Sustainable and Efficient Maritime Operations

Sustainable and Efficient Maritime Operations

Guest Editors

Jiaguo Liu
Jun Ye

MDPI

Basel • Beijing • Wuhan • Barcelona • Belgrade • Novi Sad • Cluj • Manchester

Guest Editors
Jiaguo Liu
School of Maritime
Economics and Management
Dalian Maritime University
Dalian
China

Jun Ye
College of Business
Administration
Ningbo University of Finance
& Economics
Ningbo
China

Editorial Office
MDPI AG
Grosspeteranlage 5
4052 Basel, Switzerland

This is a reprint of the Special Issue, published open access by the journal *Journal of Marine Science and Engineering* (ISSN 2077-1312), freely accessible at: https://www.mdpi.com/journal/jmse/special_issues/OA401FOM51.

For citation purposes, cite each article independently as indicated on the article page online and as indicated below:

Lastname, A.A.; Lastname, B.B. Article Title. *Journal Name* **Year**, *Volume Number*, Page Range.

ISBN 978-3-7258-7044-8 (Hbk)
ISBN 978-3-7258-7045-5 (PDF)
https://doi.org/10.3390/books978-3-7258-7045-5

Contents

About the Editors

Jiaguo Liu

Jiaguo Liu is the Head of the Department of Business Administration, Dalian Maritime University, and the Director of the Institute for the Development of Ports and Shipping in Liaoning Province, China. His research expertise lies in the fields of Maritime Supply Chain Management, Maritime Transport, and Maritime Economics. Dr. Liu has made significant contributions to the academic community, having authored over 150 peer-reviewed articles in prestigious journals, including *Transportation Research Part A/B/D/E*, *Computers & Industrial Engineering*, *International Journal of Logistics*, *International Journal of Production Research*, *IEEE Transactions on Engineering Management*, and *Maritime Policy & Management*. His research endeavors have been funded by prominent organizations such as the National Natural Science Foundation of China and the National Development and Reform Commission.

Jun Ye

Jun Ye is an Associate Professor at Ningbo University of Finance & Economics, and a Top-notch Talent of Ningbo, China. Her research expertise lies in the fields of green shipping, maritime policy, and global supply chain management. Dr. Ye has presided over several research projects, including those of the National Natural Science Foundation of China and the National Social Science Fund of China. She has published more than 20 papers in prestigious peer-reviewed journals indexed by SCI, SSCI, and CSSCI, including *Transportation Research Part B/D*, *Ocean Engineering*, *Ocean & Coastal Management*, and *Chinese Journal of Management Science*, among which 3 papers are ESI Highly Cited Papers.

Preface

This Reprint brings together a selection of contributions examining sustainable and efficient maritime operations, a topic of growing importance for both academic research and industry practice. The maritime sector is undergoing a profound transformation driven by environmental pressures, technological innovation, regulatory developments, and evolving operational demands. Against this backdrop, this Reprint aims to provide a structured overview of recent research efforts that address these challenges from complementary perspectives.

The scope of this Reprint spans multiple dimensions of maritime operations, including decarbonization strategies, alternative fuels, digitalization, intelligent decision support, port efficiency, maritime communications, safety management, and system resilience. Rather than viewing these themes in isolation, the collected studies highlight the interconnected nature of technological, environmental, and managerial factors shaping maritime systems.

The motivation for compiling this Reprint lies in the need to consolidate emerging knowledge and to facilitate dialogue across disciplinary boundaries. Sustainable and efficient maritime operations require insights from engineering, economics, operations research, data science, and policy analysis. By presenting diverse yet related contributions, this Reprint seeks to support researchers, practitioners, and policymakers interested in understanding both methodological advances and practical implications.

This Reprint is addressed to scholars and professionals engaged in maritime studies, transportation research, logistics, and sustainability analysis. It is hoped that the works presented here will stimulate further investigation and contribute to the development of more sustainable, resilient, and intelligently managed maritime operations.

Jiaguo Liu and Jun Ye
Guest Editors

Journal of
Marine Science
and Engineering

MDPI

Editorial

Sustainable and Efficient Maritime Operations

Jiaguo Liu [1,*] and Jun Ye [2]

[1] School of Maritime Economics and Management, Dalian Maritime University, Dalian 116026, China
[2] College of Business Administration, Ningbo University of Finance & Economics, Ningbo 315175, China
* Correspondence: liujiaguo@gmail.com

1. Introduction

Maritime transport plays a basic role in the global economy. Most international trade in goods, energy, and raw materials still depends on shipping [1]. In recent years, however, maritime operations have become harder to manage. Environmental regulations are expanding, new technologies are entering practice, and global supply chains are facing repeated disruptions [2]. These changes mean that sustainability and efficiency are no longer long-term goals in isolation: they now affect everyday decisions in shipping and port operations.

The current changes in the maritime sector are not driven by one single factor [3,4]. Digital tools, automation, and environmental requirements are developing at the same time, and they often affect the same operational activities. Data analysis and artificial intelligence are now used in traffic management, voyage planning, and port logistics. Autonomous vessels are also moving closer to real operation, even if many limits remain. At the same time, stricter emission targets are pushing companies to test alternative fuels, modify existing ships, and adjust routine operations [5–7]. Recent accidents and supply chain problems have also shown that safety and system stability cannot be treated as secondary issues.

With these challenges in mind, we organized this Special Issue, entitled "Sustainable and Efficient Maritime Operations." Thirteen papers were selected, covering shipping, ports, maritime communications, and safety topics. The studies differ in methods and focus, but they all deal with practical problems related to environmental pressure, technological change, and operational risk in today's maritime industry.

2. Contributions of the Papers in This Special Issue

The 13 papers (see the list of contributions) in this Special Issue examine sustainable and efficient maritime operations from different research angles. Together, they reflect current interests and concerns in the field, as summarized in the following.

Several papers explore the transition to alternative fuels and carbon reduction strategies. Charvalos et al. (Contribution 1) examine different marine fuels for bulk carriers, comparing environmental and economic outcomes. Their results provide practical guidance for selecting fuels under stricter emission regulations. Díaz-Secades et al. (Contribution 2) focus on older cruise vessels, weighing scrubbers against low-sulfur fuels to show the trade-offs in retrofitting fleets. Zhang et al. (Contribution 3) discuss ship-based carbon capture, utilization, and storage (CCUS), highlighting technical limits and operational challenges. Wang et al. (Contribution 4) analyze fuel choices for dual-fuel vessels under FuelEU Maritime regulations, offering actionable advice for compliance and investment decisions. Finally, Wang et al. (Contribution 5) investigate decarbonization potential across dual-fuel propulsion systems using life-cycle assessment aligned with IMO requirements.

 https://doi.org/10.3390/jmse14030294

Another group of papers addresses digital tools, automation, and safety management. Shin and Jung (Contribution 6) look at the human element of automated waterways. Their machine-learning model aims to improve Vessel Traffic Service Operator (VTSO) recruitment, as Maritime Autonomous Surface Ships (MASS) share the same traffic routes as conventional vessels. Lee et al. (Contribution 7) focus on energy efficiency in maritime communications, proposing a two-dimensional Thompson sampling method for uplink transmissions from ocean buoys. Wu et al. (Contribution 8) tackle scheduling in smart coal ports, combining a multi-agent proximal policy optimization (MAPPO) method with an adaptive micro-evolutionary algorithm, resulting in measurable efficiency gains. Yin et al. (Contribution 9) integrate an improved Human Factor Analysis and Classification System with Bayesian networks to better understand accident causation and more effectively support safety management.

Port sustainability and performance evaluation are addressed from complementary perspectives. González-Cancelas et al. (Contribution 10) analyze smart port development in Spain using the Blue Ocean Strategy, demonstrating how Port 4.0 technologies can simultaneously enhance competitiveness and sustainability. In parallel, Zhang and Gu (Contribution 11) propose an improved epsilon-based measure model that incorporates heterogeneous abatement capacities to evaluate port environmental efficiency more realistically, offering valuable methodological support for differentiated environmental regulation and port-level decarbonization strategies.

Finally, two papers review broader trends and emerging directions. Djoumessi et al. (Contribution 12) provide a bibliometric review of technological innovation in the maritime sector, identifying established trends and research gaps. Xue et al. (Contribution 13) examine applications of machine learning in maritime safety research, focusing on autonomous shipping and highlighting current hotspots and future directions.

3. Outlook

Taken together, the papers in this Special Issue clarify several important insights for current research on sustainable and efficient maritime operations.

Firstly, the contributions collectively show that improvements in maritime performance increasingly arise from the integration of digital intelligence with operational and safety management, rather than from standalone technological solutions. Although data-driven models and artificial intelligence are proving effective in areas such as traffic management, port operations, and risk analysis, their successful application depends on data quality, system interoperability, and transparent human–machine interaction. As a result, human expertise remains a central component of maritime decision-making.

Secondly, this Special Issue highlights that maritime decarbonization is inherently a system-level challenge. Studies on alternative fuels, emission abatement technologies, and environmental efficiency consistently demonstrate that environmental outcomes are shaped by interactions among technology choices, operational strategies, regulatory constraints, and economic conditions. Life-cycle and system-wide perspectives are therefore essential for evaluating the long-term feasibility and effectiveness of decarbonization pathways.

Thirdly, the collected papers reinforce the growing importance of resilience as a concept that links efficiency, sustainability, and risk management. Increasing exposure to extreme weather events, geopolitical disruptions, and operational uncertainties underscores the need to strengthen the adaptive capacity of vessels, ports, and maritime networks. Future research and practice will need to address sustainability and efficiency alongside resilience to support the maritime sector's transition toward more robust, intelligent, and sustainable operations.

https://doi.org/10.3390/jmse14030294

Acknowledgments: The Guest Editors would like to thank all the authors for their valuable contributions to this Special Issue, as well as the editorial staff for their professional assistance.

Conflicts of Interest: The authors declare no conflicts of interest.

List of Contributions:

1. Charvalos, G.; Tzakis, A.; Arvanitis, A.; Peppa, S.; Papadopoulos, C. Environmental and Economic Assessment of Alternative Marine Fuels for Bulk Carriers: A Comparative Analysis of Handymax, Panamax and Supramax Vessels. *J. Mar. Sci. Eng.* **2025**, *13*, 1757. https://doi.org/10.3390/jmse13091757.
2. Díaz-Secades, L.A.; Baptista, L.; Pereira, S. Case Study on Emissions Abatement Strategies for Aging Cruise Vessels: Environmental and Economic Comparison of Scrubbers and Low-Sulphur Fuels. *J. Mar. Sci. Eng.* **2025**, *13*, 1454. https://doi.org/10.3390/jmse13081454.
3. Zhang, N.; Wang, X.; Wang, C.; Peng, Z.; Chen, Y.; Liu, L.; Chen, L.; Hu, J.; Wei, W.; Li, X. Design and Innovative Application of Ship CCUS Technology Under the Requirements of Green Shipping. *J. Mar. Sci. Eng.* **2025**, *13*, 1157. https://doi.org/10.3390/jmse13061157.
4. Wang, C.; Peng, Z.; Yang, J.; Zhang, N.; Li, K.; Li, X. Optimal Alternative Fuel Selection for Dual-Fuel Ships under FuelEU Maritime Regulations: Environmental and Economic Assessment. *J. Mar. Sci. Eng.* **2025**, *13*, 1105. https://doi.org/10.3390/jmse13061105.
5. Wang, Z.; Fan, S.; Han, Z. Decarbonation Effects of Mainstream Dual-Fuel Power Schemes Focus on IMO Mandatory Regulation and LCA Method. *J. Mar. Sci. Eng.* **2025**, *13*, 847. https://doi.org/10.3390/jmse13050847.
6. Shin, G.-h.; Jung, M. Preparing VTS for the MASS Era: A Machine Learning-Based VTSO Recruitment Model. *J. Mar. Sci. Eng.* **2025**, *13*, 2127. https://doi.org/10.3390/jmse13112127.
7. Lee, K.J.; Jo, J.-H.; Cho, S.; Kwon, K.-W.; Kim, D. Two-Dimensional Thompson Sampling for Joint Beam and Power Control for Uplink Maritime Communications. *J. Mar. Sci. Eng.* **2025**, *13*, 2034. https://doi.org/10.3390/jmse13112034.
8. Wu, Y.; He, S.; Tang, H.; Long, Z.; Xiang, A. Integrated Scheduling of Handling and Spraying Operations in Smart Coal Ports: A MAPPO-Driven Adaptive Micro-Evolutionary Algorithm Framework. *J. Mar. Sci. Eng.* **2025**, *13*, 1840. https://doi.org/10.3390/jmse13101840.
9. Yin, X.; Yan, Y.; Wang, J.; Zhao, H.; Wu, Q.; Xu, Q. Data-Driven Analysis of the Causal Chain of Waterborne Traffic Accidents: A Hybrid Framework Based on an Improved Human Factors Analysis and Classification System with a Bayesian Network. *J. Mar. Sci. Eng.* **2025**, *13*, 393. https://doi.org/10.3390/jmse13030393.
10. González-Cancelas, N.; Guil López, J.J.; Vaca-Cabrero, J.; Camarero-Orive, A. Toward Smart and Sustainable Port Operations: A Blue Ocean Strategy Approach for the Spanish Port System. *J. Mar. Sci. Eng.* **2025**, *13*, 872. https://doi.org/10.3390/jmse13050872.
11. Zhang, J.; Gu, G. An Evaluation of Port Environmental Efficiency Considering Heterogeneous Abatement Capacities: Integrating Weak Disposability into the Epsilon-Based Measure Model. *J. Mar. Sci. Eng.* **2025**, *13*, 2064. https://doi.org/10.3390/jmse13112064.
12. Djoumessi, A.; Tei, A.; Ferrari, C. The Adoption of Technological Innovations in the Maritime Industry: A Bibliometric Review. *J. Mar. Sci. Eng.* **2025**, *13*, 1484. https://doi.org/10.3390/jmse13081484.
13. Xue, J.; Yang, P.; Li, Q.; Song, Y.; van Gelder, P.H.A.J.M.; Papadimitriou, E.; Hu, H. Machine Learning in Maritime Safety for Autonomous Shipping: A Bibliometric Review and Future Trends. *J. Mar. Sci. Eng.* **2025**, *13*, 746. https://doi.org/10.3390/jmse13040746.

References

1. Gu, B.; Liu, J. A Systematic Review of Resilience in Maritime Transport. *Int. J. Logist. Res. Appl.* **2025**, *28*, 257–278. [CrossRef]
2. Liu, J.; Wu, J.; Gong, Y. Maritime Supply Chain Resilience: From Concept to Practice. *Comput. Ind. Eng.* **2023**, *182*, 109366. [CrossRef]
3. Vidović, T.; Šimunović, J.; Radica, G.; Penga, Ž. Systematic Overview of Newly Available Technologies in the Green Maritime Sector. *Energies* **2023**, *16*, 641. [CrossRef]
4. Xiao, G.; Pan, L.; Lai, F. Application, Opportunities, and Challenges of Digital Technologies in the Decarbonizing Ship Industry: A Bibliometric Analysis. *Front. Mar. Sci.* **2025**, *12*, 1523267. [CrossRef]

https://doi.org/10.3390/jmse14030294

5. Ye, J.; Chen, J.; Shi, J.; Jiang, X.; Zhou, S. Novel synergy mechanism for carbon emissions abatement in shipping decarbonization. *Transp. Res. Part D Transp. Environ.* **2024**, *127*, 104059. [CrossRef]
6. Zou, Y.; Xiao, G.; Li, Q.; Biancardo, S.A. Intelligent Maritime Ship: A Bibliometric Analysis of Internet Technologies and Automated Port Infrastructure Applications. *J. Mar. Sci. Eng.* **2025**, *13*, 979. [CrossRef]
7. Wang, L.; Hsu, H.H. IoT Technology in Maritime Logistics Management: Exploration of Data Analysis Methods. *Discov. Internet Things* **2025**, *5*, 66. [CrossRef]

https://doi.org/10.3390/jmse14030294

Journal of
*Marine Science
and Engineering*

MDPI

Review

The Adoption of Technological Innovations in the Maritime Industry: A Bibliometric Review

Armand Djoumessi, Alessio Tei and Claudio Ferrari *

Italian Centre of Excellence on Logistics, Transport and Infrastructures (CIELI), University of Genoa, Via Vivaldi 5, 16126 Genoa, Italy; thedjoumessima@gmail.com (A.D.); alessio.tei@unige.it (A.T.)
* Correspondence: claudio.ferrari@unige.it

Abstract

The adoption of technological innovations in the maritime industry is of interest to business, policy, and academic communities. In the last group, this interest has translated into the publication of a large but scattered literature, making it difficult to compare findings and identify the dynamics, structures, and patterns that might inform future research. A comprehensive review of past research on this topic might help achieve this. To date, no such review has been carried out, which is an important gap in the literature that this paper contributes to bridging. Two bibliometric review techniques—co-citation analysis of cited references and bibliographic coupling of documents—are applied to 171 journal articles published between 1999 and February 2025 to answer the following questions: 1. What is the knowledge base of this literature? 2. What are the recent research trends (research fronts) in this literature? The analysis reveals that research on "shore power" dominates both the knowledge base and research fronts. Other key research themes centre on "autonomous shipping", "blockchain", and "alternative fuels". Based on these results, implications for future research are drawn.

Keywords: technological innovations; innovation adoption; maritime industry; knowledge base; research fronts; bibliometrics

1. Introduction

As the maritime industry faces mounting pressure to meet the International Maritime Organization's 2050 decarbonisation targets, shipowners and port operators are rapidly exploring technological innovations such as shore power, alternative fuels, and artificial intelligence [1,2]. The introduction of technological innovations has always led to significant changes in the structure and organisation of the shipping industry [3].

For example, the introduction of the steam engine and later of diesel propulsion into the industry ensured faster delivery of goods [4]. Similarly, the adoption of the container ensured that more goods could be transported efficiently [5,6], and the adoption of larger ships ensured that the final customer could pay a lower price [7,8]. Some technological innovations have also been adopted into the industry for safety and environmental purposes—for example, the adoption of double-hull oil tankers to reduce their negative environmental impact in terms of oil spills [9,10].

In general, as more technologies that could be applied to the maritime industry have become available, operators have been faced with the challenge of selecting the right one for their specific case, especially as some of these technologies could be substitutes for each other. From a research point of view, this has resulted in a surge of publications, as authors from various disciplines have approached this issue from different perspectives. For example, the application of smart technologies [11,12], blockchain technologies [13–16], autonomous technologies [17–19], and environmentally friendly technologies [20–22] in the maritime industry has been explored.

Depending on the technology studied, the perspective taken, and the goal of the research, studies on this topic have been published beyond maritime- and transport-focused journals—for example, in behaviour-focused journals (e.g., [12]), energy-focused journals (e.g., [21,23]), environment-focused journals [24,25], and strategic management-focused journals [10,26], just to name a few. Although this has enriched the body of knowledge on the topic, it has also resulted in a very scattered literature, making it difficult to compare findings and identify the dynamics, structures, and patterns. A comprehensive review of past research on this topic might help identify such patterns. To date, no such review has been carried out, which is an important gap in the literature that this paper contributes to bridging.

The purpose of this study is thus to bring together and provide a comprehensive review of the growing literature on the adoption of technological innovations in the maritime industry. Though there are many types of reviews (e.g., systematic, narrative, critical, scoping, bibliometric, etc.), we are particularly interested in this first attempt to bring this literature together to provide a macroscopic perspective. As such, a bibliometric literature approach seems appropriate. A core benefit of bibliometric studies is that they allow the identification of influential authors, documents, sources, etc. in a field of studies as well as associations and similarities between them, which in turn provide researchers with a solid foundation for positioning their contribution and identifying new lines of research [27,28]. As argued by Ref. [29], in the context of maritime research, carrying out bibliometric reviews is important because they bridge the gap between researchers and practitioners, especially in the current context of sustainability transitions in the maritime industry.

In line with bibliometric studies in other fields [30–32], the following research questions will guide this study:

- Q1. What is the knowledge base of this literature?
- Q2. What are the recent research trends (research fronts) in this literature?

Two bibliometric techniques—co-citation analysis and bibliographic coupling (further explained in Section 3.4)—are used to answer the above questions. The dataset consists of peer-reviewed journal articles on the adoption of technological innovations in the maritime industry published between 1999 and February 2025. Of note, 1999 represents the year the earliest article relevant on the topic was identified, while February 2025 (the dataset was downloaded on 2 December 2025) represents the month and year the data used in this article were downloaded.

By carrying out a literature review on the adoption of technological innovations, this paper contributes to the increasing literature on maritime bibliometric studies in the following ways. First, it explores an important but rather unexplored topic in the context of maritime bibliometric studies. In their recent review of maritime bibliometric studies, Ref. [29] has called for such studies. In the current context of maritime decarbonisation, it is anticipated that more technological innovations will be adopted as shipowners and charterers strive to meet the International Maritime Organization (IMO) regulation requirements and that more research will be conducted [33–36]. Taking stock of past research on the adoption of technological innovations in the maritime industry is necessary, as it will lay a strong foundation upon which future research on the topic can be built. Second, it demonstrates that the knowledge base of this literature is composed of seven thematic areas at various stages of development. Third, by capturing the seven research fronts on the different technologies discussed in the extant literature, the paper will benefit researchers wanting to orientate themselves in this field. Finally, the paper also contributes to increasing the number of maritime bibliometric studies using co-citation analysis and bibliographic coupling, whose use has so far been limited [29].

The remainder of this paper is structured as follows. Section 2 lays the theoretical foundation of this study. Section 3 presents the research design and methodology, followed by the findings in Section 4. Finally, Section 5 concludes and provides some trajectories for future research.

2. Theoretical Background: Technological Innovations and Their Adoption Process

2.1. How Should Technological Innovation Be Understood in This Paper?

According to Ref. [37], the term "technological innovation" might be characterised as an inclusive concept (in the sense that it can be used to mean different things depending on the context) whose meaning has evolved over the years. In essence, it has served and continues to serve different purposes in the literature. It has often been used as a theoretical concept, process, good, or service. As an example, in their review article on the literature on technological innovation, Ref. [38] defines technological innovation as "a procedure that contains the interplay of multiple sources, a multidimensional concept difficult to measure". This rather broad definition leaves the interpretation to the reader. In fact, many authors using the phrase shy away from defining it, choosing rather to give examples. For example, in the chapter titled "Defining technological innovation", Ref. [39] fails to provide a definition of the term, demonstrating the difficulty that might be associated with doing so. It is therefore critical to clarify how this phrase/term is used in this paper.

It can be argued that the difficulty to define technological innovation stems from the fact that it is composed of two words, "technological" and "innovation" [37]. It is therefore useful to clarify what each of these words means in the context of this research before providing an understanding of how the two combined should be understood in this research. Starting with "technological", there seems to be a consensus in the literature that it relates to or involves technology [37,39]. The Cambridge online dictionary defines technology as "a method by which science is used for practical purposes, new machinery and equipment that has been developed using scientific knowledge or processes" [40]. It can be inferred that a technology is a scientific method (e.g., automated container terminal), a tangible tool (e.g., electric tugboat), or an intangible tool (e.g., blockchain). This is how technology will be understood in this paper.

Turning now to "innovation", the other word in the phrase "technological innovation", Ref. [37] refers to it as a "buzzword, a magic word that seeps into almost every sentence, a panacea for every socio-economic problem". Although multitudes of definitions of "innovation" have been provided [41,42], there is a consensus in the literature that it includes an element of newness to the world, industry, organisation, etc. [43]. This paper focuses on maritime industry, and thus on innovations that are new to organisations involved in this industry. As for what is new, various typologies of innovations have been provided [44,45]. In this paper, the focus is on technology, and thus technological innovation as type of innovation. The degree of newness of an innovation (radical, incremental) is also often discussed in the literature (e.g., [46,47]). In this paper, both will be considered without distinction in the literature explored.

In this paper, therefore, technological innovation refers to a new scientific method, or a tangible or intangible tool. They are adopted by users to either achieve competitiveness or compliance with sustainability regulations. For an in-depth discussion on this term, refer to Refs. [37,38]. Various theories of technological innovations have been developed, but a discussion of them is beyond the scope of this paper (see Refs. [48,49] for details on these).

Since our focus is on technology adoption, we will focus in the next section on the most used theory to explain the adoption of technological innovation.

2.2. Adoption Lifecycle of Technological Innovations

Ref. [50] provides a review of the major models and theories related to users' adoption of technology. Their findings show that the most important of these theories as per the number of cited documents is the Innovation Diffusion theory (also called the Diffusion of Innovation (DOI) theory by Ref. [51]), which will be briefly discussed in this paper because it has recently been applied in the maritime context by Ref. [52] to study the adoption of LNG and electric batteries by ferries in Norway. The DOI theory distinguishes five end-user categories in the technology adoption lifecycle spread over a normally distributed bell curve [51,53,54].

Innovators are generally more eager to experiment with a new technology as soon as possible to learn about its potential monetary returns. Motivated by profits they can capture by adopting the technology before their competitors, early adopters are willing to take the risk of investing in an immature technology [55]. According to Refs. [54,56], there are chasms or cracks in the bell curve representing adoption barriers (in form of various types of uncertainties and risks) between different groups of adopters, the most significant of which is between early adopters and the early majority, preventing the new technology from taking off. It should be noted, however, that the absence of chasms in the curve between other groups is not synonymous with their non-existence but rather with the fact that they are of lower magnitude, and thus crossed by adopters [57]. The early majority group (nearly a third of the market) is risk averse and conservative, and adopts technological innovation only when it has been proven. The late majority (also a third of the market) is much more risk averse and will only be committed after most technological and economic uncertainties associated with the new technology have been resolved. Finally, the laggards resist new technologies and only adopt them when compelled to.

3. Research Design and Methods

We relied on the literature review approach proposed by Ref. [58] (p. 102). According to these authors, despite their differences, all reviews can be conducted following eight steps: (1) formulating the research problem, (2) developing and validating the review protocol, (3) searching the literature, (4) screening for inclusion, (5) assessing quality, (6) extracting data, (7) analysing and synthesising data, and (8) reporting the findings. For the purpose of this review, some of these steps are combined, as shown in Figure 1. Steps 1, 2, 3, and 4 are elaborated upon in Section 3 of the paper, and Step 5 is in Section 4 of this paper.

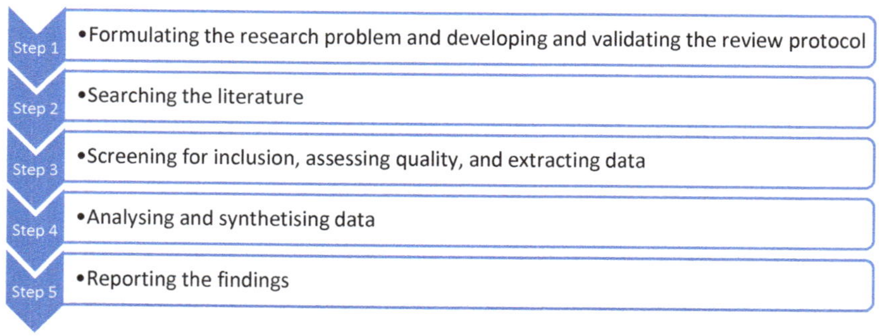

Figure 1. Research framework.

3.1. Formulating the Research Problem and Developing and Validating the Review Protocol

In this step, the topic investigated is defined and justified, and the type of review that will be conducted is chosen. In this paper, the topic of interest is the adoption of technological innovations in the maritime industry. This topic is especially relevant with the current push toward the decarbonisation and digitalisation of the industry, which will necessitate the adoption of many technologies and potentially drive more research on this topic. To the best of our knowledge, there has been no attempt to bring together the extant literature on the topic. Doing so is critical and will lay a strong foundation on which future research on the topic will build and identify issues requiring further investigations. According to Ref. [58], a question modelled on the topic can be used to define keywords and relevant synonyms that will be used for identifying relevant articles on the topic. Here, it is important to stress that the sole purpose of the question framed in this step is to retrieve the relevant literature. As such, it is not to be confused with the main research questions (Q1 and Q2, mentioned in the introduction, which this paper attempts to answer).

The use of synonyms in the question that is modelled on the topic to retrieve the data is particularly relevant in the context of the maritime industry, where several terminologies are often used to refer to the same concept (i.e., the maritime industry is often just referred to as shipping, and ships can also be referred to as vessels). Accordingly, the following question was used to identify relevant articles on the topic:

- What are the drivers and barriers to the adoption of technological innovations in the maritime industry?

Various technological innovations in the maritime industry have been analysed through the lens of their drivers and barriers in the literature (e.g., [21,22,59–62]. This justifies why "barriers" and "drivers" were included in the question. After defining the question used to retrieve the literature, Refs. [58,63] recommend dissecting it into concept domains that will be used to search the literature. Accordingly, the domains of the research questions are "drivers", "barriers", "adoption", "technological innovations", and "maritime industry".

As pointed out by Ref. [58], the domains can be extended by synonyms, abbreviations, other alternative spelling, or related terms, or even broken down into other words. We argue that in addition to this, the researcher's knowledge of the language/terms/vocabulary commonly used in their discipline for the industry addressed in the study and of previous known studies on the topic can also direct the selection of alternative keyword domains [63]. For example, although the word "factors" does not appear in the research question, it is often used in the maritime industry/shipping literature as a generic term encompassing both drivers and barriers (e.g., [19,64–66]). Therefore, it will be used as a domain during the search process in combination with other domains. The same reasoning goes for the selection of the words "challenges" and "constraints", which often are associated with the barriers [67].

Regarding the domain "technological innovations", since "technological" is a qualifier for innovations, it is more appropriate in the context of this research to break this domain into the related terms "technology" and "innovations", which are widely used in the maritime literature (e.g., [4,17,68,69]. Table 1 shows the results of the extended domains.

Table 1. Domains and their extensions used in the data selection search.

Domains Derived from the Research Questions	Synonyms Provided by the Cambridge Thesaurus (Online)	Synonyms Potentially Relevant to this Research	Relevant Related Terms
Drivers	Chauffeurs, jockeys, motorists, operators Trainers	-	Factors Triggers Enablers
Barriers	Hindrances, checks, difficulties Restrictions, hurdles, obstacles Limitations, hazards, stumbling blocks Impediments	Obstacles Impediments Hindrances Limitations	Challenges Constraints
Adoption	Acceptance, approval, enactment Endorsement, maintenance Ratification, selection	Acceptance	Deployment Implementation
Technological innovations	High-tech, industrial, mechanical Professional, scholarly, scientific Special, specialised, vocational	-	Technology AND innovations
Maritime industry	-	Shipping industry	Shipping, port, ship

3.2. Searching the Literature

The different domains combinations were used as inputs in the "Advanced search TITLE-ABS-KEY" function of the Scopus database. It was selected because it covers most peer-reviewed academic research data in the fields relevant to this research (science, technology, and social sciences) and has also been used for conducting reviews in maritime research [17,70]. Separate search combinations were carried out depending on whether the retrieved documents focus on drivers or barriers. An example combination with drivers is [TITLE-ABS-KEY (("Drivers" OR "factors" OR "triggers" OR "enablers") AND "technology*" AND ("adoption" OR "deployment" OR "implementation" OR "acceptance") AND (innovation) AND ("shipping" OR "port" OR ("maritime industry") OR "ship")). An example with barriers is [TITLE-ABS-KEY (("Barriers" OR "obstacles" OR "challenges" OR "impediments" OR "hindrances" OR "constraints" OR "limitations") AND ("adoption" OR "deployment" OR "implementation" OR "acceptance") AND ("technology*") AND innovation* AND ("shipping" OR "port" OR "ship" OR "maritime industry"))]. The search results for all combinations led to the identification of 2207 documents, while the keywords combinations used for searching papers are shown in Appendix A.

3.3. Screening for Inclusion, Assessing Quality, and Extracting Data

The documents selected for further analysis were limited to academic journal articles (916), and those published prior to 2023 (612) had to have been cited at least once. In total, 519 articles were selected, whose abstracts were read to ascertain their aims, the questions they were attempting to answer, and the main findings. When these were not very clear, the articles were read in full, as recommended by Ref. [70]. A total of 348 articles unrelated to research questions were deleted, leaving 171 journal articles deemed relevant for analysis. The data are analysed next.

3.4. Analysing and Synthetising Data

This paper adopts a bibliometric approach. Compared with traditional reviews (most often qualitative), bibliometric reviews enable researchers to replicate their review results. The advantage of bibliometric methods when used correctly is their quantitative rigour and ability to provide succent visualisations of large quantities of data [71]. Given the relatively large quantity of data to analyse (171 articles), they are suitable for this study.

They are increasingly being used to carry out reviews on topics and journals in social sciences [30,71–73] and are increasingly being applied in maritime transport research [29,74].

The choice of the software was based on its ability to provide clarity for the question being answered and its ability to support data imported from Scopus and Microsoft Excel. The software VOSviewer 1.6.20 was used to perform the analysis [75]. A key strength of this software is its ability to show network, overlay, and density visualisations.

In general, bibliometric techniques are used for performance assessment and science mapping [76,77]. The former is used when the purpose is to describe the performance of individual/groups of articles, authors, institutions, countries, etc. in a body of knowledge. The latter is performed when the aim is to identify dynamics, structures, and patterns in the data analysed. There are various types of science mapping techniques suitable for specific questions (see Refs. [71,78] for a detailed account of these techniques). Given the questions this paper attempts to answer, two bibliometric techniques will be used: co-citation analysis and bibliographic coupling.

Both techniques are centred on intertextual relationships among published documents, which are established based on the referencing behaviour of authors under the assumption that these relationships are an indication of textual similarity between the coupled or co-cited documents [28]. Two documents are said to be bibliographically coupled if they have at least one reference in common and co-cited if they appear on the same reference list [79]. Each of these techniques has some strengths and weaknesses, as shown in Table 2. It can be inferred from Table 2 that the two techniques are complementary. This justifies why both are used in this paper.

Table 2. Summary of bibliographic coupling and co-citation analysis.

Bibliometric Techniques	Occurs When	Key Features	Merits	Critique
Bibliographic coupling	Two documents have at least one reference in common	(i) Overlapping bibliographies (ii) Measure of association between two citing documents	(a) Forward looking (b) Best suited for mapping current trends and future priorities (research fronts)	(a) Static because number of references shared by two documents does not change over time (b) Two articles may reference a completely unrelated subject matter in the third
Co-citation	Two documents are included in the same reference list	(a) Similarity relationship between two cited documents (b) Applied to references of documents rather than the document itself	(i) Dynamic approach given that co-citation frequencies increase over time (ii) Best indicator of subject similarity (iii) Best suited for mapping the intellectual heritage on the basis of highly cited documents	(1) Not suitable for clustering smaller niche specialties formed by less cited documents (2) Biased towards the past (backward looking)

Given the goal of the paper and the research questions of interest, bibliographic coupling in this paper will only use documents as units of analysis (to answer question Q2), though sources, authors, organisations, and countries could also be used. In the same vein, though sources and authors can also be used as units of analysis, co-citation in this paper will only be applied to references (to answer Q1). A graphical representation of co-citation and bibliographic coupling as applied in this paper is shown in Figure 2.

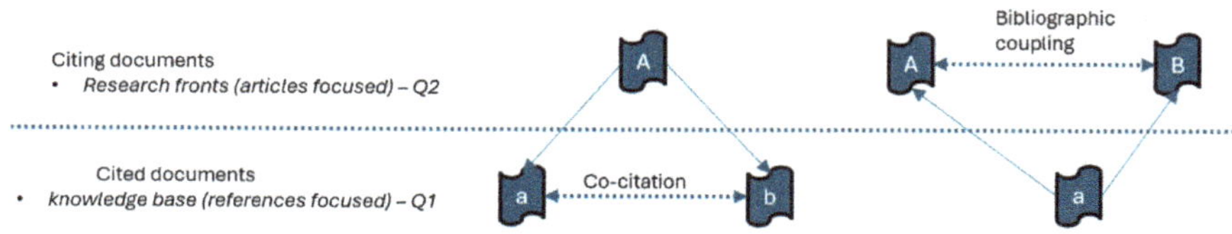

Figure 2. A graphic illustration of co-citation and bibliographic coupling. Adapted from Ref. [28] (p. 429).

4. Reporting the Findings

4.1. Descriptive Analysis

In this section, the annual scientific production, the most influential documents, and the authors and countries involved in research on the adoption of technological innovations in the maritime industry are discussed.

As can be seen in Figure 3, interest in the adoption of technological innovations in the maritime industry picked up in 2019 and has been increasing ever since, with 2024 being the most productive year. The 171 articles analysed were authored by 544 authors, with an average value of 3.68 authors per paper. Out of these, only 19 were single authored, reflecting a high level of collaboration. It can be seen from the corresponding author data in Figure 4 that authors from China dominate this research, both in terms of national collaboration, expressed as single-country production (SCP), and international collaboration, expressed as multiple-country production (MCP), followed by the UK, the USA, Norway, Italy, Sweden, Australia, Germany, and Hong Kong. For Germany, Italy, and Sweden, multiple-country collaborations overtake single-country productions.

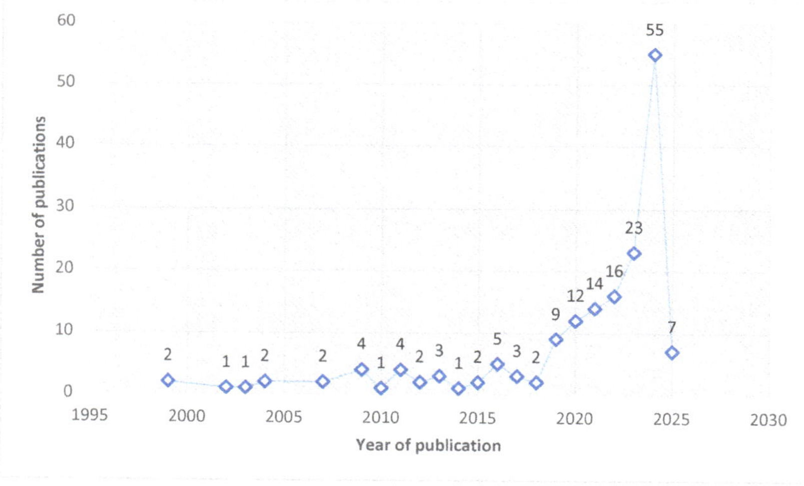

Figure 3. Annual scientific production between 1999 and 2025.

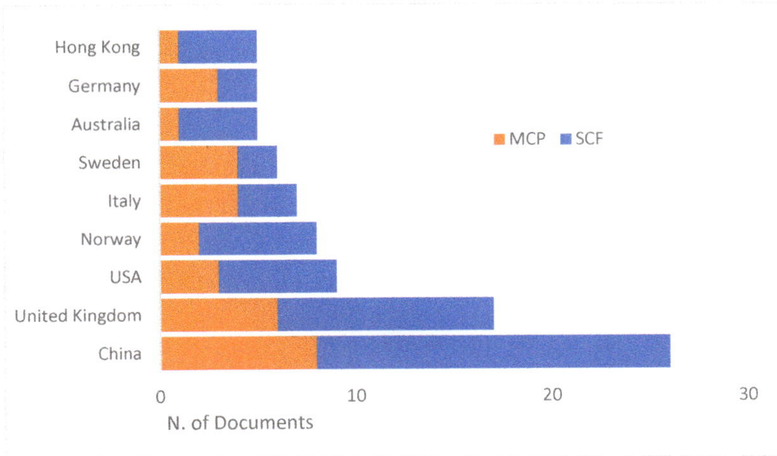

Figure 4. Countries of the authors.

A snapshot linking the top 15 authors (AU) that have contributed the most to this research, the sources in which they have published (SO), and the keywords (DE) used is shown in the three-field plot in Figure 5. The three-field plot was built based on the concept of Sankey diagrams. The keywords shown in Figure 5 reflect the topic being explored. Of note, when considering all 544 authors, a total 632 keywords were used, with those that were used at least five (5) times shown in Table 3. The sources in which the examined articles are published, shown in Figure 5, cover maritime business- and transportation-related issues in their scope. The 171 articles were published in 105 different sources covering various disciplines and topics related to the adoption of technological innovation, with a focus on the maritime industry, including technology, innovation, pollution, the environment, etc. Sources in which there were at least three (3) articles on the topic are shown in Table 4. The diversity of sources and keywords used reflects the multidisciplinary nature of the topic under consideration in this paper.

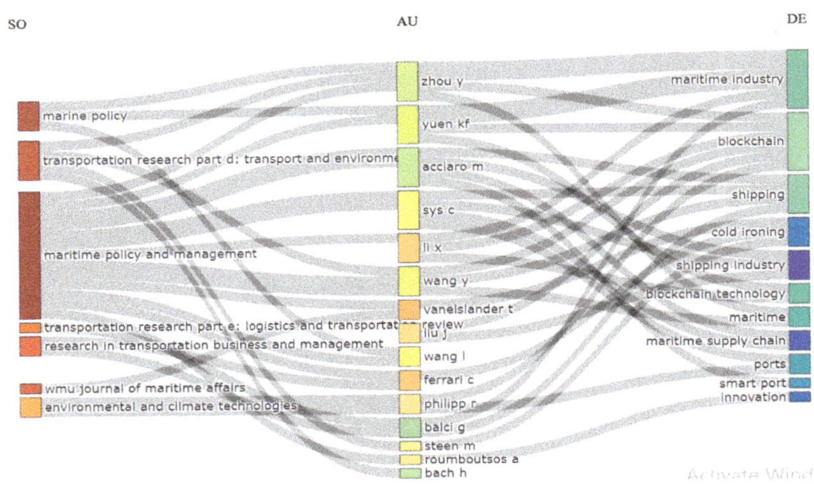

Figure 5. Three-field plot connecting sources, authors, and keywords.

Table 3. Author keywords occurring at least five times.

Keywords	Frequency
Blockchain (or blockchain technology)	25
Shipping	17
Maritime industry	15
Seaports	13
Shore power	10
Digital transformation	9
Autonomous ships	9
Supply chain management	8
Decarbonisation	7
Container	7
Sustainability	6
Smart port	6
Innovation	5
Maritime supply chain	5
Maritime transportation	5

Table 4. Sources contributing at least 3 articles.

Rank	Sources	2024 Impact Factor	Nr. of Documents
1	Maritime Policy & Management	3.6	11
2	Marine Policy	3.7	9
3	Sustainability	3.3	6
4	Transportation Research Part D: Transport and Environment	7.7	6
5	WMU Journal of Maritime Affairs	2.4	6
6	Research in Transportation Business and Management	4.4	5
7	Technological Forecasting and Social Change	13.3	4
8	Journal of Marine Science and Engineering	2.8	4
9	Maritime Business Review	2.79	4
10	Transport Policy	5.3	3
11	Transportation Research Part E: Logistics and Transportation Review	8.8	3

The most cited articles (with a citation count greater than 90) among the 171 selected articles are shown in Table 5. Ref. [80] is by far the most cited and also the oldest. Though published just two years ago, Ref. [81] has the highest number of citations per year (54). In the majority of the titles of these top cited papers, the technological innovation(s) they deal with are mentioned. Though it is beyond the scope of this paper to discuss the many technological innovations discussed in the 171 articles individually, they can be broadly grouped into seven categories: digital, autonomous, energy-efficient ships, information and communication, machinery and equipment, tracing and tracking, and industrial production. Examples of each category are given in Table 6.

Table 5. Top cited articles (citation ≥ 90).

Authors	Title	Cited by	Citations per Year
Giannopoulos (2004) [80]	The application of information and communication technologies in transport	201	9.57
Zis (2019) [52]	Prospects of cold ironing as an emissions reduction option	167	27.83
Schinas and Butler (2016) [82]	Feasibility and commercial considerations of LNG-fueled ships	161	17.89
Lu et al. (2007) [83]	Application of structural equation modeling to evaluate the intention of shippers to use Internet services in liner shipping	135	7.5
Ahmad et al. (2021) [84]	Blockchain applications and architectures for port operations and logistics management	130	32.5
Tan and Sidhu (2022) [85]	Review of RFID and IoT integration in supply chain management	129	43
Greve (2009) [10]	Bigger and safer: The diffusion of competitive advantage	125	7.81
Zhou et al. (2020) [86]	The key challenges and critical success factors of blockchain implementation: Policy implications for Singapore's maritime industry	121	24.20
Gausdal et al. (2018) [60]	Applying blockchain technology: Evidence from Norwegian companies	119	17
Baldi et al. (2020) [87]	The role of solid oxide fuel cells in future ship energy systems	113	22.60
Liu et al. (2023) [81]	Blockchain technology in maritime supply chains: applications, architecture and challenges	108	54
Balci and Surucu-Balci (2021) [88]	Blockchain adoption in the maritime supply chain: Examining barriers and salient stakeholders in containerized international trade	101	25.25
Bach et al. (2020) [20]	Implementing maritime battery-electric and hydrogen solutions: A technological innovation systems analysis	93	18.60
Hogg and Ghosh (2016) [85]	Autonomous merchant vessels: examination of factors that impact the effective implementation of unmanned ships	93	10.33

Table 6. Examples of technological innovations discussed in the 171 papers.

Group of Technologies	Examples of Types of Technologies	Example of References per Type
Digital	Blockchain, digital twins, digital platforms	[84–90]
Autonomous	Autonomous ships, automated container terminals, remote automated vehicles	[91–93]
Energy efficient ships	Shore power, alternative fuels, electric tugboats, scrubber, wind-assisted propulsion technologies	[23,94–96]
Information and communication	Internet, Internet of Things (IOT), 5G, augmented reality	[97–100]
Machinery and equipment	Wearable safety devices, ship bridge, simulator, stack train, diesel engine, air compressor, dehumidifier, inverter welding machine, PV, welding machine, fast ferries, steel hull, double-hull tankers, post-Panamax container ship, indented berth	[10,12,26,101–105]
Tracing and tracking	RFID, X-ray scanning, asset tracking system	[106–108]
Industrial production	Advanced manufacturing technology, additive manufacturing	[109,110]

Regarding the institutions that contributed the most (see Table 7), universities in Asia dominate, followed by European universities, with Shanghai Maritime University as the institution that contributed the most. Regarding the most productive authors, European and Asian share dominance, as reflected in Table 8, with Vanelslander, Thierry, being the most productive author, affiliated with the University of Antwerp.

Table 7. Institutions that contributed the most (≥4 articles).

Affiliation	Articles
Shanghai Maritime University	10
University of Antwerp	8
Dalian Maritime University	7
The Hong Kong Polytechnic University	7
University of the Aegean	6
National Taiwan University	5
Norwegian University of Science and Technology	5
Delft University of Technology	4
Nanyang Technological University	4
University of Genoa	4

Table 8. Most productive authors (≥3 articles).

Authors	Articles
Vanelslander, T.	4
Ferrari, C.	3
Li, X.	3
Liu, J.	3
Philipp, R.	3
Roumboutsos, A.	3
Steen, M.	3
Sys, C.	3
Wang, L.	3
Wang, Y.	3
Yuen, K.F.	3

Overall, the descriptive analysis has provided the following insights regarding this literature: the increased interest reflected in the number of papers published in recent years (since 2019), the high number of technological innovations studied, the dominance of Asian institutions in terms of the number of documents they have contributed, and the dominance of Asian and European authors in terms of productivity. The global nature of this literature is evidenced by the number of multiple-country collaborations. Furthermore, the variety of journals in which this literature is published also demonstrates that this research is approached from different perspectives and is multidisciplinary.

4.2. Knowledge Base of the Literature on the Adoption of Technological Innovations in the Maritime Industry (Q1)

As already explained in Section 3.4 a reference co-citation analysis was carried out to identify the knowledge base of this literature. In line with the principles of co-citation also explained above, a key assumption of this method is that the content of two documents frequently cited together (i.e., frequently appearing together in the reference list of other documents) is related in the sense that they share some textual similarities [28,111]. Of note, the analysis was conducted on the 9119 references of the 171 journal articles identified. The analysis was carried out using the VOSviewer 1.6.20 software with the purpose of identifying networks of documents often co-cited together. The mathematical underpinnings of co-citation analysis are beyond the scope of this paper and can be found in Ref. [75]. In a nutshell, the software uses normalising, mapping, and clustering algorithms to build networks of interconnected documents based on intertextual similarities and (manually) defined citation of cited reference thresholds. As a general rule, the higher the threshold set for a citation of a cited reference, the fewer documents are connected.

The best results obtained in the context of this paper after running the analysis are based on a minimum citation threshold of 3 of a cited reference, with 44 out the 9119 references interconnected and organised into seven clusters, as shown in Figures 6 and 7.

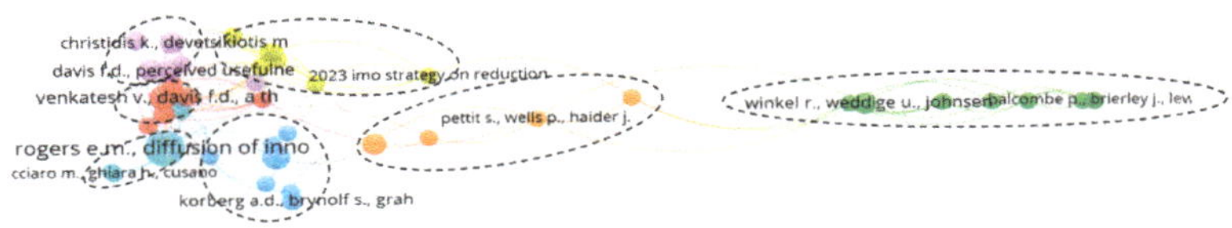

Figure 6. Network visualization of reference co-citation analysis.

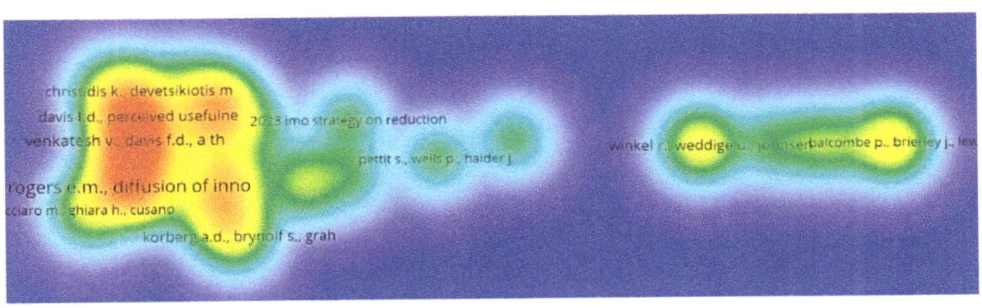

Figure 7. Density visualisation of reference co-citation analysis.

Each sphere in Figure 6 represents a cited reference. The size of the sphere is proportional to the number of citations of a cited reference. The lines between the spheres represent the strength of the interconnection link between the cited references. Finally, spheres of the same colour clustered together represent cited references with the highest levels of similarities between them (and are assumed to be dealing with similar themes or topics). As can be seen in Figure 7, the stronger the link between two cited references, the denser the visualisation. Documents with stronger links are represented in different colours (the strongest links are shown in red and in green the weakest ones), as shown in Figure 7.

Given these results, an attempt is made to name the different clusters of the knowledge base of this literature based on the documents they contain (Figure 6). It is important to note that an in-depth analysis of the content of each of the documents in each of the clusters will necessitate a different type of literature review (e.g., systematic review) and is therefore beyond the scope of this paper. In this first attempt to bring together this rather scattered literature on the adoption of technological innovations in the maritime industry, our aim was to use a robust and objective approach to organise this literature and provide a macro perspective. This is why we opted for a bibliometric approach. The only criteria used for naming the clusters was therefore limited to the titles of the documents since the software clusters together based on intertextual similarities. As also argued by Refs. [112,113], authors choose the titles of their research carefully, and they are indicative of the content of the articles. It is also worth noting that the names proposed were derived based on the data and could be considered as "ideal types" rather than rigid categorizations.

- Cluster 1 (green cluster): Shore power as technological innovation for maritime decarbonisation

The documents belonging to this cluster are shown in Table 9. It can be inferred from the titles of these documents that the broadest topic in this cluster is maritime decarbonisation, with a focus on shore power (or cold ironing) as a technological innovation solution. This cluster is therefore named "shore power as technological innovation for maritime decarbonisation". Issues related to cost, design, regulations, policies, and environmental and economic benefits are addressed by the references in this cluster. These documents are the main documents currently informing research on the adoption of shore power as a decarbonisation solution.

Table 9. Documents in the green cluster.

Documents	Titles
[114]	How to decarbonise international shipping: Options for fuels, technologies and policies
[115]	Cost benefit calculation tool onshore power supply, (2016)
[116]	Technical design aspects of harbour area grid for shore to ship power: State of the art and future solutions
[117]	Environmental assessment and regulatory aspects of cold ironing planning for a maritime route in the Adriatic Sea
[118]	Shore power management for maritime transportation: status and perspectives
[119]	Shore side electricity in Europe: potential and environmental benefits
[94]	Policy implementation barriers and economic analysis of shore power promotion in China
[120]	Is cold ironing hot enough? An actor focus perspective of on shore power supply (ops) at Copenhagen's harbour

- Cluster 2 (red cluster): Autonomous shipping and digital technologies

The documents belonging to this cluster are shown in Table 10. This is a rather heterogeneous cluster in the sense that the focus is on autonomous shipping and digital technologies. As such, it is named "autonomous shipping and digital technologies"; the issues analysed relate to their implementation and their impact in the industry. The main theoretical perspective informing these studies is the technology acceptance model.

Table 10. Documents in the red cluster.

Documents	Titles
[121]	Digitization in maritime logistics—What is there and what is missing?
[91]	Autonomous merchant vessels: examination of factors that impact the effective implementation of unmanned ships
[122]	Technological trajectories and scenarios in seaport digitalization
[123]	Analyzing the economic benefit of unmanned autonomous ships: an exploratory cost-comparison between an autonomous and a conventional bulk carrier
[124]	A theoretical extension of the technology acceptance model: four longitudinal field studies
[125]	User acceptance of information technology: Toward a unified view
[126]	Shipping innovation
[127]	Towards the assessment of potential impact of unmanned vessels on maritime transportation safety

- Cluster 3 (blue cluster): Alternative fuels as a decarbonisation solution

The documents in this cluster are shown in Table 11. This is another heterogeneous cluster that nonetheless can be named "alternative fuels as a decarbonisation solution" given that two documents deal with the topic. The presence of a blockchain-related paper in the cluster could be due to the fact two documents dealing with unrelated topics can both appear in the reference list of more than one document. This is a weakness of the bibliometric approach issued, though it remains a valid and robust review method.

Table 11. Documents in the blue cluster.

Documents	Titles
[128]	Agency theory: An assessment and review
[129]	Alternative marine fuels: Prospects based on multi-criteria decision analysis involving Swedish stakeholders
[130]	Techno-economic assessment of advanced fuels and propulsion systems in future fossil-free ships
[131]	Blockchain technology and the sustainable supply chain: Theoretically exploring adoption barriers

- Cluster 4 (yellow cluster): Blockchain technology applications in maritime supply chains and logistics

The documents included in this cluster are shown in Table 12. Blockchain is undoubtedly the technological innovation at the centre, with a focus on applications in maritime supply chains and logistics. This justifies why it is named "blockchain technologies applications in maritime supply chains and logistics".

Table 12. Documents in the yellow cluster.

Documents	Titles
[132]	Blockchain in global supply chains and cross border trade: a critical synthesis of the state-of-the-art, challenges and opportunities
[133]	The architectural design requirements of a blockchain-based port community system
[13]	Blockchain adoption from the Shipping industry: An empirical study
[134]	Blockchain's roles in meeting key supply chain management objectives
[135]	Blockchain technology implementation in logistics

- Cluster 5 (light purple cluster): Blockchain technology applications to information communication technologies and transport sustainability.

The documents in this cluster are included in Table 13. Blockchain is the technological innovation addressed, with a focus on applications in information and communication technologies, and transport sustainability, which justifies the chosen name. Research on this thematic area appears to be informed by the user acceptance theory.

Table 13. Documents in the light purple cluster.

Documents	Titles
[136]	Blockchains and smart contracts for the internet of things
[137]	User acceptance of computer technology: A comparison of two theoretical models
[138]	Perceived usefulness, perceived ease of use, and user acceptance of information technology
[60]	Applying blockchain technology: Evidence from Norwegian companies
[139]	Improving maritime transport sustainability using blockchain-based information exchange

- Cluster 6 (deep sky blue): Blockchain applications in maritime digitalisation

The documents included in this cluster are shown in Table 14. Blockchain is also at the centre stage, with a focus on applications in maritime shipping digitalisation, as reflected in the chosen name. This technology appears to be discussed in the context of innovation diffusion.

Table 14. Documents in the deep sky-blue cluster.

Documents	Titles
[140]	Energy management in seaports: A new role for port authorities
[51]	Diffusion of innovations
[141]	Markets and hierarchies: analysis and antitrust implications
[142]	Maritime shipping digitalization: Blockchain-based technology applications, future improvements, and intention to use

- Cluster 7 (orange cluster): Shipping sustainability transitions

The documents included in this cluster are shown in Table 15. The context of these papers is the broader topic of shipping sustainability, with some documents focusing on liquefied natural gas (LNG), regulations, and the conditions for a second socio-technical transition in shipping to occur.

Table 15. Documents in the orange cluster.

Documents	Titles
[143]	Improving sustainability of maritime transport through utilization of liquefied natural gas (LNG) for propulsion
[144]	Revisiting history: Can shipping achieve a second socio-technical transition for carbon emissions reduction?
[145]	Sustainability transitions in Baltic Sea shipping: Exploring the responses of firms to regulatory changes
[146]	Third IMO GHG study

The results of the co-citation analysis of the cited references revealed that the knowledge base of this literature is still relatively recent and is still being built. This is evidenced by the limited number of citations of a cited reference (3) that had to be applied for the network of 44 articles to be built. Of note, a citation threshold of 4 resulted in only 9 of the 9119 references being interconnected and a citation threshold of 5 resulting in only 5 interconnected references. This justifies why the threshold of 3 was selected.

The decrease in the number of interconnected cited references as the threshold of citation of cited reference increases reveals the following about the reviewed papers. First, the knowledge base of the literature on the studied topic constitutes non-influential documents (in the sense of not being very highly cited). A close look at the documents constituting the knowledge base reveals that the quasi-totality of them were published between 2014 and 2021. Given that citations take time to accumulate and depend on the popularity of the topic in the research community, this is understandable. With such a "recent" knowledge base, it can also be inferred that the adoption of technological innovations in the maritime industry is a recent topic, and thus worth exploring, as is being carried out in this paper. Second, the knowledge base of this literature is still in development and will be consolidated and partitioned into distinctive areas, forming the foundation of future research in these areas in the coming years.

In addition, the results also reveal that academic research on the adoption of technological innovations in the maritime industry has predominantly focused on two main technologies: shore power and blockchain. These two thematic areas are well delineated in the knowledge base. Furthermore, they also reveal that though shore power and blockchain technologies currently dominate, there are emerging thematic areas for autonomous ships, alternative fuels, digitalisation, and sustainability transitions that will be consolidated as more research is carried out in these areas. This research is accumulating and serving as the foundation for future research on these topics.

4.3. Research Fronts on the Adoption of Technological Innovations in the Maritime Industry (Q3)

The results of bibliographic coupling are reported and discussed in this section. They are based on a citation threshold of at least 15 citations. As can be seen in Figures 8–10, there are seven (7) research fronts in this literature. Each sphere in Figure 8 represents a published article. The size of each sphere is proportional to the number of citations of the article. The lines between the spheres represent the strength of the interconnection link between the coupled articles. Finally, spheres of the same colour clustered together represent coupled documents with the highest levels of similarities between them (with overlapping bibliographies). As can be seen in Figure 9, the stronger the link between the coupled documents, the denser the visualisation. The period within which the coupled documents were published is shown in Figure 10.

Similar to what has been carried out with the clusters, hereafter an attempt is made to name the different clustered research fronts based on their constituting documents (Figure 8). As for the knowledge base, the naming is based on the titles of the documents, with a focus on the technological innovation examined. Also, the names proposed are based on the data and could be considered "ideal types" rather than rigid categorisations.

- Research front 1 (red cluster): Adoption of shore power as a technological innovation for maritime decarbonisation

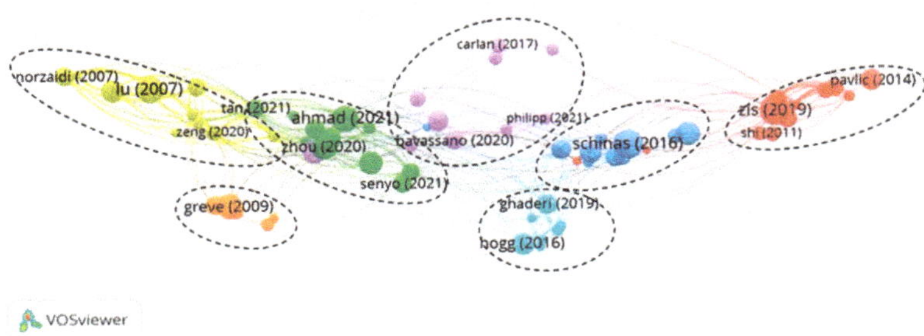

Figure 8. Network visualisation of bibliographic coupling results (based on a citation threshold of at least 15) In red are the strongest link.

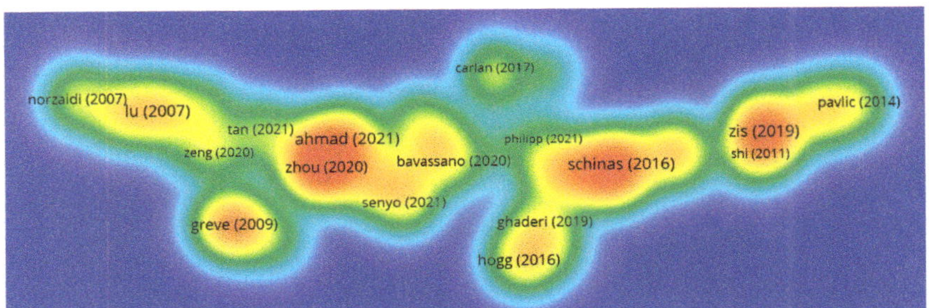

Figure 9. Density visualisation of bibliographic coupling results.

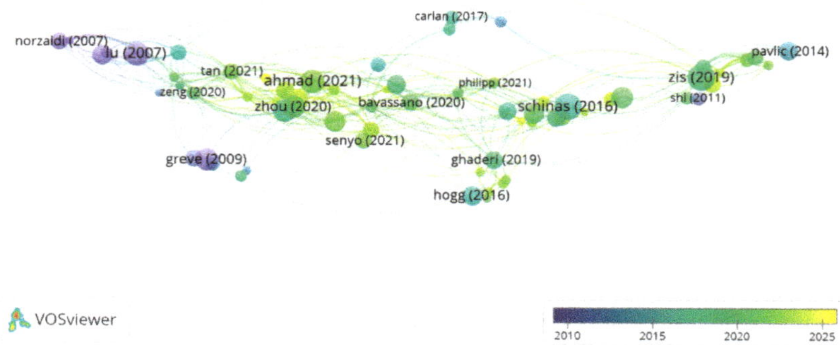

Figure 10. Overlay visualisation of bibliographic coupling results.

The documents included in this cluster are shown in Table 16, in which it can be seen that the main topic for shore power is decarbonisation, with a focus on shore power/cold ironing (7 of the 12 documents dealt with this topic). The adoption of shore power is thus a core trend in this literature and justifies the name chosen. Issues explored in relation to shore power include the role of government subsidies, barriers to adoption, industry-related opinions, emissions reduction potential, and policies that facilitate implementation. The appearance of other decarbonisation technological solutions such as green ammonia, biofuels, and sustainable port infrastructure is understandable, as the authors might have referred to them as examples of other solutions when discussing shore power. On the other hand, the appearance of RFID and automated container terminals in the table is a limitation of bibliographic coupling. As explained by Ref. [147], the bibliographies of two articles can overlap regarding unrelated subject matters.

Table 16. Documents in the red cluster.

Documents	Titles
[148]	Electrification of onshore power systems in maritime transportation towards decarbonization of ports: A review of the cold ironing technology
[149]	A roadmap to alternative fuels for decarbonising shipping: The case of green ammonia
[150]	The global trends of automated container terminal: a systematic literature review
[151]	Optimization of shore power deployment in green ports considering government subsidies
[152]	Sustainable port infrastructure, practical implementation of the green port concept
[61]	Critical barriers to the introduction of shore power supply for green port development: case of Djibouti container terminals
[106]	RFID technology and its application to port-based container logistics
[85]	Adoption of biofuels for marine shipping decarbonization: A long-term price and scalability assessment
[153]	Identifying industry-related opinions on shore power from a survey in China
[154]	A bilevel hybrid economic approach for optimal deployment of onshore power supply in maritime ports
[94]	Policy implementation barriers and economic analysis of shore power promotion in China
[155]	Prospects of cold ironing as an emissions reduction option

- Research front 2 (green cluster): Blockchain technology applications in maritime supply chains and logistics

The documents included in this cluster are shown in Table 17, which highlights how blockchain is the main technological innovation discussed, with an emphasis on its application in maritime supply chains and logistics.

Table 17. Documents in the green cluster.

Documents	Titles
[84]	Blockchain applications and architectures for port operations and logistics management
[88]	Blockchain adoption in the maritime supply chain: Examining barriers and salient stakeholders in containerized international trade
[156]	Technological drivers of seaports' business model innovation: An exploratory case study on the port of Barcelona
[157]	Managing a blockchain-based platform ecosystem for industry-wide adoption: The case of TradeLens
[81]	Blockchain technology in maritime supply chains: applications, architecture and challenges
[158]	Blockchain Adoption for Sustainable Supply Chain Management: Economic, Environmental, and Social Perspectives
[159]	Blockchain Technology for tracking and tracing containers: model and conception
[160]	Blockchain adoption in container shipping: An empirical study on barriers, approaches, and recommendations
[90]	Digital platformisation as public sector transformation strategy: A case of Ghana's paperless port
[161]	Assessing Blockchain Technology application for freight booking business: a case study from Technology Acceptance Model perspective
[86]	The key challenges and critical success factors of blockchain implementation: Policy implications for Singapore's maritime industry

- Research front 3 (blue cluster): Adoption of energy efficient ship technologies

Documents belonging to this research front are shown in Table 18. As evidenced in the titles of the documents, the focus is on energy efficient ship technologies. Examples of technologies analysed from an adoption perspective include hydrogen, electric batteries, biofuels, biogas, LNG, and solid oxide fuel cells. This is a rather homogeneous front, with all documents dealing with the same topic.

Table 18. Documents in the blue cluster.

Documents	Titles
[20]	Implementing maritime battery-electric and hydrogen solutions: A technological innovation systems analysis
[162]	Blending new and old in sustainability transitions: Technological alignment between fossil fuels and biofuels in Norwegian coastal shipping
[87]	The role of solid oxide fuel cells in future ship energy systems
[163]	Adopting different wind-assisted ship propulsion technologies as fleet retrofit: An agent-based modeling approach
[59]	Drivers for and barriers to biogas use in manufacturing, road transport and shipping: a demand-side perspective
[164]	Fuel cell and hydrogen in maritime application: A review on aspects of technology, cost and regulations
[165]	Biofuel as an alternative shipping fuel: technological, environmental and economic assessment
[21]	Drivers and barriers for the large-scale adoption of hydrogen fuel cells by Nordic shipping companies
[22]	Wind technologies: Opportunities and barriers to a low carbon shipping industry
[166]	The impact of split incentives on energy efficiency technology investments in maritime transport
[82]	Feasibility and commercial considerations of LNG-fueled ships

- Research front 4 (yellow cluster): Adoption of ICT and tracking technologies

Documents belonging to this research front are shown in Table 19, the majority of which deal with the adoption of ICTs, followed by tracing and tracking technologies. This justifies the name chosen: "adoption of ICT and tracking technologies".

Table 19. Documents in the yellow cluster.

Documents	Titles
[167]	Critical factors affecting the adoption of container security service: The shippers' perspective
[168]	The facilitating role of IT systems for legal compliance: the case of port community systems and container Verified Gross Mass (VGM)
[13]	Blockchain adoption from the Shipping industry: An empirical study
[12]	Development and validation of a technology acceptance model for safety-enhancing, wearable locating systems
[83]	Application of structural equation modeling to evaluate the intention of shippers to use Internet services in liner shipping
[169]	Intranet usage and managers' performance in the port industry
[170]	A study of RFID adoption for vehicle tracking in a container terminal
[171]	RFID benefits, costs, and possibilities: The economical analysis of RFID deployment in a cruise corporation global service supply chain
[172]	Factors that affect acceptance and use of information systems within the Maritime industry in developing countries: The case of Ghana
[173]	The adoption of open platform for container bookings in the maritime supply chain

- Research front 5 (light purple cluster): Innovation adoption in seaports

Documents in this cluster are shown in Table 20. The emphasis is on innovation adoption in seaports, with blockchain, liquefied biogas, and LNG being discussed. As such, the cluster is named "innovation adoption in seaports".

Table 20. Documents in the light purple cluster.

Documents	Titles
[174]	Are the innovation processes in seaport terminal operations successful?
[175]	Innovation in the maritime sector: aligning strategy with outcomes
[105]	How to turn an innovative concept into a success? An application to seaport-related innovation
[176]	Blockchain: How shipping industry is dealing with the ultimate technological leap
[177]	Digital innovation in the port sector: Barriers and facilitators
[60]	Applying Blockchain technology: Evidence from Norwegian companies
[178]	Blockchain for LBG Maritime Energy Contracting and Value Chain Management: A Green Shipping Business Model for Seaports
[179]	Blockchain and Smart Contracts for Entrepreneurial Collaboration in Maritime Supply Chains
[180]	Towards Green and Smart Seaports: Renewable Energy and Automation Technologies for Bulk Cargo Loading Operations
[181]	The role of port authorities in the development of LNG bunkering facilities in North European ports

- Research front 6 (deep sky-blue cluster): Adoption of maritime autonomous ships

Documents in this cluster are shown in Table 21 below, and all deal with autonomous surface ships. Aspects explored concern public perception, implementation factors, challenges, and regulations.

Table 21. Documents in the deep sky-blue cluster.

Documents	Titles
[17]	Assessing innovation in transport: An application of the Technology Adoption (TechAdo) model to Maritime Autonomous Surface Ships (MASS)
[182]	Autonomous technologies in short sea shipping: trends, feasibility and implications
[183]	An exploratory investigation of public perceptions towards autonomous urban ferries
[91]	Autonomous merchant vessels: examination of factors that impact the effective implementation of unmanned ships
[184]	Maritime Autonomous Surface Ships: Problems and Challenges Facing the Regulatory Process
[18]	Autonomous ships: A study of critical success factors
[185]	Regulatory framework analysis for the unmanned inland waterway vessel

- Research front 7 (orange cluster): Adoption of machinery and equipment

Documents in this cluster are shown in Table 22, with the adoption of machinery and equipment being a core topic. As such, it is named "adoption of machinery and equipment".

Table 22. Documents in the orange cluster.

Documents	Titles
[109]	Digitalizing the maritime industry: A case study of technology acquisition and enabling advanced manufacturing technology
[10]	Bigger and safer: The diffusion of competitive advantage
[26]	Fast and expensive: The diffusion of a disappointing innovation
[186]	Environmental innovation and the role of stakeholder collaboration in West Coast port gateways
[187]	'Information communication technology' innovation in a non-high technology sector: achieving competitive advantage in the shipping industry

The bibliographic coupling analysis revealed that the extant literature on the adoption of technology comprises seven (7) research fronts: adoption of shore power as a technological innovation for maritime decarbonisation, blockchain technology applications in maritime supply chains and logistics, adoption of energy efficient ship technologies, adoption of ICT and tracking technologies, innovation adoption in seaports, and adoption of machinery and equipment. Energy efficient ship technologies (among which shore power is the most studied technology), blockchain, and autonomous ships are of most interest to researchers. This can be explained by the drive towards a cleaner maritime industry. Indeed, though maritime autonomous surface ships and blockchain technologies do not directly contribute to decarbonisation, their adoption has the potential to contribute to more sustainable maritime shipping [188,189].

4.4. Linking the Knowledge Base to Research Fronts

As explained in Section 3.4, co-citation of references and bibliographic coupling were chosen because they are two complementary techniques. The former allowed us to identify seven thematic areas that constitute the knowledge base of the literature under consideration, while the latter allowed us to identify seven research fronts. It was observed that shore power is a well-delineated thematic area of the knowledge base, as well as a key research front. Second, though still a nascent thematic area of the knowledge base, the adoption of autonomous surface ships is a well-delineated research front. Third, blockchain applications in maritime supply chains and logistics is also both a thematic knowledge base area and a research front. Fourth, the knowledge base of the research fronts adoption of ship energy efficient technologies, adoption of ICT and tracking technologies, innovation adoption in seaports, and adoption of machinery and equipment expansion (i.e., as research continues to investigate these areas) will develop in the future.

Given the period covered (1999–2025), the dominance of the knowledge base and of the research fronts based on recent technological popularity trends such as shore power, blockchain, autonomous shipping, and alternative fuels could appear intriguing at first sight. However, a look at Figure 3 reveals that publications on the topic explored pick up in 2016, when these technologies were becoming popular in the maritime domain. In the face of tightening environmental regulations by the IMO, these technologies are the main ones being considered by the industry as solutions for emissions reduction. They have thus attracted much interest from researchers, shifting the balance in the 171 articles towards these recent trends.

5. Conclusions and Trajectories for Future Research

Research on the adoption of technological innovations in the maritime industry has grown considerably in the last five years, and this trend is expected to continue in the current context of maritime decarbonisation. The purpose of this paper was to bring together and synthesise this growing literature. Using performance analysis and science mapping, we analysed 171 journal articles published on the topic. The results of the performance analysis showed the dominance of Asian institutions in terms of the number of documents they have contributed, and the dominance of Asian and European authors in terms of productivity. The global nature of this literature was evidenced by the number of multiple-country collaborations. Through a co-citation analysis of 9119 cited references, it was shown that the extant literature on the adoption of technological innovations in the maritime industry is built on seven thematic knowledge base areas in different stages of development. Finally, through bibliographic coupling of 171 journal articles, it was shown that seven research fronts dominate in this literature. The link between the knowledge base and the research fronts was also established.

The main contributions of this paper are fourfold. First, being the first attempt to bring together this scattered and transdisciplinary literature, it brings clarity and highlights the knowledge base and the research fronts. As such, it lays a strong foundation upon which researchers looking to orientate themselves on the topic can build [28]. Second, by using two objective approaches (co-citation of cited references and bibliographic coupling) to identify the knowledge base and the research fronts of this literature, it contributes to theoretical advancement [190]. Third, its practical contribution relates to the objective assessment and reporting productivity, and impact-related indicators. Fourth, having identified thematic areas of the knowledge base and the research fronts of this literature, it provides editors with a tool they can rely upon to call for special issues to increase the representativeness of themes identified in this paper in their journals. Moreover, the paper also sheds light on the technological innovation adopted in the maritime industry. It has been shown that though many technological innovations have been adopted, the focus of the academic community has been on just a few of these, namely blockchain, shore power, and alternative fuels. More research on other technologies is needed.

This research has the following limitations. The first relates to the bibliometric techniques used. Though bibliometric studies involve rigorous and replicable quantitative approaches, some studies (e.g., [191,192]) have argued that they only provide a macro perspective and miss key insights that can only be revealed through in-depth qualitative approaches. Though this is true to some extent, for the purpose of this paper, the methods used are justified given the large number of documents (171) and cited references analysed (9119). Since the techniques used cannot identify keys issues related to the settings, methods, and types of data used in the analysed papers, they have nonetheless allowed the identification of documents forming the backbone of research on the topic examined, which is an important step for future research. Traditional review approaches might thus be used in future research to complement our results. Such studies might look into the methodology used in the papers constituting the knowledge base and the research fronts of the literature identified in this paper. Another limitation of the techniques used in this paper is that they are citation- and trend-driven. As explained in Section 3.4, while bibliographic coupling favours recent publications, reference co-citation analysis favours older works. The implication is that some relevant research that had not accumulated enough citations might have been missed, given that citation thresholds had to be set. Other studies might also replicate our study using alternative citation thresholds. A further limitation is that the documents analysed were limited to journal articles. Though important documents in the form of conference papers, book chapters, industry reports, or others might have been missed, the choice of focusing only on peer-reviewed journal articles ensured the documents analysed were vetted by the relevant academic community and that the analysis was conducted on documents of high quality.

Author Contributions: Conceptualisation, A.D.; methodology, A.D.; validation, A.T. and C.F.; formal analysis, A.D.; investigation, A.D.; resources, A.T. and C.F.; data curation, A.D., A.T. and C.F.; writing—original draft preparation, A.D. and A.T.; writing—review and editing, A.D., A.T. and C.F.; visualisation, A.D.; supervision, A.T. and C.F.; project administration, A.T.; funding acquisition, A.T. All authors have read and agreed to the published version of the manuscript.

Funding: This research was carried out during the European project Optiwise, funded by HORIZON EUROPE (Climate, Energy & Mobility) with grant agreement No. 101056769.

Conflicts of Interest: The authors declare no conflicts of interest. The funders had no role in the design of the study; in the collection, analyses, or interpretation of data; in the writing of the manuscript; or in the decision to publish the results.

Appendix A

Keywords combinations used to retrieve articles from Scopus.

Keywords Combinations in Title, Abstract, and Keyword	Articles Retrieved from Scopus
Drivers	
TITLE-ABS-KEY (("Drivers" OR "factors" OR "triggers" OR "enablers") AND ("technology*") AND ("adoption" OR "deployment" OR "implementation" OR "acceptance") AND (innovation) AND ("shipping" OR "port" OR "maritime industry" OR "ship")) AND (LIMIT-TO (SRCTYPE, "j")) AND (LIMIT-TO (LANGUAGE, "English")) AND (LIMIT-TO (DOCTYPE, "ar"))	29
TITLE-ABS-KEY (("Drivers" OR "factors" OR "triggers" OR "enablers") AND "technology*" AND ("adoption" OR "deployment" OR "implementation" OR "acceptance") AND ("shipping" OR "port" OR ("maritime industry") OR "ship")) AND (LIMIT-TO (SRCTYPE, "j")) AND (LIMIT-TO (DOCTYPE, "ar") AND (LIMIT-TO (LANGUAGE, "English"))	271
TITLE-ABS-KEY ((drivers OR factors OR triggers OR enablers) AND innovation AND (adoption OR deployment OR implementation OR acceptance) AND (shipping OR port OR ("maritime industry") OR ship) AND (LIMIT-TO (DOCTYPE, "ar")) AND (LIMIT-TO (SRCTYPE, "j")) AND (LIMIT-TO (LANGUAGE, "English"))	55

Keywords Combinations in Title, Abstract, and Keyword	Articles Retrieved from Scopus
TITLE-ABS-KEY ((drivers OR factors OR triggers OR enablers) AND (technology OR innovation) AND (adoption OR deployment OR implementation OR acceptance) AND (shipping OR port OR ("maritime industry") OR ship) AND (LIMIT-TO (LANGUAGE, "English")) AND (LIMIT-TO (DOCTYPE, "ar")) AND (LIMIT-TO (SRCTYPE, "j"))	297
Barriers	
TITLE-ABS-KEY (("Barriers" OR "obstacles" OR "challenges" OR "impediments" OR "hindrances" OR "constraints" OR "limitations") AND ("adoption" OR "deployment" OR "implementation" OR "acceptance") AND ("technology*") AND innovation* AND ("shipping" OR "port" OR "ship" OR "maritime industry") AND (LIMIT-TO (DOCTYPE, "ar")) AND (LIMIT-TO (LANGUAGE, "English")) AND (LIMIT-TO (SRCTYPE, "j"))	47
TITLE-ABS-KEY ((barriers OR obstacles OR challenges OR impediments OR hindrances OR constraints OR limitations) AND technology AND (adoption OR deployment OR implementation OR acceptance) AND (shipping OR port OR ship OR ("maritime industry"))) AND (LIMIT-TO (DOCTYPE, "ar") AND (LIMIT-TO (SRCTYPE, "j")) AND (LIMIT-TO (LANGUAGE, "English"))	457
TITLE-ABS-KEY ((barriers OR obstacles OR challenges OR impediments OR hindrances OR constraints OR limitations) AND innovation AND (adoption OR deployment OR implementation OR acceptance) AND (shipping OR port OR ship OR ("maritime industry"))) AND (LIMIT-TO (DOCTYPE, "ar") AND (LIMIT-TO (LANGUAGE, "English")) AND (LIMIT-TO (SRCTYPE, "j"))	87
TITLE-ABS-KEY ((barriers OR obstacles OR challenges OR impediments OR hindrances OR constraints OR limitations) AND (innovation OR technology) AND (adoption OR deployment OR implementation OR acceptance) AND (shipping OR port OR ship OR ("maritime industry"))) AND (LIMIT-TO (SRCTYPE, "j")) AND (LIMIT-TO (DOCTYPE, "ar")) AND (LIMIT-TO (LANGUAGE, "English"))	497

References

1. Chalfant, J.; Kite-Powell, H.; Bonfiglio, L.; Chryssostomidis, C. Decarbonization of the Cargo Shipping Fleet. *J. Ship Prod. Des.* **2022**, *38*, 199–214. [CrossRef]
2. Romano, A.; Yang, Z. Decarbonisation of shipping: A state of the art survey for 2000–2020. *Ocean Coast. Manag.* **2021**, *214*, 105936. [CrossRef]
3. King, J. Technology and the course of shipping. *Ocean Coast. Manag.* **2001**, *44*, 567–577. [CrossRef]
4. Strez, C. The dynamics of propulsion technology adoption in the maritime industry: A systems dynamics model of technology transition. In Proceedings of the 16th Annual General Assembly and Conference of the International Association of Maritime Universities, Rijeka, Croatia, 7–10 October 2015.
5. Coşar, A.K.; Demir, B. Shipping inside the box: Containerization and trade. *J. Int. Econ.* **2018**, *114*, 331–345. [CrossRef]
6. Notteboom, T.; Rodrigue, J.-P. The future of containerization: Perspectives from maritime and inland freight distribution. *GeoJournal* **2009**, *74*, 7–22. [CrossRef]
7. Haralambides, H.E. Gigantism in container shipping, ports and global logistics: A time-lapse into the future. *Marit. Econ. Logist.* **2019**, *21*, 1–60. [CrossRef]
8. Tran, N.K.; Haasis, H.-D. An empirical study of fleet expansion and growth of ship size in container liner shipping. *Int. J. Prod. Econ.* **2015**, *159*, 241–253. [CrossRef]
9. Glen, D. Modelling the impact of double hull technology on oil spill numbers. *Marit. Policy Manag.* **2010**, *37*, 475–487. [CrossRef]
10. Greve, H.R. Bigger and safer: The diffusion of competitive advantage. *Strateg. Manag. J.* **2009**, *30*, 1–23. [CrossRef]
11. Buhalis, D.; Papathanassis, A.; Vafeidou, M. Smart cruising: Smart technology applications and their diffusion in cruise tourism. *J. Hosp. Tour. Technol.* **2022**, *13*, 626–649. [CrossRef]
12. Kwee-Meier, S.T.; Bützler, J.E.; Schlick, C. Development and validation of a technology acceptance model for safety-enhancing, wearable locating systems. *Behav. Inform. Technol.* **2016**, *35*, 394–409. [CrossRef]
13. Kapnissis, G.; Vaggelas, G.K.; Leligou, H.C.; Panos, A.; Doumi, M. Blockchain adoption from the Shipping industry: An empirical study. *Marit. Transp. Res.* **2022**, *3*, 100058. [CrossRef]
14. Lin, H.-F. Blockchain adoption in the maritime industry: Empirical evidence from the technological-organizational-environmental framework. *Marit. Policy Manag.* **2023**, *51*, 1474–1496. [CrossRef]

15. Orji, I.J.; Kusi-Sarpong, S.; Huang, S.; Vazquez-Brust, D. Evaluating the factors that influence blockchain adoption in the freight logistics industry. *Transp. Res. Part E Logist. Transp. Rev.* **2020**, *141*, 102025. [CrossRef]
16. Todd, P. Electronic bills of lading, blockchains and smart contracts. *Int. J. Law Inf. Technol.* **2019**, *27*, 339–371. [CrossRef]
17. Fonseca, T.; Lagdami, K.; Schröder-Hinrichs, J.-U. Assessing innovation in transport: An application of the Technology Adoption (TechAdo) model to Maritime Autonomous Surface Ships (MASS). *Transp. Policy* **2021**, *114*, 182–195. [CrossRef]
18. Li, X.; Yuen, K.F. Autonomous ships: A study of critical success factors. *Marit. Econ. Logist.* **2022**, *24*, 228–254. [CrossRef]
19. Park, Y.-J.; Jeong, Y.-J.; An, Y.-S.; Ahn, J.-K. Analyzing the Factors Influencing the Intention to Adopt Autonomous Ships Using the TOE Framework and DOI Theory. *J. Navig. Port. Res.* **2022**, *46*, 134–144.
20. Bach, H.; Bergek, A.; Bjørgum, Ø.; Hansen, T.; Kenzhegaliyeva, A.; Steen, M. Implementing maritime battery-electric and hydrogen solutions: A technological innovation systems analysis. *Transp. Res. Part D Transp. Environ.* **2020**, *87*, 102492. [CrossRef]
21. Latapí, M.; Davíðsdóttir, B.; Jóhannsdóttir, L. Drivers and barriers for the large-scale adoption of hydrogen fuel cells by Nordic shipping companies. *Int. J. Hydrogen Energy* **2023**, *48*, 6099–6119. [CrossRef]
22. Rehmatulla, N.; Parker, S.; Smith, T.; Stulgis, V. Wind technologies: Opportunities and barriers to a low carbon shipping industry. *Mar. Policy* **2017**, *75*, 217–226. [CrossRef]
23. Rojon, I.; Dieperink, C. Blowin' in the wind? Drivers and barriers for the uptake of wind propulsion in international shipping. *Energy Policy* **2014**, *67*, 394–402. [CrossRef]
24. Schwartz, H.; Gustafsson, M.; Spohr, J. Emission abatement in shipping-is it possible to reduce carbon dioxide emissions profitably? *J. Clean. Prod.* **2020**, *254*, 120069. [CrossRef]
25. Shi, J.; Zhu, Y.; Feng, Y.; Yang, J.; Xia, C. A Prompt Decarbonization Pathway for Shipping: Green Hydrogen, Ammonia, and Methanol Production and Utilization in Marine Engines. *Atmosphere* **2023**, *14*, 584. [CrossRef]
26. Greve, H.R. Fast and expensive: The diffusion of a disappointing innovation. *Strateg. Manag. J.* **2011**, *32*, 949–968. [CrossRef]
27. Ferreira, F.A. Mapping the field of arts-based management: Bibliographic coupling and co-citation analyses. *J. Bus. Res.* **2018**, *85*, 348–357. [CrossRef]
28. Vogel, R.; Güttel, W.H. The dynamic capability view in strategic management: A bibliometric review. *Int. J. Manag. Rev.* **2013**, *15*, 426–446. [CrossRef]
29. Dragović, A.; Zrnić, N.; Dragović, B.; Dulebenets, M.A. A comprehensive review of Maritime Bibliometric Studies (2014–2024). *Ocean Eng.* **2024**, *311*, 118917. [CrossRef]
30. Ferrigno, G.; Crupi, A.; Di Minin, A.; Ritala, P. 50+ years of R&D Management: A retrospective synthesis and new research trajectories. *R&D Manag.* **2023**, *53*, 900–926.
31. Faraji, O.; Asiaei, K.; Rezaee, Z.; Bontis, N.; Dolatzarei, E. Mapping the conceptual structure of intellectual capital research: A co-word analysis. *J. Innov. Knowl.* **2022**, *7*, 100202. [CrossRef]
32. Xu, Z.; Ge, Z.; Wang, X.; Skare, M. Bibliometric analysis of technology adoption literature published from 1997 to 2020. *Technol. Forecast. Soc. Change* **2021**, *170*, 120896. [CrossRef]
33. Bai, X.; Hou, Y.; Yang, D. Choose clean energy or green technology? Empirical evidence from global ships. *Transp. Res. Part E Logist. Transp. Rev.* **2021**, *151*, 102364. [CrossRef]
34. IMO. *IMO Approves Net-Zero Regulations for Global Shipping*; International Maritime Organization (IMO): London, UK, 2025; Available online: https://www.imo.org/en/mediacentre/pressbriefings/pages/imo-approves-netzero-regulations.aspx (accessed on 1 May 2025).
35. Sheng, D.; Jiang, J.; Wang, H.; Tan, Z.; Wang, Y. Optimal compliance choices for ocean carriers under the sulphur regulation. *Transp. Res. Part D Transp. Environ.* **2023**, *116*, 103639. [CrossRef]
36. Zhang, X.; Bao, Z.; Ge, Y.-E. Investigating the determinants of shipowners' emission abatement solutions for newbuilding vessels. *Transp. Res. Part D Transp. Environ.* **2021**, *99*, 102989. [CrossRef]
37. Godin, B. Technological innovation: On the origins and development of an inclusive concept. *Technol. Cult.* **2016**, *57*, 527–556. [CrossRef]
38. Akbari, M.; Khodayari, M.; Khaleghi, A.; Danesh, M.; Padash, H. Technological innovation research in the last six decades: A bibliometric analysis. *Eur. J. Innov. Manag.* **2021**, *24*, 1806–1831. [CrossRef]
39. Vaughan, J. Defining technological innovation. *Lib. Technol. Rep.* **2013**, *49*, 10–46.
40. Cambridge Dictionary. Technology. 2005. Available online: https://dictionary.cambridge.org/dictionary/english/technology (accessed on 4 May 2025).
41. Baregheh, A.; Rowley, J.; Sambrook, S. Towards a multidisciplinary definition of innovation. *Manag. Decis.* **2009**, *47*, 1323–1339. [CrossRef]
42. Edison, H.; Bin Ali, N.; Torkar, R. Towards innovation measurement in the software industry. *J. Syst. Softw.* **2013**, *86*, 1390–1407. [CrossRef]
43. Johannessen, J.A.; Olsen, B.; Lumpkin, G.T. Innovation as newness: What is new, how new, and new to whom? *Eur. J. Innov. Manag.* **2001**, *4*, 20–31. [CrossRef]

44. OECD. *The Oslo Manual*; OECD/Euro-Stat: Paris, France, 2005.
45. Varis, M.; Littunen, H. Types of innovation, sources of information and performance in entrepreneurial SMEs. *Eur. J. Innov Manag.* **2010**, *13*, 128–154. [CrossRef]
46. Du Plessis, M. The role of knowledge management in innovation. *J. Know Manag.* **2007**, *11*, 20–29. [CrossRef]
47. Ettlie, J.E.; Bridges, W.P.; O'keefe, R.D. Organization strategy and structural differences for radical versus incremental innovation *Manag. Sci.* **1984**, *30*, 682–695. [CrossRef]
48. Butler, J.E. Theories of Technological Innovation as Useful Tools for Corporate Strategy. *Strateg. Manag. J.* **1988**, *9*, 15–29. [CrossRef]
49. Wei, X.; Liu, R.; Chen, W. Meta theories of technological innovation based on the analysis of classic texts. *Heliyon* **2023**, *9*, e16779. [CrossRef]
50. Yadegari, M.; Mohammadi, S.; Masoumi, A.H. Technology adoption: An analysis of the major models and theories. *Technol. Anal. Strateg.* **2022**, 1096–1110. [CrossRef]
51. Rogers, E. *Diffusions of Innovations*; Free Press: New York, NY, USA, 1995.
52. Laribi, S.; Guy, E. Marine energy transition with LNG and electric batteries: A technological adoption analysis of Norwegian ferries. *Marit. Bus. Rev.* **2023**, *8*, 80–96. [CrossRef]
53. Mansfield, E. Technical change and the rate of imitation. *Econometrica* **1961**, *29*, 741–766. [CrossRef]
54. Moore, G. *Crossing the Chasm: Marketing and Selling Technology Products to Mainstream Customers*; HarperCollins: New York, NY, USA, 1999.
55. Frattini, F.; Bianchi, M.; De Massis, A.; Sikimic, U. The role of early adopters in the diffusion of new products: Differences between platform and nonplatform innovations. *J. Prod. Innov. Manag.* **2014**, *31*, 466–488. [CrossRef]
56. Park, S.; Yoon, S.-H. Separating early-adopters from the majority: The case of Broadband Internet access in Korea. *Technol. Forecast. Soc. Change* **2005**, *72*, 301–325. [CrossRef]
57. Rycroft, R.W. Time and technological innovation: Implications for public policy. *Technol. Soc.* **2006**, *28*, 281–301. [CrossRef]
58. Xiao, Y.; Watson, M. Guidance on conducting a systematic literature review. *J. Plan. Educ. Res.* **2019**, *39*, 93–112. [CrossRef]
59. Dahlgren, S.; Kanda, W.; Anderberg, S. Drivers for and barriers to biogas use in manufacturing, road transport and shipping: A demand-side perspective. *Biofuels* **2022**, *13*, 177–188. [CrossRef]
60. Gausdal, A.H.; Czachorowski, K.V.; Solesvik, M.Z. Applying blockchain technology: Evidence from Norwegian companies. *Sustainability* **2018**, *10*, 1985. [CrossRef]
61. Radwan, M.E.; Chen, J.; Wan, Z.; Zheng, T.; Hua, C.; Huang, X. Critical barriers to the introduction of shore power supply for green port development: Case of Djibouti container terminals. *Clean. Technol. Environ.* **2019**, *21*, 1293–1306. [CrossRef]
62. Zhang, X.; Lam, J.S.L. A fuzzy Delphi-AHP-TOPSIS framework to identify barriers in big data analytics adoption: Case of maritime organizations. *Marit. Policy Manag.* **2019**, *46*, 781–801. [CrossRef]
63. Kitchenham, B.; Brereton, O.P.; Budgen, D.; Turner, M.; Bailey, J.; Linkman, S. Systematic literature reviews in software engineering–a systematic literature review. *Inform. Softw. Technol.* **2009**, *51*, 7–15. [CrossRef]
64. Li, X.; Zhou, Y.; Yuen, K.F. Blockchain implementation in the maritime industry: Critical success factors and strategy formulation. *Marit. Policy Manag.* **2022**, *51*, 304–322. [CrossRef]
65. Prussi, M.; Scarlat, N.; Acciaro, M.; Kosmas, V. Potential and limiting factors in the use of alternative fuels in the European maritime sector. *J. Clean. Prod.* **2021**, *291*, 125849. [CrossRef]
66. Wan, C.; Yan, X.; Zhang, D.; Yang, Z. Analysis of risk factors influencing the safety of maritime container supply chains. *Int. J. Ship Trans. Log.* **2019**, *11*, 476–507. [CrossRef]
67. Damian, S.E.; Wong, L.A.; Shareef, H.; Ramachandaramurthy, V.K.; Chan, C.K.; Moh, T.S.Y.; Tiong, M.C. Review on the challenges of hybrid propulsion system in marine transport system. *J. Energy Storage* **2022**, *56*, 105983. [CrossRef]
68. Searcy, T. Harnessing the wind: A case study of applying Flettner rotor technology to achieve fuel and cost savings for Fiji's domestic shipping industry. *Mar. Policy* **2017**, *86*, 164–172. [CrossRef]
69. Wiśnicki, B.; Wagner, N.; Wołejsza, P. Critical areas for successful adoption of technological innovations in sea shipping–the autonomous ship case study. *Innovation* **2021**, *37*, 582–608. [CrossRef]
70. Okoli, C.; Schabram, K. A Guide to Conducting a Systematic Literature Review of Information Systems Research. Available online: https://ssrn.com/abstract=1954824 (accessed on 10 January 2025).
71. Zupic, I.; Čater, T. Bibliometric Methods in Management and Organization. *Organ. Res. Methods* **2015**, *18*, 429–472. [CrossRef]
72. Dabić, M.; Marzi, G.; Vlačić, B.; Daim, T.U.; Vanhaverbeke, W. 40 years of excellence: An overview of Technovation and a roadmap for future research. *Technovation* **2021**, *106*, 102303. [CrossRef]
73. Nesari, M.; Naghizadeh, M.; Ghazinoori, S.; Manteghi, M. The evolution of socio-technical transition studies: A scientometric analysis. *Technol. Soc.* **2022**, *68*, 101834. [CrossRef]
74. Li, K.X.; Li, M.; Zhu, Y.; Yuen, K.F.; Tong, H.; Zhou, H. Smart port: A bibliometric review and future research directions. *Transp. Res. Part E Logist. Transp. Rev.* **2023**, *174*, 103098. [CrossRef]

75. van Eck, N.J.; Waltman, L. Visualizing Bibliometric Networks. In *Measuring Scholarly Impact: Methods and Practice*; Ding, Y., Rousseau, R., Wolfram, D., Eds.; Springer International Publishing: Basel, Switzerland, 2014; pp. 285–320.
76. Aria, M.; Cuccurullo, C. Bibliometrix: An R-tool for comprehensive science mapping analysis. *J. Informetr.* **2017**, *11*, 959–975. [CrossRef]
77. Cobo, M.J.; López-Herrera, A.G.; Herrera-Viedma, E.; Herrera, F. An approach for detecting, quantifying, and visualizing the evolution of a research field: A practical application to the Fuzzy Sets Theory field. *J. Informetr.* **2011**, *5*, 146–166. [CrossRef]
78. Donthu, N.; Kumar, S.; Mukherjee, D.; Pandey, N.; Lim, W.M. How to conduct a bibliometric analysis: An overview and guidelines. *J. Bus. Res.* **2021**, *133*, 285–296. [CrossRef]
79. Boyack, K.W.; Klavans, R. Co-citation analysis, bibliographic coupling, and direct citation: Which citation approach represents the research front most accurately? *J. Am. Soc. Inf. Sci. Tec.* **2010**, *61*, 2389–2404. [CrossRef]
80. Giannopoulos, G.A. The application of information and communication technologies in transport. *Eur. J. Oper. Res.* **2004**, *152*, 302–320. [CrossRef]
81. Liu, J.; Zhang, H.; Zhen, L. Blockchain technology in maritime supply chains: Applications, architecture and challenges. *Int. J. Prod. Res.* **2023**, *61*, 3547–3563. [CrossRef]
82. Schinas, O.; Butler, M. Feasibility and commercial considerations of LNG-fueled ships. *Ocean Eng.* **2016**, *122*, 84–96. [CrossRef]
83. Lu, C.-S.; Lai, K.-h.; Cheng, T.C.E. Application of structural equation modeling to evaluate the intention of shippers to use Internet services in liner shipping. *Eur. J. Oper. Res.* **2007**, *180*, 845–867. [CrossRef]
84. Ahmad, R.W.; Hasan, H.; Jayaraman, R.; Salah, K.; Omar, M. Blockchain applications and architectures for port operations and logistics management. *Res. Transp. Bus. Manag.* **2021**, *41*, 100620. [CrossRef]
85. Tan, W.C.; Sidhu, M.S. Review of RFID and IoT integration in supply chain management. *Oper. Res. Perspect.* **2022**, *9*, 100229. [CrossRef]
86. Zhou, Y.; Soh, Y.S.; Loh, H.S.; Yuen, K.F. The key challenges and critical success factors of blockchain implementation: Policy implications for Singapore's maritime industry. *Mar. Policy* **2020**, *122*, 104265. [CrossRef]
87. Baldi, F.; Moret, S.; Tammi, K.; Maréchal, F. The role of solid oxide fuel cells in future ship energy systems. *Energy* **2020**, *194*, 116811. [CrossRef]
88. Balci, G.; Surucu-Balci, E. Blockchain adoption in the maritime supply chain: Examining barriers and salient stakeholders in containerized international trade. *Transp. Res. Part E Logist. Transp. Rev.* **2021**, *156*, 102539. [CrossRef]
89. Koilo, V. Unlocking the sustainable value with digitalization: Views of maritime stakeholders on business opportunities. *Probl. Perspect. Manag.* **2024**, *22*, 401–417. [CrossRef]
90. Senyo, P.K.; Effah, J.; Osabutey, E.L.C. Digital platformisation as public sector transformation strategy: A case of Ghana's paperless port. *Technol. Forecast. Soc. Change* **2021**, *162*, 120387. [CrossRef]
91. Hogg, T.; Ghosh, S. Autonomous merchant vessels: Examination of factors that impact the effective implementation of unmanned ships. *Aust. J. Marit. Ocean Aff.* **2016**, *8*, 206–222. [CrossRef]
92. Majoral, G.; Reyes, A.; Saurí, S. Lessons from Reality on Automated Container Terminals: What Can Be Expected from Future Technological Developments? *Transp. Res. Rec.* **2023**, *2678*, 401–415. [CrossRef]
93. Hadley, M.; Pourzanjani, M. How remote is remote pilotage? *WMU J. Marit. Aff.* **2003**, *2*, 181–197. [CrossRef]
94. Yin, M.; Wang, Y.; Zhang, Q. Policy implementation barriers and economic analysis of shore power promotion in China. *Transp. Res. Part D Transp. Environ.* **2020**, *87*, 102506. [CrossRef]
95. Devarapali, S.; Manske, A.; Khayamim, R.; Jacobs, E.; Li, B.; Elmi, Z.; Dulebenets, M.A. Electric tugboat deployment in maritime transportation: Detailed analysis of advantages and disadvantages. *Marit. Bus. Rev.* **2024**, *9*, 263–291. [CrossRef]
96. Ho, T.C.; Lee, H.S. An Analysis of the Technology, Service Quality, and Relevance for CSBC Corporation: Taiwan's Installation of Scrubber Systems. *Sustainability* **2023**, *15*, 5641. [CrossRef]
97. Pisanias, N.; Willcocks, L. Understanding slow Internet adoption: 'Infomediation'; in ship-broking markets. *J. Inf. Technol.* **1999**, *14*, 399–413. [CrossRef]
98. Singh, S.; Lakshay; Pratap, S.; Jauhar, S.K. Unveiling barriers to IoT adoption in the maritime freight industry. *Int. J. Syst. Assur. Eng. Manag.* **2024**, *15*, 1–11.
99. Potter, A.; Wang, Y.; Naim, M. Scaling-up 5G adoption in smart ports: Barriers and enablers. *Marit. Policy Manag.* **2024**, *52*, 517–534. [CrossRef]
100. O'Keeffe, V.; Jang, R.; Manning, K.; Trott, R.; Howard, S.; Hordacre, A.-L.; Spoehr, J. Forming a view: A human factors case study of augmented reality collaboration in assembly. *Ergonomics* **2024**, *67*, 1828–1844. [CrossRef]
101. Danielsen, B.-E.; Lützhöft, M.; Haavik, T.K.; Johnsen, S.O.; Porathe, T. "Seafarers should be navigating by the stars": Barriers to usability in ship bridge design. *Cogn. Technol. Work* **2022**, *24*, 675–691. [CrossRef]
102. Kim, T.E.; Sharma, A.; Bustgaard, M.; Gyldensten, W.C.; Nymoen, O.K.; Tusher, H.M.; Nazir, S. The continuum of simulator-based maritime training and education. *WMU J. Marit. Aff.* **2021**, *20*, 135–150. [CrossRef]

103. Stapleton, A.; Hanna, J.B. Technological Innovation Adoption: An Empirical Investigation of Steamship Line Sales Force Integration. *Transp. J.* **2002**, *41*, 5–22.
104. Liu, J.; Liao, R.; Dong, F.; Huang, C.; Li, H.; Liu, J.; Zhao, T. Low-carbon technology selection and carbon reduction potential assessment in the shipbuilding industry with dynamically changing grid emission factors. *J. Clean. Prod.* **2024**, *441*, 140707. [CrossRef]
105. Arduino, G.; Aronietis, R.; Crozet, Y.; Frouws, K.; Ferrari, C.; Guihéry, L.; Kapros, S.; Kourounioti, I.; Laroche, F.; Lambrou; et al. How to turn an innovative concept into a success? An application to seaport-related innovation. *Res. Transp. Econ.* **2013**, *42*, 97–107. [CrossRef]
106. Shi, X.; Tao, D.; Voß, S. RFID Technology and its Application to Port-Based Container Logistics. *J. Org. Comp. Elect. Com.* **2011**, *21*, 332–347. [CrossRef]
107. Min, H.; Park, J.-W.; Lim, Y.-K.; So, A.; Cho, Y.K. Challenges and opportunities for implementing X-ray scanning technology at the Korean hub ports. *Int. J. Logist. Syst. Manag.* **2016**, *25*, 513–531. [CrossRef]
108. Choi, H.; Baek, Y.; Lee, B. Design and implementation of practical asset tracking system in container terminals. *Int. J. Precis. Eng. Man.* **2012**, *13*, 1955–1964. [CrossRef]
109. Ellingsen, O.; Aasland, K.E. Digitalizing the maritime industry: A case study of technology acquisition and enabling advanced manufacturing technology. *J. Eng. Technol. Manag.* **2019**, *54*, 12–27. [CrossRef]
110. Kostidi, E.; Nikitakos, N.; Progoulakis, I. Additive manufacturing and maritime spare parts: Benefits and obstacles for the end-users. *J. Mar. Sci. Eng.* **2021**, *9*, 895. [CrossRef]
111. Small, H. Co-citation in the scientific literature: A new measure of the relationship between two documents. *J. Am. Soc. Inform. Sci.* **1973**, *24*, 265–269. [CrossRef]
112. Bowman, D.; Kinnan, S. Creating effective titles for your scientific publications. *VideoGIE* **2018**, *3*, 260–261. [CrossRef]
113. Tullu, M.S. Writing the title and abstract for a research paper: Being concise, precise, and meticulous is the key. *Saudi J. Anaesth.* **2019**, *13* (Suppl. S1), S12–S17. [CrossRef]
114. Balcombe, P.; Brierley, J.; Lewis, C.; Skatvedt, L.; Speirs, J.; Hawkes, A.; Staffell, I. How to decarbonise international shipping: Options for fuels, technologies and policies. *Energy Convers. Manag.* **2019**, *182*, 72–88. [CrossRef]
115. 'T Hoen, M.; den Boer, E. *Cost Benefit Calculation Tool Onshore Power Supply*; C. Delft: Valle de los Molinos, Mexico, 2016.
116. Kumar, J.; Kumpulainen, L.; Kauhaniemi, K. Technical design aspects of harbour area grid for shore to ship power: State of the art and future solutions. *Int. J. Electr. Power* **2019**, *104*, 840–852. [CrossRef]
117. Piccoli, T.; Fermeglia, M.; Bosich, D.; Bevilacqua, P.; Sulligoi, G. Environmental assessment and regulatory aspects of cold ironing planning for a maritime route in the Adriatic Sea. *Energies* **2021**, *14*, 5836. [CrossRef]
118. Qi, J.; Wang, S.; Peng, C. Shore power management for maritime transportation: Status and perspectives. *Marit. Transp. Res.* **2020**, *1*, 100004. [CrossRef]
119. Winkel, R.; Weddige, U.; Johnsen, D.; Hoen, V.; Papaefthimiou, S. Shore side electricity in Europe: Potential and environmental benefits. *Energy Policy* **2016**, *88*, 584–593. [CrossRef]
120. Zanetti, S.L. Is Cold Ironing Hot Enough? An Actor Focus Perspective of on Shore Power Supply (OPS) at Copenhagen's Harbour. Master's Thesis, Lund University, Lund, Sweden, 2013.
121. Fruth, M.; Teuteberg, F. Digitization in maritime logistics-What is there and what is missing? *Cogent Bus. Manag.* **2017**, *4*, 1411066. [CrossRef]
122. Inkinen, T.; Helminen, R.; Saarikoski, J. Technological trajectories and scenarios in seaport digitalization. *Res. Transp. Bus. Manag.* **2021**, *41*, 100633. [CrossRef]
123. Kretschmann, L.; Burmeister, H.-C.; Jahn, C. Analyzing the economic benefit of unmanned autonomous ships: An exploratory cost-comparison between an autonomous and a conventional bulk carrier. *Res. Transp. Bus. Manag.* **2017**, *25*, 76–86. [CrossRef]
124. Venkatesh, V.; Davis, F.D. A theoretical extension of the technology acceptance model: Four longitudinal field studies. *Manag. Sci.* **2000**, *46*, 186–204. [CrossRef]
125. Venkatesh, V.; Morris, M.G.; Davis, G.B.; Davis, F.D. User acceptance of information technology: Toward a unified view. *MIS Q.* **2003**, *27*, 425–478. [CrossRef]
126. Wijnolst, N.; Wergeland, T.; Levander, K. *Shipping Innovation*; IOS Press: Amsterdam, The Netherlands, 2009.
127. Wróbel, K.; Montewka, J.; Kujala, P. Towards the assessment of potential impact of unmanned vessels on maritime transportation safety. *Reliab. Eng. Syst. Safe* **2017**, *165*, 155–169. [CrossRef]
128. Eisenhardt, K.M. Agency theory: An assessment and review. *Acad. Manag. Rev.* **1989**, *14*, 57–74. [CrossRef]
129. Hansson, J.; Månsson, S.; Brynolf, S.; Grahn, M. Alternative marine fuels: Prospects based on multi-criteria decision analysis involving Swedish stakeholders. *Biomass Bioenergy* **2019**, *126*, 159–173. [CrossRef]
130. Korberg, A.D.; Brynolf, S.; Grahn, M.; Skov, I.R. Techno-economic assessment of advanced fuels and propulsion systems in future fossil-free ships. *Renew. Sust. Energy Rev.* **2021**, *142*, 110861. [CrossRef]

131. Kouhizadeh, M.; Saberi, S.; Sarkis, J. Blockchain technology and the sustainable supply chain: Theoretically exploring adoption barriers. *Int. J. Prod. Econ.* **2021**, *231*, 107831. [CrossRef]
132. Chang, Y.; Iakovou, E.; Shi, W. Blockchain in global supply chains and cross border trade: A critical synthesis of the state-of-the-art, challenges and opportunities. *Int. J. Prod. Res.* **2020**, *58*, 2082–2099. [CrossRef]
133. Irannezhad, E. The architectural design requirements of a blockchain-based port community system. *Logistics* **2020**, *4*, 30. [CrossRef]
134. Kshetri, N. Blockchain's roles in meeting key supply chain management objectives. *Int. J. Inform. Manag.* **2018**, *39*, 80–89. [CrossRef]
135. Tijan, E.; Aksentijević, S.; Ivanić, K.; Jardas, M. Blockchain technology implementation in logistics. *Sustainability* **2019**, *11*, 1185. [CrossRef]
136. Christidis, K.; Devetsikiotis, M. Blockchains and smart contracts for the internet of things. *IEEE Access* **2016**, *4*, 2292–2303. [CrossRef]
137. Davis, F.D.; Bagozzi, R.P.; Warshaw, P.R. User acceptance of computer technology: A comparison of two theoretical models. *Manag. Sci.* **1989**, *35*, 982–1003. [CrossRef]
138. Davis, F.D. Perceived usefulness, perceived ease of use, and user acceptance of information technology. *MIS Q.* **1989**, *13*, 319–340. [CrossRef]
139. Jović, M.; Tijan, E.; Žgaljić, D.; Aksentijević, S. Improving maritime transport sustainability using blockchain-based information exchange. *Sustainability* **2020**, *12*, 8866. [CrossRef]
140. Acciaro, M.; Ghiara, H.; Cusano, M.I. Energy management in seaports: A new role for port authorities. *Energy Policy* **2014**, *71*, 4–12. [CrossRef]
141. Williamson, O.E. *Markets and Hierarchies: Analysis and Antitrust Implications: A Study in the Economics of Internal Organization*; Free Press: New York, NY, USA, 1975.
142. Yang, C.-S. Maritime shipping digitalization: Blockchain-based technology applications, future improvements, and intention to use. *Transp. Res. Part E Logist. Transp. Rev.* **2019**, *131*, 108–117. [CrossRef]
143. Burel, F.; Taccani, R.; Zuliani, N. Improving sustainability of maritime transport through utilization of Liquefied Natural Gas (LNG) for propulsion. *Energy* **2013**, *57*, 412–420. [CrossRef]
144. Pettit, S.; Wells, P.; Haider, J.; Abouarghoub, W. Revisiting history: Can shipping achieve a second socio-technical transition for carbon emissions reduction? *Transp. Res. Part D Transp. Environ.* **2018**, *58*, 292–307. [CrossRef]
145. Stalmokaitė, I.; Yliskylä-Peuralahti, J. Sustainability transitions in Baltic Sea shipping: Exploring the responses of firms to regulatory changes. *Sustainability* **2019**, *11*, 1916. [CrossRef]
146. Smith, T.; Jalkanen, J.; Anderson, B.; Corbett, J.; Faber, J.; Hanayama, S.; O'keeffe, E.; Parker, S.; Johansson, L.; Aldous, L. *Third IMO Greenhouse Gas Study*; International Maritime Organization: London, UK, 2014.
147. Garfield, E. From bibliographic coupling to co-citation analysis via algorithmic. In *A Citationist's Tribute to Belver C. Griffith*; Lazerow Lecture presented at Drexel University: Philadelphia, PA, USA, 2001.
148. Abu Bakar, N.N.; Bazmohammadi, N.; Vasquez, J.C.; Guerrero, J.M. Electrification of onshore power systems in maritime transportation towards decarbonization of ports: A review of the cold ironing technology. *Renew. Sustain. Energy Rev.* **2023**, *178*, 113243. [CrossRef]
149. Balci, G.; Phan, T.T.N.; Surucu-Balci, E.; Iris, Ç. A roadmap to alternative fuels for decarbonising shipping: The case of green ammonia. *Res. Transp. Bus. Manag.* **2024**, *53*, 101100. [CrossRef]
150. Kon, W.K.; Abdul Rahman, N.S.F.; Md Hanafiah, R.; Abdul Hamid, S. The global trends of automated container terminal: A systematic literature review. *Marit. Bus. Rev.* **2021**, *6*, 206–233. [CrossRef]
151. Lu, H.; Huang, L. Optimization of shore power deployment in green ports considering government subsidies. *Sustainability* **2021**, *13*, 1640. [CrossRef]
152. Pavlic, B.; Cepak, F.; Sucic, B.; Peckaj, M.; Kandus, B. Sustainable port infrastructure, practical implementation of the green port concept. *Therm. Sci.* **2014**, *18*, 935–948. [CrossRef]
153. Wang, J.; Zhong, M.; Wang, T.; Ge, Y.E. Identifying industry-related opinions on shore power from a survey in China. *Transp. Policy* **2023**, *134*, 65–81. [CrossRef]
154. Wang, L.; Liang, C.; Shi, J.; Molavi, A.; Lim, G.; Zhang, Y. A bilevel hybrid economic approach for optimal deployment of onshore power supply in maritime ports. *Appl. Energy* **2021**, *292*, 116892. [CrossRef]
155. Zis, T.P.V. Prospects of cold ironing as an emissions reduction option. *Transp. Res. Part A Pol. Pract.* **2019**, *119*, 82–95. [CrossRef]
156. Henríquez, R.; Martínez de Osés, F.X.; Martínez Marín, J.E. Technological drivers of seaports' business model innovation: An exploratory case study on the port of Barcelona. *Res. Transp. Bus. Manag.* **2022**, *43*, 100803. [CrossRef]
157. Jovanovic, M.; Kostić, N.; Sebastian, I.M.; Sedej, T. Managing a blockchain-based platform ecosystem for industry-wide adoption: The case of TradeLens. *Technol. Forecast. Soc. Change* **2022**, *184*, 121981. [CrossRef]

158. Munir, M.A.; Habib, M.S.; Hussain, A.; Shahbaz, M.A.; Qamar, A.; Masood, T.; Sultan, M.; Mujtaba, M.A.; Imran, S.; Hasan, M. et al. Blockchain Adoption for Sustainable Supply Chain Management: Economic, Environmental, and Social Perspectives. *Front. Energy Res.* **2022**, *10*, 899632. [CrossRef]
159. Nasih, S.; Arezki, S.; Gadi, T. Blockchain Technology for tracking and tracing containers: Model and conception. *Data Metadata* **2024**, *3*, 373. [CrossRef]
160. Nguyen, S.; Chen, P.S.L.; Du, Y. Blockchain adoption in container shipping: An empirical study on barriers, approaches, and recommendations. *Mar. Policy* **2023**, *155*, 105724. [CrossRef]
161. Tan, W.K.A.; Sundarakani, B. Assessing Blockchain Technology application for freight booking business: A case study from Technology Acceptance Model perspective. *J. Glob. Oper. Strateg. Sourc.* **2021**, *14*, 202–223. [CrossRef]
162. Bach, H.; Mäkitie, T.; Hansen, T.; Steen, M. Blending new and old in sustainability transitions: Technological alignment between fossil fuels and biofuels in Norwegian coastal shipping. *Energy Res. Soc. Sci.* **2021**, *74*, 101957. [CrossRef]
163. Chica, M.; Hermann, R.R.; Lin, N. Adopting different wind-assisted ship propulsion technologies as fleet retrofit: An agent-based modeling approach. *Technol. Forecast. Soc. Change* **2023**, *192*, 122559. [CrossRef]
164. Fu, Z.; Lu, L.; Zhang, C.; Xu, Q.; Zhang, X.; Gao, Z.; Li, J. Fuel cell and hydrogen in maritime application: A review on aspects of technology, cost and regulations. *Sustain. Energy Technol. Assess.* **2023**, *57*, 103181. [CrossRef]
165. Kesieme, U.; Pazouki, K.; Murphy, A.; Chrysanthou, A. Biofuel as an alternative shipping fuel: Technological, environmental and economic assessment. *Sustain. Energy Fuels* **2019**, *3*, 899–909. [CrossRef]
166. Rehmatulla, N.; Smith, T. The impact of split incentives on energy efficiency technology investments in maritime transport. *Energy Policy* **2020**, *147*, 111721. [CrossRef]
167. Chao, S.L.; Lin, P.S. Critical factors affecting the adoption of container security service: The shippers' perspective. *Int. J. Prod. Econ.* **2009**, *122*, 67–77. [CrossRef]
168. Fedi, L.; Lavissiere, A.; Russell, D.; Swanson, D. The facilitating role of IT systems for legal compliance: The case of port community systems and container Verified Gross Mass (VGM). *Supply Chain Forum* **2019**, *20*, 29–42. [CrossRef]
169. Norzaidi, M.D.; Chong, S.C.; Murali, R.; Salwani, M.I. Intranet usage and managers' performance in the port industry. *Ind. Manag. Data Syst.* **2007**, *107*, 1227–1250. [CrossRef]
170. Ting, S.L.; Wang, L.X.; Ip, W. A study on RFID adoption for vehicle tracking in container terminal. *J. Ind. Eng. Manag.* **2012**, *5*, 22–52. [CrossRef]
171. Véronneau, S.; Roy, J. RFID benefits, costs, and possibilities: The economical analysis of RFID deployment in a cruise corporation global service supply chain. *Int. J. Prod. Econ.* **2009**, *122*, 692–702. [CrossRef]
172. Wiafe, I.; Koranteng, F.N.; Tettey, T.; Kastriku, F.A.; Abdulai, J.-D. Factors that affect acceptance and use of information systems within the Maritime industry in developing countries: The case of Ghana. *J. Syst. Inf. Technol.* **2020**, *22*, 21–45. [CrossRef]
173. Zeng, F.; Chan, H.K.; Pawar, K. The adoption of open platform for container bookings in the maritime supply chain. *Transp. Res. Part E Logist. Transp. Rev.* **2020**, *141*, 102019. [CrossRef]
174. Acciaro, M.; Ferrari, C.; Lam, J.S.L.; Macario, R.; Roumboutsos, A.; Sys, C.; Tei, A.; Vanelslander, T. Are the innovation processes in seaport terminal operations successful? *Marit. Policy Manag.* **2018**, *45*, 787–802. [CrossRef]
175. Acciaro, M.; Sys, C. Innovation in the maritime sector: Aligning strategy with outcomes. *Marit. Policy Manag.* **2020**, *47*, 1045–1063. [CrossRef]
176. Bavassano, G.; Ferrari, C.; Tei, A. Blockchain: How shipping industry is dealing with the ultimate technological leap. *Res. Transp. Bus. Manag.* **2020**, *34*, 100428. [CrossRef]
177. Carlan, V.; Sys, C.; Vanelslander, T.; Roumboutsos, A. Digital innovation in the port sector: Barriers and facilitators. *Compet. Regul. Netw. Ind.* **2017**, *18*, 71–93. [CrossRef]
178. Philipp, R. Blockchain for LBG Maritime Energy Contracting and Value Chain Management: A Green Shipping Business Model for Seaports. *Environ. Clim. Technol.* **2020**, *24*, 329–349. [CrossRef]
179. Philipp, R.; Prause, G.; Gerlitz, L. Blockchain and Smart Contracts for Entrepreneurial Collaboration in Maritime Supply Chains. *Transp. Telecommun.* **2019**, *20*, 365–378. [CrossRef]
180. Philipp, R.; Prause, G.; Olaniyi, E.O.; Lemke, F. Towards Green and Smart Seaports: Renewable Energy and Automation Technologies for Bulk Cargo Loading Operations. *Environ. Clim. Technol.* **2021**, *25*, 650–665. [CrossRef]
181. Wang, S.; Notteboom, T. The role of port authorities in the development of LNG bunkering facilities in North European ports. *WMU J. Marit. Aff.* **2015**, *14*, 61–92. [CrossRef]
182. Ghaderi, H. Autonomous technologies in short sea shipping: Trends, feasibility and implications. *Transp. Rev.* **2019**, *39*, 152–173. [CrossRef]
183. Goerlandt, F.; Pulsifer, K. An exploratory investigation of public perceptions towards autonomous urban ferries. *Safety Sci.* **2022**, *145*, 105496. [CrossRef]
184. Issa, M.; Ilinca, A.; Ibrahim, H.; Rizk, P. Maritime Autonomous Surface Ships: Problems and Challenges Facing the Regulatory Process. *Sustainability* **2022**, *14*, 15630. [CrossRef]

185. Nzengu, W.; Faivre, J.; Pauwelyn, A.-S.; Bolbot, V.; Lien Wennersberg, L.A.; Theotokatos, G. Regulatory framework analysis for the unmanned inland waterway vessel. *WMU J. Marit. Aff.* **2021**, *20*, 357–376. [CrossRef]
186. Hall, P.V.; O'Brien, T.; Woudsma, C. Environmental innovation and the role of stakeholder collaboration in West Coast port gateways. *Res. Transp. Econ.* **2013**, *42*, 87–96. [CrossRef]
187. Poulis, E.; Poulis, K.; Dooley, L. 'Information communication technology' innovation in a non-high technology sector: Achieving competitive advantage in the shipping industry. *Serv. Ind. J.* **2013**, *33*, 594–608. [CrossRef]
188. Alamoush, A.S.; Ölçer, A.I.; Ballini, F. Drivers, opportunities, and barriers, for adoption of Maritime Autonomous Surface Ships (MASS). *J. Int. Marit. Saf. Environ. Aff. Ship.* **2024**, *8*, 2411183. [CrossRef]
189. Plomaritou, E.; Jeropoulos, S. The digitalisation in chartering business: Special reference to the role of e-bill of lading in the bulk and liner markets. *J. Ship Trade* **2022**, *7*, 28. [CrossRef]
190. Mukherjee, D.; Lim, W.M.; Kumar, S.; Donthu, N. Guidelines for advancing theory and practice through bibliometric research. *J. Bus. Res.* **2022**, *148*, 101–115. [CrossRef]
191. Breslin, D.; Bailey, K. Expanding the conversation through 'Debate Essays' and 'Review Methodology'papers. *Int. J. Manag. Rev.* **2020**, *22*, 219–221. [CrossRef]
192. Post, C.; Sarala, R.; Gatrell, C.; Prescott, J.E. Advancing theory with review articles. *J. Manag. Stud.* **2020**, *57*, 351–376. [CrossRef]

Journal of
Marine Science and Engineering

MDPI

Review

Machine Learning in Maritime Safety for Autonomous Shipping: A Bibliometric Review and Future Trends

Jie Xue [1,*], Peijie Yang [1], Qianbing Li [1], Yuanming Song [1], P. H. A. J. M. van Gelder [2], Eleonora Papadimitriou [2] and Hao Hu [1,3]

1 School of Ocean and Civil Engineering, Shanghai Jiao Tong University, Shanghai 200240, China; yangpeijie21@sjtu.edu.cn (P.Y.); qianbing_li@sjtu.edu.cn (Q.L.); ymsong2024@sjtu.edu.cn (Y.S.); hhu@sjtu.edu.cn (H.H.)

2 Safety and Security Science Group (S3G), Faculty of Technology, Policy and Management, Delft University of Technology, 2628 CD Delft, The Netherlands; p.h.a.j.m.vangelder@tudelft.nl (P.H.A.J.M.v.G.); e.papadimitriou@tudelft.nl (E.P.)

3 State Key Laboratory of Ocean Engineering, Shanghai Jiao Tong University, Shanghai 200240, China

* Correspondence: j.xue@sjtu.edu.cn

Abstract: Autonomous vessels are becoming paramount to ocean transportation, while they also face complex risks in dynamic marine environments. Machine learning plays a crucial role in enhancing maritime safety by leveraging its data analysis and predictive capabilities. However, there has been no review grounded in bibliometric analysis in this field. To explore the research evolution and knowledge frontier in the field of maritime safety for autonomous shipping, a bibliometric analysis was conducted using 719 publications from the Web of Science database, covering the period from 2000 up to May 2024. This study utilized VOSviewer, alongside traditional literature analysis methods, to construct a knowledge network map and perform cluster analysis, thereby identifying research hotspots, evolution trends, and emerging knowledge frontiers. The findings reveal a robust cooperative network among journals, researchers, research institutions, and countries or regions, underscoring the interdisciplinary nature of this research domain. Through the review, we found that maritime safety machine learning methods are evolving toward a systematic and comprehensive direction, and the integration with AI and human interaction may be the next bellwether. Future research will concentrate on three main areas: evolving safety objectives towards proactive management and autonomous coordination, developing advanced safety technologies, such as bio-inspired sensors, quantum machine learning, and self-healing systems, and enhancing decision-making with machine learning algorithms such as generative adversarial networks (GANs), hierarchical reinforcement learning (HRL), and federated learning. By visualizing collaborative networks, analyzing evolutionary trends, and identifying research hotspots, this study lays a groundwork for pioneering advancements and sets a visionary angle for the future of safety in autonomous shipping. Moreover, it also facilitates partnerships between industry and academia, making for concerted efforts in the domain of USVs.

Keywords: maritime safety; machine learning; autonomous shipping; bibliometrics; evolutionary trends

1. Introduction

In recent years, the concept of autonomous shipping has gained momentum on a global scale [1,2]. Leading nations such as Norway, China, and the US, alongside some companies such as Kongsberg Maritime, have been investing in R&D for autonomous vessels. Marked by the deployment of unmanned vessels capable of independent navigation and decision-making, it represents a significant advancement in the maritime industry. This transformative technology is poised to revolutionize various aspects of maritime operations, promising increased efficiency, reduced operational costs, and potentially safer voyages [3]. At the heart of this technological evolution lies the integration of machine learning (ML), a branch of artificial intelligence (AI) that equips autonomous systems with the ability to learn from data and improve their performance over time [4].

The importance of autonomous shipping extends beyond operational efficiency to encompass broader economic, environmental, and safety considerations. Economically, autonomous ships promise to lower operational costs associated with crew salaries, accommodation, and insurance premiums [5]. By combining the ship cost model and carbon pricing, it is proven that the significant economic benefit of autonomous ships was outdoing conventional ones [6]. Environmentally, they have the potential to optimize routes, reduce fuel consumption, and mitigate the ecological impact of maritime transport [7]. From a safety perspective, the concept of autonomous shipping has evolved from traditional piloted vessels to autonomous or unmanned ships equipped with advanced sensors, communication systems, and decision-making algorithms [8]. The primary drivers behind the development of autonomous shipping include the pursuit of operational efficiencies, reduction of human error, and compliance with increasingly stringent environmental regulations. Moreover, these technologies offer the possibility of enhanced situational awareness [9], collision avoidance, and rapid response capabilities in emergency situations, thereby bolstering overall maritime safety.

To achieve all-sided effects, machine learning has played a pivotal role in enabling the autonomy of maritime vessels and promoting maritime safety. In the domain of maritime transport, machine learning interprets real-time information about vessel position, environmental conditions, nearby obstacles, and other critical parameters. This capability enables autonomous ships to make informed decisions autonomously, such as route planning, speed adjustment, and collision avoidance, based on predictive analytics and probabilistic modeling. Currently, numerous meticulous studies have been conducted in the past years, covering various perspectives of maritime safety. The application of ML in the safety of autonomous shipping encompasses various domains, including but not limited to:

(1) Collision Avoidance: Besides some machine learning algorithms such as deep reinforcement learning (DRL) [10,11] or deep neural networks (DNN), several integrated methods were also used, such as an improved reinforcement learning with the knowledge transfer (KT) method [12], rule-guided vision supervised learning (RGVSL) [13], the multi-agent deep reinforcement learning (MADRL) [14], a distributed multi-USV navigation method based on deep reinforcement learning (DRL) [15], and novel Artificial Potential Fields (APFs) to perform obstacle and collision avoidance in marine environments [16]. These algorithms utilize supervised learning techniques to classify and predict collision risks based on real-time vessel dynamics and environmental conditions.

(2) Path planning: Machine learning models employ techniques such as deep reinforcement learning (DRL) [17,18], Deep Q-learning network [19], and Dynamic path guidance method [20,21] to optimize vessel routes. These models integrate data on weather forecasts, sea conditions, and historical voyage patterns to dynamically adjust routes and maximize efficiency while ensuring safety.

(3) Emergency Response: For emergency response, risk assessment and management come forth. Collision risk metric [22], deep learning multi-model integration method [23], and Bayesian belief network [24] were mainly used to analyze historical incident data, environmental factors, and human factors to assess risk levels and inform proactive risk

management strategies in autonomous shipping operations. After assessment, machine learning-driven emergency response systems utilize deep convolutional neural network (CNN) [25], deep learning [26], Bayesian networks [27], OCSVM [28] or ensemble learning (EL) [29] to prioritize emergency actions. These models integrate real-time data on vessel status, weather forecasts, and emergency protocols to simulate emergency scenarios and recommend optimal response strategies for crew safety and environmental protection.

(4) Environmental Adaptation: Machine learning algorithms such as lightweight volumetric convolutional neural network (VCNN) [30], deep learning recognition method [31], Learning-Without-Forgetting (LWF) approach [32], and ensemble learning methods [33] optimize vessel operations for environmental sustainability, a more accurate estimate of the environment [34,35], and overall safety. These models analyze data on emissions, fuel consumption, and regulatory requirements to dynamically adjust operational parameters and minimize the ecological footprint of maritime activities.

The aforementioned studies illustrate the diverse applications of machine learning (ML) in enhancing maritime safety, particularly in the context of autonomous shipping. With the increasing complexity of operational scenarios in unmanned surface vessels (USVs), ML technologies have been widely adopted to address critical challenges, including collision avoidance, route optimization, real-time risk assessment, and environmental adaptation. Specific techniques such as deep reinforcement learning (DRL), convolutional neural networks (CNNs), Bayesian networks, and generative adversarial networks (GANs) have demonstrated significant potential in enabling autonomous decision-making, predictive analytics, and adaptive control systems for maritime operations.

Despite these advancements, the field still faces several pressing questions that warrant further investigation: (1) What are the key research hotspots regarding ML applications in maritime safety for autonomous shipping? (2) Which institutions and collaboration networks have driven progress in ML-based maritime safety research, and how do they interact? (3) How have machine learning methods evolved to address the unique demands of maritime safety in recent years? (4) What emerging trends and technologies are shaping the future of ML in this domain?

Furthermore, while the interest in integrating ML into autonomous shipping continues to grow, according to the search results of WoS, there is a notable absence of a systematic literature review and featured bibliometric analysis regarding the maritime safety of autonomous shipping in the context of machine learning (ML). This gap highlights the need for a systematic examination of current research to identify trends, evaluate progress, and guide future innovations in this critical field. Hence, a bibliometric analysis is necessary to track the field's development and trends in the past years.

This paper aims to offer a comprehensive overview of the literature concerning the application of machine learning in ensuring the maritime safety of autonomous shipping. Emphasizing technological advancements and current research status, the study intends to meticulously analyze existing scholarly works to (1) map influential journals, authors, and keywords and their collaboration networks; (2) discuss prominent trends, challenges, and future prospects in this field; and (3) enhance understanding of the evolving field of machine learning and offer enlightening references to researchers in this field.

The remainder of the present paper is organized as follows. Firstly, Section 2 presents the data extraction and exclusion criteria and bibliometric analysis method and tool. Secondly, the results for the temporal and spatial distribution of publications, influential journals, author cooperation networks, citation and co-citation networks, and research subject categories, hotspots, and trends are demonstrated in Section 3. Thirdly, Section 4 details the discussions both for the bibliometric results and selected literature. Finally, the concluding remarks are addressed in Section 5.

2. Materials and Methods

2.1. Data Sources

The scientific publications analyzed in this study were retrieved on 31 May 2024 from Science Citation Index Expanded (SCI-EXPANDED), Social Sciences Citation Index (SSCI), Conference Proceedings Citation Index-Science (CPCI-S), and Conference Proceedings Citation Index-Social Science and Humanities (CPCI-SSH) citation index databases in the Web of Science (WoS) Core Collection, which are the most commonly used, authoritative, and the highest data quality databases for bibliometric analysis research domain [36–39].

The analysis for the topic "machine learning for maritime safety of autonomous shipping" mainly contains two parts: "machine learning" and "maritime safety of autonomous shipping". Moreover, the process of data analysis is the basis for conducting machine learning, and the safety of the autonomous shipping research areas also relates to various safety-related domains; thus, the search topic includes some strings/keywords, which can be seen in Table A1 in Appendix A.

Moreover, Boolean operators are often used in advanced search functions of database information. This paper uses Boolean operators "AND" and "OR" to combine the logical relationships between different field tags and different query sets (keywords) to create our query when searching in WoS. The iteration of the search on the website is shown in Figure 1.

Additionally, to provide the accurate retrieval, we also utilize double quotation marks for phrase retrieval, i.e., the exact phrase in double quotation marks would be searched as a whole when it is retrieved. For instance, when searching for papers related to autonomous ships. If you directly use the "autonomous ship" as the search term to retrieve the related publications, the obtained publications contain the autonomous ship research but also contain some research not related to autonomous ships, such as the research on autonomous cars or ordinary ships. When searching with "autonomous ship", the two words "autonomous" and "ship" in the search results need to appear in groups, which meets our search aims and ensures the search accuracy.

Furthermore, this paper uses wildcard characters which represent unknown characters in the literature search. Familiar wildcard characters include the asterisk (*), question mark (?), and dollar sign ($). The asterisk (*) represents any character group, including null characters, e.g., "safe*" can be used to search for "safety" or "safely", etc.; Similarly, "ship*" for "ships" or "shipping", etc.; "mari*" for "marine" or "maritime", etc. The question mark (?) represents any character, which is very useful for searching when the last character is uncertain, e.g., "los?" can be used to search for "lose", "loss" or "lost". The dollar sign ($) means zero or one character. It is usually used to find the English and American spellings of the same word, e.g., "man$euver" can be used to find "maneuver" or "manoeuver". The use of the above wildcard character symbols greatly avoids the related search missing and improves the search efficiency. The explanation of the related substitute words is shown at the bottom of Table A1 in Appendix A and the schematic drawing of the search is shown in Figure 1.

Although the Web of Science database covers publications from 1900 through 2024, very few relevant studies on unmanned vessels existed prior to 2000 due to the nascent state of computing power and machine-learning methods required for autonomous navigation. Therefore, this study only collected the publications after 2000. The language of publications was limited to English, the currently most widely utilized academic language. Moreover, as document types, the types of articles, review articles, and proceedings papers were included in our database.

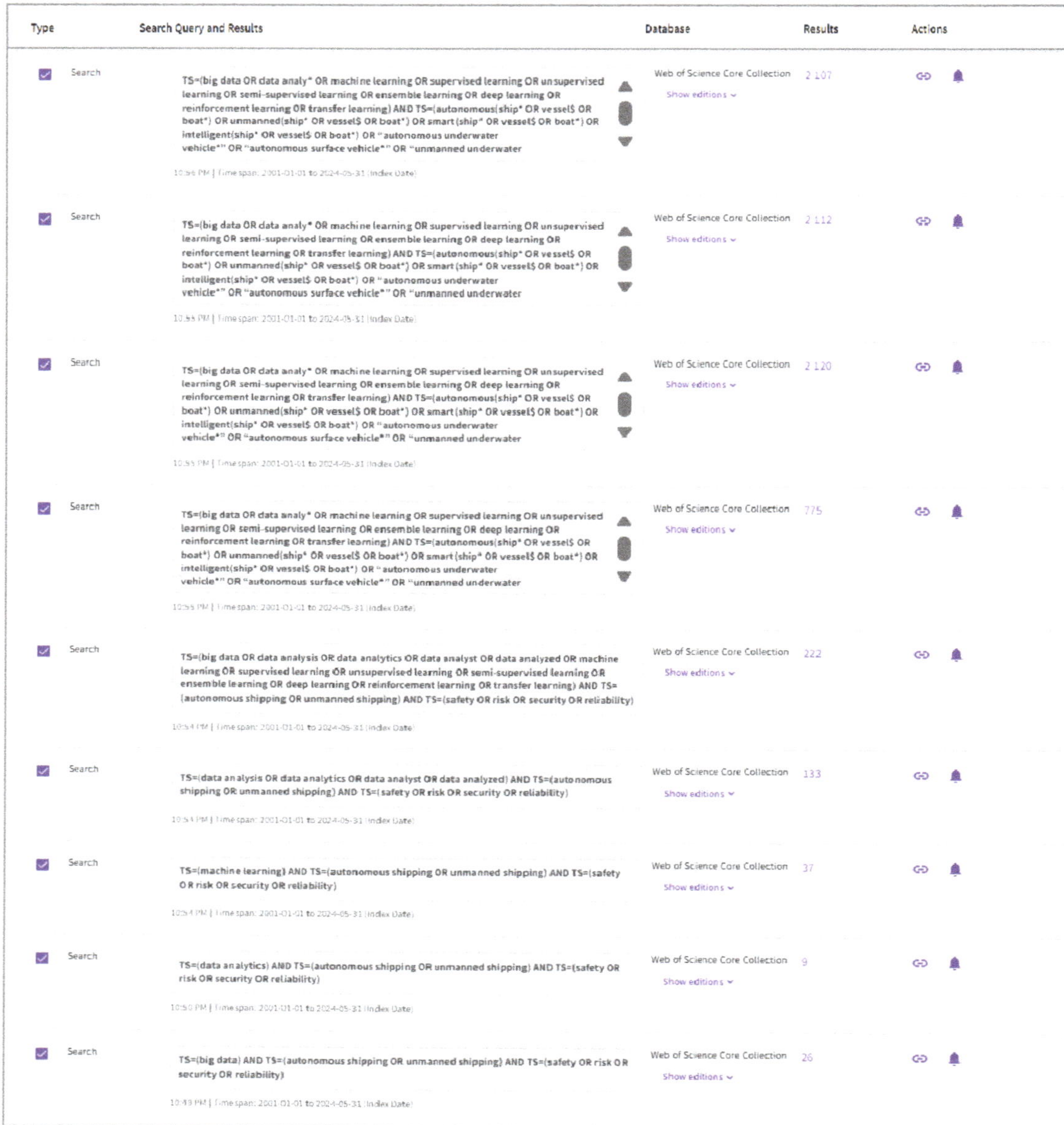

Figure 1. Data retrieval strategies in Web of Science. Note: TS = Topic; DT = Document Types; SCI-EXPANDED = Science Citation Index Expanded; SSCI = Social Sciences Citation Index; CPCI-S = Conference Proceedings Citation Index-Science; CPCI-SSH = Conference Proceedings Citation Index-Social Science and Humanities; The detailed description for each retrieval step and the presentation of the Boolean operators, double quotation marks, and wildcard characters are shown in Table A1 and the following corresponding context.

Moreover, due to the limitations of the search of WoS, it is necessary to further manually exclude documents in this dataset that do not address the explored topic, such as autonomous cars, autonomous drones, etc. The manual filtering was mainly conducted based on the titles and abstracts of the papers; for the specific paper, the paper's full text is also carefully overviewed. Note that the publications that only utilized the autonomous unmanned vessel as a tool to conduct research and do not promote the development of AUV technologies (i.e., the technology of target recognition, obstacle avoidance, navigation, and communication, etc.) are manually excluded.

Finally, 719 publications on machine learning for maritime safety of autonomous shipping are obtained for further bibliometric analysis. These documents included 2404 authors affiliated with 824 institutions in 66 countries, which obtained 11,124 citations in total (an average of 15.47 citations per document) and were published in 399 source journals and proceedings, citing 8218 references (Table 1). Additionally, there are three types of documents, including article = 453 (63.00%); proceedings paper = 251 (34.91%); review = 15 (2.09%), respectively. The full search records and cited references are saved as plain text. The framework of the search process and the steps to obtain the final analyzed dataset are shown in Figure 2.

Table 1. Data extraction and exclusion criteria.

Criteria	TS	TND	AU	SJP	SC	NI	TC	CR
Quantity	31 May 2001–31 May 2024	719	2404	399	66	824	11,124	8218

Note: TS = time span; TND = total number of documents; AU = authors; SJP = source journals and proceedings; SC = source countries; NI = number of institutions/organizations; TC = total number of citations; CR = cited references.

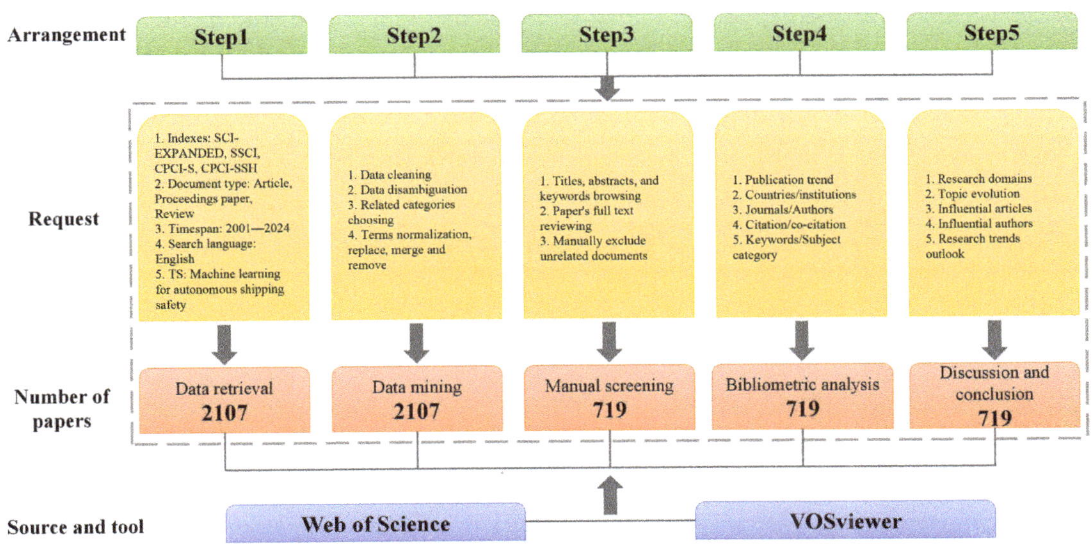

Figure 2. Methodological flowchart of the process of document analysis.

2.2. Bibliometric Analysis Method and Tool

Bibliometric analysis delivers a structural overview of a specific research domain by publications, authors, countries, institutions, journals, and keywords in terms of descriptive statistics, network analysis, and frequency analysis, etc. [40–42]. Additionally, bibliometric research is used to assimilate various viewpoints found in the research and provides a static, transparent, and systematic research perspective [40]. It enables the literature review and analysis to be more simple, objective, understandable, reproducible, and visual user-friendly [41,43].

Nowadays, bibliometric analysis has become popular, and more than 30 free-used tools have already been developed for bibliometric mapping, and visualization of similarity viewer (VOSviewer) software is one of the most internationally renowned analysis tools [44,45]. VOSviewer was developed by Dutch scholars van Eck and Waltman from the Advanced Bibliometric Methods group of the Centre for Science and Technology Studies of Leiden University [46]. It has become an effective auxiliary analysis tool for scientometric mapping research. Using advanced bibliometric methods and visualization, VOSviewer facilitates citation, collaboration, and keyword co-occurrence analyses for comprehensive mapping of a research field [47]. Moreover, several papers have already applied VOSviewer to do bibliometric mapping analysis in safety-related topics, e.g., output distributions and topic maps of safety-related journals [48], safety journal identification [49], safety culture [50], construction safety [51,52], process safety [53], domino effect [54], laboratory safety in universities [55], and road safety research [56], etc.

Therefore, in the present research, a bibliometric analysis method is employed, and the bibliometric mapping tool VOSviewer program is performed to graphically map the material and visually represent the bibliometric analysis and statistics results so as to facilitate the corresponding visual interpretations. Moreover, along with VOSviewer, the data analysis function of WoS was also utilized in this study. The analysis results mainly focus on the following aspects, i.e., temporal and spatial distribution of publications, influential journals, author collaboration networks, citation and co-citation networks, research subject categories and topics, etc.

3. Results

3.1. Temporal Distribution of Publications

A statistical analysis of the annual trends of publications in a particular field can be used to explore and understand the research domain's development from the temporal distribution [57]. The preliminary search in Section 2.1, as the comparable result of the eighth and ninth steps in Table A1 of Appendix A, shows that the productions in the WoS's core collection database had only records after the 2000s.

To more comprehensively analyze and understand the development process of machine learning technology utilized in the autonomous shipping safety research area, Figure 3 is shown to manifest the trend and variation of publication. The entire period's development process is divided into three stages: the initial germination stage (annual increased number of publications is not more than five), the initial growth stage (annual increased number of publications is more than five and not more than 25), and the rapid development stage (annual increased number of publications is more than 25), respectively, as can be seen from Figure 3 and Table 2.

Initial germination stage (2001–2010): In WoS core collection databases, the explored studies' documentation began until 2001. As shown in Table 2, prior to 2010, there were few documents on the relevant studies; the number of publications only accounts for 4.44% of the number of documents. In addition, the primary type of documents is the proceeding paper, accounting for 84.38% of the number of documents published in the initial germination stage. Moreover, the annual increased number of publications is not more than five, shown as the gray curve in Figure 3.

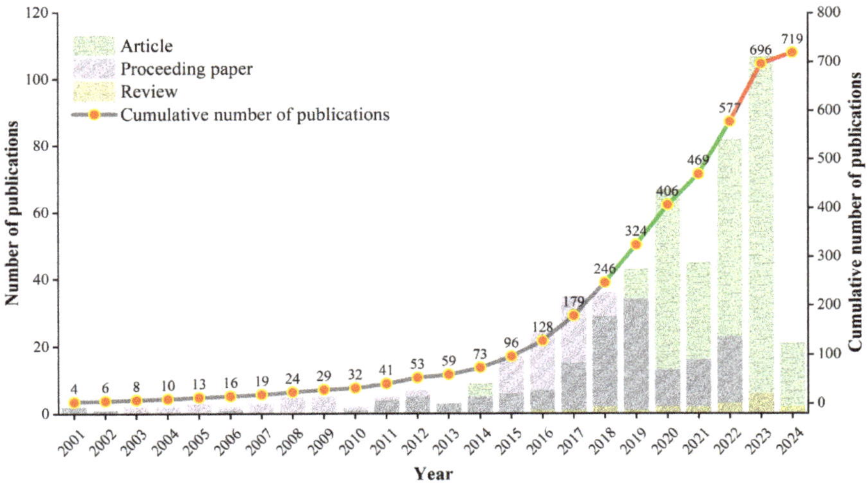

Figure 3. Annual trends of publications in maritime safety of autonomous shipping.

Initial growth stage (2011–2015): The initial growth period for the explored research area spans from 2011 to 2015, and the number of publications accounts for 8.90% of the total documents' number (Table 2). Shown as the green curve in Figure 3, although there were still a few publications on the relevant study in the period, the cumulative number of publications continues to increase with a low growth rate, and the annual increased number of publications is more than five and not more than 25. Additionally, the variation tendency of the publications' number indicates that the research in the area of machine learning for maritime safety of autonomous shipping has attracted the attention of scholars year by year.

Rapid development stage (2016 and beyond): After 2016, with the rapid development of computer technology and big data technology, machine learning methods have been widely used in autonomous shipping safety. Shown as the red curve in Figure 3, the number of publications exceeded 100 for the first time in 2016, and the annual increased number of publications is more than 25. In addition, the volume of publications on the explored studies had shown a rapid growth tendency (the number of publications in 2024 has not been fully counted yet due to there still being several months left in this year when the statistics were conducted); thus, this stage is named the rapid development stage. At this stage, in terms of the document's type, the number of journal articles (416) exceeds the number of proceeding papers (187) for the first time, and the number of publications accounts for 83.86% of the total documents' number (Table 2).

Table 2. The distributions of document types in various stages.

Stage	Time Span	Classification Criteria	Number of Articles	Proportion (%)	Number of Proceeding Paper	Proportion (%)	Number of Reviews	Proportion (%)	Total Number	Proportion (%)
Initial germination stage	2001–2010	0 < AINP ≤ 5	5	0.69%	27	3.76%	0	0	32	4.44%
Initial growth stage	2011–2015	5 < AINP ≤ 25	27	3.76%	37	5.15%	0	0	64	8.90%
Rapid development stage	2016 and beyond	AINP > 25	416	57.86%	187	26.01%	20	2.78%	603	83.86%
Total	-	-	448	62.31%	251	34.91%	20	2.78%	719	100.00%

Note: AINP = Annual increased number of publications.

3.2. Spatial Distribution of Publications

3.2.1. Publications Distribution in Pattern of Countries/Regions

Through the statistics for the authors' affiliations, the spatial distribution of various countries/regions and organizations for the explored publications was investigated. Due to the cooperative publication, one document might be conducted by more than one country and thus might belong to two or more countries, hence finally with a total of 937 records in this part. As shown in Figure 4, the academic publications' geographic distribution covers 64 countries/regions. Among them, 30 are from Europe, 19 from Asia, 3 from Africa (South Africa, Morocco, and Egypt), 3 from North America (USA, Canada, and Mexico), 3 from South America (Brazil, Argentina, and Colombia), 3 from Oceania (Australia, New Zealand, and New Caledonia), and 4 countries from Europe and Asia (Russia, Turkey, Azerbaijan, and United Arab Emirates). Note that the documents from England (60), Scotland (13), and Northern Ireland (2) merged to belong to the United Kingdom (UK, 75 documents) in Figure 4. There are 19 countries/regions (29.69%) publishing ten or more documents, and 32 countries/regions have produced less than three publications in the explored research domain. It is noteworthy that there are 303 documents from European countries, accounting for 32.34% of the total number of publications. Moreover, Figure 4 illustrates that among the top ten productive countries/regions of the explored domain during 2001–2024, China (341, 36.39%) is the most productive country, followed by the USA (101, 10.78%), the UK (75, 8.00%), South Korea (48, 5.12%), Germany (24, 2.56%), Spain (24, 2.56%), Italy (22, 2.35%), Australia (19, 2.03%), and Portugal (16, 1.71%), respectively.

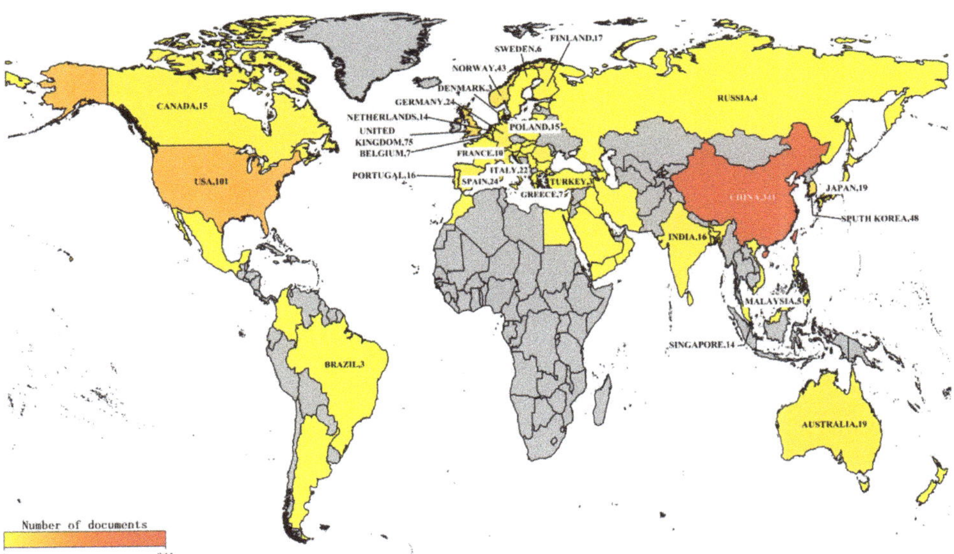

Figure 4. The number of publications distributed by countries/regions.

3.2.2. International Cooperation and Influential Research Institutions

The cooperation relation of countries/regions in the explored field was also visualized and analyzed by utilizing cooperation networks analysis to investigate affiliated countries and institutes through the VOSviewer software, shown in Figure 5a. It should be noted that a node is apportioned to each co-author of a publication in the networks. The nodes' color presents the average time for the publications of each country [57]. The nodes' size denotes the related publications' number, and the thickness of the links indicates the international collaborations' degree [54,58], i.e., the larger the node is, the more critical the country/region is, and the thicker the line is, the closer the cooperation relationship between countries/regions.

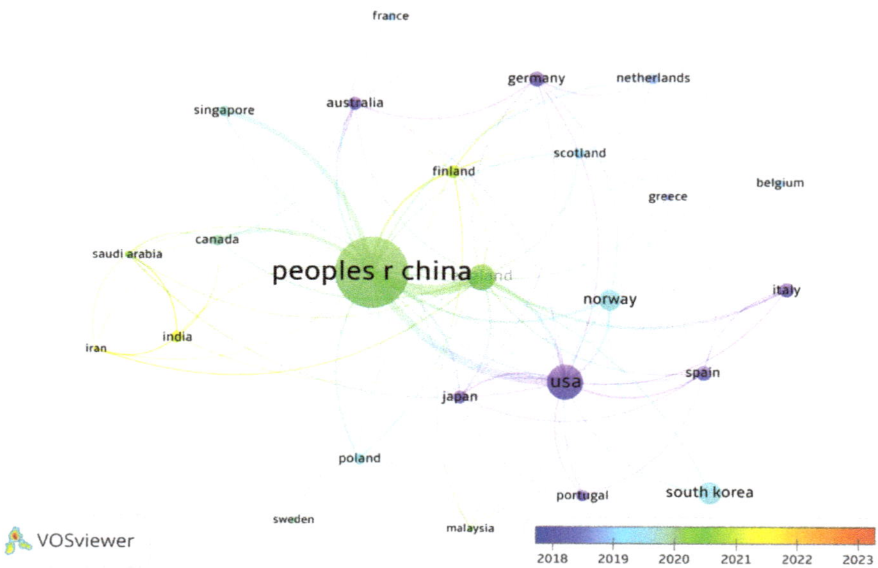

(**a**)

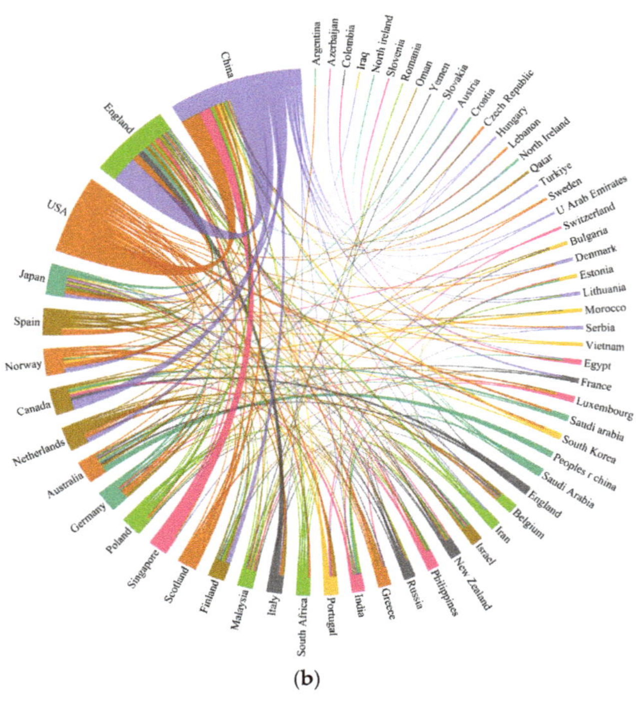

(**b**)

Figure 5. (**a**) Cooperation network and the timespan of the publications of various countries/regions; (**b**) Cooperation network among countries/regions.

Figure 5a indicates that China (21 links) and the USA (17 links) have been at the forefront and play the predominant roles in the explored machine learning for maritime safety of autonomous shipping studies. Meanwhile, Figure 5a,b show that China collaborates with the USA, Singapore, Norway, the Netherlands, Canada, Australia, Spain, and England, etc. Similarly, the close collaboration countries/regions with the USA are China, England, Spain, Italy, South Africa, Saudi Arabia, and Norway, etc. In addition, among the close collaboration countries/regions, China and the USA have the closest cooperation and research relationship in the explored research domain (with the link strength of 16). Moreover, as shown in Figure 5a, in terms of the publication time for the studies, Iran, India, and Finland are relatively active research countries recently and have the latest output in the explored field. Furthermore, the USA is among the first countries to carry out the related research and published the first journal paper in 2001 [59] according to the investigated results from the core database of the WoS. Generally, more international collaboration needs to be promoted and enhanced to share knowledge globally in the future.

According to the bibliometric analysis for authors' affiliation, 793 institutions were recognized. The top 10 affiliated research institutions that contribute to the publications on the explored domain are presented in Table 3. It can be seen that the institutions are mainly from China, for instance, Dalian Maritime University, Wuhan University of Technology, Ocean University of China, Harbin Engineering University, Chinese Academy of Sciences, Shanghai Jiaotong University, and Shanghai Maritime University. Among them, Dalian Maritime University (58, 8.07%) is the most productive and active institution related to the explored domain in China as well as globally, followed by Wuhan University of Technology (44, 6.12%) and Norwegian University of Science and Technology (26, 3.62%). Furthermore, Wuhan University of Technology has the highest total link strength (81), and Wuhan University of Technology also has the most links (55) with other institutions.

Table 3. The top ten research institutions in the explored research domain.

Rank	Institution	Country/Region	Links	TLS	NP	P (%)	TC	APY	AC
1	Dalian Maritime Univ.	China	53	68	58	8.07%	1161	2020.21	20.02
2	Wuhan Univ. Technol.	China	55	81	44	6.12%	1013	2020.30	23.02
3	Harbin Engn Univ.	China	17	21	30	4.17%	449	2019.10	14.97
4	Norwegian Univ Sci and Technol.	Norway	29	36	26	3.62%	407	2019.75	15.65
5	Ocean Univ China.	China	20	27	23	3.20%	266	2018.52	11.57
6	Chinese Acad. Sci.	China	25	34	18	2.50%	211	2020.61	11.72
7	Shanghai Jiaotong Univ.	China	14	26	18	2.50%	488	2021.06	27.11
8	Shanghai Maritime Univ.	China	19	20	13	1.81%	317	2020.00	24.38
9	Liverpool John Moores Univ.	UK	13	19	12	1.67%	169	2023.17	14.08
10	Univ Southampton.	UK	14	16	10	1.39%	276	2018.60	27.60

Note: TLS = total link strength; NP = number of publications; P (%) = the proportion of NP/TND; TND = total number of documents; TC = total number of citations; APY = average publications year; AC= average citations = TC/NP.

Moreover, the Norwegian University of Science and Technology (Norway), Liverpool John Moores University (UK), and University of Southampton (UK) also ranked in the top ten for the number of publications. It should be noted that the highest number of cited documents in the explored field is from Dalian Maritime University, with 1161 citations and an average of 20.02 citations. Additionally, except for the institutions that only published 1 document, the University of Southampton has the latest average publication year of 2023.17.

The VOSviewer software is utilized to establish a cooperative network map for the primary research institutions in the explored domain. Figure 6 demonstrates the interactions through joint publications among 98 institutions (with more than 3 publications, finally including 74 associated institutions) with 314 links (with the full counting method). It can be found that Dalian Maritime University is located in the middle of the cooperation network with the most prominent node, indicating that it has the greatest number of publications and broad cooperation with other institutions. Moreover, Wuhan University of Technology (55 links in total) has the most frequent cooperation with other institutions.

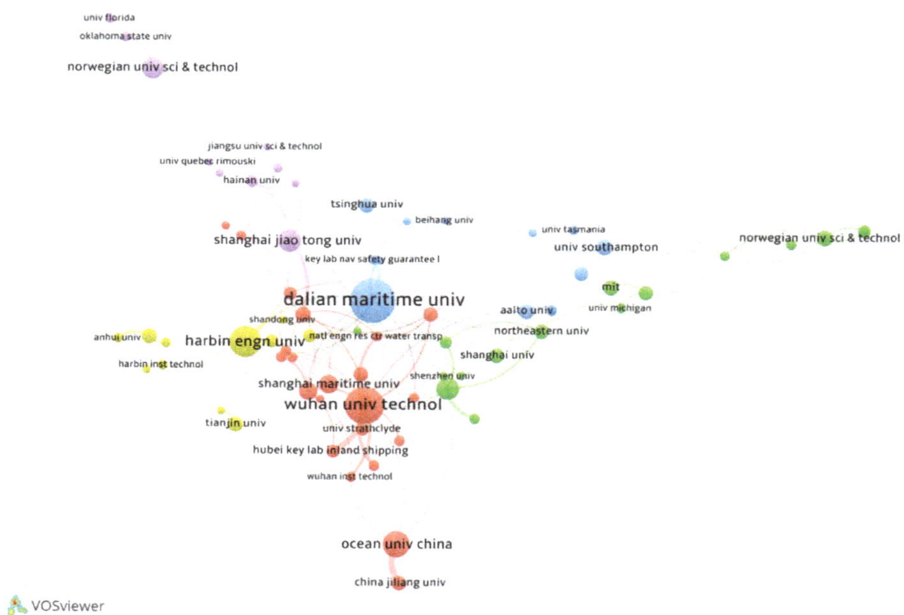

Figure 6. Cooperation network of institutions for the explored research domain.

According to the results of various classifications with different colors, the global cooperation network among different research institutions for machine learning in maritime safety of autonomous shipping studies has been initially formed. However, institutions primarily collaborate at the national level, with limited cross-continental partnerships. Scholars only carried out a certain degree of cooperation at a relatively low level, basically cooperation within each similar institution or an individual country. However, institutions need to be encouraged to cooperate in exchanging knowledge among scholars to obtain and promote the output for the highest research quality [60].

It should be noted that when utilizing the VOSViewer, the node's size presents the number of publications, the node's color reflects the cluster to which the node belongs, and the links between nodes denote cooperation between countries, respectively. In addition, the larger the node's size is, the greater the number of publications is, and the thicker the line is, the closer the cooperation relationship between institutions. Moreover, to clearly illustrate the cooperative relationship between the prominent institutions, 24 unassociated institutions were excluded when drawing Figure 6 in the software.

3.3. Influential Journal Analysis

Between 2001 and 2024, 399 publication sources contributed to the field. Table 4 summarizes the ten most prolific journals (≥3 publications), led by Ocean Engineering (68 papers, 9.45% of the total; 1396 citations; average 20.53 citations; total link strength = 220; Q1 in Engineering), followed by the Journal of Marine Science and Engineering (36 papers; highest mean publication year = 2022.44; impact factor = 2.7; 5-year IF = 2.8) and IEEE Access (29 papers). Other top outlets include Sensors (24), Applied Sciences-Basel (14), IEEE Journal of Oceanic Engineering (14), IEEE Transactions on Intelligent Transportation Systems (14), IEEE Transactions on Neural Networks and Learning Systems (8), Reliability Engineering and System Safety (8), and Applied Ocean Research (7)—all ranking Q1 in their primary research categories. Engineering, oceanography, and Computer Science emerge most frequently as subject areas among these journals. It should be noted that the specific research area category and the corresponding quartile are given if the classification is different in the subdivision area according to the journal citation reports from WoS; then the detailed information is separately marked.

In addition, as shown in Table 4, though the top 10 most prolific journals are analyzed; however, there is a significant difference in the number of citations of the specific journal. To further investigate the most influential journal related to the research domain of machine learning in autonomous shipping safety, mapping the sources with regard to the co-citations is depicted in Figure 7.

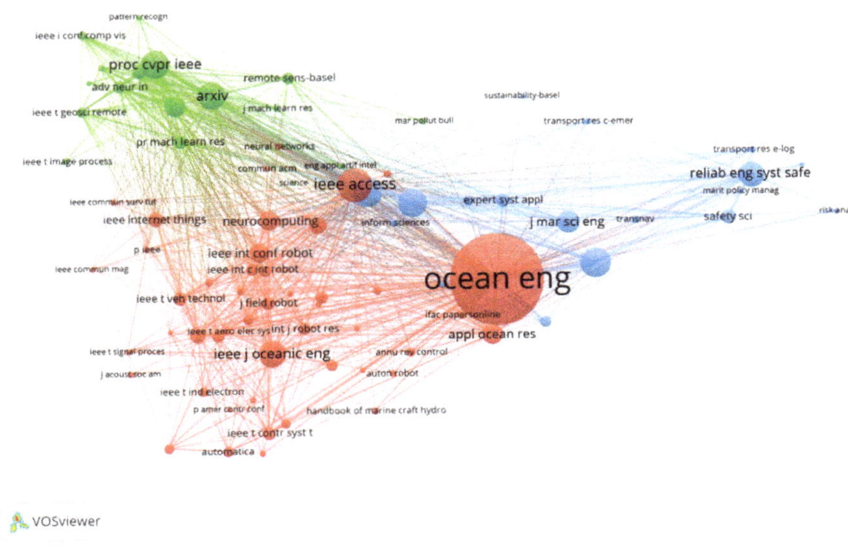

Figure 7. The mapping of the sources with regards to the co-citations.

The mutual relations between the journal sources are usually visualized by utilizing this type of bibliometric network analysis [36]. In addition, the minimum value of citations of a source is settled as 20. Herein, 67 of the 4774 sources meet the threshold. The weights are settled by the citations' number. Thus, four clusters are recognized using the full counting method for sources and the association strength method for data normalization in VOSviewer software.

The top five co-cited journals in this domain are Ocean Engineering (1563 citations), IEEE Access (438), the Journal of Navigation (387), Sensors (367), and IEEE Transactions on Intelligent Transportation Systems (358).

Table 4. The top 10 most prolific journals published on maritime safety of autonomous shipping research from 1981 to 2020.

Rank	Journal Title	Links	TLS	NP	P (%)	TC	APY	AC	IF	5 Year IF	Journal Category	Corresponding Quartile Rank
1	Ocean Engineering	58	220	68	9.45%	1396	2021.76	20.53	4.6	4.8	Engineering Civil; Engineering Marine; Engineering Ocean; Oceanography	Q1; Q1; Q1; Q1
2	Journal of Marine Science and Engineering	29	114	36	5.00%	271	2022.44	7.53	2.7	2.8	Engineering Marine; Engineering Ocean; Oceanography	Q1; Q2; Q2
3	IEEE Access	28	65	29	4.03%	635	2019.72	21.90	3.4	3.7	Computer Science; Electrical and Electronic; Telecommunication	Q2; Q2; Q2
4	Sensors	28	69	24	3.34%	507	2020.83	21.13	3.4	3.7	Chemistry; Electrical and Electronic; Instruments and Instrumentation	Q2; Q2; Q2
5	Applied Sciences	8	8	14	1.95%	143	2020.93	10.21	2.5	2.7	Chemistry; Engineering; Materials Science; Physics	Q2; Q1; Q3; Q2
6	IEEE Journal of Oceanic Engineering	8	8	14	1.95%	412	2015.64	29.43	3.8	4.2	Engineering Civil; Electrical and Electronic; Engineering Ocean; Oceanography	Q1; Q2; Q2; Q1
7	IEEE Transactions on Intelligent Transportation Systems	22	40	14	1.95%	327	2022.21	23.36	7.9	8.3	Engineering Civil; Electrical and Electronic; Transportation Science and Technology	Q1; Q1; Q1
8	IEEE Transactions on Neural Networks and Learning Systems	16	20	8	1.11%	243	2022.13	30.38	10.2	10.4	Artificial Intelligence; Hardware and Architecture; Theory and Methods; Electrical and Electronic	Q1; Q1; Q1; Q1
9	Reliability Engineering and System Safety	10	21	8	1.11%	217	2021.63	27.13	9.4	8.1	Engineering Industrial; Operations Research and Management Science	Q1; Q1
10	Applied Ocean Research	22	58	7	0.97%	250	2021.86	35.71	4.3	4.1	Engineering Ocean; Oceanography	Q1; Q1

Note: JT = journal title; TLS = total link strength; NP = number of publications; P (%) = the proportion of NP/TND; TND = total number of documents; TC = total number of citations; APY = average publications year; AC = average citations = TC/NP; IF = impact factor in 2019; JC = journal category; QR = corresponding quartile rank.

Figure 7 identifies five thematic co-citation clusters. The largest (red) cluster centers on ocean and marine engineering, robotics, and machine learning—anchored by Ocean Engineering and including Applied Ocean Research, IEEE ICRA, and Neural Networks. A green cluster focuses on computer vision and pattern recognition, led by CVPR proceedings and linked to IEEE TPAMI, Remote Sensing, and IJCV. The blue cluster emphasizes safety and system reliability, featuring the Journal of Navigation, Reliability Engineering and System Safety, and Accident Analysis and Prevention. The fourth cluster covers information technology and automation (e.g., IEEE Access, Automatica), while the fifth cluster concentrates on signal processing and control theory in maritime contexts, represented by the IEEE Journal of Oceanic Engineering and IEEE Transactions on Signal Processing.

3.4. Cooperation Network Analysis for the Authors

A total of 2597 authors have contributed to research on machine learning applications for maritime safety in autonomous shipping. Table A2 (Appendix A) identifies the top 20 authors by publication output and citation impact.

Bo He of Ocean University of China is the most prolific author, with 14 publications (1.95% of the corpus), while Zhouhua Peng of Dalian Maritime University attains the highest average citation rate (58 citations per paper). Furthermore, the most average number of citations (AC) is 58 (from Zhouhua Peng, Dalian Maritime University, China), and the average number of citations varies from 2.29 to 58 in this table.

In citation impact analysis, Weidong Zhang (Shanghai Jiaotong University) emerges as the most cited author (373 total citations; average 41.44 citations per paper). Furthermore, the average number of citations ranges from 6.25 to 48 in the table.

Across both productivity and citation metrics, Yuanchang Liu demonstrates the greatest collaborative influence, exhibiting the highest co-authorship network centrality (96 links; total link strength = 1803).

Figure 8a,b depict author co-authorship and co-citation networks, respectively, with node size proportional to publication or citation volume, link thickness denoting collaboration strength, and color clusters indicating distinct collaborative communities. Only authors with a minimum of four publications (49 out of 2597) meet the threshold for inclusion in these network visualizations.

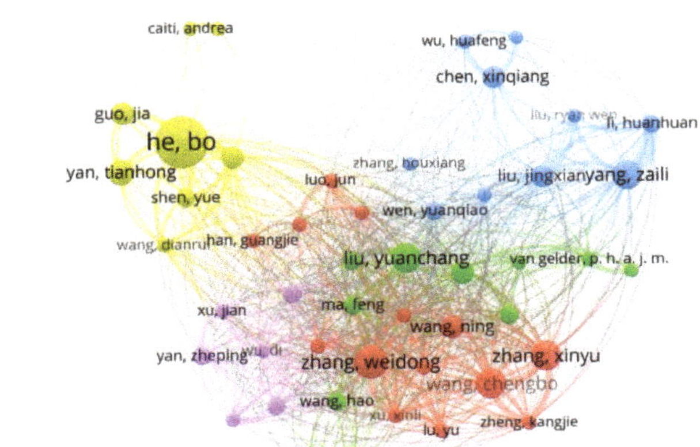

(a)

Figure 8. *Cont.*

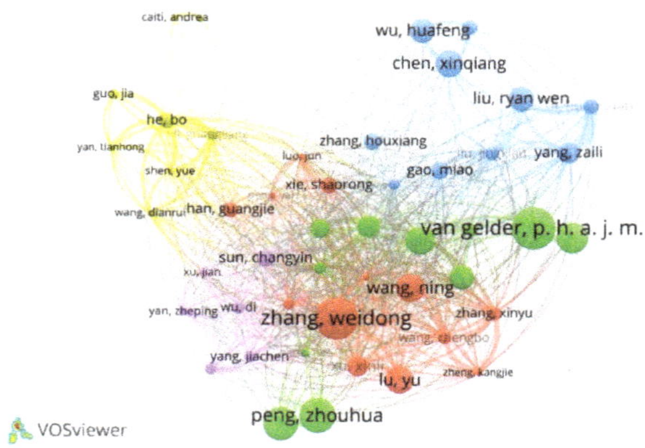

(b)

Figure 8. (a) Bibliographic coupling network of the authors for the investigated topic based on the weights of the number of documents; (b) Bibliographic coupling network of the authors for the investigated topic based on the weights of the number of citations.

3.5. Citation and Co-Citation Network Analysis for the Publications

3.5.1. Publications Citation Analysis

The publication citation analysis demonstrates the process of analyzing the number of citations that the documents of the explored domain have been cited by other publications through the database of WoS [58], and it is a method to measure the specific publication's quality and influence [61]. The most cited publication is authored by Peng et al. [62] in IEEE Transactions on Industrial Electronics (257 citations). Additionally, two publications authored by Huang et al. [63] and Cheng and Zhang [64], which are published in Safety Science (239 citations) and Neurocomputing (160 citations), rank as the second and third most frequently cited publications, respectively.

Among the three publications with the highest citation, Peng et al. proposed a neurodynamic-based output feedback scheme for distributed containment maneuvering of marine vessels guided by multiple parameterized paths without using velocity measurements [62]. This innovative approach, which operates without using velocity measurements, exemplifies the advanced control systems essential for the autonomous navigation and coordination of maritime vehicles guided by multiple parameterized paths, thereby enhancing their operational efficacy and reliability. Huang et al. offered a comprehensive overview of collision prevention techniques based on the three fundamental processes of determining evasive solutions: motion prediction, conflict detection, and conflict resolution [63]. Its thorough analysis and innovative approach have made it a cornerstone in the field of autonomous maritime navigation, significantly influencing subsequent research and providing a foundational framework for developing advanced collision avoidance systems. Cheng and Zhang introduced a novel Concise Deep Reinforcement Learning Obstacle Avoidance (CDRLOA) algorithm [64]. This method utilized the robust deep Q-networks architecture to smoothly address the usability challenges posed by the complex control laws inherent in traditional analytical approaches.

Note that since most of the influential works' research topics are cross fused with the research hotspots in Section 3.6, the related literature productions are not discussed and analyzed in detail in this section.

3.5.2. Publications Co-Citation Analysis

Publication co-citation analysis identifies pairs of documents that are frequently cited together, thereby quantifying their thematic similarity and revealing underlying narrative structures within a research domain. VOSviewer enables reference-level co-citation mapping, where nodes represent cited documents and node size reflects citation frequency. In this study, 59 of the 20,280 references, each cited at least 15 times across 719 retrieved publications, constitute the co-citation network (Figure 9).

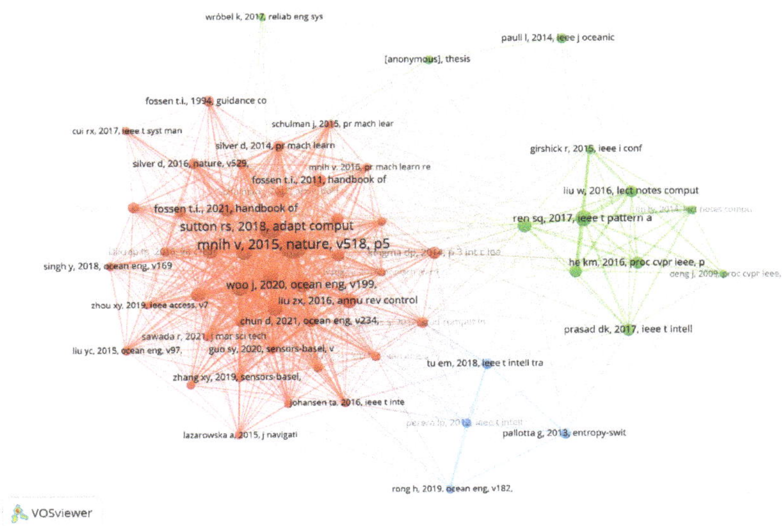

Figure 9. Map of co-cited reference network in explored research domain.

Figure 9 shows the references co-citation map for the documents cited more than 15 times by the explored publications. Herein, 59 of the 20,280 references cited by 719 retrieved documents on the explored domain meet the threshold. It can be seen that four clusters are recognized from 59 references in the map of the co-cited reference network, which demonstrates various research themes in the explored research domain. The blue cluster mainly focuses on trajectory prediction and AIS data application [65–68]. The document authored by Pallotta et al. [68] is the core and representative publication with the most citations (21), and Tu et al. [65] has the most links (33). Moreover, the red cluster mainly emphasizes machine learning methods, especially deep reinforcement learning [64,69–71]. The document authored by Mnih et al. [71] is the most influential publication, with the most citations (74) and links (52). In addition, the green cluster mainly concentrates on object detection and image processing [72–76]. Among them, the document authored by Ren et al. [72] has the most number of citations (35), and the document authored by Krizhevsky et al. [75] has the most number of links (28).

3.6. Research Subject Categories, Hot Topics, and Trends

3.6.1. Subject Categories Analysis

A total of 719 publications indexed in the Web of Science belong to 71 distinct subject categories, reflecting the inherently multidisciplinary character of machine learning research for maritime safety in autonomous shipping. As shown in Figure 10, the predominant categories are Engineering, Electrical and Electronic (222 publications; 30.88%), Oceanography (190; 26.43%), Marine Engineering (180; 25.04%), Ocean Engineering (175;

24.34%), and Civil Engineering (115; 16.00%), with substantial representation also observed in Automation and Control Systems, Computer Science (artificial intelligence), Computer Science (Information Systems), Telecommunications, and Robotics, each comprising between 8.8% and 12.9% of the rest. In general, the explored research is apparently multidisciplinary; moreover, the researchers range from scientists and engineers to marine ecologists. Note that one publication may belong to different categories.

Figure 10. The distribution of the explored documents into categories as various research areas.

3.6.2. Influential Authors' Research Interests and Domains

To figure out the main research topics relevant to the explored domain, the bibliographic coupling analysis is carried out for the 719 publications. Figure 11 demonstrates the 56 publications with at least 50 citations. Each node represents the specific publication, and the size of the node indicates the number of publication citations. Bibliographic coupling happens when two publications cite the same third publication and is denoted by the link curve between specific publications. As shown in Figure 11, there are five colors in the network, which indicate five clusters, and each cluster's topic can be recognized and summarized by the title and abstract of the documents. The red cluster contains most publications (14) and mainly concentrates on obstacle detection, object detection, and trajectory prediction. Moreover, the theme of the green cluster (10) focuses on path planning and optimization under deep reinforcement learning, and the blue cluster (9) presents control methods for collision avoidance and navigation. In addition, the yellow cluster (7) is related to the research topic of the intelligent decision-making method and system, while that in the purple cluster (7) represents the amelioration and enhancement of sensors, ship localization, and ship surveillance. The results indicate that the aforementioned topics attract the most attention in the explored research domain.

Generally, scholars with a considerable amount of highly cited publications dominate the research field's development direction and methodological trends and significantly influence the domain. Additionally, high-yield authors also usually have a relatively deep understanding of the research frontiers in the specific field and often have new outputs. Therefore, identifying and analyzing influential scholars from the highly cited and high-yielding levels allows obtaining insights and bringing inspiration into understanding the development of hotspots and trends in the explored research domain.

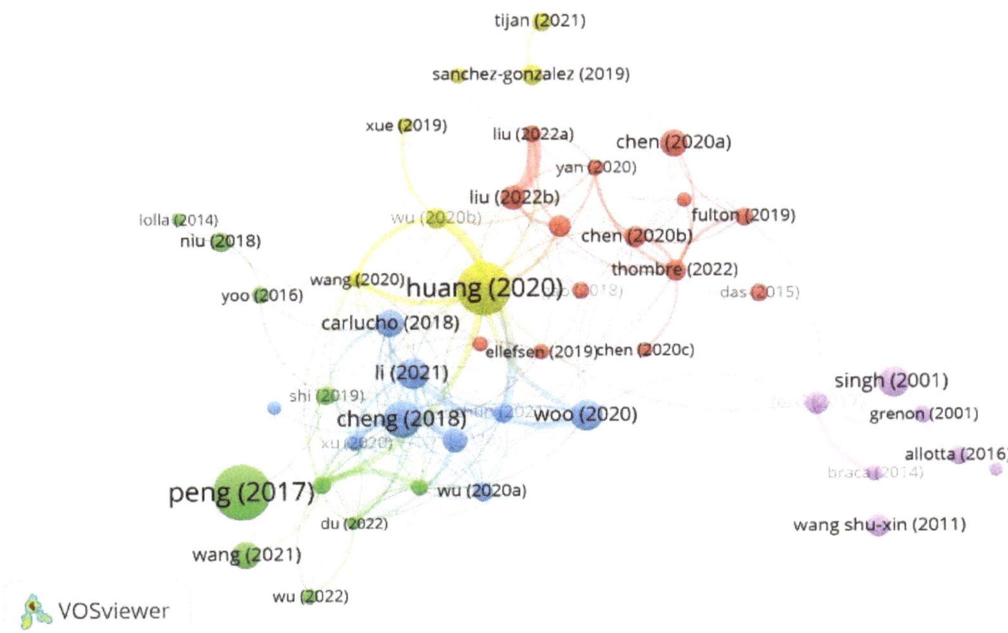

Figure 11. Bibliographic coupling network between publications on the explored domain.

Hence, in this paper, to further analyze the main research hot topics and trends, the top three authors in the top 20 most productive authors and the top 20 most cited authors are selected, respectively. Moreover, considering the average citations (AC) comprehensively and objectively reflect the number of publications and the influence of the author in a specific research field [77], the top three authors are also identified based on AC from the top 20 most productive and most cited authors, as presented in Table A2. Recognizing that citation counts alone may favor researchers with longer academic careers, the average publication year (APY) is also incorporated as an additional metric to account for the emerging influence of younger researchers. Specifically, APY, or average publication year, is a metric used to reflect the temporal distribution of an author's publications. It is calculated as the weighted average of the publication years of all works by an author, considering the publication count as the weighting factor.

By integrating these criteria, a total of 10 influential authors with 76 publications, accounting for 10.57% of the 719 publications, are identified. The selected authors' main research interests are then summarized based on information obtained from Google Scholar, ResearchGate, and affiliated universities or institutions, as shown in Table 5.

Table 5. Top 10 most influential authors in the explored domain and their research interests.

No.	Author	Country	Institution	NP	P (%)	TC	APY	AC	Main Research Interests
1	Zhang, Weidong	China	Shanghai Jiao Tong University	9	1.25%	**373**	2021.56	41.44	Image restoration; intelligent agriculture; deep learning; image processing and computer vision; control theory and pattern recognition theory and their applications in USV/UAV/AUV
2	Van Gelder, P. H. A. J. M.	Netherlands	Delft University of Technology	4	0.56%	370	2019.25	**92.50**	Risk analysis and optimization of systems; processes and structures; infrastructure safety; statistical modelling of high impact low probability (HILP)
3	Peng, Zhouhua	China	Dalian Maritime University	5	0.70%	290	2019.60	58.00	Guidance, control, and coordination of unmanned surface vehicles; multi-vehicle systems; unmanned surface vehicles; formation control; neural networks
4	He, Bo	China	Ocean University of China	**14**	**1.95%**	98	2018.14	7.00	Mobile robots; unmanned vehicles; precise navigation, and control and communication; AUV design and applications; AUV SLAM (simultaneous localization and mapping); AUV control; machine learning
5	Liu, Yuanchang	England	University College London	8	1.11%	155	2020.63	19.38	Autonomous system; artificial intelligence; marine robotics; statistical machine learning; automation and autonomy; guidance and control of intelligent and autonomous vehicles
6	Wang, Chengbo	China	Dalian Maritime University	8	1.11%	69	2022.25	8.63	Maritime autonomous surface ships; collision avoidance; decision-making; deep reinforcement learning
7	Yang, Zaili	England	Liverpool John Moores University	8	1.11%	149	**2023.25**	18.63	Maritime transport; risk analysis; analysis and modelling of safety; resilience and sustainability of transport networks; maritime and logistics systems
8	Zhang, Xinyu	China	Dalian Maritime University	8	1.11%	77	2021.88	9.63	Traffic organization optimization; intelligent navigation of USV; analysis and integration of maritime big data; three-dimensional maritime supervision methods; port traffic capability simulation

Table 5. *Cont.*

No.	Author	Country	Institution	NP	P (%)	TC	APY	AC	Main Research Interests
9	Yan, Xinping	China	Wuhan University of Technology	6	0.83%	216	2017.50	36.00	Intelligent transport system key technologies; energy efficiency management of vessel; marine system design and control; vessel condition monitoring and fault diagnosis; maritime safety; tribology and safety
10	Wang, Dan	China	Dalian Maritime University	4	0.56%	274	2018.75	68.50	Marine vehicle control; unmanned surface vehicles; multi-agent system control; tracking control; linear multiagent systems

Note: NP = number of publications; P (%) = the proportion of NP/TND; TND = total number of documents; TC = total number of citations; APY = average publications year; AC = average citations = TC/NP.

Note that the authors Yuanchang Liu, Chengbo Wang, Zaili Yang, and Xinyu Zhang all have 8 publications when choosing the top 3 authors among the top 20 most productive authors, so they are all listed in Table 5. Moreover, as shown in Table A2, the author Xinping Yan belongs to the top 3 authors based on average citations in the top 20 most productive authors. In addition, the fourth author, Dan Wang, belongs to the top 3 authors based on average citations in the top 20 most cited authors and is selected into the table list.

As shown in Table 5, the most cited author is Weidong Zhang from Shanghai Jiaotong University with 373 citations, followed by Van Gelder, P. H. A. J. M. (370) from Delft University of Technology, respectively. In addition, Bo He from the Ocean University of China is the most productive author with 14 publications, followed by Weidong Zhang (9) from Shanghai Jiao Tong University, Yuanchang Liu (8) from University College of London, Chengbo Wang (8) from Dalian Maritime University, Zaili Yang (8) from Liverpool John Moores University, and Xinyu Zhang (8) from Dalian Maritime University, respectively. Moreover, Zaili Yang is also the author with the largest average publications year (2023.25), which indicates the publications relatively have new updates. Additionally, Van Gelder, P. H. A. J. M. is also the author with the highest average citations of 92.50 per paper, followed by Dan Wang (68.50) from Dalian Maritime University and Zhijun Chen (62.33) from Wuhan University of Technology. Thus, the results reveal that their research has received extensive attention from other scholars.

From the primary research interest shown in Table 5, we could conclude that the influential authors mainly focus on AUV/USV reliability and safety-related topics in maritime industries, as well as advancements in intelligent systems for maritime applications. Key areas of focus include image processing, computer vision, control theory, and their applications in USV, UAV, and AUV; risk analysis and optimization of systems, processes, and infrastructures; guidance, control, and coordination of unmanned surface vehicles and multi-vehicle systems; and formation control using neural networks.

Additionally, these authors are keen on exploring autonomous systems design, precise navigation, communication, AUV control and design, and simultaneous localization and mapping (SLAM). By analyzing their interested fields, it can be inferred that machine learning, deep learning, artificial intelligence, and deep reinforcement learning are critical for enhancing decision-making and collision avoidance in marine robotics and autonomous vehicles. Energy efficiency management, vessel condition monitoring, fault diagnosis, tribology, and multi-agent system control also contribute to the safety of robust and efficient unmanned and intelligent systems in complex maritime environments.

3.6.3. Research Fields Identification and Analysis

Bibliometric analysis through keywords in a specific period has been used and developed for several years, and it has proved to be a very helpful and efficient way of discovering and revealing scientific research trends and hotspots [78–80]. Therefore, the core content and research topics of the papers can be summarized and represented by keywords. In addition, we can use the text mining technology through VOSviewer to infer the topics and trends in a specific research field from the text strings in keywords, titles, and abstracts of the documents in the database [77]. Thereby, in this paper, to investigate and explore the research fields and trends, the keywords of documents are visualized by utilizing the co-occurrence analysis module in VOSviewer software. The "All keywords" function, containing "Author keywords" and "Keywords plus", is employed. Figure 12 demonstrates the keywords co-occurrence map where the minimum number of occurrences of a keyword is 10 among the 719 retrieved publications. Herein there are 50 of the 2624 keywords, i.e., nodes, that meet the threshold (for example, among the 2624 keywords, 2080 keywords appeared only once while 152 keywords appeared more than 5 times). Note that a node presents the term gotten from the publications related to the explored research domain. Moreover, the node's size denotes the co-occurrence frequencies for the analyzed keywords [54].

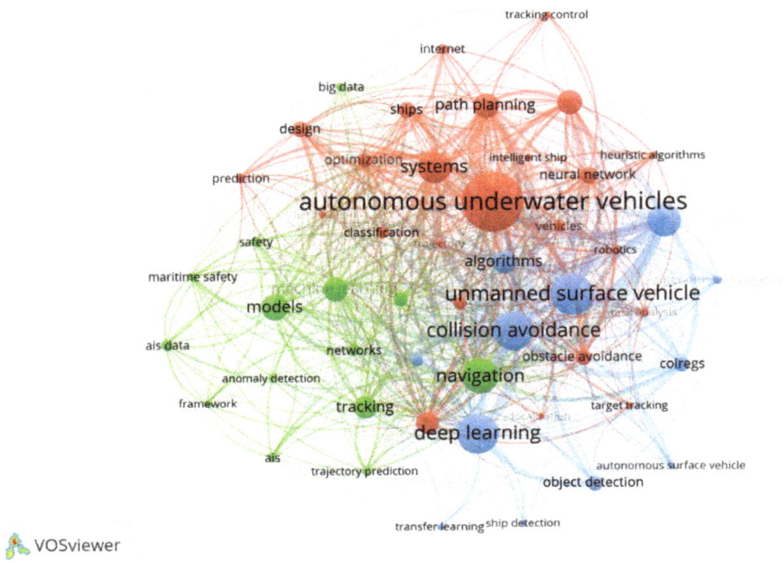

Figure 12. Keywords co-occurrence analysis for the explored research domain.

As shown in Figure 12, there are three clusters. The blue cluster concentrates mainly on new technologies and methods, including machine learning, deep learning, autonomous navigation systems, intelligent collision avoidance, and object detection, etc. In addition, the red cluster is associated with the applications of AUV and other types of marine robotics, covering prediction, localization, optimization, target tracking, etc. Furthermore, the green cluster mainly focuses on the foundation of AUV, including models, AIS data, networks, big data, artificial intelligence, etc.

3.6.4. Research Frontier and Evolutionary Process

Table A3 in the Appendix A summarizes the top 30 keywords in the explored research domain, which appeared most frequently in each period. Apparently, the most co-occurring keyword during all the investigated period is "autonomous underwater vehicle (AUV)", with 131 times, followed by "collision avoidance (83)", "deep learning (79)", "unmanned surface vessel (77)", "navigation (68)", "system (62)", "deep reinforcement learning (57)". Furthermore, "reinforcement learning (45)", "machine learning (43)", "model (43)", "tracking (41)", "path planning (39)", "algorithm (33)", etc., which are also the main keywords and research hotspots of the explored domain. These keywords are also highlighted in the density map, as shown in Figure 13a. Note that, for the data processing of this part, the keywords with the same meaning are summarized and replaced by the keyword that appears most frequently; e.g., the keyword "unmanned surface vehicle" is used to represent maritime autonomous surface ship (MASS), unmanned surface vessel, and autonomous surface vehicle (ASV). Similarly, the autonomous underwater vehicle represents the unmanned underwater vehicle/unmanned undersea vehicle (UUV), and the autonomous ship represents the keywords, such as autonomous vessel, smart ship, intelligent ship, unmanned ship, and autonomous sailboat, etc.

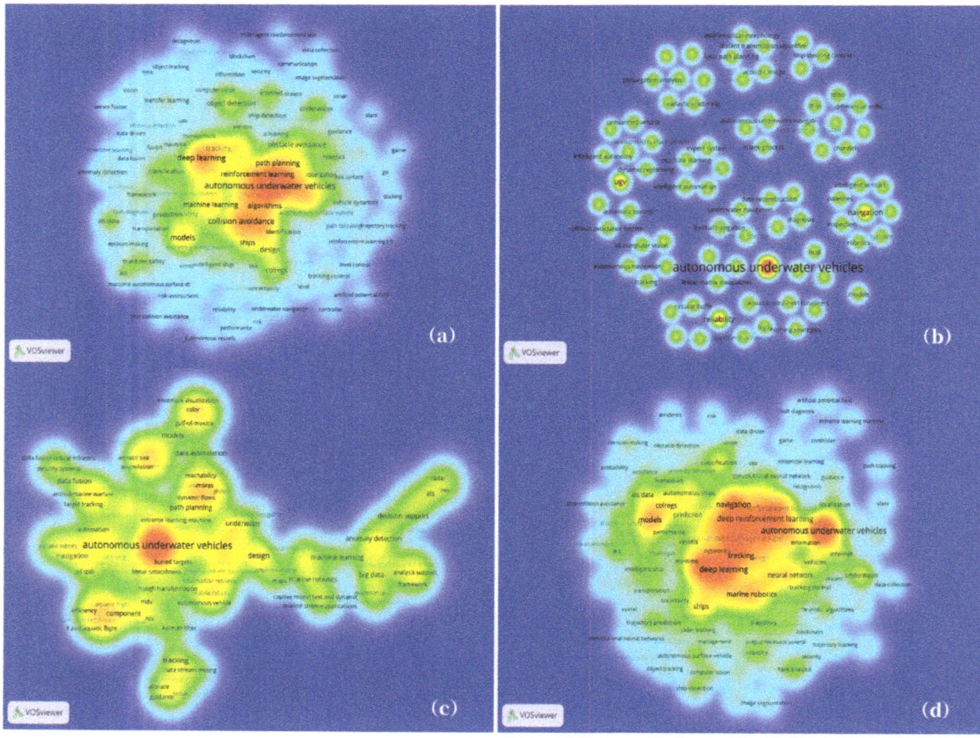

Figure 13. Heat maps of keywords in the explored domain during various time periods. (**a**) All time; (**b**) 2000–2010; (**c**) 2011–2015; (**d**) 2016–2024.

To further understand and analyze the research theme and development context and trend on the explored research domain in specific various periods. In this paper, the explored documents are separated into three parts based on the publication time: (i) 2001–2010, (ii) 2011–2015, and (iii) 2016–May 2024. Moreover, the research topics and main machine learning research methods during various periods are recognized and analyzed from keywords and presented and visualized through the density map function of VOSviewer (Figures 13 and 14). Note that the colors from green to red in the heat maps represent the occurrence frequency for various keywords, and the distance between two keywords demonstrates the association strength in the publications.

Period of 2001–2010: Figure 13b and Table A3 in the Appendix A present the hot keywords from 2001 to 2010. Obviously, the most occurring keyword in this period is "autonomous underwater vehicle", with 6 times, followed by "reliability (2)" and "navigation (2)", etc.

Period of 2011–2015: The keywords for the explored research domain from 2011 to 2015 are illustrated in Figure 13c. As shown in Table A3 in Appendix A, the top 5 frequent keywords are "autonomous underwater vehicle (20)", "design (5)", "system (4)", and "tracking (3)", etc.

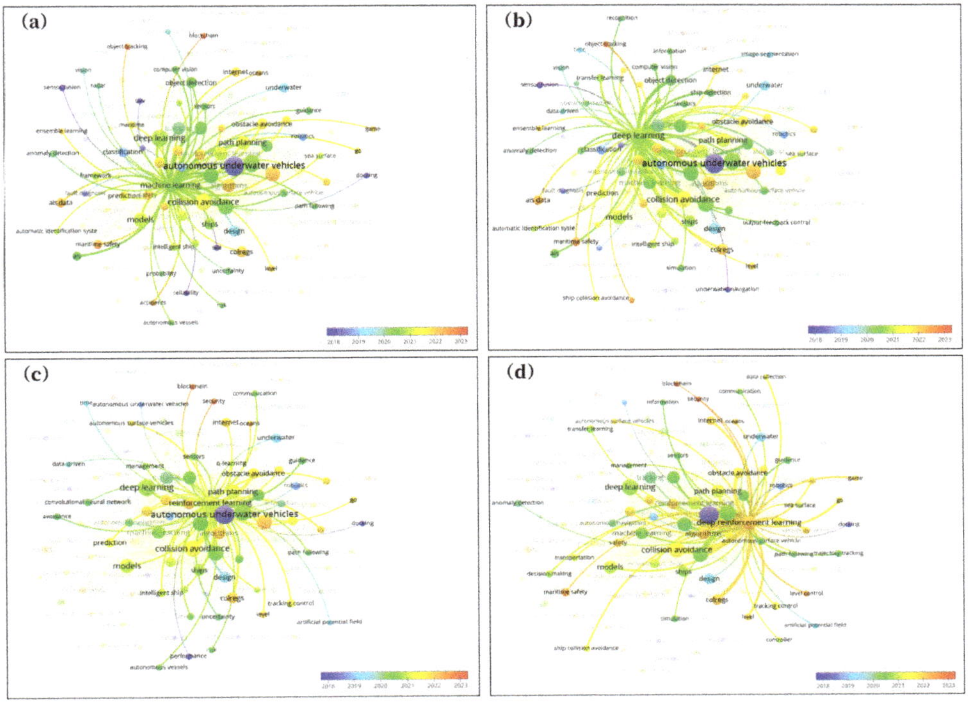

Figure 14. Co-occurrence analysis networks for the main machine learning methods used in the safety and security of maritime shipping. (**a**) Machine learning; (**b**) deep learning; (**c**) reinforcement learning; (**d**) deep reinforcement learning.

Period of 2016–2024: From 2016 to 2024, a total of 623 publications offered 2343 keywords, 248 of which appeared more than 3 times. The explored research boomed in this period. The hot keywords are "autonomous underwater vehicle (99)", "unmanned surface vehicle (80)", "deep learning (79)", "collision avoidance (76)", "navigation (64)", "deep reinforcement learning (63)", "system (55)", "models (46)", "path planning (44)", and "reinforcement learning (44)", etc., as shown in Figure 13d and Table A3 in the Appendix A.

By comparison among the three periods, the research in machine learning for the safety of autonomous shipping has progressed significantly over the past two decades. Between 2001 and 2010, the foundational work focused on autonomous underwater vehicles, emphasizing keywords such as reliability, navigation, and unmanned ground vehicles. This period laid the groundwork for understanding the basic capabilities required for autonomous navigation and system reliability. From 2011 to 2015, the research emphasis expanded to include design, system integration, and tracking technologies, highlighting the increasing complexity of autonomous systems. During this time, machine learning techniques were employed to enhance system robustness and decision-making processes.

From 2016 onwards, there was a notable shift towards incorporating advanced machine learning methodologies to improve safety and operational efficiency. Between 2016 and 2024, the field witnessed a significant boom in publications, with hot keywords such as deep learning, collision avoidance, and deep reinforcement learning emerging as focal points. This era marked the integration of sophisticated AI algorithms to address challenges in navigation, path planning, and real-time decision-making. Furthermore, the incorporation of technologies such as unmanned surface vehicles and sensor fusion indicated a trend towards multifunctionality and the amalgamation of diverse data sources. By utilizing satellite data, computer vision, and deep neural networks, researchers have enhanced the communication and intelligent decision-making capabilities of autonomous ships, ensuring safer maritime operations. These developments reflect a broader trend towards leveraging machine learning to achieve more reliable and autonomous maritime systems, setting new standards for safety in the industry.

To facilitate exploring and understanding the main machine learning methods used in the safety and security of maritime shipping, the co-occurrence analysis networks for the corresponding keywords are demonstrated in Figure 14. Additionally, four terms, i.e., "machine learning", "deep learning", "reinforcement learning", and "deep reinforcement learning", are selected so that the strongest related items are highlighted, and then the according words that show their respective application are conducted in Table A4 in the Appendix A.

According to Table A4 in the Appendix A, machine learning (ML), deep learning (DL), reinforcement learning (RL), and deep reinforcement learning (DRL) each have unique applications in autonomous shipping, dictated by their capabilities and the nature of the tasks they handle. Overall, these methods all aim to improve maritime safety by predicting potential hazards, optimizing routes, and enabling autonomous navigation, ultimately reducing the risk of accidents.

Machine learning is widely employed in structured data tasks such as fault diagnosis, classification, and anomaly detection, where its reliance on manual feature engineering and simpler models proves effective. It is also instrumental in applications such as path planning, navigation, and optimization, which benefit from its capacity to model deterministic tasks and optimize outcomes in well-structured environments. In contrast, deep learning excels at processing unstructured data, such as ship detection, image segmentation, and sensor fusion, where the automatic extraction of features from high-dimensional datasets is critical. Its versatility extends to more advanced domains, including autonomous ships, marine robotics, and AUV operations, where deep learning's ability to process sensory inputs and adapt to dynamic environments makes it indispensable. Reinforcement learning, with its focus on decision-making and adaptive learning through interaction with the environment, is well-suited for applications involving path following, obstacle avoidance, and tasks that require continuous adjustments to changing conditions. Deep reinforcement learning combines the strengths of DL and RL and is particularly effective in scenarios that require both perception and action, such as collision avoidance and multi-agent coordination. These applications are especially critical in complex maritime environments, where the integration of real-time data and strategic decision-making is essential for safe and efficient operations.

Given the unique capabilities and applications of each method, deep reinforcement learning (DRL) emerges as the most suitable approach for autonomous shipping. Its ability to integrate perception and decision-making, adapt to dynamic environments, and handle complex, multi-agent interactions make it the optimal choice for achieving safe, efficient, and intelligent ship operations in the future.

4. Discussion

4.1. Bibliometric Analysis Findings

4.1.1. Past and Current Trends

As is depicted in Figure 3, the evolution of research in maritime safety and autonomous shipping can be divided into 3 stages. Before 2010, the focus was primarily on foundational technologies for autonomous underwater vehicles (AUVs), emphasizing reliability and navigation. This period featured limited publications, mainly conference papers, indicating a nascent interest in exploring the fundamental capabilities of autonomous systems.

However, after the period, research expanded to include design, system integration, and tracking technologies. There was a moderate increase in publications, reflecting growing scholarly attention and a shift towards improving decision-making processes within autonomous systems. After 2016, there has been a significant rise in the use of advanced machine learning techniques, such as deep learning and reinforcement learning, to address challenges in navigation and collision avoidance. The number of publications has increased dramatically.

The publication trend in maritime safety and autonomous shipping is expected to continue increasing and reach a new breakthrough due to several factors: advancements in AI, a sustainability focus, industry-academia collaboration, and regulatory developments.

4.1.2. The Features of Social Structure

Regarding the geographic distribution of research on machine learning applications in maritime safety and autonomous shipping, the field is notably led by contributions from China, the USA, and European countries. Among these, China has emerged as the most prolific contributor, with its dominance reflecting strong governmental support, substantial investment in maritime technology, and a well-established academic infrastructure that promotes innovation. The USA and European countries also play significant roles, driven by their advanced technological capabilities and emphasis on addressing global maritime challenges. This distribution underscores a globally collaborative effort to advance the integration of machine learning in autonomous maritime technologies, as evidenced by the analysis of publication outputs and international cooperation networks in this study.

With regard to notable international collaborations, the cooperation, particularly between China and the USA, as well as with countries such as the UK, Norway, and Australia, underscores the importance of cooperative efforts in addressing global maritime challenges. These collaborations facilitate the exchange of knowledge and expertise, driving the development of more comprehensive and effective solutions. Furthermore, international cooperation is significantly reflected in institutional networks. Dalian Maritime University, Wuhan University of Technology, and Shanghai Jiaotong University serve as pivotal hubs in fostering collaboration between regional and international institutions within their respective clusters. This highlights their robust capacity to engage in international cooperation and resource integration. According to the VOSviewer analysis, Dalian Maritime University occupies a central position in the cooperation network, as evidenced by its prominent node, which reflects its substantial publication output and extensive collaboration with other institutions. Moreover, Wuhan University of Technology exhibits the highest number of cooperative links, emphasizing its active engagement with diverse institutions. These findings provide compelling evidence of the strong mutual cooperation links among Chinese institutions, underscoring their domestic strength and potential for further expanding international collaboration.

Regarding key institutions, Dalian Maritime University, Wuhan University of Technology, and Shanghai Jiaotong University play pivotal roles in advancing research, showcasing China's strong academic presence in this field. The concentration of research in these institutions suggests a well-established academic network that supports maritime safety and autonomous shipping. The presence of leading universities and research centers indicates a commitment to fostering innovation and advancing the capabilities of autonomous maritime systems.

When it comes to the influential authors, Weidong Zhang from Shanghai Jiao Tong University, Bo He from Ocean University of China, Zaili Yang from Liverpool John Moores

University, and Xinping Yan from Wuhan University of Technology have emerged as the leading figures. It is intriguing to note that these authors are affiliated with institutions renowned for their strong engineering programs, providing them with cutting-edge resources and collaborative environments that have fueled their research endeavors. Moreover, these authors have not only exhibited remarkable productivity (Bo he with the highest individual publication) and extensive influence (Van Gelder, P. H. A. J. M. with the highest average citation) but have also demonstrated significant activity (Zaili Yang with the latest publication year), particularly in the last three years.

4.1.3. Citation and Co-Citation Network Summary

In citation analysis, key works by Peng et al. [62], Huang et al. [63], and Cheng et al. [64] are highly cited, reflecting their foundational contributions to advanced control systems, collision avoidance, and reinforcement learning applications. The high citation count of these publications highlights their pivotal role in shaping current research directions and methodologies, serving as essential references for subsequent studies. The co-citation network reveals four main clusters focusing on trajectory prediction, machine learning methods, object detection, and system reliability. The clustering of co-cited documents highlights the interconnected nature of research topics, emphasizing the multidisciplinary approach necessary for advancements in maritime safety. The strong co-citation ties between foundational works illustrate the integrated research efforts driving innovation.

4.2. Comparison Analysis of Research Methods

Figure 15 presents an overview of the principal machine learning techniques and models employed in maritime safety for autonomous shipping from 2000 to 2024. This figure is based on data extracted from abstracts and keywords of 719 articles, compiled through a meticulous manual review and de-duplication process. The statistical analysis of methods used over different time periods reveals that although there is some significant variation between past and present methodologies, many traditional machine learning methods still hold their unshakable position. The development of technology has enriched their application scenarios and improved their accuracy.

(1) 2000–2005: Foundational Techniques

Between 2000 and 2005, the focus was on establishing foundational techniques that could support the nascent field of autonomous shipping. Reinforcement learning and artificial neural networks were among the early methods explored, emphasizing control systems and basic decision-making processes. While these methods provided a solid basis for understanding risks, they were often constrained by limited data availability and challenges in incorporating human factors and expressing uncertainty.

(2) 2006–2010: Enhanced Methods for Improved Precision

From 2006 to 2010, the focus shifted towards more sophisticated machine learning methods that could enhance precision and address complex maritime scenarios. Techniques such as deep learning (DL) and sensor fusion were introduced to improve data processing capabilities and accuracy. The period also saw the integration of Fuzzy Fault Trees and Neural Network-based Case-based Q Learning (NCQL), which provided more robust frameworks for dealing with uncertain data and decision-making processes. Principal Component Analysis (PCA) and ARTMAP models were employed for dimensionality reduction and adaptive pattern recognition. Furthermore, unsupervised machine learning methods such as clustering, Modular Network Self-Organizing Maps (MNSOM), and Radial Basis Function (RBF) neural networks were used to detect patterns and anomalies in data. Innovative approaches such as Distant Transmission (DT) algorithms, Self-learning

Behavior Agents, and Fuzzy Logic Representation were also explored to enhance system autonomy and adaptability in maritime environments.

Figure 15. Major methods and models used in maritime safety for autonomous shipping.

(3) 2011–2015: Emphasis on Data-Driven Approaches

The period from 2011 to 2015 marked a significant shift towards data-driven approaches in maritime safety, leveraging more advanced machine learning techniques to gain deeper insights into complex maritime systems. Extreme Learning Machines (ELM) and Support Vector Machines (SVM) became popular for their ability to efficiently handle large datasets and improve classification accuracy. Hidden Markov Models were utilized to model temporal dependencies and sequential data, providing a robust framework for predictive analysis. The integration of OpenGL and VC++ facilitated the visualization of complex data structures and simulations. Techniques such as K-Nearest Neighbors (KNN) and ID3 contributed to enhanced pattern recognition and decision tree analyses, while Lasso Model Predictive Control (MPC) was used for optimization in predictive modeling. Risk evaluation models and Gaussian Mixture Models (GMM) were developed to improve probabilistic assessments of maritime safety. Furthermore, situation analysis approaches and data mining techniques were employed to extract actionable insights from vast amounts of data. Controlled Lagrangian Particle Tracking (CLPT) and model-free reinforcement learning algorithms such as SARSA (lambda) were also explored to enhance real-time decision-making and adaptive learning capabilities in autonomous maritime systems.

(4) 2016–2024: Integration of Advanced Machine Learning and Multidisciplinary Methods

From 2016 to 2024, advancements in maritime safety for autonomous shipping were driven by key machine learning techniques such as deep reinforcement learning (DRL) and Long Short-Term Memory (LSTM) networks, which enhanced decision-making and predictive capabilities. Explainable AI (XAI) methods, including SHAP, improved transparency and trust in autonomous systems. Ensemble methods such as Random Forests (RF) and Genetic Algorithms (GA) increased model accuracy and resilience. Additionally, Transformers and Spatio-Temporal Graph Convolutional Networks (STGCNs) offered advanced modeling of spatial and temporal dependencies, while transfer learning and multi-agent deep reinforcement learning (MADRL) facilitated adaptive and collaborative decision-making. Overall, this period marked a shift towards more sophisticated and integrated approaches, reflecting the convergence of machine learning advancements with maritime safety needs. These methodologies have substantially enhanced the capability to ensure safe and efficient autonomous shipping, aligning with the industry's goal of reducing human intervention while maintaining high safety standards.

All in all, the evolution of methods in autonomous shipping safety from 2000 to 2024 highlights a continuous advancement toward more sophisticated, data-driven, and multidisciplinary approaches. While foundational methods remain relevant, the integration of cutting-edge technologies and methodologies has significantly enhanced the accuracy and applicability of risk assessments in maritime transport.

4.3. Future Research Trends

The advancement of machine learning for enhancing maritime safety in autonomous shipping has undergone substantial evolution. This bibliometric analysis delves into pertinent literature from the last 24 years to offer a thorough understanding of machine learning's role in this field. Over the years, research in maritime safety utilizing machine learning has progressed through three distinct phases. Initially, the focus shifted from conventional maritime practices to autonomous ships, with maritime autonomous surface ships (MASS) acting as an essential intermediary step. Following this, the research emphasis broadened to include three pivotal areas: risk assessment, modeling of ship behavior, and sophisticated decision-making techniques. Additionally, machine learning methodologies have advanced through three stages: the mathematical model stage, the rule-based algorithm stage, and the intelligent algorithm stage.

This evolution underscores the critical integration of traditional maritime research with modern machine learning techniques. By leveraging decades of accumulated knowledge and insights, substantial progress has been made toward achieving safe, efficient, and intelligent autonomous ship operations. These advancements promise to transform maritime safety, setting new standards for the future of autonomous shipping.

In the future, for the safety of intelligent ships under machine learning, further exploration can be carried out from the following 3 aspects.

4.3.1. Trends in Safety Objectives

The traditional objectives of maritime safety have predominantly focused on reactive measures, such as compliance with regulations and responding to incidents. However, with advancements in machine learning and AI, the safety objectives are shifting in new directions. To seek this change, the findings presented in this section are derived from the selected literature included in this study. Specifically, the inspiration for identifying the evolving safety objectives in the maritime domain stems from the co-occurrence analysis of keywords and the citation network mapping. By examining recurring terms in the keyword co-occurrence network in the exact literature, we traced how the research focus has shifted over time as follows.

(1) Proactive safety management: As MASS evolves to the unmanned ship stage, the factors affecting marine accidents need to focus on four factors: ship factors, environmental factors, management factors, and technical factors [81]. The new objective is to predict and prevent incidents before they occur. Machine learning models analyze historical and real-time data to identify patterns that indicate potential safety hazards, allowing for preemptive actions. For instance, by continuously monitoring weather conditions, traffic density, and mechanical health, these systems can dynamically update risk profiles and adjust operations accordingly to mitigate risks. A case in point can be illustrated in MLDet [82], a deep learning method that aims to clear and recycle obstacle wastes in the way. In the future, a systematic multi-scale collision risk estimation approach should be developed and upgraded to capture traffic conflict patterns under different spatial scales [83].

(2) Human-machine synergy: Human factors are always the key element of unsafe behaviors [84], and some research has provided insight into human interactions in remote ship operations and consequently facilitates performance improvement actions [85]. For example, it was found that seafarers tend to experience more anxiety when dealing with emergency situations, while marine pilots experience more anxiety during multi-ship encounter periods [86]. However, as artificial intelligence develops in a fast manner, emotional AI is a groundbreaking trend that focuses on the well-being of human crew members still present on autonomous ships. By monitoring physiological data, voice tones, and facial expressions, emotional AI can assess stress and fatigue levels, providing timely interventions to prevent human errors. This ensures that the remaining human operators are always in optimal condition to manage any critical situations.

(3) Autonomous collaboration: Swarm intelligence is a novel objective where multiple autonomous ships operate as a cohesive unit, sharing data and insights in real-time. Currently, swarm intelligence has been used in multi-ship path planning [87] and collision avoidance decision-making systems [88,89]. This collective intelligence enhances situational awareness and decision-making, reducing the likelihood of accidents and improving overall safety. For this reason, swarm intelligence is able to significantly enhance the capabilities of unmanned ships in several innovative ways in the future. These include coordinated search and rescue operations and disaster response, where vessels autonomously distribute search patterns and respond rapidly to emergencies, improving efficiency and reducing response times. Moreover, swarm-based ships can perform fleet defense and security patrols, maintaining maritime safety with dynamic patrols and coordinated responses. These vessels can also support collaborative scientific research missions, enabling comprehensive data collection and exploration, since automatic ship alarm systems, coastal radars, and coastal cameras are not alone sufficient equipment to build maritime awareness [90]. Finally, swarm intelligence can act as "police" to assist in maritime traffic management, improving safety, reducing congestion, and optimizing operations in busy waterways.

4.3.2. Trends in Safety Technology Based on Machine Learning

Machine learning technologies are evolving to meet the sophisticated needs of autonomous maritime operations, introducing new safety mechanisms:

(1) Natural Language Processing (NLP) for Multilingual Operations: NLP enhances communication between ship systems and human operators, especially during emergencies. AI-powered systems can interpret voice commands, generate detailed incident reports, and translate complex maritime regulations into actionable instructions, facilitating more effective management of incidents.

(2) Bio-Inspired Sensors: Autonomous ships are expected to extensively rely on perception sensors for situation awareness and safety during challenging operations, such as reactive collision avoidance [91]. Up to now, some new sensors have been devised in combination with machine learning [92,93]. As we look to the future, the evolution of these technologies suggests the potential for even more sophisticated solutions, such as bio-inspired sensors. By mimicking the sensory capabilities of animals such as bats and dolphins, these sensors can offer enhanced detection and navigation capabilities. Machine learning will play a critical role in this advancement, enabling these bio-inspired sensors to adapt to complex maritime environments and challenges. Thus, as sensor technology continues to advance, the combination of machine learning and bio-inspired designs may set new standards for autonomy and precision in marine navigation.

(3) Quantum Machine Learning: Quantum computing has been proven to excel in factorization issues and unordered search problems due to its capability of quantum parallelism [94]. So far, quantum technology has been used in malware detection on IoT-enabled maritime transportation systems [95], co-communication protocols of underwater sensor networks [96], etc. By leveraging quantum computing, machine learning algorithms can process and analyze vast amounts of data at unprecedented speeds. This capability is crucial for real-time safety applications, such as predicting sudden environmental changes or mechanical failures.

(4) Self-Healing Systems: Based on proactive safety management, future autonomous ships could leverage machine learning to significantly enhance their safety capabilities by incorporating self-healing materials and systems. Machine learning algorithms can continuously monitor the ship's structural integrity and detect anomalies or potential damage in real time. For example, if a hull breach is detected, the ship's machine learning system can analyze the severity and location of the damage, predict potential risks, and automatically activate repair mechanisms. These advanced materials can seal the breach autonomously while the ship navigates safely to a port for more extensive repairs. Additionally, machine learning can optimize the repair process by learning from past incidents, improving the system's efficiency and effectiveness over time. This approach not only minimizes human intervention but also ensures a higher level of safety and reliability in maritime operations.

4.3.3. Trends in Machine Learning Algorithms

Machine learning algorithms are continuously advancing, providing new possibilities for maritime safety:

In terms of maritime safety management, machine learning algorithms, such as LSTM, RNN, NLP, RL, UL, DRL, STMGCN, SVM, ANN, and so on, have been widely used for vessel trajectory prediction [97,98], ship behavior [99,100], fault diagnosis of marine diesel engines [101], collision avoidance [102,103], visual recognition [104,105], and environmental perception [106,107]. For instance, Jiang et al. proposed a diagnostic network combining a wide convolutional neural network (WDCNN) and an extreme learning machine (ELM) to precisely conduct AUV actuator fault diagnosis [108]. A double-gated recurrent unit neural network (GRU-RNN) [103] was constructed to learn unmanned surface vehicle (USV) collision avoidance decisions from the extracted data of successful encounters of ships. This GRU-RNN model contributes significantly to the increased efficiency and safety of sea operations. These above technologies can facilitate the deep application of maritime data and improve the accuracy of risk assessment models, which may still be one of the directions for future research in the field of maritime safety.

Additionally, some newly burgeoning machine learning methods in other fields are proposed and suggested for future development.

(1) Generative Adversarial Networks (GANs) for Simulation: GANs are emerging unsupervised deep learning models known for their ability to generate realistic samples that are used to amplify a number of failures within training datasets [109]. This method has been used in machinery monitoring [109] and multi-ship trajectory prediction [87,110]. GANs are expected to create realistic simulations of maritime environments and potential hazards, which are used to train machine learning models under a wide range of scenarios. This ensures that autonomous ships are prepared for even the most unexpected events.

(2) Hierarchical Reinforcement Learning (HRL): HRL has been designed to improve training speed and resolve planning conflicts within the formation [111]. It also shows excellence in high learning speed and performance [112]. This approach is believed to become particularly suited for managing the complex tasks faced by autonomous vessels. By breaking down operations into smaller, manageable sub-tasks, hierarchical reinforcement learning allows for more precise and effective decision-making processes.

(3). Explainable AI (XAI): As AI systems become more integral to safety-critical operations, there is a growing demand for transparency. To remedy this problem, a number of explainable AI methods have been presented, such as SHAP and LIME [113], and they have been widely used in the field of wireless communications [114] and autonomous driving systems [115–117]. Moreover, intelligent software [118] can also serve as an important way in the development of XAI, with the goal of greatly improving the service efficiency and the shipping security of maritime transport industries. Despite the fact that research about explainable AI in autonomous shipping is limited in number, explainable AI provides insights into how decisions are made by machine learning models, enhancing trust and facilitating regulatory approval.

(4). Federated Learning: In the latest research, federated learning is testified to be a novel machine learning method that can guarantee collaborative learning among all ships while preserving ships' privacy [119]. It is promising that this decentralized approach enables the training of models across multiple ships without sharing raw data. Federated learning enhances the ability of machine learning systems to generalize across diverse maritime conditions while maintaining data privacy and security.

4.3.4. Summary for Future Prospect

The future of machine learning in maritime safety is boundless. With continuous advancements in AI, autonomous vessels will become increasingly capable of operating safely and efficiently with minimal human oversight. The future of maritime safety in autonomous shipping is being reshaped by advanced machine learning technologies. This evolution involves a shift from reactive to proactive safety management, where AI predicts and prevents incidents, enhancing human-machine synergy and enabling swarm intelligence among ships. Key advancements in technology include NLP for multilingual operations, bio-inspired sensors for enhanced navigation, and quantum machine learning for rapid data processing. Explainable AI and federated learning further enhance transparency and collaboration while ensuring data privacy. The mentioned future trend evolution is depicted in Figure 16.

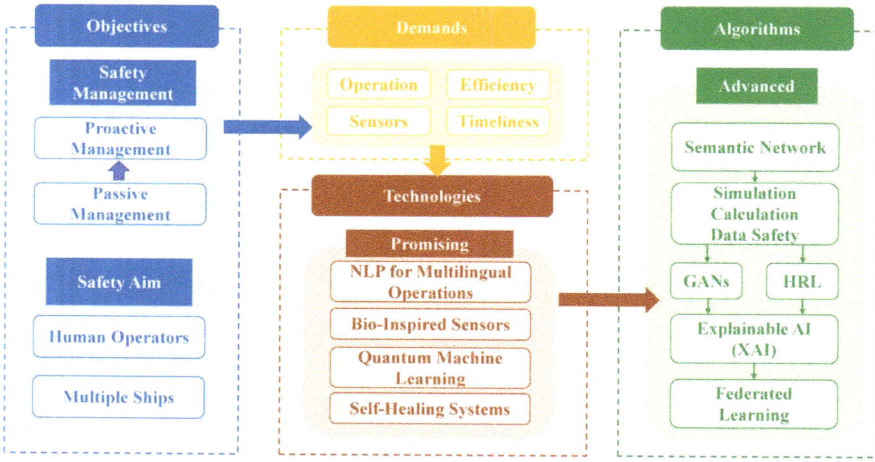

Figure 16. The trend analysis of machine learning for maritime safety.

Overall, extant literature has primarily concentrated on technological innovation and data analytics, yet the transformative impact of these advancements is contingent upon their systemic adoption within the maritime industry. Such adoption necessitates comprehensive institutional changes encompassing regulatory frameworks, organizational culture, commercial structures, and stakeholder trust—beyond mere technical feasibility. Critically, cybersecurity constitutes the foremost challenge to autonomous shipping, vulnerabilities in digital networks, data integrity, and control systems present significant threats to vessel safety, cargo security, and industry credibility. A recent comprehensive review of maritime cybersecurity research highlights not only a rapid escalation in both publicly reported and likely underreported cyber incidents, including ransomware, denial-of-service attacks, and sophisticated supply-chain compromises, but also a fragmented research landscape that lacks holistic, empirically validated frameworks for risk assessment and resilience [120]. Accordingly, future research should prioritize rigorous examination of cybersecurity within autonomous maritime contexts, including threat modeling, risk assessment methodologies, and the development of legal and regulatory mechanisms addressing data ownership, breach liability, and incident response. Realizing the full potential of autonomous shipping will require the establishment of robust international standards, interdisciplinary collaboration, and continual innovation in cyber-resilience to safeguard operational integrity and foster trust across the maritime ecosystem.

4.3.5. Bias and Limitations

Despite the valuable insights provided by this study, it is crucial to acknowledge several limitations inherent in the methodology and data sources utilized. Firstly, the exclusive reliance on the Web of Science (WoS) database for bibliometric analysis, while justified by its comprehensive coverage and high data quality, may have inadvertently excluded relevant publications from other databases such as Scopus or Google Scholar. This limitation could introduce a selection bias, potentially omitting significant contributions to the field of maritime safety and autonomous shipping, particularly from emerging or niche journals not indexed in WoS. Future research should consider integrating multiple databases to provide a more holistic view of the field.

Secondly, this study focused exclusively on English-language publications, as English remains the dominant medium of academic discourse. The exclusion of these works may limit the comprehensiveness of the analysis and overlook region-specific advancements or perspectives in maritime safety and machine learning.

Thirdly, the chosen time frame for the analysis, covering publications from 2000 to 2024, could influence the identified trends and patterns. While this period captures the rise of machine learning applications in autonomous shipping, earlier foundational research or very recent publications may not have been adequately represented.

Lastly, the design of the search query, including the selection of keywords and Boolean operators, may have unintentionally excluded relevant studies or broadened the scope to include less relevant papers. This methodological limitation highlights the need for more precise query designs and iterative refinements in future bibliometric studies.

5. Conclusions

In this paper, to provide a detailed overview of the key contributions in this research domain, a comprehensive bibliometric analysis of machine learning for maritime safety of autonomous shipping was conducted by utilizing 719 publications sourced from the WoS database. The analysis entailed the exploration of publication trends and the identification of influential authors, institutions, countries, notable articles, journals, disciplines, references, and keywords. Additionally, deeper insights were demonstrated of the international cooperation in ML research, explaining it from the perspectives of countries, institutions, and authors. Furthermore, we used a dual-function word analysis to identify prevalent keywords and the variation of them, thereby uncovering emerging research topics and trends. Finally, three future directions for the advancement of machine learning in maritime safety were brought forth, which provide ideas for the development of marine research.

The implications of these findings are significant for various stakeholders. For academic researchers, the results offer valuable insights into the current landscape of research in autonomous shipping and maritime safety, elucidating both the strengths and areas needing further exploration. This information can aid scholars in identifying potential collaborators for educational and research projects in the field of autonomous maritime systems. Such partnerships have the potential to advance not only sustainable marine technology development but also enhance related industrial sectors on a global scale.

Author Contributions: Conceptualization, J.X.; methodology, J.X. and P.Y.; software, J.X., P.Y. and Q.L.; validation, J.X., Y.S. and H.H.; resources, J.X., P.H.A.J.M.v.G., Y.S. and H.H.; data curation, J.X., E.P., P.Y. and Q.L.; writing—original draft preparation, J.X. and P.Y.; review and editing, J.X., P.H.A.J.M.v.G., E.P. and H.H.; visualization, J.X. and P.Y.; supervision, J.X., P.H.A.J.M.v.G. and H.H.; funding acquisition, J.X. All authors have read and agreed to the published version of the manuscript.

Funding: This work was supported by the National Natural Science Foundation of China (Grant No. 52201413); the Natural Science Foundation of Science and Technology Commission of Shanghai Municipality (Grant No. 23ZR1434400); the Soft Science Research Program of Science and Technology Commission of Shanghai Municipality (Grant No. 24692105500); the Opening Foundation of Key Laboratory of Safety and Risk Management on Transport Infrastructures for Ministry of Transport (Grant No. 2023KFKT015); the Oceanic Interdisciplinary Program of Shanghai Jiao Tong University (Grant No. SL2022PT107); and the Startup Fund for Young Faculty at Shanghai Jiao Tong University (Grant No. 22X010503469).

Data Availability Statement: The raw data supporting the conclusions of this article will be made available by the authors on request.

Conflicts of Interest: The authors declare that they have no known competing financial interests or personal relationships that could have appeared to influence the work reported in this paper.

Appendix A

Table A1. The keywords and strings search in WoS.

Step	Keywords/Strings/Topics	Number of Papers
1	TS = (big data) AND TS = (autonomous shipping OR unmanned shipping) AND TS = (safety OR risk OR security OR reliability) Indexes = SCI-EXPANDED, SSCI, CPCI-S, CPCI-SSH; Timespan = 1 January 2001–31 May 2024	26
2	TS = (data analytics) AND TS = (autonomous shipping OR unmanned shipping) AND TS = (safety OR risk OR security OR reliability) Indexes = SCI-EXPANDED, SSCI, CPCI-S, CPCI-SSH; Timespan = 1 January 2001–31 May 2024	9
3	TS = (machine learning) AND TS = (autonomous shipping OR unmanned shipping) AND TS = (safety OR risk OR security OR reliability) Indexes = SCI-EXPANDED, SSCI, CPCI-S, CPCI-SSH; Timespan = 1 January 2001–31 May 2024	37
4	TS = (data analysis OR data analytics OR data analyst OR data analyzed) AND TS = (autonomous shipping OR unmanned shipping) AND TS = (safety OR risk OR security OR reliability) Indexes = SCI-EXPANDED, SSCI, CPCI-S, CPCI-SSH; Timespan = 1 January 2001–31 May 2024	133
5	TS = (big data OR data analysis OR data analytics OR data analyst OR data analyzed OR machine learning OR supervised learning OR unsupervised learning OR semi-supervised learning OR ensemble learning OR deep learning OR reinforcement learning OR transfer learning) AND TS = (autonomous shipping OR unmanned shipping) AND TS = (safety OR risk OR security OR reliability) Indexes = SCI-EXPANDED, SSCI, CPCI-S, CPCI-SSH; Timespan = 1 January 2001–31 May 2024	222
6	TS = (big data OR data analy* OR machine learning OR supervised learning OR unsupervised learning OR semi-supervised learning OR ensemble learning OR deep learning OR reinforcement learning OR transfer learning) AND TS = (autonomous(ship* OR vessel$ OR boat*) OR unmanned(ship* OR vessel$ OR boat*) OR smart(ship* OR vessel$ OR boat*) OR intelligent(ship* OR vessel$ OR boat*) OR "autonomous underwater vehicle*" OR "autonomous surface vehicle*" OR "unmanned underwater vehicle*" OR "unmanned surface vehicle*" OR "autonomous marine robotic vehicle*" OR "unmanned marine robotic vehicle*" OR "underwater robotic vehicle*" OR "surface robotic vehicle*" OR "robotic underwater vehicle*" OR "robotic surface vehicle*" OR "untethered underwater vehicle" OR "untethered surface vehicle") AND TS = (safety OR risk OR security OR reliability) Indexes = SCI-EXPANDED, SSCI, CPCI-S, CPCI-SSH; Timespan = 1 January 2001–31 May 2024	775
7	TS = (big data OR data analy* OR machine learning OR supervised learning OR unsupervised learning OR semi-supervised learning OR ensemble learning OR deep learning OR reinforcement learning OR transfer learning) AND TS = (autonomous(ship* OR vessel$ OR boat*) OR unmanned(ship* OR vessel$ OR boat*) OR smart(ship* OR vessel$ OR boat*) OR intelligent(ship* OR vessel$ OR boat*) OR "autonomous underwater vehicle*" OR "autonomous surface vehicle*" OR "unmanned underwater vehicle*" OR "unmanned surface vehicle*" OR "autonomous marine robotic vehicle*" OR "unmanned marine robotic vehicle*" OR "underwater robotic vehicle*" OR "surface robotic vehicle*" OR "robotic underwater vehicle*" OR "robotic surface vehicle*" OR "untethered underwater vehicle" OR "untethered surface vehicle") AND TS = (safe* OR risk* OR secur* OR reliab* OR resilience* OR emergen* OR danger* OR hazard* OR maintainab* OR los$ OR accident* OR incident* OR colli* OR encounter* OR ground* OR sink* OR list* OR capsiz* OR dragg* OR contact* OR damag* OR COLREG* OR fire* OR explosion* OR wind* OR "human factor*" OR marine* OR maritime* OR "maritime tra*" OR "maritime transportation system*") Indexes = SCI-EXPANDED, SSCI, CPCI-S, CPCI-SSH; Timespan = 1 January 2001–31 May 2024	2120

Table A1. *Cont.*

Step	Keywords/Strings/Topics	Number of Papers
8	TS = (big data OR data analy* OR machine learning OR supervised learning OR unsupervised learning OR semi-supervised learning OR ensemble learning OR deep learning OR reinforcement learning OR transfer learning) AND TS = (autonomous(ship* OR vessel$ OR boat*) OR unmanned(ship* OR vessel$ OR boat*) OR smart(ship* OR vessel$ OR boat*) OR intelligent(ship* OR vessel$ OR boat*) OR "autonomous underwater vehicle*" OR "autonomous surface vehicle*" OR "unmanned underwater vehicle*" OR "unmanned surface vehicle*" OR "autonomous marine robotic vehicle*" OR "unmanned marine robotic vehicle*" OR "underwater robotic vehicle*" OR "surface robotic vehicle*" OR "robotic underwater vehicle*" OR "robotic surface vehicle*" OR "untethered underwater vehicle" OR "untethered surface vehicle") AND TS = (safe* OR risk* OR secur* OR reliab* OR resilience* OR emergen* OR danger* OR hazard* OR maintainab* OR los$ OR accident* OR incident* OR colli* OR encounter* OR ground* OR sink* OR list* OR capsiz* OR dragg* OR contact* OR damag* OR COLREG* OR fire* OR explosion* OR wind* OR "human factor*" OR marine* OR maritime* OR "maritime tra*" OR "maritime transportation system*") AND DT = (Article OR Proceedings Paper OR Review) Indexes = SCI-EXPANDED, SSCI, CPCI-S, CPCI-SSH; Timespan = 1 January 2001–31 May 2024; Search language = English	2107
9	TS = (big data OR data analy* OR machine learning OR supervised learning OR unsupervised learning OR semi-supervised learning OR ensemble learning OR deep learning OR reinforcement learning OR transfer learning) AND TS = (autonomous(ship* OR vessel$ OR boat*) OR unmanned(ship* OR vessel$ OR boat*) OR smart(ship* OR vessel$ OR boat*) OR intelligent(ship* OR vessel$ OR boat*) OR "autonomous underwater vehicle*" OR "autonomous surface vehicle*" OR "unmanned underwater vehicle*" OR "unmanned surface vehicle*" OR "autonomous marine robotic vehicle*" OR "unmanned marine robotic vehicle*" OR "underwater robotic vehicle*" OR "surface robotic vehicle*" OR "robotic underwater vehicle*" OR "robotic surface vehicle*" OR "untethered underwater vehicle" OR "untethered surface vehicle") AND TS = (safe* OR risk* OR secur* OR reliab* OR resilience* OR emergen* OR danger* OR hazard* OR maintainab* OR los$ OR accident* OR incident* OR colli* OR encounter* OR ground* OR sink* OR list* OR capsiz* OR dragg* OR contact* OR damag* OR COLREG* OR fire* OR explosion* OR wind* OR "human factor*" OR marine* OR maritime* OR "maritime tra*" OR "maritime transportation system*") AND DT = (Article OR Proceedings Paper OR Review) Manually screened by relevance Indexes = SCI-EXPANDED, SSCI, CPCI-S, CPCI-SSH; Timespan = 1 January 2001–31 May 2024; Search language = English	719

Note: TS = Topic; DT = Document Types; SCI-EXPANDED = Science Citation Index Expanded; SSCI = Social Sciences Citation Index; CPCI-S = Conference Proceedings Citation Index- Science; CPCI-SSH = Conference Proceedings Citation Index- Social Science and Humanities; the following words which combined with wildcard (* and $) contain but not only the following words: analy* = analysis/analyze/analytics/analyst/analytical; ship* = ship/ships/shipping; vessel$ = vessel/vessels; boat* = boat/boats/boating; vehicle* = vehicle/vehicles/vehicules/vehicular; safe* = safe/safety/safely; risk* = risk/risks/risky; secur* = secure/security; reliab* = reliability/reliable/reliably; resilience* = resilience/resiliences; emergen* = emergency/emergence/emergencies/emergent; danger* = danger/dangers/dangerous/dangerously; hazard* = hazard/hazards/hazardous; maintainab* = maintainability/maintainable; los$ = loss/lose/lost; accident* = accident/accidents/accidentally/accidental; incident* = incident/incidents/incidental/incidentally; colli* = collision//collide/collided/colliding/collisions; encounter* = encounter/encounters/encountered; ground* = ground/grounds/grounded/grounding/groundings; sink* = sink/sinks/sinking; list* = list/lists/listing; capsiz* = capsize/capsized/capsizes/capsizing; dragg* = dragged/dragging/draggable; contact* = contact/contacts; damag* = damage/damaged/damages/damaging; colreg* = colreg/colregs; fire* = fire/fires; explosion* = explosion/explosion; wind* = wind/winds/windy; marine* = mine/mines; maritime* = maritime/maritimes; "maritime tra*" = "maritime traffic"/"maritime transport"/"maritime transportation"/"maritime trade".

Table A2. Top 20 authors with most publications and citations on machine learning for maritime safety of autonomous shipping research.

Rank	Top 20 Most Productive Authors									Top 20 Most Citation Authors								
	Author	Country	Links	TLS	NP	P (%)	TC	APY	AC	Author	Country	Links	TLS	NP	P (%)	TC	APY	AC
1	He, Bo	China	93	1542	14	1.95%	98	2018.14	7.00	Zhang, Weidong	China	84	2918	9	1.25%	373	2021.56	41.44
2	Zhang, Weidong	China	84	2918	9	1.25%	373	2021.56	41.44	Van Gelder, P. H. A. J. M.	Netherlands	78	1066	4	0.56%	370	2019.25	92.50
3	Liu, Yuanchang	England	96	1803	8	1.11%	155	2020.63	19.38	Peng, Zhouhua	China	61	865	5	0.70%	290	2019.60	58.00
4	Wang, Chengbo	China	76	1715	8	1.11%	69	2022.25	8.63	Wang, Dan	China	60	719	4	0.56%	274	2018.75	68.50
5	Yang, Zaili	England	64	3673	8	1.11%	149	2023.25	18.63	Wu, Chaozhong	China	46	628	4	0.56%	246	2019.25	61.50
6	Zhang, Xinyu	China	90	1670	8	1.11%	77	2021.88	9.63	Yan, Xinping	China	79	1232	6	0.83%	216	2017.50	36.00
7	Yan, Tianhong	China	56	598	7	0.97%	16	2017.43	2.29	Wang, Ning	China	86	981	6	0.83%	215	2021.83	35.83
8	Chen, Xinqiang	China	49	803	6	0.83%	205	2021.33	34.17	Chen, Xinqiang	China	49	803	6	0.83%	205	2021.33	34.17
9	Guo, Jia	China	32	576	6	0.83%	51	2018.50	8.50	Lu, Yu	China	73	792	4	0.56%	203	2019.00	50.75
10	Li, Guangliang	China	80	997	6	0.83%	52	2019.17	8.67	Chen, Zhijun	China	41	418	3	0.42%	187	2019.33	62.33
11	Liu, Jingxian	China	84	1326	6	0.83%	83	2021.00	13.83	Liu, Ryan Wen	China	58	912	4	0.56%	185	2020.50	46.25
12	Wang, Ning	China	86	981	6	0.83%	215	2021.83	35.83	Wu, Huafeng	China	45	548	4	0.56%	181	2021.00	45.25
13	Yan, Xinping	China	79	1232	6	0.83%	216	2017.50	36.00	Li, Zhixiong	China	80	491	5	0.70%	175	2016.60	35.00
14	Li, Huanhuan	China	62	2974	5	0.70%	98	2023.20	19.60	Liang, Maohan	China	45	530	3	0.42%	169	2019.67	56.33
15	Li, Zhixiong	China	80	491	5	0.70%	175	2016.60	35.00	Yang, Yongsheng	China	42	434	3	0.42%	168	2020.00	56.00
16	Ma, Feng	China	76	1278	5	0.70%	66	2019.80	13.20	Liu, Yuanchang	England	96	1803	8	1.11%	155	2020.63	19.38

Table A2. *Cont.*

Rank	Top 20 Most Productive Authors									Top 20 Most Citation Authors								
	Author	Country	Links	TLS	NP	P (%)	TC	APY	AC	Author	Country	Links	TLS	NP	P (%)	TC	APY	AC
17	Peng, Zhouhua	China	61	865	5	0.70%	290	2019.60	58.00	Wang, Yang	China	73	622	3	0.42%	150	2020.67	50.00
18	Shen, Yue	China	74	692	5	0.70%	20	2019.40	4.00	Yang, Zaili	England	64	3673	8	1.11%	149	2023.25	18.63
19	Sun, Changyin	China	85	1182	5	0.70%	113	2022.80	22.60	Xu, Xinli	China	79	1540	4	0.56%	138	2021.75	34.50
20	Wang, Hao	China	62	353	5	0.70%	42	2021.00	8.40	Zhang, Mingyang	China	71	617	3	0.42%	135	2021.67	45.00

Note: TLS = total link strength; NP = number of publications; P (%) = the proportion of NP/TND; TND = total number of documents; TC = total number of citations; APY = average publications year; AC = average citations = TC/NP.

Table A3. Top 30 keywords for the explored research domain in terms of various periods.

Rank	All Time		2000–2010	
	Keyword	**Frequency**	**Keyword**	**Frequency**
1	AUV	131	AUV	6
2	Collision avoidance	83	Navigation	2
3	Deep learning	79	Reliability	2
4	USV	77	UGV	2
5	Navigation	68	3D computer vision	1
6	System	62	Acoustic image	1
7	Deep reinforcement learning	57	Acoustic ultrashort baseline system	1
8	Reinforcement learning	45	Automatic control	1
9	Machine learning	43	Autonomous navigation	1
10	Model	43	Autonomous underwater navigation	1
11	Tracking	41	AUV docking	1
12	Path planning	39	Avoidance	1
13	Algorithm	33	Basis expansion model	1
14	Ship	29	Channels	1
15	Design	28	Circular buffe	1
16	Object detection	26	Collision avoidance system	1
17	Obstacle avoidance	26	Data logging	1
18	COLREGS	25	Data reconstruction	1
19	Optimization	25	Detection	1
20	Artificial Intelligence	24	Diagnosis	1
21	Classification	21	Differential detection	1
22	Marine vehicles	21	Differential OSTBC	1
23	Marine robotics	19	Distant transmission algorithm	1
24	AIS data	18	DSSS	1
25	Internet	16	Dynamic replanning	1
26	Prediction	16	Expert system	1
27	Safety	16	Fault detection	1
28	AIS	15	Fuzzy fault tree	1
29	Autonomous navigation	14	Homing strategies	1
30	Big data	14	Image process	1
	2011–2015		2016–2024	
	Keyword	Frequency	Keyword	Frequency
	AUV	20	AUV	99
	Design	5	USV	80
	Systems	4	Deep learning	79
	Tracking	3	Collision avoidance	76
	Acoustic communication	3	Navigation	64
	AIS	2	Deep reinforcement learning	63
	Anomaly detection	2	System	55
	Big data	2	Models	46
	Component	2	Path planning	44

Table A3. *Cont.*

Rank	All Time		2000–2010	
	Keyword	**Frequency**	**Keyword**	**Frequency**
	Data assimilation	2	Reinforcement learning	44
	Data fusion	2	Algorithms	42
	Decision support	2	Machine learning	40
	Docking	2	Marine robotics	38
	Instantaneous angular speed	2	Tracking	37
	Intelligent systems	2	Neural network	29
	Machine learning	2	Ship	28
	Marine robotics	2	Object detection	26
	Maritime domain awareness	2	COLREGS	25
	Models	2	Obstacle avoidance	25
	Navigation	2	Optimization	25
	Path planning	2	Artificial Intelligence	24
	Underwater	2	Design	23
	Underwater communication	2	Classification	20
	UUV	2	Networks	20
	3 axis aquatic flight	1	Autonomous ships	19
	Accumulation	1	Underwater	19
	Acoustic navigation	1	Vehicles	18
	Adriatic sea	1	AIS data	17
	Air launch	1	Safety	16
	AIS data	1	Internet	15

Note: AUV = autonomous underwater vehicle; USV = unmanned surface vehicle; AIS = automatic identification system; UGV = unmanned ground vehicle; UUV = unmanned underwater vehicle.

Table A4. Summary of application field about four intelligent methods.

Number	Machine Learning	Deep Learning	Reinforcement Learning	Deep Reinforcement Learning
1	Sensor fusion	Sensor fusion	-	-
2	-	Ship detection	-	-
3	-	Image segmentation	-	-
4	Reinforcement learning	Reinforcement learning	-	Reinforcement learning
5	Optimization	Optimization	-	Optimization
6	Object detection	Object detection	-	-
7	-	Machine learning	-	Machine learning
8	-	Transfer learning	-	Transfer learning
9	-	-	COLREGs	COLREGs
10	-	-	Autonomous navigation	Autonomous navigation
11	-	Simulation	Simulation	Simulation
12	-	-	-	decision-making
13	Path planning	Path planning	Path planning	Path planning
14	-	-	Underwater vehicle	-
15	Fault diagnosis	Fault diagnosis	-	-
16	-	-	-	Data collection
17	Classification	Classification	-	-
18	Prediction	Prediction	Prediction	-
19	-	Marine robotics	-	Marine robotics
20	AUV	AUV	AUV	AUV
21	AIS data	AIS data	-	-
22	Anomaly detection	Anomaly detection	-	Anomaly detection

Table A4. *Cont.*

Number	Machine Learning	Deep Learning	Reinforcement Learning	Deep Reinforcement Learning
23	Computer vision	Computer vision	-	-
24	Navigation	Navigation	Navigation	Navigation
25	Robotics	Robotics	Robotics	Robotics
26	Tracking	Tracking	Tracking	Tracking
27	Models	Models	Models	Models
28	AIS	AIS	-	-
29	System	System	System	System
30	Uncertainty	-	Uncertainty	-
31	Design	Design	Design	Design
32	Neural network	Neural network	Neural network	Neural network
33	Algorithm	Algorithm	Algorithm	Algorithm
34	Marine vehicles	Marine vehicles	Marine vehicles	-
35	Deep learning	-	Deep learning	Deep learning
36	Autonomous ship	Autonomous ship	Autonomous ship	Autonomous ship
37	USV	USV	USV	USV
38	-	Convolutional neural Network	Convolutional neural Network	-
39	Collision avoidance	Collision avoidance	Collision avoidance	Collision avoidance
40	Sensors	Sensors	Sensors	Sensors
41	Obstacle avoidance	Obstacle avoidance	Obstacle avoidance	Obstacle avoidance
42	Path following	-	Path following	Path following
43	Ensemble learning	Ensemble learning	-	-
44	Maritime safety	Maritime safety	-	Maritime safety
45	-	Vision	-	-
46	Blockchain	-	Blockchain	Blockchain
47	-	-	Management	Management

References

1. Olapoju, O. Autonomous ships, port operations, and the challenges of African ports. *Marit. Technol. Res.* **2023**, *5*, 260194. [CrossRef]
2. Li, H.H.; Yang, Z.L. Incorporation of AIS data-based machine learning into unsupervised route planning for maritime autonomous surface ships. *Transp. Res. Part E-Logist. Transp. Rev.* **2023**, *176*, 103171. [CrossRef]
3. Chou, C.-C.; Wang, C.-N.; Hsu, H.-P. A novel quantitative and qualitative model for forecasting the navigational risks of Maritime Autonomous Surface Ships. *Ocean Eng.* **2022**, *248*, 110852. [CrossRef]
4. Tan, A.Y.N.; Loh, H.S.; Hsieh, C.-H.; Lopez, M.C.R. Adoption of digital technologies in the maritime industry: Insights from Singapore. *Marit. Technol. Res.* **2025**, *7*, 275821. [CrossRef]
5. Zhou, X.Y.; Liu, Z.J.; Wang, F.W.; Wu, Z.L. A system-theoretic approach to safety and security co-analysis of autonomous ships. *Ocean Eng.* **2021**, *222*, 108569. [CrossRef]
6. Jovanovic, I.; Percic, M.; Korican, M.; Vladimir, N.; Fan, A.L. Investigation of the Viability of Unmanned Autonomous Container Ships under Different Carbon Pricing Scenarios. *J. Mar. Sci. Eng.* **2022**, *10*, 1991. [CrossRef]
7. Tang, W.S.; Roman, D.; Dickie, R.; Robu, V.; Flynn, D. Prognostics and Health Management for the Optimization of Marine Hybrid Energy Systems. *Energies* **2020**, *13*, 4676. [CrossRef]
8. Kim, D.; Kim, J.S.; Kim, J.H.; Im, N.K. Development of ship collision avoidance system and sea trial test for autonomous ship. *Ocean Eng.* **2022**, *266*, 113120. [CrossRef]
9. Qiao, D.L.; Liu, G.Z.; Lv, T.Z.; Li, W.; Zhang, J. Marine Vision-Based Situational Awareness Using Discriminative Deep Learning: A Survey. *J. Mar. Sci. Eng.* **2021**, *9*, 397. [CrossRef]
10. Xie, W.D.; Gang, L.H.; Zhang, M.H.; Liu, T.; Lan, Z.X. Optimizing Multi-Vessel Collision Avoidance Decision Making for Autonomous Surface Vessels: A COLREGs-Compliant Deep Reinforcement Learning Approach. *J. Mar. Sci. Eng.* **2024**, *12*, 372. [CrossRef]
11. Wang, Y.; Xu, H.X.; Feng, H.; He, J.H.; Yang, H.J.; Li, F.; Yang, Z. Deep reinforcement learning based collision avoidance system for autonomous ships. *Ocean Eng.* **2024**, *292*, 116527. [CrossRef]

12. Wang, C.B.; Wang, N.; Gao, H.B.; Wang, L.H.; Zhao, Y.Z.; Fang, M.X. Knowledge transfer enabled reinforcement learning for efficient and safe autonomous ship collision avoidance. *Int. J. Mach. Learn. Cybern.* **2024**, *15*, 3715–3731. [CrossRef]
13. Zheng, K.J.; Zhang, X.Y.; Wang, C.B.; Li, Y.K.; Cui, J.L.; Jiang, L.L. Adaptive collision avoidance decisions in autonomous ship encounter scenarios through rule-guided vision supervised learning. *Ocean Eng.* **2024**, *297*, 117096. [CrossRef]
14. Niu, Y.H.; Zhu, F.X.; Wei, M.X.; Du, Y.F.; Zhai, P.Y. A Multi-Ship Collision Avoidance Algorithm Using Data-Driven Multi-Agent Deep Reinforcement Learning. *J. Mar. Sci. Eng.* **2023**, *11*, 2101. [CrossRef]
15. Xue, D.L.; Wu, D.F.; Yamashita, A.S.; Li, Z.X. Proximal policy optimization with reciprocal velocity obstacle based collision avoidance path planning for multi-unmanned surface vehicles. *Ocean Eng.* **2023**, *273*, 114005. [CrossRef]
16. Jadhav, A.K.; Pandi, A.R.; Somayajula, A. Collision avoidance for autonomous surface vessels using novel artificial potential fields. *Ocean Eng.* **2023**, *288*, 116011. [CrossRef]
17. Peng, X.; Han, F.L.; Xia, G.H.; Zhao, W.Y.; Zhao, Y.M. Autonomous Obstacle Avoidance in Crowded Ocean Environment Based on COLREGs and POND. *J. Mar. Sci. Eng.* **2023**, *11*, 1320. [CrossRef]
18. Waltz, M.; Paulig, N.; Okhrin, O. 2-level reinforcement learning for ships on inland waterways: Path planning and following. *Expert Syst. Appl.* **2025**, *274*, 126933. [CrossRef]
19. Li, L.Y.; Wu, D.F.; Huang, Y.Q.; Yuan, Z.-M. A path planning strategy unified with a COLREGS collision avoidance function based on deep reinforcement learning and artificial potential field. *Appl. Ocean Res.* **2021**, *113*, 102759. [CrossRef]
20. Wang, D.; Chen, H.M.; Lao, S.H.; Drew, S. Efficient Path Planning and Dynamic Obstacle Avoidance in Edge for Safe Navigation of USV. *IEEE Internet Things J.* **2024**, *11*, 10084–10094. [CrossRef]
21. Degorre, L.; Fossen, T.I.; Delaleau, E.; Chocron, O. A Virtual Reference Point Kinematic Guidance Law for 3-D Path-Following of Autonomous Underwater Vehicles. *IEEE Access* **2024**, *12*, 109822–109831. [CrossRef]
22. Waltz, M.; Okhrin, O. Spatial-temporal recurrent reinforcement learning for autonomous ships. *Neural Netw.* **2023**, *165*, 634–653. [CrossRef]
23. Wang, M.M.; Wang, Y.F.; Cui, E.H.; Fu, X.J. A novel multi-ship collision probability estimation method considering data-driven quantification of trajectory uncertainty. *Ocean Eng.* **2023**, *272*, 113825. [CrossRef]
24. Guo, C.Q.; Haugen, S.; Utne, I.B. Risk assessment of collisions of an autonomous passenger ferry. *Proc. Inst. Mech. Eng. Part O-J. Risk Reliab.* **2023**, *237*, 425–435. [CrossRef]
25. Thum, G.W.; Tang, S.H.; Ahmad, S.A.; Alrifaey, M. Toward a Highly Accurate Classification of Underwater Cable Images via Deep Convolutional Neural Network. *J. Mar. Sci. Eng.* **2020**, *8*, 924. [CrossRef]
26. Kim, J.-y.; Lee, T.-h.; Lee, S.-h.; Lee, J.-j.; Lee, W.-k.; Kim, Y.-j.; Park, J.-w. A Study on Deep Learning-Based Fault Diagnosis and Classification for Marine Engine System Auxiliary Equipment. *Processes* **2022**, *10*, 1345. [CrossRef]
27. Zhao, X.Y.; Yuan, H.W.; Yu, Q. Autonomous Vessels in the Yangtze River: A Study on the Maritime Accidents Using Data-Driven Bayesian Networks. *Sustainability* **2021**, *13*, 9985. [CrossRef]
28. Tan, Y.H.; Tian, H.; Jiang, R.Z.; Lin, Y.J.; Zhang, J.D. A comparative investigation of data-driven approaches based on one-class classifiers for condition monitoring of marine machinery system. *Ocean Eng.* **2020**, *201*, 107174. [CrossRef]
29. Das, D.B.; Birant, D. GASEL: Genetic algorithm-supported ensemble learning for fault detection in autonomous underwater vehicles. *Ocean Eng.* **2023**, *272*, 113844. [CrossRef]
30. Pereira, M.I.; Claro, R.M.; Leite, P.N.; Pinto, A.M. Advancing Autonomous Surface Vehicles: A 3D Perception System for the Recognition and Assessment of Docking-Based Structures. *IEEE Access* **2021**, *9*, 53030–53045. [CrossRef]
31. Jian, J.; Sun, Z.; Sun, K. An Intelligent Automatic Sea Forecasting System Targeting Specific Areas on Sailing Routes. *Sustainability* **2024**, *16*, 1117. [CrossRef]
32. Dalal, S.; Seth, B.; Radulescu, M.; Cilan, T.F.; Serbanescu, L. Optimized Deep Learning with Learning without Forgetting (LwF) for Weather Classification for Sustainable Transportation and Traffic Safety. *Sustainability* **2023**, *15*, 6070. [CrossRef]
33. Chen, X.Q.; Liu, S.H.; Zhao, J.S.; Wu, H.F.; Xian, J.F.; Montewka, J. Autonomous port management based AGV path planning and optimization via an ensemble reinforcement learning framework. *Ocean Coast. Manag.* **2024**, *251*, 107087. [CrossRef]
34. Cope, S.; Zetterlind, V.; Tougher, B.; Easterday, K. Using machine learning to optimize autonomous tracking of vessels by marine radar. In Proceedings of the OCEANS Conference, San Diego, CA, USA, 20–23 September 2021; pp. 1–6.
35. Zhou, X. Spatial risk assessment of maritime transportation in offshore waters of China using machine learning and geospatial big data. *Ocean Coast. Manag.* **2024**, *247*, 106934. [CrossRef]
36. Gil, M.; Wrobel, K.; Montewka, J.; Goerlandt, F. A bibliometric analysis and systematic review of shipboard Decision Support Systems for accident prevention. *Saf. Sci.* **2020**, *128*, 104717. [CrossRef]
37. Li, Z.H.; Zhang, D.; Han, B.; Wan, C.P. Risk and reliability analysis for maritime autonomous surface ship: A bibliometric review of literature from 2015 to 2022. *Accid. Anal. Prev.* **2023**, *187*, 107090. [CrossRef]
38. Karatug, C.; Arslanoglu, Y.; Soares, C.G. Review of maintenance strategies for ship machinery systems. *J. Mar. Eng. Technol.* **2023**, *22*, 233–247. [CrossRef]

39. Chaal, M.; Ren, X.; BahooToroody, A.; Basnet, S.; Bolbot, V.; Banda, O.A.V.; Van Gelder, P. Research on risk, safety, and reliability of autonomous ships: A bibliometric review. *Saf. Sci.* **2023**, *167*, 106256. [CrossRef]
40. Singh, S.W.; Dhir, S.L.; Das, V.M.; Sharma, A. Bibliometric overview of the Technological Forecasting and Social Change journal: Analysis from 1970 to 2018. *Technol. Forecast. Soc. Change* **2020**, *154*, 119963. [CrossRef]
41. Aria, M.; Cuccurullo, C. bibliometrix: An R-tool for comprehensive science mapping analysis. *J. Informetr.* **2017**, *11*, 959–975. [CrossRef]
42. Sun, Y.T.; Grimes, S. The emerging dynamic structure of national innovation studies: A bibliometric analysis. *Scientometrics* **2016**, *106*, 17–40. [CrossRef]
43. Askun, V.; Cizel, R. Twenty Years of Research on Mixed Methods. *J. Mix. Methods Stud.* **2020**, *1*, 28–43. [CrossRef]
44. Bukar, U.A.; Sayeed, M.S.; Razak, S.F.A.; Yogarayan, S.; Amodu, O.A.; Mahmood, R.A.R. A method for analyzing text using VOSviewer. *MethodsX* **2023**, *11*, 102339. [CrossRef] [PubMed]
45. Van Eck, N.J.; Waltman, L. Citation-based clustering of publications using CitNetExplorer and VOSviewer. *Scientometrics* **2017**, *111*, 1053–1070. [CrossRef]
46. Van Eck, N.J.; Waltman, L. Software survey: VOSviewer, a computer program for bibliometric mapping. *Scientometrics* **2010**, *84*, 523–538. [CrossRef]
47. Li, J.; Goerlandt, F.; Reniers, G. Mapping process safety: A retrospective scientometric analysis of three process safety related journals (1999–2018). *J. Loss Prev. Process Ind.* **2020**, *65*, 104141. [CrossRef]
48. Li, J.; Hale, A. Output distributions and topic maps of safety related journals. *Saf. Sci.* **2016**, *82*, 236–244. [CrossRef]
49. Li, J.; Hale, A. Identification of, and knowledge communication among core safety science journals. *Saf. Sci.* **2015**, *74*, 70–78. [CrossRef]
50. Van Nunen, K.; Li, J.; Reniers, G.; Ponnet, K. Bibliometric analysis of safety culture research. *Saf. Sci.* **2018**, *108*, 248–258. [CrossRef]
51. Jin, R.Y.; Zou, P.X.W.; Piroozfar, P.; Wood, H.; Yang, Y.; Yan, L.B.; Han, Y. A science mapping approach based review of construction safety research. *Saf. Sci.* **2019**, *113*, 285–297. [CrossRef]
52. Akram, R.; Thaheem, M.J.; Nasir, A.R.; Ali, T.H.; Khan, S. Exploring the role of building information modeling in construction safety through science mapping. *Saf. Sci.* **2019**, *120*, 456–470. [CrossRef]
53. Amin, M.T.; Khan, F.; Amyotte, P. A bibliometric review of process safety and risk analysis. *Process Saf. Environ. Prot.* **2019**, *126*, 366–381. [CrossRef]
54. Li, J.; Reniers, G.; Cozzani, V.; Khan, F. A bibliometric analysis of peer-reviewed publications on domino effects in the process industry. *J. Loss Prev. Process Ind.* **2017**, *49*, 103–110. [CrossRef]
55. Yang, Y.F.; Reniers, G.; Chen, G.H.; Goerlandt, F. A bibliometric review of laboratory safety in universities. *Saf. Sci.* **2019**, *120*, 14–24. [CrossRef]
56. Zou, X.; Yue, W.L.; Vu, H.L. Visualization and analysis of mapping knowledge domain of road safety studies. *Accid. Anal. Prev.* **2018**, *118*, 131–145. [CrossRef]
57. Liu, H.; Chen, H.L.; Hong, R.; Liu, H.G.; You, W.J. Mapping knowledge structure and research trends of emergency evacuation studies. *Saf. Sci.* **2020**, *121*, 348–361. [CrossRef]
58. Tao, J.; Yang, F.Q.; Qiu, D.Y.; Reniers, G. Analysis of safety leadership using a science mapping approach. *Process Saf. Environ. Prot.* **2020**, *140*, 244–257. [CrossRef]
59. Grenon, G.; An, P.E.; Smith, S.M.; Healey, A.J. Enhancement of the inertial navigation system for the Morpheus autonomous underwater vehicles. *IEEE J. Ocean. Eng.* **2001**, *26*, 548–560. [CrossRef]
60. Aldieri, L.; Kotsemir, M.; Vinci, C.P. The impact of research collaboration on academic performance: An empirical analysis for some European countries. *Socio-Econ. Plan. Sci.* **2018**, *62*, 13–30. [CrossRef]
61. Wang, Q.; Wang, J.P.; Xue, M.M.; Zhang, X.F. Characteristics and Trends of Ocean Remote Sensing Research from 1990 to 2020: A Bibliometric Network Analysis and Its Implications. *J. Mar. Sci. Eng.* **2022**, *10*, 373. [CrossRef]
62. Peng, Z.H.; Wang, J.; Wang, D. Distributed Containment Maneuvering of Multiple Marine Vessels via Neurodynamics-Based Output Feedback. *IEEE Trans. Ind. Electron.* **2017**, *64*, 3831–3839. [CrossRef]
63. Huang, Y.M.; Chen, L.Y.; Chen, P.F.; Negenborn, R.R.; van Gelder, P.H.A.J.M. Ship collision avoidance methods: State-of-the-art. *Saf. Sci.* **2020**, *121*, 451–473. [CrossRef]
64. Cheng, Y.; Zhang, W.D. Concise deep reinforcement learning obstacle avoidance for underactuated unmanned marine vessels. *Neurocomputing* **2018**, *272*, 63–73. [CrossRef]
65. Tu, E.M.; Zhang, G.H.; Rachmawati, L.; Rajabally, E.; Huang, G.B. Exploiting AIS Data for Intelligent Maritime Navigation: A Comprehensive Survey From Data to Methodology. *IEEE Trans. Intell. Transp. Syst.* **2018**, *19*, 1559–1582. [CrossRef]
66. Rong, H.; Teixeira, A.P.; Soares, C.G. Ship trajectory uncertainty prediction based on a Gaussian Process model. *Ocean Eng.* **2019**, *182*, 499–511. [CrossRef]

67. Perera, L.P.; Oliveira, P.; Soares, C.G. Maritime Traffic Monitoring Based on Vessel Detection, Tracking, State Estimation, and Trajectory Prediction. *IEEE Trans. Intell. Transp. Syst.* **2012**, *13*, 1188–1200. [CrossRef]
68. Pallotta, G.; Vespe, M.; Bryan, K. Vessel Pattern Knowledge Discovery from AIS Data: A Framework for Anomaly Detection and Route Prediction. *Entropy* **2013**, *15*, 2218–2245. [CrossRef]
69. Zhao, L.M.; Roh, M.I. COLREGs-compliant multiship collision avoidance based on deep reinforcement learning. *Ocean Eng.* **2019**, *191*, 106436. [CrossRef]
70. Woo, J.; Kim, N. Collision avoidance for an unmanned surface vehicle using deep reinforcement learning. *Ocean Eng.* **2020**, *199*, 107001. [CrossRef]
71. Mnih, V.; Kavukcuoglu, K.; Silver, D.; Rusu, A.A.; Veness, J.; Bellemare, M.G.; Graves, A.; Riedmiller, M.; Fidjeland, A.K.; Ostrovski, G.; et al. Human-level control through deep reinforcement learning. *Nature* **2015**, *518*, 529–533. [CrossRef]
72. Ren, S.Q.; He, K.M.; Girshick, R.; Sun, J. Faster R-CNN: Towards Real-Time Object Detection with Region Proposal Networks. *IEEE Trans. Pattern Anal. Mach. Intell.* **2017**, *39*, 1137–1149. [CrossRef] [PubMed]
73. Redmon, J.; Divvala, S.; Girshick, R.; Farhadi, A. You Only Look Once: Unified, Real-Time Object Detection. In Proceedings of the 2016 IEEE Conference on Computer Vision and Pattern Recognition (CVPR), Las Vegas, NV, USA, 27–30 June 2016; pp. 779–788.
74. Liu, W.; Anguelov, D.; Erhan, D.; Szegedy, C.; Reed, S.; Fu, C.Y.; Berg, A.C. SSD: Single Shot MultiBox Detector. In Proceedings of the Computer Vision—ECCV 2016, Amsterdam, The Netherlands, 11–14 October 2016; pp. 21–37.
75. Krizhevsky, A.; Sutskever, I.; Hinton, G.E. ImageNet Classification with Deep Convolutional Neural Networks. *Commun. ACM* **2017**, *60*, 84–90. [CrossRef]
76. He, K.M.; Zhang, X.Y.; Ren, S.Q.; Sun, J. Deep Residual Learning for Image Recognition. In Proceedings of the 2016 IEEE Conference on Computer Vision and Pattern Recognition (CVPR), Las Vegas, NV, USA, 27–30 June 2016; pp. 770–778.
77. Yang, Y.F.; Chen, G.H.; Reniers, G.; Goerlandt, F. A bibliometric analysis of process safety research in China:Understanding safety research progress as a basis for making China's chemical industry more sustainable. *J. Clean. Prod.* **2020**, *263*, 121433. [CrossRef]
78. Wang, Q.; Yang, Z.G.; Yang, Y.; Long, C.L.; Li, H. A bibliometric analysis of research on the risk of engineering nanomaterials during 1999–2012. *Sci. Total Environ.* **2014**, *473–474*, 483–489. [CrossRef]
79. Chiu, W.-T.; Huang, J.-S.; Ho, Y.-S. Bibliometric analysis of severe acute respiratory syndrome-related research in the beginning stage. *Scientometrics* **2004**, *61*, 69–77. [CrossRef]
80. Li, L.-L.; Ding, G.H.; Feng, N.; Wang, M.-H.; Ho, Y.-S. Global stem cell research trend: Bibliometric analysis as a tool for mapping of trends from 1991 to 2006. *Scientometrics* **2009**, *80*, 39–58. [CrossRef]
81. Cao, Y.H.; Wang, X.L.; Yang, Z.L.; Wang, J.; Wang, H.H.; Liu, Z.J. Research in marine accidents: A bibliometric analysis, systematic review and future directions. *Ocean Eng.* **2023**, *284*, 115048. [CrossRef]
82. Ma, D.L.; Wei, J.; Li, Y.; Zhao, F.; Chen, X.; Hu, Y.C.; Yu, S.S.; He, T.H.; Jin, R.H.; Li, Z.Z.; et al. MLDet: Towards efficient and accurate deep learning method for Marine Litter Detection. *Ocean Coast. Manag.* **2023**, *243*, 106765. [CrossRef]
83. Xin, X.R.; Liu, K.Z.; Loughney, S.; Wang, J.; Li, H.H.; Ekere, N.; Yang, Z.L. Multi-scale collision risk estimation for maritime traffic in complex port waters. *Reliab. Eng. Syst. Saf.* **2023**, *240*, 109554. [CrossRef]
84. Zhang, D.; Han, Z.P.; Zhang, K.; Zhang, J.F.; Zhang, M.; Zhang, F. Use of Hybrid Causal Logic Method for Preliminary Hazard Analysis of Maritime Autonomous Surface Ships. *J. Mar. Sci. Eng.* **2022**, *10*, 725. [CrossRef]
85. Kari, R.; Gaspar, H.M.; Gausdal, A.H.; Morshedi, M. Human Interactions Framework for Remote Ship Operations. In Proceedings of the 26th Mediterranean Conference on Control and Automation (MED), Zadar, Croatia, 19–22 June 2018; pp. 581–587.
86. Shi, K.; Weng, J.X.; Fan, S.Q.; Yang, Z.L.; Ding, H.F. Exploring seafarers' emotional responses to emergencies: An empirical study using a shiphandling simulator. *Ocean Coast. Manag.* **2023**, *243*, 106736. [CrossRef]
87. Chen, P.F.; Yang, F.K.; Mou, J.M.; Chen, L.Y.; Li, M.X. Regional ship behavior and trajectory prediction for maritime traffic management: A social generative adversarial network approach. *Ocean Eng.* **2024**, *299*, 117186. [CrossRef]
88. Guan, W.; Luo, W.Z.; Cui, Z.W. Intelligent decision-making system for multiple marine autonomous surface ships based on deep reinforcement learning. *Rob. Auton. Syst.* **2024**, *172*, 104587. [CrossRef]
89. Wang, T.F.; Wu, Q.; Zhang, J.F.; Wu, B.; Wang, Y. Autonomous decision-making scheme for multi-ship collision avoidance with iterative observation and inference. *Ocean Eng.* **2020**, *197*, 106873. [CrossRef]
90. Simola, J.; Poyhonen, J.; Acad Conf, L.T.D. Emerging Cyber risk Challenges in Maritime Transportation. In Proceedings of the 17th International Conference on Cyber Warfare and Security (ICCWS), State Univ New York, Albany, NY, USA, 17–18 March 2022; pp. 306–314.
91. Lee, P.; Theotokatos, G.; Boulougouris, E. Robust Decision-Making for the Reactive Collision Avoidance of Autonomous Ships against Various Perception Sensor Noise Levels. *J. Mar. Sci. Eng.* **2024**, *12*, 557. [CrossRef]
92. Vandavasi, B.N.J.; Venkataraman, H.; Gidugu, A.R. Machine learning-based electro-magnetic field guided localization technique for autonomous underwater vehicle homing. *Ocean Eng.* **2023**, *280*, 114692. [CrossRef]
93. Kim, H.; Kim, D.; Lee, S.-M. Marine Object Segmentation and Tracking by Learning Marine Radar Images for Autonomous Surface Vehicles. *IEEE Sens. J.* **2023**, *23*, 10062–10070. [CrossRef]

94. Tychola, K.A.; Kalampokas, T.; Papakostas, G.A. Quantum Machine Learning-An Overview. *Electronics* **2023**, *12*, 2379. [CrossRef]
95. Maray, M.; Alghamdi, M.; Alrayes, F.S.; Alotaibi, S.S.; Alazwari, S.; Alabdan, R.; Al Duhayyim, M. Intelligent metaheuristics with optimal machine learning approach for malware detection on IoT-enabled maritime transportation systems. *Expert Syst.* **2022**, *39*, e13155. [CrossRef]
96. Ma, H.Y.; Teng, J.K.; Hu, T.; Shi, P.; Wang, S.M. Co-communication Protocol of Underwater Sensor Networks with Quantum and Acoustic Communication Capabilities. *Wirel. Pers. Commun.* **2020**, *113*, 337–347. [CrossRef]
97. Liu, R.W.; Liang, M.H.; Nie, J.T.; Lim, W.Y.B.; Zhang, Y.; Guizani, M. Deep Learning-Powered Vessel Trajectory Prediction for Improving Smart Traffic Services in Maritime Internet of Things. *IEEE Trans. Netw. Sci. Eng.* **2022**, *9*, 3080–3094. [CrossRef]
98. Murray, B.; Perera, L.P. A dual linear autoencoder approach for vessel trajectory prediction using historical AIS data. *Ocean Eng.* **2020**, *209*, 107478. [CrossRef]
99. Gao, M.; Shi, G.Y.; Li, S. Online Prediction of Ship Behavior with Automatic Identification System Sensor Data Using Bidirectional Long Short-Term Memory Recurrent Neural Network. *Sensors* **2018**, *18*, 4211. [CrossRef] [PubMed]
100. Murray, B.; Perera, L.P. Ship behavior prediction via trajectory extraction-based clustering for maritime situation awareness. *J. Ocean. Eng. Sci.* **2022**, *7*, 1–13. [CrossRef]
101. Kowalski, J.; Krawczyk, B.; Wozniak, M. Fault diagnosis of marine 4-stroke diesel engines using a one-vs-one extreme learning ensemble. *Eng. Appl. Artif. Intell.* **2017**, *57*, 134–141. [CrossRef]
102. Meyer, E.; Heiberg, A.; Rasheed, A.; San, O. COLREG-Compliant Collision Avoidance for Unmanned Surface Vehicle Using Deep Reinforcement Learning. *IEEE Access* **2020**, *8*, 165344–165364. [CrossRef]
103. Shi, J.-h.; Liu, Z.-j. Deep Learning in Unmanned Surface Vehicles Collision-Avoidance Pattern Based on AIS Big Data with Double GRU-RNN. *J. Mar. Sci. Eng.* **2020**, *8*, 682. [CrossRef]
104. Pan, M.Y.; Liu, Y.S.; Cao, J.Y.; Li, Y.; Li, C.; Chen, C.-H. Visual Recognition Based on Deep Learning for Navigation Mark Classification. *IEEE Access* **2020**, *8*, 32767–32775. [CrossRef]
105. Zurowietz, M.; Langenkaemper, D.; Hosking, B.; Ruhl, H.A.; Nattkemper, T.W. MAIA-A machine learning assisted image annotation method for environmental monitoring and exploration. *PLoS ONE* **2018**, *13*, e0207498. [CrossRef]
106. Rawson, A.; Brito, M.; Sabeur, Z.; Tran-Thanh, L. A machine learning approach for monitoring ship safety in extreme weather events. *Saf. Sci.* **2021**, *141*, 105336. [CrossRef]
107. Fan, S.Q.; Yang, Z.L. Towards objective human performance measurement for maritime safety: A new psychophysiological data-driven machine learning method. *Reliab. Eng. Syst. Saf.* **2023**, *233*, 109103. [CrossRef]
108. Jiang, Y.; Feng, C.; He, B.; Guo, J.; Wang, D.R.; Lv, P.F. Actuator fault diagnosis in autonomous underwater vehicle based on neural network. *Sens. Actuators A-Phys.* **2021**, *324*, 112668. [CrossRef]
109. Yigin, B.; Celik, M. A Prescriptive Model for Failure Analysis in Ship Machinery Monitoring Using Generative Adversarial Networks. *J. Mar. Sci. Eng.* **2024**, *12*, 493. [CrossRef]
110. Fang, Z.; Jiang, D.; Huang, J.; Cheng, C.; Sha, Q.; He, B.; Li, G.L. Autonomous underwater vehicle formation control and obstacle avoidance using multi-agent generative adversarial imitation learning. *Ocean Eng.* **2022**, *262*, 112182. [CrossRef]
111. Wei, X.W.; Wang, H.; Tang, Y.X. Deep hierarchical reinforcement learning based formation planning for multiple unmanned surface vehicles with experimental results. *Ocean Eng.* **2023**, *286*, 115577. [CrossRef]
112. Kawano, H.; Ura, T.; Ieee, I. Fast reinforcement learning algorithm for motion planning of non-holonomic Autonomous Underwater Vehicle in disturbance. In Proceedings of the IEEE/RSJ International Conference on Intelligent Robots and Systems (IROS 2002), Lausanne, Switzerland, 30 September–4 October 2002; pp. 903–908.
113. Gjaerum, V.B.; Rorvik, E.-L.H.; Lekkas, A.M. Approximating a deep reinforcement learning docking agent using linear model trees. In Proceedings of the European Control Conference (ECC), Delft, The Netherlands, 29 June–2 July 2021; pp. 1465–1471.
114. Gizzini, A.K.; Medjahdi, Y.; Ghandour, A.J.; Clavier, L. Towards Explainable AI for Channel Estimation in Wireless Communications. *IEEE Trans. Veh. Technol.* **2024**, *73*, 7389–7394. [CrossRef]
115. Zablocki, E.; Ben-Younes, H.; Pérez, P.; Cord, M. Explainability of Deep Vision-Based Autonomous Driving Systems: Review and Challenges. *Int. J. Comput. Vis.* **2022**, *130*, 2425–2452. [CrossRef]
116. Nogueira, C.; Fernandes, L.; Fernandes, J.N.D.; Cardoso, J.S. Explaining Bounding Boxes in Deep Object Detectors Using Post Hoc Methods for Autonomous Driving Systems. *Sensors* **2024**, *24*, 516. [CrossRef]
117. Nazat, S.; Arreche, O.; Abdallah, M. On Evaluating Black-Box Explainable AI Methods for Enhancing Anomaly Detection in Autonomous Driving Systems. *Sensors* **2024**, *24*, 3515. [CrossRef]
118. Li, D. Development of Intelligent Marine Traffic Service Application Software for Haikou Bay. *J. Coast. Res.* **2019**, *94*, 490–494. [CrossRef]

119. Hammedi, W.; Brik, B.; Senouci, S.M. Federated Deep Learning-Based Framework to Avoid Collisions Between Inland Ships. In Proceedings of the 18th IEEE International Wireless Communications and Mobile Computing Conference (IWCMC), Dubrovnik, Croatia, 30 May–3 June 2022; pp. 967–972.
120. Harish, A.V.; Tam, K.; Jones, K. Literature review of maritime cyber security: The first decade. *Marit. Technol. Res.* **2024**, 7, 273805. [CrossRef]

Disclaimer/Publisher's Note: The statements, opinions and data contained in all publications are solely those of the individual author(s) and contributor(s) and not of MDPI and/or the editor(s). MDPI and/or the editor(s) disclaim responsibility for any injury to people or property resulting from any ideas, methods, instructions or products referred to in the content.

Journal of *Marine Science and Engineering*

MDPI

Article

Preparing VTS for the MASS Era: A Machine Learning-Based VTSO Recruitment Model

Gil-ho Shin [1] and Min Jung [2,*]

[1] Korea Coast Guard, Busan 49112, Republic of Korea; newroadtiger@gmail.com
[2] Korea Institute of Maritime and Fisheries Technology, Busan 49111, Republic of Korea
* Correspondence: seamini@naver.com

Abstract

As the maritime industry transitions toward Maritime Autonomous Surface Ships (MASS), Vessel Traffic Service Operators (VTSOs) face new challenges in managing mixed traffic of conventional and autonomous vessels. Effective VTSO selection is becoming increasingly critical for maritime safety, yet current recruitment processes rely on subjective methods that limit objective evaluation of candidate suitability. This study presents the first machine learning-based classification model for VTSO recruitment. Eight features were defined, including sea service experience, navigation career, education, certifications, and language proficiency. Due to limited access to actual recruitment data, expert-validated simulated datasets were constructed through labeling by 40 maritime professionals and density estimation-based augmentation. Four algorithms were compared, with XGBoost achieving 94.6% F1-score. Feature importance analysis revealed TOEIC score as the most critical predictor, followed by seafaring career, with 3–4 years of experience identified as optimal. These findings indicate that English proficiency for communication with shore remote control centers and practical maritime experience for assessing autonomous vessel behaviors constitute core VTSO competencies in the MASS era. The proposed model demonstrates potential to improve subjective recruitment methods by discovering quantifiable competency patterns, offering a pathway toward data-driven, standardized, and transparent decision-making for enhanced maritime safety.

Keywords: Vessel Traffic Services (VTS); Maritime Autonomous Surface Ships (MASS); Vessel Traffic Service Operator (VTSO); machine learning; Human-in-the-Loop (HITL)

1. Introduction

The maritime industry is undergoing a transformative shift with the emergence of Maritime Autonomous Surface Ships (MASS). As autonomous and remotely controlled vessels move toward commercial deployment in the coming years, they will inevitably share waterways with conventional manned ships, creating mixed traffic environments that pose unprecedented challenges for maritime traffic management [1,2]. The introduction of MASS fundamentally alters the established human-to-human interaction between vessel operators and shore-based traffic controllers, as the absence of seafarers on board eliminates traditional communication and coordination practices that have been the foundation of maritime safety for decades [1].

In this evolving maritime landscape, the role and responsibility of VTSOs become increasingly critical, as they must not only manage mixed traffic scenarios but also serve as the primary human decision-makers in overseeing vessels with varying degrees of autonomy, thereby necessitating enhanced competencies and more rigorous recruitment criteria. In recent decades, the growth of the global merchant fleet has led to increased congestion and complexity in maritime traffic. This is particularly pronounced in coastal regions, straits, and adjacent waterways. Paradoxically, maritime accidents have decreased by half over the past decade, which can be attributed to the implementation of Vessel Traffic Services (VTS) systems in conflict-prone areas [3].

VTS serves as a cornerstone of modern maritime traffic safety management, playing a crucial role in ensuring safe navigation of vessels in ports and coastal waters [4]. The significance of effective maritime traffic management extends beyond safety to encompass broader sustainability dimensions. Research demonstrates that professional navigation oversight, such as maritime pilotage, directly contributes to emission reduction, spill avoidance, and marine biodiversity protection while optimizing operational efficiency [5].

Furthermore, given that maritime logistics operations within seaports account for approximately 60% of port CO_2 emissions [6], the role of VTS in managing vessel movements efficiently becomes critical not only for safety but also for environmental sustainability objectives. VTS have become a cornerstone in maintaining the safety and efficiency of maritime navigation. As noted in the IALA VTS Manual [7], over 500 VTS centers are currently in operation worldwide, and their significance continues to increase in parallel with the persistent growth of global maritime traffic. VTSO perform complex and specialized tasks including collision avoidance between vessels, provision of navigational safety information, and emergency response. Their expertise and decision-making capabilities directly impact maritime safety.

Research findings indicating that 60 to 90% of maritime accidents are attributed to human factors [8] further emphasize the importance of maritime professionals such as VTSO. Moreno et al. [3] highlighted that VTSO serve as the critical link connecting safety and efficiency within maritime systems. Particularly, as VTSO must simultaneously control multiple vessels in real-time while performing complex situational assessments and decision-making, their professional competencies become decisive factors determining the safety and efficiency of maritime traffic.

However, current VTSO recruitment processes primarily rely on traditional methods centered on subject-specific written examinations and interviews. These methods present limitations in objectively and systematically evaluating candidates' actual job suitability. Such subjective evaluation methods heavily depend on the experience and intuition of hiring personnel, making it difficult to establish consistent selection criteria and carrying the risk of missing excellent talent or selecting unsuitable personnel. These limitations of traditional evaluation methods are further exacerbated in the MASS era. In MASS operational environments, VTSOs must manage complex traffic scenarios involving mixed fleets of autonomous and manned vessels.

This requires multifaceted competencies beyond conventional navigation knowledge and experience, including understanding of autonomous systems, remote monitoring capabilities, and international communication skills. Traditional subjective evaluation methods are inadequate for systematically identifying and predicting the complex interactions and nonlinear relationships among these multiple competency factors. Therefore, a data-driven approach capable of objectively analyzing diverse candidate characteristics and discovering hidden patterns and competency combinations is essential. Recently, machine learning-based personnel selection systems have been introduced across various industries, significantly improving recruitment objectivity and prediction accuracy.

In the maritime field, modern approaches are being attempted, such as Pekdas et al. [9] applying machine learning to maritime professional recruitment using the Human-in-the-Loop (HITL) approach and achieving 86% prediction accuracy. The HITL approach is a methodology where domain experts directly participate in data labeling and model validation during the machine learning model training process, systematically incorporating professional knowledge. This is particularly effective in securing high accuracy and reliability that are difficult to achieve through automated labeling alone in complex and specialized domains. The effectiveness of this approach has been demonstrated across various fields: Butler et al. [10] applied expert manual correction to micro-expression recognition datasets, Zhang et al. [11] developed entity extraction methods utilizing human input, Bartolo et al. [12] improved reading comprehension model performance through iterative expert annotation, and Fan et al. [13] enhanced detection accuracy by integrating user intervention into network anomaly detection systems.

These studies commonly demonstrate that direct expert labeling and validation are essential for performance improvement in complex domains. Research on personnel selection has been actively conducted in the maritime field, though it has primarily focused on seafarer recruitment. Pekdas et al. [9] applied machine learning to maritime professional recruitment, Celik et al. [14] proposed a systematic model for captain selection, and Kartal et al. [15] conducted comparative analysis of multinational navigators. However, these studies all targeted seafarers working at sea, and research on VTSO recruitment, who control vessel traffic from shore, remains absent. VTSOs possess fundamentally different working environments and required competencies compared to seafarers. While seafarers board vessels directly to perform navigation duties, VTSOs monitor and control multiple vessels simultaneously from shore-based control centers using various electronic equipment such as radar, Automatic Identification System (AIS), and Closed-Circuit Television (CCTV).

Therefore, there are fundamental limitations in directly applying existing seafarer recruitment research results to VTSO recruitment. This study aims to address this research gap by predicting VTSO recruitment suitability through a binary classification model combining machine learning and HITL approaches, utilizing candidate features such as sea service experience, navigation career, education, certifications, and language proficiency.

However, accessing actual VTSO recruitment datasets, including detailed information of both successful and unsuccessful candidates, is practically impossible due to privacy regulations and institutional policies protecting personal information. This constraint necessitated the construction of expert-validated simulated datasets, where 40 current VTS professionals with over 5 years of control experience participated in the HITL labeling process to ensure data reliability and realism.

The key novelties of this research are threefold: (1) it represents the first application of machine learning-based prediction models specifically designed for shore-based VTSO recruitment, distinct from existing studies focused on seafarer selection; (2) it systematically integrates domain expert knowledge through HITL labeling with 40 current VTS professionals to ensure prediction reliability; and (3) it addresses limited data challenges through KDE-based augmentation while preserving the statistical characteristics of realistic recruitment scenarios.

Section 2 reviews previous research on personnel selection in maritime and other fields and identifies research gaps. Section 3 provides detailed explanations of data collection, preprocessing, inter-variable correlation analysis, and augmentation processes. Section 4 presents the methodology of the proposed machine learning model. Section 5 analyzes experimental results, and Section 6 conducts in-depth discussion of the results. Finally, Section 7 presents research conclusions and future research directions.

2. Literature Review

This literature review section analyzes existing studies related to personnel selection in the maritime sector and examines current research trends. Traditionally, multi-criteria decision making (MCDM) methodologies have been predominantly utilized in the maritime field, with statistical modeling and machine learning approaches beginning to be introduced in recent years. To systematically identify research gaps, existing studies are analyzed across two key methodological dimensions: feature design (general vs. maritime-specific competencies) and algorithmic approaches (MCDM vs. machine learning methods). Table 1 demonstrates the methodological diversity of personnel selection research in the maritime sector, summarizing the key characteristics and contributions of each study. Subsequently, machine learning-based personnel selection studies from other industries are reviewed to explore their applicability to the maritime field. Finally, through comprehensive analysis of existing literature, research gaps are identified, and the necessity of this study is presented.

Table 1. Summary of Personnel Selection Studies in Maritime Industry.

Author(s)	Study Focus	Features	Methodology
Pekdas et al. [9]	Maritime professionals recruitment	Experience, References, English proficiency, GPA, Experience, Professional knowledge, MMPI-I scales	Decision Tree, Random Forest, Gradient Boosted Trees, Naive Bayes, PNN
Kartal et al. [15]	Multinational officers of the watch comparison	Cost-related (wage, social security fees, other costs), Cultural properties, Education, Professional properties	Fuzzy Analytic Hierarchy Process (FAHP)
Koutra et al. [16]	Maritime executive personnel selection	Trustworthiness, Responsibility, Decision making, Team spirit, Communication skills, Time management, Foreign languages, Computer skills	AHP + Correspondence Analysis (CA)
Wang and Yeo [17]	Foreign seafarer supply country selection for Korean flag vessels	Total crew costs, Seafarer qualifications, Government relations, Communication (language competence, cultural adaptation), Supply ability for well-trained seafarers, Government support from supply countries	Delphi, Fuzzy AHP, TOPSIS
Celik et al. [14]	Master selection for merchant ships	Occupational information, Professional discipline, Leadership & coaching, Personality characteristics	Analytic Network Process (ANP)
Ding and Liang [18]	Seafarer employment choice behavior	Crew cost, Competence & efficiency (knowledge, skills, communication, physical/psychological conditions), Quality standard system (STCW95 compliance)	Binary Logit Model

2.1. Machine Learning-Based Personnel Selection Studies in Other Industries

Machine learning-based personnel selection research has been actively conducted across various industries, primarily developing around text-based matching and HITL approaches. In text-based matching research, Wang et al. [19] proposed the PJFCANN model combining co-attention neural networks and graph neural networks to utilize past successful recruitment cases for current hiring decisions. Wang and Zhu [20] compared advanced deep learning architectures including BERT and Bi-LSTM-CRF, emphasizing the importance of selecting appropriate models based on dataset characteristics rather than model complexity.

HITL-based research focuses on the systematic integration of expert knowledge. Pessach et al. [21] achieved 86% prediction accuracy by combining Variable-Order Bayesian Networks (VOBN) with HITL methodology, while Al-Quhfa et al. [22] systematically compared nine machine learning algorithms including K-NN, SVM, Random Forest, and Multi-Layer Perceptron, confirming that Random Forest achieved the highest performance at 92.8%. According to Wu et al.'s [23] comprehensive survey, systematically integrating domain expert knowledge into data processing and model training stages can simultaneously improve both model performance and reliability. These studies commonly demonstrate the superiority of ensemble methods, the importance of integrating domain expertise through HITL approaches, and the effectiveness of utilizing multiple data sources.

2.2. *Applicability to Maritime Field and Research Gaps*

Most existing studies have targeted general corporate environments or IT sectors, with Pekdas et al. [9] being the first to attempt machine learning-based personnel selection in the maritime field, though limited to seafarer recruitment. The methodological insights from these existing studies are applicable to VTSO recruitment. The superiority of ensemble methods such as Random Forest and neural networks can be applied to evaluating complex maritime professional competencies, while HITL methodologies for integrating domain expertise are useful for systematizing maritime experts' experience.

Moreover, the importance of combining multiple data sources and building interpretable models can contribute to ensuring transparency and reliability in VTSO recruitment. Despite these applicabilities, existing research cannot resolve fundamental limitations in VTSO recruitment. In terms of job characteristics, VTSOs require highly specialized competencies distinct from general jobs studied in existing research, including real-time multi-vessel monitoring, radar/AIS system operation, and complex maritime situation assessment. Regarding evaluation characteristics, unlike the general personality traits or standardized job experiences utilized in existing studies, VTSOs require maritime-specific competencies as core elements, including maritime expertise, sea-going experience, navigation certificates, and maritime safety knowledge.

Most importantly, VTS work is directly linked to maritime safety. Therefore, prediction models require much higher accuracy and reliability than those for general corporate recruitment. To address these research gaps, this study proposes the development of machine learning models that reflect the unique job characteristics and maritime-specific competencies of VTSOs. We aim to systematically integrate maritime experts' domain knowledge through an HITL approach to construct a VTSO-specific binary classification model based on sea-going experience, navigation career, maritime certificates, and other relevant factors.

3. Data Preprocessing

This study developed a systematic dataset that combines HITL approaches with data augmentation techniques for developing machine learning models that reflect the realistic characteristics of VTSO recruitment. The problem was formulated as a binary classification task for pass/fail decisions, reflecting the high competition rates of actual Korean VTSO recruitment, and a labeling process was implemented that systematically incorporates domain knowledge from maritime experts. As illustrated in Figure 1, the data construction process consists of four sequential stages, each designed to address specific methodological challenges.

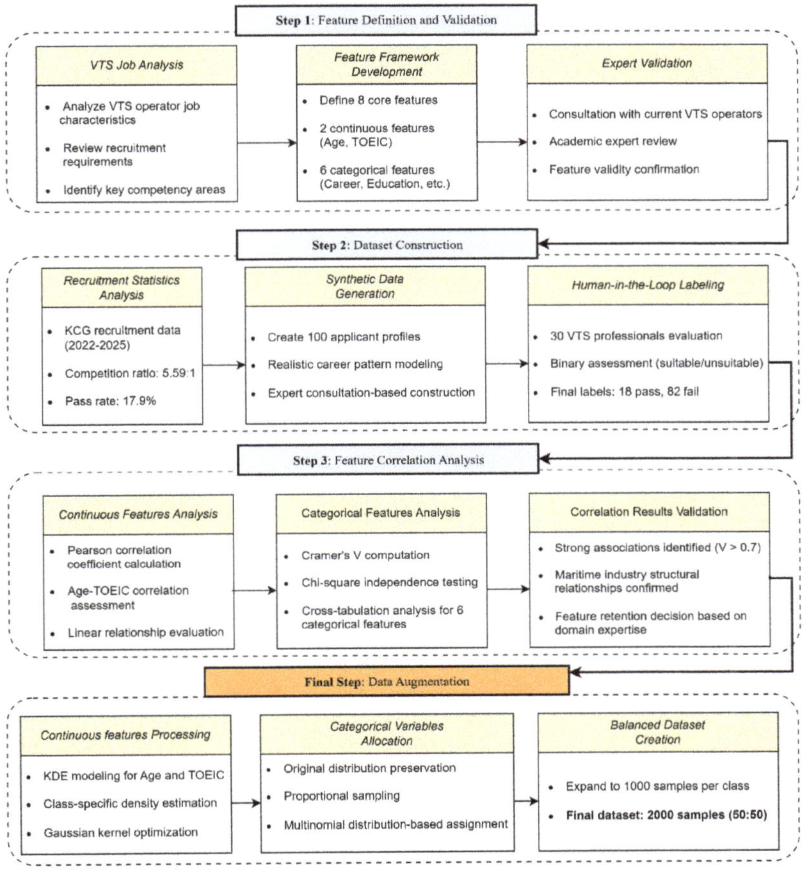

Figure 1. Data preprocessing workflow for VTSO recruitment binary classification model.

First, Feature Definition identifies eight key features that capture both maritime-specific competencies (License, Career, Position) and general qualifications (Age, Education, TOEIC Score). Second, HITL Labeling employs 40 maritime experts from multiple VTS centers to evaluate 100 candidate profiles, with the top 18% labeled as successful to reflect actual Korean VTS statistics (17.9% pass rate). Third, Correlation Analysis applies Pearson correlation coefficients for continuous variables and Cramer's V for categorical variables to confirm feature independence while identifying structural relationships in maritime career progression. Fourth, KDE-based Augmentation expands the original 100 samples to 2000 balanced samples (1000 successful, 1000 unsuccessful) through Kernel Density Estimation for continuous variables and proportional allocation for categorical variables, maintaining statistical characteristics while addressing class imbalance.

This systematic workflow ensures data quality and realism while enabling effective machine learning model training.

3.1. Dataset Construction

Eight key features were defined by analyzing the job characteristics and recruitment requirements of VTSO. The feature design was configured to comprehensively reflect the elements evaluated in actual recruitment processes, and their validity was verified through consultation with current VTSOs and maritime academics. Table 2 presents the detailed definitions and characteristics of each feature.

Table 2. Feature Definitions and Characteristics for VTSO Dataset.

Category	Feature	Type	Range/Classification	Description
Basic Information	Age	Continuous	18–60 years	Applicant's age
	Education Level	Categorical	High school or below/Bachelor's/Master's or above	Final education level (3 levels)
	Educational Institution	Categorical	Maritime university/Non-maritime university/Maritime-related high school/Maritime training institute	Classification of maritime-related educational institutions
Maritime Professional Experience	Officer License Grade	Categorical	1st/2nd/3rd/4th	Maritime officer license grade (1st is highest)
	Navigation Career	Categorical	0/0~1 year/1~3 years/3~5 years/5~10 years/10+ years	Actual navigation experience period (5 intervals)
	Last Position	Categorical	Cadet/3rd Officer/2nd Officer/Chief Officer/Master	Final position aboard ship (5 levels)
Competency Indicators	VTS Course Completion	Binary	Completed/Not completed	Completion of VTS-related courses at educational institution
	TOEIC Score	Continuous	550–985 points	Certified English proficiency test score

Realistic career progression models were applied to reflect systematic career development patterns in the maritime field. The data incorporated standard career advancement pathways in the Korean maritime industry: high school graduates begin sea service at age 19 and progress sequentially through 4th → 3rd → 2nd → 1st class positions, while university graduates start at age 23 and advance through 3rd → 2nd → 1st class positions. These patterns faithfully reflect the actual licensing acquisition processes and promotion systems in the maritime industry, ensuring data realism.

3.2. Human-in-the-Loop (HITL) Labeling

To construct a realistic dataset, VTSO recruitment statistics from the Korea Coast Guard's competitive recruitment for general civil servants were analyzed for 2022–2025. Over three years, a total of 481 candidates applied, with 86 being selected, recording an average competition ratio of 5.59:1 and a pass rate of 17.9% [24]. Reflecting this realistic recruitment ratio, 100 virtual candidate profiles were constructed through consultation with current VTSO and maritime field professors, with 18 candidates (18%) designated as successful and 82 candidates (82%) as unsuccessful, implementing a realistic class distribution. For VTSO recruitment, determining candidate suitability requires advanced professional knowledge in maritime safety, vessel operations, and traffic control.

Therefore, an HITL labeling process utilizing the collective intelligence of domain experts was applied to achieve high accuracy and reliability that would be difficult to attain through automated labeling alone. The expert group comprised 40 current professionals with over 5 years of control experience from Busan Port VTS, Busan New Port VTS, Jeju Port VTS, and Mokpo Port VTS, selected through direct field visits. The selected experts possessed the professional knowledge necessary for candidate suitability assessment, having accurate understanding of the complexity and required competencies of actual VTS operations.

During the labeling process, each expert reviewed the profiles of 100 candidates and performed binary evaluation as either "suitable for recruitment" or "unsuitable for recruitment." The evaluation results from 40 experts were aggregated to calculate the Approval Count for each candidate, and reflecting the actual recruitment rate (17.9%), the top 18 candidates by approval count were labeled as successful (1), with the remaining 82 candidates labeled as unsuccessful (0). To assess inter-rater reliability, we computed Fleiss' Kappa among the 40 domain experts, yielding $\kappa = 0.258$ ($p < 0.001$), indicating fair agreement according to Landis and Koch [25]. While the fair-level Kappa reflects moderate variability in expert judgments, it demonstrates the inherent complexity and multifaceted nature of VTSO recruitment decisions. The statistically significant result ($p < 0.001$) confirms that expert evaluations were systematic rather than random, and our collective intelligence approach—using aggregated approval counts rather than requiring unanimous consensus—effectively captures this diversity of professional perspectives to generate reliable training labels. This collective intelligence-based labeling approach minimized individual expert subjective bias and enabled the generation of objective and reliable labels.

3.3. Feature Correlation Analysis

Prior to modeling, quantitative correlation analysis between features was conducted to verify the presence of strong associations among variables. According to feature types, Pearson correlation coefficients were applied to continuous variables and Cramer's V was applied to categorical variables for quantitative assessment of associations. The Pearson correlation coefficient measures the strength of linear relationships between two continuous variables X and Y, defined as follows:

$$r = \frac{\sum_{i=1}^{n_{cont}} (X_i - \overline{X})(Y_i - \overline{Y})}{\sqrt{\sum_{i=1}^{n_{cont}} (X_i - \overline{X})^2 \sum_{i=1}^{n_{cont}} (Y_i - \overline{Y})^2}} \quad (1)$$

where X_i and Y_i represent the variable values of the i-th observation, $\overline{X}$ and $\overline{Y}$ are the means of each variable, and n_{cont} is the number of observations for continuous variable pairs. For example, the analysis between Age and TOEIC Score yielded $r = -0.028$, indicating a very weak correlation that was not statistically significant, confirming that these two features were mutually independent. Meanwhile, associations between categorical variables were measured using Cramer's V, which is calculated based on the chi-square statistic as follows:

$$V = \sqrt{\frac{\chi^2}{n_{cat} \cdot min(k-1, r-1)}} \quad (2)$$

where χ^2 is the chi-square statistic obtained through independence testing between two categorical variables, n_{cat} is the total sample size in the contingency table, and k and r represent the number of categories for each variable, respectively. Figure 2 presents the results of this feature independence analysis. The analysis revealed strong associations between Seafaring Career and Last Position ($V = 0.755$) and between Education Level and Educational Institution ($V = 0.709$).

This reflects the industry's personnel structure where career advancement naturally leads to position promotion and education level is naturally associated with institutional type. Additionally, License showed moderate associations with other maritime professional features ($V = 0.356 \sim 0.623$), while Completion of VTS Courses demonstrated weak associations with all features ($V \leq 0.176$), confirming its independence. Consequently, although strong associations exist between some features, these can be interpreted as structural relationships inherent to the maritime industry. Since each feature provides unique information for VTSO suitability prediction, inclusion of all features in the model was deemed appropriate.

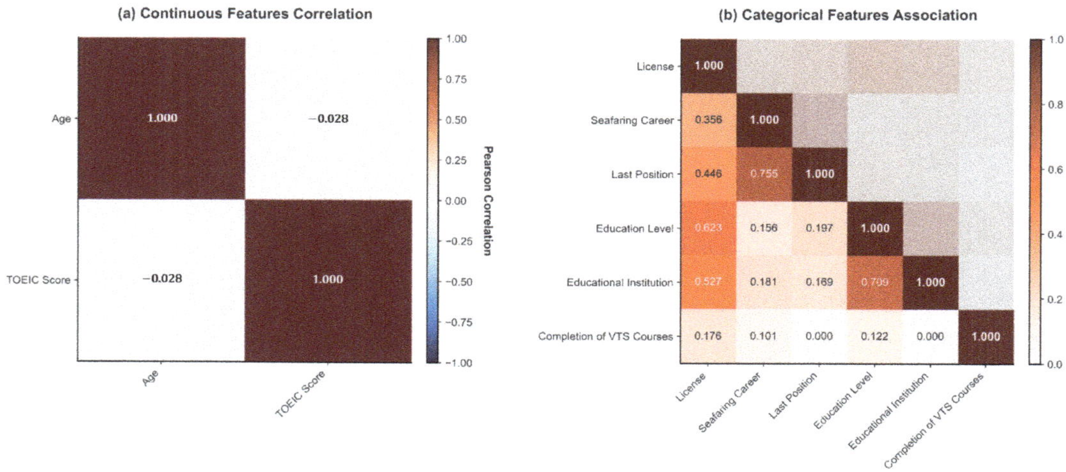

Figure 2. Feature independence analysis for continuous and categorical features.

3.4. Data Augmentation

The original dataset of 100 samples that completed correlation review presented two critical limitations for machine learning model training: insufficient data volume and severe class imbalance (18 successful vs. 82 unsuccessful candidates). To address these issues, differentiated data augmentation strategies were applied according to variable types. Among various data augmentation techniques, we selected Kernel Density Estimation (KDE) over alternatives such as Synthetic Minority Over-sampling Technique (SMOTE) and Bayesian sampling methods based on the following rationale.

SMOTE [26], which generates synthetic samples by interpolating between neighboring minority class instances, is primarily designed for purely continuous feature spaces and exhibits limitations when dealing with mixed-type datasets containing both continuous and categorical variables [27]. Since our VTSO dataset comprises six categorical features (Education Level, Educational Institution, License, Seafaring Career, Last Position, VTS Course Completion) and only two continuous features (Age, TOEIC Score), SMOTE's interpolation-based approach would require complex adaptations for categorical features, potentially introducing unrealistic feature combinations that violate domain constraints inherent to maritime career progression.

Bayesian sampling methods [28], while theoretically elegant, require specification of prior distributions and assumptions about the underlying data generation process. Given our limited sample size ($n = 100$) and the absence of established prior knowledge about VTSO candidate distributions, Bayesian approaches would introduce additional uncertainty through potentially misspecified priors. In contrast, KDE offers a non-parametric approach that directly estimates probability density functions from observed data without imposing distributional assumptions, making it particularly suitable for small datasets with unknown underlying distributions [29]. Moreover, KDE enables independent augmentation of continuous variables while preserving their marginal distributions, allowing us to maintain the proportional allocation strategy for categorical variables based on observed frequencies in the original dataset. KDE is a non-parametric statistical method that estimates probability density functions using kernel functions from given data, defined as follows:

$$\hat{f}(x) = \frac{1}{nh}\sum_{i=1}^{n} K\left(\frac{x - x_i}{h}\right) \quad (3)$$

where K is the kernel function, h is the bandwidth, and n is the number of data points. The augmentation process was specifically applied to continuous variables while preserving the statistical characteristics of the original data. Separate KDE models were constructed for each class (successful/unsuccessful) to capture different distribution patterns. For bandwidth selection, we applied Silverman's rule of thumb [30]:

$$h = 0.9 \times min(\sigma, \tfrac{IQR}{1.34}) \times n^{-1/5} \quad (4)$$

where σ is the standard deviation and IQR is the interquartile range. This yielded bandwidth values of $h_{Age} = 2.09$ years for Age and $h_{TOEIC} = 40.20$ points for TOEIC Score in the successful candidate class ($n = 18$), and $h_{Age} = 1.99$ years and $h_{TOEIC} = 33.79$ points in the unsuccessful candidate class ($n = 82$). These bandwidth values enable smooth density estimation while avoiding over-smoothing that would obscure important distributional characteristics.

New values were sampled from the estimated probability density functions, with Age constrained to integer values within the 18–60 year range and TOEIC Score limited to 5-point increments within the 400–990 point range to maintain realistic score granularity consistent with actual TOEIC assessment protocols. Categorical variables were proportionally allocated according to the distribution patterns of the original data. For example, if a specific career segment comprised 40% of successful candidates, the same proportion was allocated to the augmented successful candidate samples to maintain realistic feature combinations that reflect actual maritime career progression pathways.

This proportional allocation strategy ensures that the augmented dataset preserves the structural relationships between categorical features observed in the original data, such as the natural association between Seafaring Career and Last Position. Table 3 presents a comparison of dataset characteristics before and after augmentation.

Table 3. Comparison of Dataset Characteristics Before and After KDE-Based Augmentation.

Characteristic	Original Dataset	Augmented Dataset
Pass Samples	18 (18%)	1000 (50%)
Fail Samples	82 (82%)	1000 (50%)
Total Samples	100(100%)	2000(100%)

Figure 3 visually demonstrates the distribution comparison before and after augmentation. In (a) Age distribution and (b) TOEIC Score distribution, it can be confirmed that KDE-based augmentation successfully preserved the probability density characteristics of the original data while generating new samples. Solid lines represent original data, and dotted lines represent augmented data, with distribution characteristics of successful (red) and unsuccessful (blue) candidates consistently maintained after augmentation. (c) Categorical variable distribution preservation shows that license grade distributions were augmented while maintaining original proportions within each class. (d) Class balance improvement confirms that the severe imbalance of the original data was improved to complete balance through augmentation.

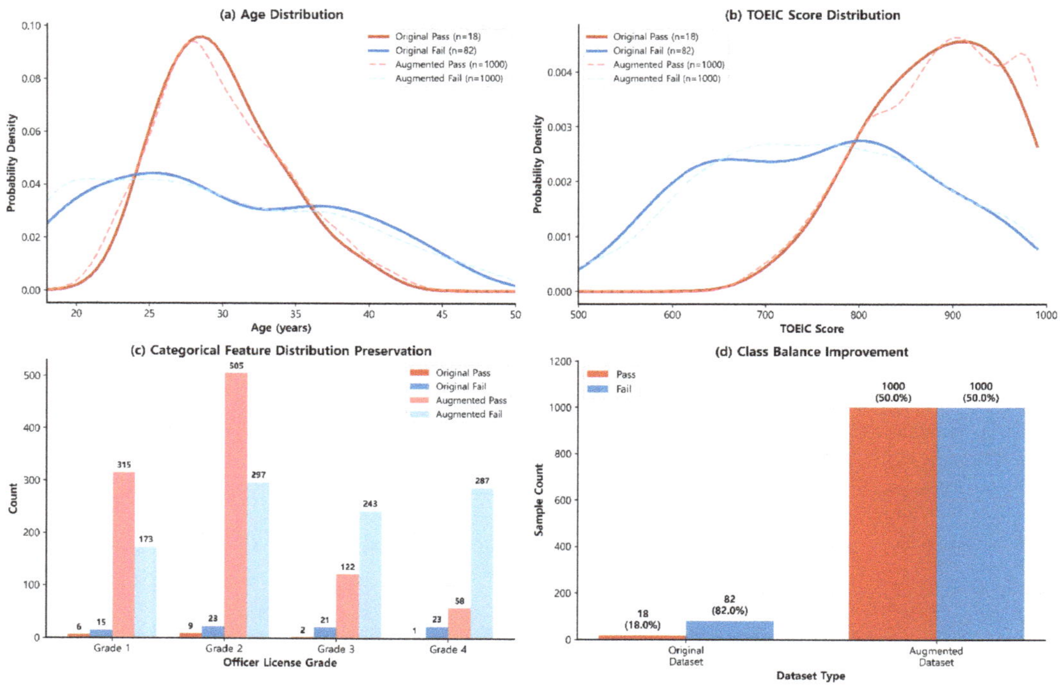

Figure 3. Distribution comparison before and after KDE-based data augmentation.

Through this systematic augmentation process, as shown in Table 3, the original 18 successful samples were expanded to 1000 and the 82 unsuccessful samples were expanded to 1000, constructing a final dataset of 2000 completely balanced samples.

4. Methodology

This section implements and compares four machine learning algorithms to solve the binary classification problem for predicting VTSO recruitment suitability. The algorithm selection criteria considered: (1) superior performance on structured tabular data, (2) effectiveness for binary classification problems, (3) interpretability and feature importance analysis support, and (4) robustness in limited data environments. To facilitate understanding, we present definitions for common variables and notations used throughout this section:

(1) n: Total count of training instances
(2) $y_i, \hat{y}_i$: Actual value (y_i) and predicted value ($\hat{y}_i$) of the i-th sample
(3) K: Tree quantity in ensemble architectures
(4) $\mathcal{L}$: Objective or loss criterion function

4.1. XGBoost

XGBoost is an efficient implementation of the gradient boosting framework developed by Chen and Guestrin [31], widely used in various machine learning problems due to its high predictive performance and computational efficiency [32]. For VTSO recruitment data, the objective function of XGBoost is defined as follows:

$$obj = \sum_{i=1}^{n} L(y_i, \hat{y}_i) + \sum_{k=1}^{K} \Omega(f_k) \tag{5}$$

In this context, $\Omega(f_k)$ denotes the penalty component that quantifies the structural complexity of tree k. The mathematical formulation of this penalty component $\Omega(f_k)$ is expressed as:

$$\Omega(f_k) = \gamma T + \frac{1}{2}\lambda \sum_{j=1}^{T} w_j^2 \tag{6}$$

In this formulation, γ controls the penalty applied to leaf node quantity, T denotes the total leaf count, λ serves as the L2 regularization coefficient, and w_j indicates the j-th leaf's weight value. The learning mechanism of XGBoost operates by minimizing the defined objective. Throughout training, the objective is approximated via Taylor series expansion, facilitating effective determination of both tree architecture and leaf weight values across successive iterations. Hyperparameter tuning was conducted using grid search methodology, focusing on critical parameters such as step size for learning, maximum depth of trees, and minimum child weight thresholds [33]. A key benefit of XGBoost lies in its capability to quantify feature contribution [34]. This functionality enabled identification of the most influential variables for predicting.

4.2. *Random Forest*

Random Forest represents an ensemble approach utilizing multiple decision trees, originally introduced by Breiman [35], and has gained widespread adoption owing to its superior predictive performance, resilience to overfitting, and versatility across diverse data types [36]. The training methodology of Random Forest comprises several key steps. Initially, through bootstrap sampling, multiple subsets of the training data are randomly created, and this process is repeated to generate multiple bootstrap samples.

Subsequently, each bootstrap subset is used to construct an individual decision tree, wherein at every branching point, the best partition is determined from a randomly selected feature subset. Ultimately, the complete model is assembled by combining numerous trees into a forest structure. When making predictions, the output is derived by consolidating predictions from all constituent trees. In classification scenarios, a plurality vote scheme is employed:

$$\hat{y} = mode\{f_k(x), k = 1, \ldots, K\} \tag{7}$$

where $f_k(x)$ is the prediction of the k-th tree. Random Forest includes several key hyperparameters that control model performance. The number of trees in the forest determines the ensemble size, while the maximum depth of each tree controls model complexity. Parameters for minimum samples required for node splitting and leaf nodes affect tree structure and generalization.

The number of features considered at each split influences model diversity and performance. Similar to XGBoost, Random Forest offers the notable benefit of quantifying feature contribution [37]. This capability enabled the identification of variables exerting the most substantial influence on VTSO recruitment suitability predictions. Furthermore, Random Forest was considered appropriate for recruitment prediction within intricate maritime contexts given its proficiency in modeling non-linear patterns and multi-dimensional feature interactions.

4.3. *Multi-Layer Perceptron Neural Network*

Multi-Layer Perceptron (MLP) is a feedforward artificial neural network proposed by Rumelhart et al. [38], consisting of an input layer, one or more hidden layers, and an output layer. MLP was applied to learn complex nonlinear patterns in VTSO recruitment data. Each neuron in the neural network receives input signals, computes a weighted sum, and generates output through an activation function. This process is expressed as follows:

$$z = \sum_{i=1}^{d} w_i x_i + b, \; \hat{y}_i = g(z) \tag{8}$$

In this formulation, d represents the input feature dimension, w_i denotes the connection weight parameters, x_i indicates the input variable values, b signifies the bias parameter, and g corresponds to the activation mechanism. To address the vanishing gradient issue and enhance training efficiency, this study utilized the ReLU activation mechanism:

$$g(z) = max(0, z) \tag{9}$$

Neural network learning is performed through the backpropagation algorithm. For binary classification, the following cross-entropy loss function is used:

$$\mathcal{L} = -\frac{1}{n}\sum_{i=1}^{n} [y_i log(\sigma(\hat{y}_i)) + (1 - y_i) log(1 - \sigma(\hat{y}_i))] \tag{10}$$

where σ is the sigmoid function. The Adam algorithm [39] was used for optimization with adaptive learning rates to update weights. The main advantage of MLP in VTS recruitment prediction is its ability to effectively capture complex interactions among various features such as maritime experience, license grades, and educational background.

4.4. Support Vector Machine

Support Vector Machine (SVM) represents a robust classification methodology introduced by Vapnik [40], which aims to find the optimal separating hyperplane that maximizes the margin between classes. SVM was applied to provide robust classification performance suitable for the binary classification problem of VTSO recruitment.

SVM transforms input space data into a high-dimensional feature space to enable linear separation. The classification function of SVM using RBF kernel is defined as follows:

$$\hat{y} = sign\left(\sum_{i=1}^{n} \alpha_i y_i K(x_i, x) + b\right) \tag{11}$$

where α_i are Lagrange multipliers and $K(x_i, x)$ is the kernel function. The RBF kernel is defined as:

$$K(x_i, x_j) = exp\left(-\gamma \|x_i - x_j\|^2\right) \tag{12}$$

where γ is a parameter controlling the width of the kernel, with larger values forming more complex decision boundaries. The optimization problem of SVM is solved through the following quadratic programming:

$$\begin{gathered} \min_{\alpha} \frac{1}{2}\sum_{i=1}^{n}\sum_{j=1}^{n} \alpha_i \alpha_j y_i y_j K(x_i, x_j) - \sum_{i=1}^{n} \alpha_i \\ Subject\ to\ constraints: \; 0 \le \alpha_i \le C, \quad \sum_{i=1}^{n} \alpha_i y_i = 0 \end{gathered} \tag{13}$$

where C is a regularization parameter that balances the tolerance for classification errors and the size of the margin. In VTSO recruitment, SVM provides stable classification performance even in limited data environments and can derive interpretable results through clear decision boundaries.

4.5. Algorithm Development and Training Configuration

This study implemented and compared four machine learning algorithms for predicting VTSO recruitment suitability: XGBoost, Random Forest, MLP, and SVM with RBF kernel. The dataset configuration for each model was designed as follows:

$$D = D_{train} \cup D_{valid} \cup D_{test} \quad (14)$$

where D_{train}, D_{valid}, and D_{test} represent the training, validation, and test datasets, respectively. The complete dataset $|D|$ consists of 2000 data points, which were randomly split in an 8:1:1 ratio, resulting in 1600, 200, and 200 data points for training, validation, and testing, respectively. To address variability introduced by random splitting, 5-fold cross-validation was applied. Each dataset's data points are expressed as follows:

$$d^i_{train} = \left(x^i_{train}, y^i_{train}\right), i = 1, \ldots, 1600 \quad (15)$$

$$d^i_{valid} = \left(x^i_{valid}, y^i_{valid}\right), i = 1, \ldots, 200 \quad (16)$$

$$d^i_{test} = \left(x^i_{test}, y^i_{test}\right), i = 1, \ldots, 200 \quad (17)$$

where $y^i_{train}, y^i_{valid}, y^i_{test} \in \{0, 1\}$ represent binary labels indicating VTSO recruitment suitability. The original input feature vector consists of eight features:

$$x_i = [Age, Education\ Level, Educational\ Institution, License, Seafaring\ Career, Last\ Position, Completion\ of\ VTS\ Courses, TOEIC\ Score] \quad (18)$$

Among these, *Age* and *TOEIC Score* are continuous variables that were standardized using StandardScaler. The remaining six categorical variables were processed using ordinal encoding for ordered variables (*License, Seafaring Career, Last Position, Education Level*) and one-hot encoding for unordered variables (*Educational Institution, Completion of VTS Courses*). The hyperparameters for each model were optimized through grid search and 5-fold cross-validation based on F1-score criterion. To assess model stability and reliability, all performance metrics were evaluated using the same 5-fold cross-validation, with detailed settings presented in Table 4.

Table 4. Hyperparameter Settings for Machine Learning Models.

Model	Parameter	Value	Search Space	Description
XGBoost	n_estimators	100	[50, 100, 200]	Number of boosting rounds
	max_depth	3	[3, 6, 9]	Maximum tree depth
	learning_rate	0.1	[0.01, 0.1, 0.3]	Step size shrinkage
	min_child_weight	1	[1, 3, 5]	Minimum sum of instance weight in child
	eval_metric	logloss	N/A	Evaluation metric
Random Forest	n_estimators	50	[50, 100, 200]	Number of trees in forest
	max_depth	15	[5, 10, 15, None]	Maximum depth of trees
	min_samples_split	10	[2, 5, 10]	Minimum samples to split node
	min_samples_leaf	1	[1, 2, 4]	Minimum samples in leaf node

Table 4. *Cont.*

Model	Parameter	Value	Search Space	Description
MLP	hidden_layer_ sizes	(150, 75)	[(50,), (100,), (100,50), (150,75)]	Neurons in hidden layers
	learning_rate_init	0.01	[0.001, 0.01]	Initial learning rate
	alpha	0.01	[0.0001, 0.001, 0.01]	L2 regularization parameter
	max_iter	1000	1000	Maximum iterations
	early_stopping	True	N/A	Use early stopping
SVM	kernel	rbf	[rbf]	Kernel type
	C	0.1	[0.1, 1.0, 10.0]	Regularization parameter
	gamma	auto	[scale, auto, 0.001, 0.01]	Kernel coefficient

All experiments were conducted in the computing environment specified in Table 5.

Table 5. Hardware Specifications and Software Configuration for Model Implementation.

Component	Specification
CPU	AMD Ryzen 5 2600 Six-Core Processor 3.40 GHz
GPU	NVIDIA Geforce RTX 3060 Dual OC D6 12 GB
Memory	16 GB
Programming Language	Python 3.8.18
OS Platform	Windows 10 x64
Machine Learning Library	XGBoost 2.1.1, Scikit-learn 1.3.2
Supporting Packages	Pandas 2.0.3, Matplotlib 3.7.5, NumPy 1.24.3,

4.6. Model Assessment and Evaluation Criteria

This section outlines the validation and testing methodology employed for the models, along with the performance metrics utilized for evaluation. The validation dataset D_{valid} and test dataset D_{test} are mathematically formulated as follows:

$$D_{valid} = \{d_{i,valid} | i \in I_{valid}\}, D_{test} = \{d_{i,test} | i \in I_{test}\} \quad (19)$$

In this notation, $d_{i,valid}$ and $d_{i,test}$ denote the i-th observations within the validation and test subsets, respectively. The symbols I_{valid} and I_{test} correspond to the indexing collections for these validation and test subsets. Individual observation structures are formulated as:

$$d_{i,s} = (x_{i,s}, y_{i,s}) \quad (20)$$

Here, $x_{i,s} \in \mathbb{R}^8$ is the input feature vector (where 8 is the total number of features), and $y_{i,s} \in \{0,1\}$ is the binary label indicating VTSO recruitment suitability. The subscript s denotes either "valid" (validation) or "test" (testing). The model's predictions on the validation and test datasets are computed as follows:

$$\hat{y}_{i,valid} = f_\theta(x_{i,valid}), \hat{y}_{i,test} = f_\theta(x_{i,test}) \quad (21)$$

In this notation, f_θ denotes the optimized prediction model, with θ corresponding to the learned parameter set encompassing weight and bias values. Model performance assessment employed five key metrics: Accuracy, Precision, Recall, F1-score, and ROC AUC. Accuracy serves as the primary indicator quantifying the ratio of correct classifications relative to total predictions, thereby evaluating overall model effectiveness.

Precision quantifies the fraction of true positives among all positive classifications. Recall captures the fraction of genuine positive cases successfully detected by the model. F1-score represents the harmonic average of Precision and Recall, facilitating assessment of equilibrium between these two measures. ROC AUC evaluates the model's discriminatory capability between positive and negative categories under varying decision boundaries. The mathematical formulations for these metrics are presented below:

$$Accuracy = \frac{TP + TN}{TP + TN + FP + FN} \tag{22}$$

$$Precision = \frac{TP}{TP + FP} \tag{23}$$

$$Recall = \frac{TP}{TP + FN} \tag{24}$$

$$F1 - Score = \frac{2 \times Precision \times Recall}{Precision + Recall} \tag{25}$$

In this context, TP (True Positive) denotes instances where VTSO recruitment suitability is accurately classified, while TN (True Negative) represents instances where unsuitability is accurately identified. FP (False Positive) arises when an unsuitable candidate is erroneously classified as suitable, whereas FN (False Negative) emerges when a suitable candidate is incorrectly classified as unsuitable.

Furthermore, the ROC AUC (Receiver Operating Characteristic Area Under the Curve) metric was employed to evaluate the calibration quality of predicted probabilities. This metric offers a consolidated score capturing the model's discriminative performance across the entire spectrum of decision thresholds, computed as follows:

$$ROCAUC = \frac{1}{n_0 n_1} \sum_{i:y_i=1} \sum_{j:y_j=0} I(\hat{p}_i > \hat{p}_j) \tag{26}$$

Here, n_0 and n_1 represent the number of samples in the negative and positive classes, respectively, $\hat{p}_i$ is the predicted probability that the i-th sample belongs to the positive class, and I is the indicator function. The complete workflow for training and evaluating the four machine learning models is detailed in Algorithm 1.

Algorithm 1. Model Training and Evaluation Process for VTSO Recruitment Prediction

Input: Training data D_{train}, validation data D_{valid}, test data D_{test}
Output: Trained models $\{f^{\theta}_{XGBoost}, f^{\theta}_{RF}, f^{\theta}_{MLP}, f^{\theta}_{SVM}\}$, performance metrics
Encode categorical features using ordinal and one-hot encoding
Standardize continuous features (Age, TOEIC Score) using StandardScaler
for each model $m \in$ {XGBoost, Random Forest, MLP, SVM} **do**
    Initialize f^{θ}_{m} with hyperparameters (see Table 4)
    Initialize $cv_m \leftarrow \emptyset$
    for fold $\leftarrow$ 1 to 5 **do**
        Split D_{train} into $D^{(fold)}_{train}$ and $D^{(fold)}_{valid}$
        if m requires scaling **then**
            Scale $D^{(fold)}_{train}$ and $D^{(fold)}_{valid}$
        Train f^{θ}_{m} on $D^{(fold)}_{train}$
        Predict $\hat{y}_{valid} \leftarrow f^{\theta}_{m}\left(x^{(fold)}_{valid}\right)$
        Compute F1-score and append to cv_m
    Compute average F1-score for model m
for each model m **do**
    if m requires scaling **then**
        Scale D_{train}, D_{valid}, and D_{test}
    Train f^{θ}_{m} on D_{train}
    Predict $\hat{y}_{valid} \leftarrow f^{\theta}_{m}(x_{valid})$
    Predict $\hat{y}_{test} \leftarrow f^{\theta}_{m}(x_{test})$
    Compute performance metrics on D_{valid} and D_{test} using Equations (15)–(17)

An early stopping strategy was implemented to mitigate overfitting during the validation phase. In particular, for XGBoost and MLP architectures, the training procedure was terminated when validation loss exhibited no enhancement across 100 consecutive epochs. The aforementioned performance indicators were utilized to conduct thorough evaluation of each model's effectiveness. Throughout validation, these indicators were continuously tracked to identify the best-performing model, with identical indicators subsequently employed in the testing phase to determine final model performance.

5. Results and Analysis

This section presents the experimental results of the four machine learning models for VTSO recruitment prediction. The analysis examines model performance evaluation (Section 5.1), feature importance analysis (Section 5.2), and partial dependence plot examination (Section 5.3) to identify the optimal model and understand the key factors influencing recruitment suitability decisions.

5.1. Model Performance Evaluation

The performance of XGBoost, Random Forest, MLP, and SVM models was evaluated using accuracy, precision, recall, F1-score, and ROC AUC metrics through 5-fold cross-validation. Table 6 presents the performance results with standard deviations, demonstrating model stability and consistency across different data subsets.

Table 6. Performance Evaluation of Machine Learning Models on Test Dataset.

Model	Accuracy	Precision	Recall	F1-Score	ROC AUC
XGBoost	0.945 ± 0.008	0.932 ± 0.010	0.960 ± 0.012	0.946 ± 0.009	0.981 ± 0.007
Random Forest	0.930 ± 0.012	0.913 ± 0.015	0.950 ± 0.014	0.931 ± 0.013	0.975 ± 0.011
MLP	0.905 ± 0.015	0.878 ± 0.018	0.940 ± 0.016	0.908 ± 0.005	0.968 ± 0.013
SVM	0.930 ± 0.011	0.898 ± 0.013	0.970 ± 0.010	0.933 ± 0.009	0.988 ± 0.006

XGBoost achieved the highest performance across most metrics with 94.5% accuracy and 0.946 F1-score, while SVM demonstrated the best ROC AUC of 0.988 and highest recall of 0.970. MLP showed the lowest performance with 90.5% accuracy, indicating relatively weaker predictive capability compared to ensemble and kernel-based methods. Figure 4 shows the ROC curves for the four models.

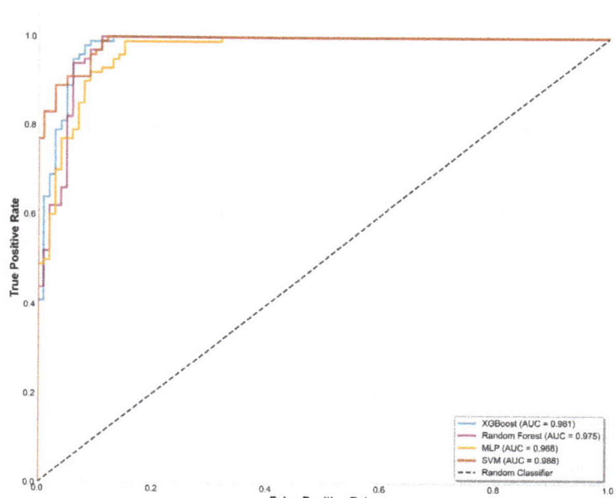

Figure 4. ROC curves comparison for four machine learning models on test dataset.

SVM achieved the highest ROC AUC value of 0.988, followed by XGBoost (0.981), Random Forest (0.975), and MLP (0.968). All models demonstrated excellent classification performance with ROC AUC values exceeding 0.968. Figure 5 presents the confusion matrices for each model. XGBoost achieved the highest F1-score of 0.946 and minimized FNs to 4, maintaining the lowest rate of missing actually suitable candidates. SVM recorded the fewest FNs with 3 cases but had a relatively high number of FPs with 11 cases. MLP showed the highest classification errors, with 13 FPs and 6 FNs.

Overall, XGBoost demonstrated the best performance in F1-score, while SVM achieved the highest ROC AUC. All models attained accuracy above 90%, confirming their effectiveness for predicting VTSO recruitment suitability.

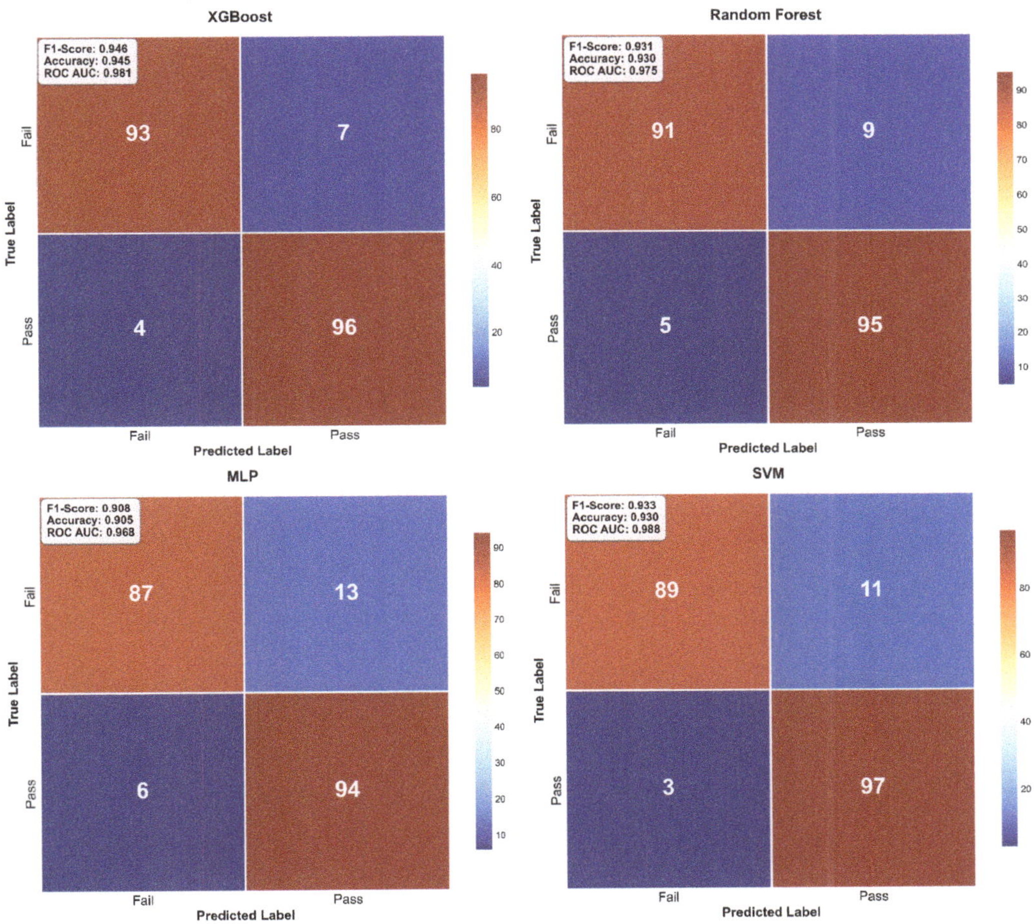

Figure 5. Confusion matrices for four machine learning models on test dataset.

5.2. Feature Importance Analysis

Feature importance analysis was conducted to identify the relative contribution of each input variable in predicting VTSO recruitment suitability across the four models. Each model's feature importance values were individually normalized to a 0–1 scale, then averaged across all models to provide a comprehensive ranking. Figure 6 shows the normalized feature importance values for each model.

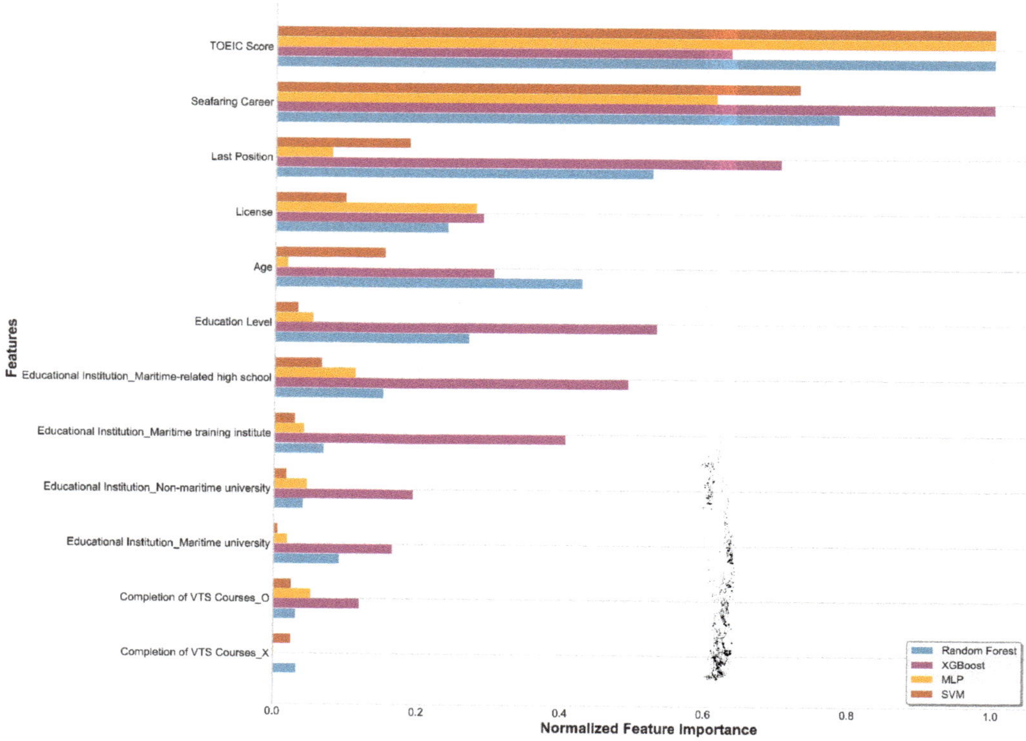

Figure 6. Feature importance comparison across four machine learning models.

As shown in Figure 6, TOEIC Score consistently demonstrated the highest importance across all models, achieving maximum normalized importance (1.0) in Random Forest, MLP, and SVM models. The average importance across all four models was 0.908, establishing it as the most critical predictor. Seafaring Career emerged as the second most important feature with an average importance of 0.782, followed by Last Position (0.373), License (0.227), and Age (0.225).

The normalized importance values reveal that Random Forest, MLP, and SVM models all identified TOEIC Score as their most influential feature, while XGBoost showed a more distributed importance pattern across features. Educational background variables (Education Level, Educational Institution) and VTS Course Completion consistently ranked among the least important factors across all models, suggesting that practical maritime experience and English proficiency are more critical determinants than formal education credentials for VTSO recruitment suitability.

5.3. Partial Dependence Plot Analysis

Partial dependence plot (PDP) analysis was conducted to quantitatively examine the individual effects of each feature on VTSO recruitment suitability prediction. Figure 7 presents the partial dependence plots for the 8 main features and 12 variables subdivided through one-hot encoding.

License demonstrated a consistent linear pattern across all models, with recruitment probability continuously increasing as the grade advances from 4th to 1st class (1st class being the highest maritime license). Seafaring Career exhibited an intriguing non-linear pattern, showing a sharp increase up to the 3–4-year experience range followed by a decline, forming an inverted U-shaped curve. Last Position showed an upward trend until the 2nd Officer position, then maintained a relatively flat pattern.

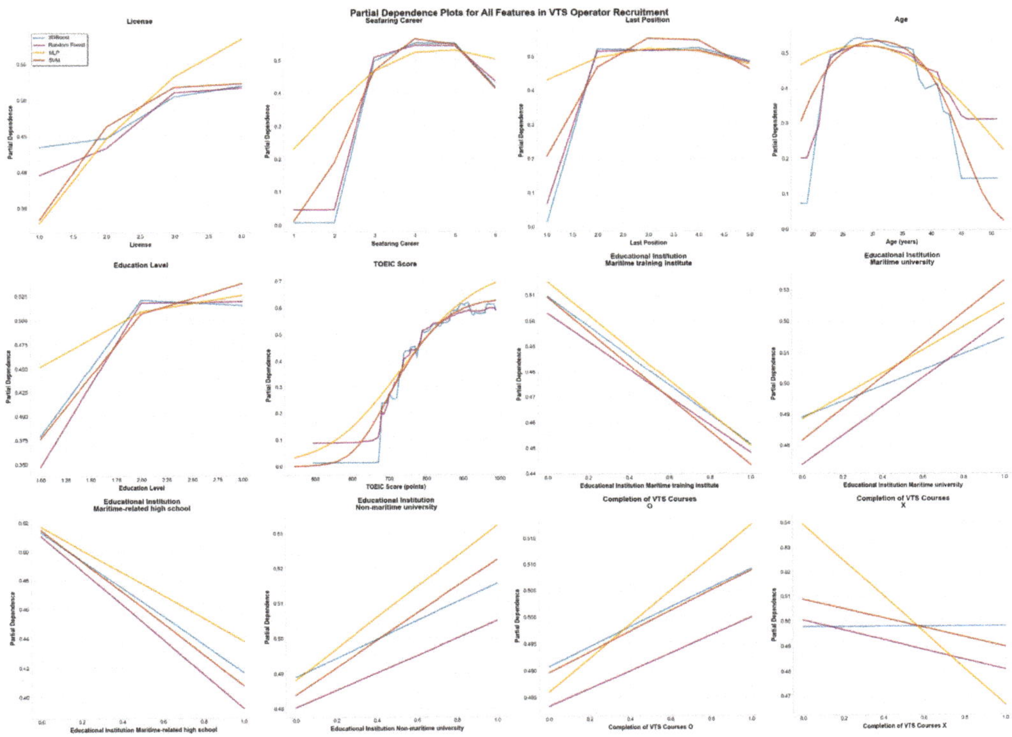

Figure 7. PDP for all features in VTSO recruitment prediction across four machine learning models.

Age displayed an inverted U-shaped distribution with recruitment probability increasing until approximately 30 years of age, then declining thereafter. Education Level showed an upward trend until the bachelor's degree level, then plateaued. TOEIC Score demonstrated a strong positive correlation across all models, with recruitment probability continuously rising as scores increased. Regarding educational institutions, both Maritime university and Non-maritime university showed upward patterns, while Maritime training institute and Maritime-related high school exhibited downward trends.

For VTS course completion, completing VTS courses (O) showed upward patterns across all models, while not completing them (X) displayed downward patterns. The specific implications of these partial dependence patterns for VTSO recruitment and their interpretation within the practical context of the maritime industry will be discussed in detail in the following section.

6. Discussion

This section analyzes the experimental results within the context of VTSO recruitment and maritime industry practices. Three key aspects are examined: model performance characteristics and their suitability for recruitment prediction, feature importance patterns and their interpretation through maritime domain expertise, and practical implications for implementing machine learning-based recruitment systems.

6.1. Model Performance Analysis

The experimental results reveal distinct performance characteristics among the four machine learning models. XGBoost achieved the highest F1-score (0.946), demonstrating strong balance between precision and recall. SVM obtained the highest ROC AUC (0.988), indicating better probabilistic calibration across all classification thresholds. This performance divergence reflects differences in model optimization objectives. XGBoost's gradient boosting framework excels at minimizing classification errors through iterative refinement.

Only 4 false negatives were produced, which is critical for avoiding rejection of suitable candidates. In contrast, SVM's margin maximization approach produces well-calibrated probability estimates, achieving the lowest false negative rate (3 cases) but with more false positives (11 cases). The strong performance of ensemble methods, particularly XGBoost and Random Forest, aligns with previous maritime personnel selection research. Pekdas et al. [9] demonstrated that XGBoost achieved the highest accuracy in seafarer recruitment prediction, followed by Random Forest. This pattern suggests the inherent suitability of ensemble methods for handling complex, multi-dimensional recruitment problems with limited data availability in the maritime domain.

Shin and Yang [41] reported similar advantages in maritime accident prediction using small-scale datasets, further supporting their robustness across different maritime applications. The choice between models involves strategic trade-offs. XGBoost's higher F1-score suggests better classification accuracy for binary pass/fail decisions, making it suitable for automated screening. SVM's superior ROC AUC indicates more reliable probability estimates, valuable for ranking candidates or setting flexible decision thresholds based on recruitment quotas. Random Forest (F1-score: 0.931, ROC AUC: 0.975) represents a robust alternative, while MLP's lower performance (F1-score: 0.908) suggests limited suitability despite its theoretical capacity for complex pattern recognition.

6.2. Feature Analysis and Maritime Context Interpretation

Feature importance and partial dependence analysis results demonstrate that international communication capabilities and appropriate levels of practical experience are key factors in VTSO recruitment. TOEIC Score achieving the highest average importance of 0.908 reflects the essential nature of VTS operations. TOEIC is a globally recognized English proficiency assessment, and effective English communication with multinational vessels in international maritime traffic control is a critical competency directly linked to maritime safety.

The consistent upward pattern of TOEIC Score across all models in partial dependence analysis indicates that improving English proficiency continuously increases recruitment probability. The average importance of 0.782 for Seafaring Career and its intriguing inverted U-shaped partial dependence pattern reflect the unique operational characteristics of the maritime industry. The pattern of sharp increase until 3–4 years of experience followed by decline suggests that experience at the Second Mate level is optimal for VTS operations. Second Mates serve as Navigation Officers aboard vessels, responsible for navigation duties, thus acquiring the core competencies required for VTS control. Conversely, from Chief Mate onwards, officers primarily handle cargo operations, reducing direct relevance to VTS work.

Additionally, excessive seafaring experience may cause recruiters to feel burdened by candidates' high-level experience or raise concerns about adaptation to shore-based control operations, resulting in declining recruitment probability. The Age pattern of increase until approximately 30 years followed by decline demonstrates the balance point between physical and cognitive abilities required for VTS operations and accumulated experience. This indicates that concentration and decision-making capabilities needed for complex multi-vessel control operations are optimized within specific age ranges. Education Level also showed an upward pattern until bachelor's level followed by plateauing, confirming the importance of systematic higher education. Analysis by educational institution revealed that Maritime university showed the highest upward pattern, attributed to systematic merchant marine officer training programs and organizational adaptability developed through collective living, which are positively perceived by evaluators.

Non-maritime university also demonstrated an upward pattern because systematic and in-depth learning experiences from four-year university education contribute to developing comprehensive thinking abilities required for VTS operations. Conversely, Maritime training institute and Maritime-related high school showed downward patterns, related to Korea's maritime officer training system. Maritime training institutes offer accelerated programs compressing four-year maritime university curricula into one year, potentially lacking theoretical depth, which may explain these results.

For VTS course completion, completing courses showed upward patterns across all models while not completing showed downward patterns, reflecting Korea's VTS education system. Currently, VTS-related courses are offered in maritime-related high schools and universities, and such theoretical foundational knowledge contributes to improving practical adaptability. Overall, feature analysis results confirm that international communication capabilities, maritime expertise with appropriate levels of sea-going experience, systematic higher education, and VTS-related specialized knowledge are key elements in VTSO recruitment.

6.3. Practical Implications and Future Directions

The results of this study provide specific practical implications for improving the efficiency and objectivity of VTSO recruitment processes. The overwhelmingly high importance of TOEIC Score (0.908) suggests the necessity of developing current general English proficiency assessment methods into forms specialized for VTS operations. Development of VTS-specific English examinations or educational programs that can evaluate specialized maritime terminology, emergency response communication, and real-time communication capabilities with multinational vessels should be considered.

The identification of Seafaring Career (0.782) as the second most critical predictor confirms that practical maritime experience remains fundamental to VTSO competency. In the emerging era of MASS, these core competencies—English proficiency and practical seafaring knowledge—will become even more vital, as VTSOs must effectively communicate with both conventional vessels and Shore Remote Control Centers (SRCC), while simultaneously applying their maritime expertise to assess autonomous vessel behaviors and manage mixed traffic environments involving both manned and unmanned vessels.

The proposed machine learning models demonstrate the potential to transform existing subjective and inconsistent recruitment methods into objective and standardized systems. The XGBoost model's 94.6% F1-score can effectively filter out unsuitable candidates in the primary screening process, reducing the number of interview candidates and significantly saving recruitment costs and time. Additionally, the SVM model's high ROC AUC (0.988) can rank candidates' suitability probabilistically, supporting recruiters in making more sophisticated selection decisions. However, this study has several important limitations. The most significant limitation is the practical constraints of data collection. Detailed information of both successful and unsuccessful candidates in actual recruitment processes is virtually impossible to obtain due to personal information protection and institutional policies.

Consequently, this study was compelled to use simulated data, but reliability was secured by constructing a dataset close to reality through rigorous validation by 40 current VTSOs and maritime experts. Additionally, this model was developed specifically for Korea's VTS recruitment system and educational framework, potentially lacking robustness for direct application in other countries. Particularly, VTS course completion reflects variables specific to Korea's educational processes.

However, core characteristics such as License, Seafaring Career, Age, and TOEIC Score are globally common maritime professional evaluation criteria, and VTS systems worldwide follow common operational standards specified by IALA, making universal application possible through appropriate localization. Future research requires model validation using actual data of successful and unsuccessful candidates through cooperation with relevant institutions. Furthermore, additional research utilizing VTS-specific English examinations developed to replace current TOEIC Scores could further improve model accuracy.

As autonomous vessels increasingly operate in coastal waters and VTS areas, future studies should explore how the feature importance identified in this study—particularly the critical role of English proficiency and seafaring experience—translates to the new operational context where VTSOs must manage mixed traffic of manned and unmanned vessels. This includes investigating whether additional features related to technical aptitude for digital navigation systems, emergency response capabilities for unmanned vessels, and ability to interpret data from Shore Remote Control Centers should be incorporated into the recruitment prediction model. Development of universal models suitable for multinational VTS systems and methodological improvements to minimize subjectivity in HITL approaches while systematically incorporating more expert opinions are also important tasks. Through such developments, it will be possible to establish standardized VTSO recruitment systems that can contribute to improving maritime safety worldwide as the maritime industry transitions toward greater automation.

7. Conclusions

This study presents the first application of machine learning models combining HITL approaches with KDE-based data augmentation for VTSO recruitment, offering a methodology to transform existing subjective hiring practices into objective and quantitative evaluation systems. Unlike previous maritime personnel selection research that primarily focused on seafarer recruitment, this study is distinguished by developing prediction models that reflect the unique operational characteristics of shore-based VTSOs.

As the maritime industry enters an era of Maritime Autonomous Surface Ships (MASS), where VTSOs will face unprecedented challenges in managing mixed traffic environments of conventional and autonomous vessels, this study provides a foundational framework for developing future VTSO selection systems that can identify candidates with the competencies necessary for these evolving operational demands. The experimental results demonstrated the superiority of ensemble methods, with XGBoost achieving optimal balanced performance. Feature importance analysis revealed that English communication ability was the overwhelmingly dominant predictor, while seafaring career exhibited a non-linear pattern with optimal recruitment probability at appropriate levels of practical experience.

These findings provide scientific evidence for establishing systematic recruitment criteria by quantitatively identifying the complexity and expertise required for international maritime traffic control. In the context of emerging MASS operations, English proficiency for communication with Shore Remote Control Centers and practical seafaring knowledge for assessing autonomous vessel behaviors will become increasingly critical as VTSOs assume expanded responsibilities as primary human decision-makers in mixed traffic scenarios.

The methodological contribution of this study lies in constructing reliable prediction models through HITL labeling utilizing domain experts' collective intelligence and KDE augmentation in limited data environments. Practically, this approach can improve recruitment process consistency and transparency, enhance the accuracy of talent selection, and contribute to administrative efficiency through pre-screening of interview candidates. However, several limitations must be acknowledged.

First, the model was developed using simulated data constructed through expert consultation rather than actual recruitment records, as real-world hiring data remains inaccessible due to privacy and institutional constraints. While the HITL approach with experienced VTSOs ensured domain validity, empirical validation with actual recruitment outcomes is necessary to confirm predictive accuracy in operational contexts. Second, the model reflects Korea-specific maritime educational systems and licensing frameworks, potentially limiting direct applicability to nations with different training pathways and regulatory structures. Third, despite rigorous expert consensus procedures, the HITL labeling process inherently retains subjective judgment, which may introduce systematic biases. Future research should address these limitations through several key directions. Collaboration with maritime authorities to collect and validate models with real-world recruitment data, cross-cultural validation studies to enhance international applicability, and incorporation of explainable AI techniques and fairness-aware machine learning approaches to improve decision transparency and address potential biases are essential.

As autonomous vessels increasingly operate in VTS areas, integrating MASS-specific competency criteria—including technical aptitude for digital navigation systems and emergency response capabilities for unmanned vessels—will be critical. Finally, developing VTS-specific English proficiency assessments that evaluate maritime terminology and emergency communication protocols would provide more targeted evaluation than general standardized tests. Through these advancements, standardized VTSO recruitment systems can be established worldwide, contributing to improved maritime safety as the industry transitions toward greater automation.

Author Contributions: Conceptualization, G.-h.S. and M.J.; methodology, G.-h.S. and M.J.; software, G.-h.S.; validation, G.-h.S. and M.J.; formal analysis, G.-h.S. and M.J.; investigation, G.-h.S.; resources, M.J.; data curation, G.-h.S.; writing—original draft preparation, G.-h.S.; writing—review and editing, G.-h.S. and M.J.; visualization, G.-h.S.; supervision, M.J.; project administration, G.-h.S. and M.J.; funding acquisition, M.J. All authors have read and agreed to the published version of the manuscript.

Funding: This research received no external funding.

Data Availability Statement: The data are available on request due to restrictions, e.g., privacy or ethics. The data presented in this study are available on request from the corresponding author. The data are not publicly available due to privacy concerns.

Conflicts of Interest: The authors declare no conflict of interest.

References

1. Chong, J.C. Impact of Maritime Autonomous Surface Ships (MASS) on VTS Operations. Master's Thesis, World Maritime University, Malmö, Sweden, 2018.
2. Kim, D.-W.; Lee, M.-K.; Park, S.-W.; Park, Y.-S. A Study on the Introduction of Maritime Autonomous Surface Ships and the Role of Vessel Traffic Service. *J. Navig. Port Res.* **2023**, *47*, 430–436. [CrossRef]
3. Moreno, F.C.; Rodríguez, R.V.; Lorente, P.S.; López, M.J.U. Relationship between human factors and a safe performance of vessel traffic service operators: A systematic qualitative-based review in maritime safety. *Saf. Sci.* **2022**, *155*, 105892. [CrossRef]
4. Shin, G.-H.; Yang, H. Vessel trajectory prediction in harbors: A deep learning approach with maritime-based data preprocessing and berthing side integration. *Ocean Eng.* **2025**, *316*, 119908. [CrossRef]
5. Issa-Zadeh, S.B.; Garay-Rondero, C.L. Maritime Pilotage and Sustainable Seaport: A Systematic Review. *J. Mar. Sci. Eng.* **2025**, *13*, 945. [CrossRef]
6. Issa-Zadeh, S.B.; Garay-Rondero, C.L. Decarbonizing Seaport Maritime Traffic: Finding Hope. *World* **2025**, *6*, 47. [CrossRef]
7. IALA. *IALA Guideline G1045 Staffing Levels at VTS Centres*; Edition 1.2; International Association of Marine Aids to Navigation and Lighthouse Authorities: Saint Germain en Laye, France, 2022; Available online: https://www.iala.int/product/g1045/ (accessed on 25 December 2024).
8. Papadimitriou, E.; Podofillini, L.; Brilakis, I.; Adjé, P.A.; Dang, V.N. Transport safety and human factors in the era of automation: What can transport modes learn from each other? *Accid. Anal. Prev.* **2020**, *144*, 105656. [CrossRef]
9. Pekdas, I.G.; Uflaz, E.; Tornacı, F.; Arslan, O.; Turan, O. Developing a machine learning-based evaluation system for the recruitment of maritime professionals. *Ocean Eng.* **2024**, *313*, 119406. [CrossRef]
10. Butler, C.; Oster, H.; Togelius, J. HITL AI for analysis of free response facial expression label sets. In Proceedings of the 20th ACM International Conference on Intelligent Virtual Agents (IVA '20), Online, 20–22 October 2020; pp. 1–8.
11. Zhang, S.; He, L.; Dragut, E.; Vucetic, S. How to invest my time: Lessons from HITL entity extraction. In Proceedings of the 25th ACM SIGKDD International Conference on Knowledge Discovery and Data Mining (KDD '19), Anchorage, AK, USA, 4–8 August 2019; pp. 2305–2313.
12. Bartolo, M.; Roberts, A.; Welbl, J.; Riedel, S.; Stenetorp, P. Beat the AI: Investigating adversarial human annotation for reading comprehension. *Trans. Assoc. Comput. Linguist.* **2020**, *8*, 662–678. [CrossRef]
13. Fan, X.; Li, C.; Yuan, X.; Dong, X.; Liang, J. An interactive visual analytics approach for network anomaly detection through smart labeling. *J. Vis.* **2019**, *22*, 955–971. [CrossRef]
14. Celik, M.; Er, I.D.; Topcu, Y.I. Computer-based systematic execution model on human resources management in maritime transportation industry: The case of master selection for embarking on board merchant ships. *Expert Syst. Appl.* **2009**, *36*, 1048–1060. [CrossRef]
15. Kartal, S.E.; Ugurlu, O.; Kaptan, M.; Arslanoglu, Y.; Wang, J.; Loughney, S. An analysis and comparison of multinational officers of the watch in the global maritime labor market. *Marit. Policy Manag.* **2019**, *46*, 757–780. [CrossRef]
16. Koutra, G.; Barbounaki, S.; Kardaras, D.; Stalidis, G. A multicriteria model for personnel selection in maritime industry in Greece. In Proceedings of the 2017 IEEE 19th Conference on Business Informatics (CBI 2017), Thessaloniki, Greece, 24–27 July 2017; Volume 1, pp. 287–294. [CrossRef]
17. Wang, Y.; Yeo, G.T. The selection of a foreign seafarer supply country for Korean flag vessels. *Asian J. Shipp. Logist.* **2016**, *32*, 221–227. [CrossRef]
18. Ding, J.F.; Liang, G.S. The choices of employing seafarers for the national shipowners in Taiwan: An empirical study. *Marit. Policy Manag.* **2005**, *32*, 123–137. [CrossRef]
19. Wang, Z.; Wei, W.; Xu, C.; Xu, J.; Mao, X.L. Person-job fit estimation from candidate profile and related recruitment history with co-attention neural networks. *Neurocomputing* **2022**, *501*, 14–24. [CrossRef]
20. Wang, Y.; Zhu, Z. The application of deep learning model in recruitment decision. *Wirel. Commun. Mob. Comput.* **2022**, *2022*, 9645830. [CrossRef]
21. Pessach, D.; Singer, G.; Avrahami, D.; Ben-Gal, H.C.; Shmueli, E.; Ben-Gal, I. Employees recruitment: A prescriptive analytics approach via machine learning and mathematical programming. *Decis. Support Syst.* **2020**, *134*, 113290. [CrossRef]
22. Al-Quhfa, H.; Mothana, A.; Aljbri, A.; Song, J. Enhancing talent recruitment in business intelligence systems: A comparative analysis of machine learning models. *Analytics* **2024**, *3*, 297–317. [CrossRef]
23. Wu, X.; Xiao, L.; Sun, Y.; Zhang, J.; Ma, T.; He, L. A survey of HITL for machine learning. *Future Gener. Comput. Syst.* **2022**, *135*, 364–381. [CrossRef]

24. Korea Coast Guard. Competition Ratios for Career-Based Competitive Recruitment of General Civil Servants. Available online: https://www.kcg.go.kr/kcg/na/ntt/selectNttList.do?mi=2797&bbsId=311 (accessed on 17 September 2025).
25. Landis, J.R.; Koch, G.G. The measurement of observer agreement for categorical data. *Biometrics* **1977**, *33*, 159–174. [CrossRef]
26. Chawla, N.V.; Bowyer, K.W.; Hall, L.O.; Kegelmeyer, W.P. SMOTE: Synthetic Minority Over-Sampling Technique. *J. Artif. Intell. Res.* **2002**, *16*, 321–357. [CrossRef]
27. Fernández, A.; García, S.; Galar, M.; Prati, R.C.; Krawczyk, B.; Herrera, F. *Learning from Imbalanced Data Sets*; Springer: Cham, Switzerland, 2018. [CrossRef]
28. Gelman, A.; Carlin, J.B.; Stern, H.S.; Rubin, D.B. *Bayesian Data Analysis*, 3rd ed.; Chapman and Hall/CRC: Boca Raton, FL, USA, 2013.
29. Scott, D.W. *Multivariate Density Estimation: Theory, Practice, and Visualization*, 2nd ed.; Wiley: New York, NY, USA, 2015.
30. Silverman, B.W. *Density Estimation for Statistics and Data Analysis*; Chapman and Hall: London, UK, 1986.
31. Chen, T.; Guestrin, C. XGBoost: A scalable tree boosting system. In Proceedings of the 22nd ACM SIGKDD International Conference on Knowledge Discovery and Data Mining, San Francisco, CA, USA, 13–17 August 2016; pp. 785–794.
32. Nielsen, D. Tree Boosting with XGBoost—Why Does XGBoost Win "Every" Machine Learning Competition? Master's Thesis, Norwegian University of Science and Technology, Trondheim, Norway, 2016.
33. Friedman, J.H. Greedy function approximation: A gradient boosting machine. *Ann. Stat.* **2001**, *29*, 1189–1232. [CrossRef]
34. Lundberg, S.M.; Lee, S.I. A unified approach to interpreting model predictions. In Proceedings of the 31st International Conference on Neural Information Processing Systems (NIPS 2017), Long Beach, CA, USA, 4–9 December 2017; pp. 4765–4774.
35. Breiman, L. Random forests. *Mach. Learn.* **2001**, *45*, 5–32. [CrossRef]
36. Liaw, A.; Wiener, M. Classification and regression by random forest. *R News* **2002**, *2*, 18–22.
37. Strobl, C.; Boulesteix, A.L.; Kneib, T.; Augustin, T.; Zeileis, A. Conditional feature importance for random forests. *BMC Bioinform.* **2008**, *9*, 307. [CrossRef]
38. Rumelhart, D.E.; Hinton, G.E.; Williams, R.J. Learning representations by backpropagating errors. *Nature* **1986**, *323*, 533–536. [CrossRef]
39. Kingma, D.P.; Ba, J. Adam: A Method for Stochastic Optimization. *arXiv* **2014**, arXiv:1412.6980.
40. Vapnik, V.N. *The Nature of Statistical Learning Theory*; Springer: New York, NY, USA, 1995.
41. Shin, G.H.; Yang, H. Maritime accident prediction in Busan port using machine learning: An integrated approach with maritime accident reports and VTS data. *Ocean Eng.* **2025**, *316*, 119968. [CrossRef]

Journal of
*Marine Science
and Engineering*

MDPI

Article

An Evaluation of Port Environmental Efficiency Considering Heterogeneous Abatement Capacities: Integrating Weak Disposability into the Epsilon-Based Measure Model

Jiewei Zhang and Gaofeng Gu *

College of Transportation Engineering, Dalian Maritime University, Dalian 116026, China;
zhangjiewei513@126.com
* Correspondence: ggu@dlmu.edu.cn

Abstract

As pivotal hubs in maritime logistics networks, ports bear a growing responsibility to harmonize economic activities with environmental stewardship. Evaluating and enhancing port environmental efficiency (PEE) is therefore imperative for maritime decarbonization and sustainability. However, conventional approaches often assume homogeneous abatement capacities across heterogeneous ports, which may distort evaluation results. To address this flaw, we develop a modified EBM-Undesirable model embedding weak disposability and non-uniform abatement factors, explicitly accounting for heterogeneity the in port's abatement capabilities. Drawing on panel data from China's major coastal ports during 2013–2022, this study further employs the Global Malmquist Index and Dagum Gini coefficient to investigate dynamic characteristics and regional disparities in PEE. Key findings reveal: (1) PEE exhibits a modest yet volatile upward trend, accompanied by pronounced inter-port divergence; (2) Total factor productivity (TFP) demonstrates sustained improvement attributable to technical efficiency advancements, yet reveals untapped potential in technological level; (3) Substantial spatial heterogeneity persists, dominated by interregional differences, though overall inequality is gradually converging. Given the observed regional disparities and technological potential, policy suggestions are proposed to advance port decarbonization, regional coordination, and maritime sustainability.

Keywords: port environmental efficiency; epsilon-based measure model; weak disposability; spatiotemporal analysis; sustainable port development

1. Introduction

Maritime transport, which carries over 80% of global trade by volume, has experienced rapid expansion in recent decades [1], thereby intensifying global energy demand and contributing to the continued rise in carbon emissions [2]. Currently, maritime transport emits around 1056 million tons of CO_2 emissions annually, contributing to more than 3% of global anthropogenic CO_2 emissions, a share that continues to rise [3]. Emissions from the shipping industry are predominantly attributed to vessels and ports. Among these, ports serve as pivotal hubs and sources of carbon emissions and play an instrumental role in the global management and control of CO_2 emissions [4]. It is estimated that ports contribute about 2% of total global CO_2 emissions, with levels rising by 5.6% to 740 million tons between 2012 and 2018. If left unchecked, projections under business-as-usual scenarios suggest a further rise of up to 50% by 2050 [5]. In this context, ports are under increasing pressure to reduce emissions and advance sustainable transition. An exclusive concentration on carbon emission reduction may be inappropriate, as overly stringent decarbonization measures could impose constraints on port productivity and hinder economic growth. Enhancing port environmental efficiency (PEE), defined as the capability of ports to maximize desirable outputs while minimizing undesirable emissions under given inputs, emerges as an effective approach to reconcile shipping-related carbon emissions and sustained economic development [6]. Scientifically evaluating port environmental performance is therefore essential to support informed policy-making and promote long-term maritime sustainability.

Given the significance of PEE, scholars have engaged in extensive discussions on its measurement. Currently, the measurement and evaluation methods for port efficiency primarily encompass Stochastic Frontier Analysis (SFA) [7] and Data Envelopment Analysis (DEA) [8]. Among these, DEA stands out as a nonparametric method, requiring no predetermined parameters, which endows it with strong applicability. Hence, it has emerged as the most commonly used method for efficiency evaluation at present [9]. Initially, most studies employed conventional models like CCR and BCC to calculate port efficiency without considering undesirable outputs, particularly CO_2 emissions [10,11]. Amid growing environmental concerns, scholars have increasingly integrated undesirable outputs into the evaluation framework. A variety of DEA models, such as the directional distance function (DDF) [12], network DEA (NDEA) [13], slack-based measure DEA (SBM-DEA) [14], and range adjusted measure DEA (RAM-DEA) [15], have been widely applied to assess PEE.

While incorporating undesirable outputs has enhanced the environmental relevance of port efficiency assessments, most existing studies largely depend on either radial or non-radial DEA models, both of which exhibit inherent limitations. Radial DEA models assume proportional adjustments in inputs and outputs while overlooking non-radial slacks, which may not hold in certain contexts [16]. In contrast, non-radial DEA models account for slack but may sacrifice original proportionality, potentially compromising efficiency estimates [17]. Therefore, a composite evaluation framework compiling the strengths of both approaches is required, enabling a more reasonable and accurate evaluation of port efficiency.

In addition, the treatment of undesirable outputs in many studies remains grounded in the strong disposability assumption. Under this assumption, undesirable outputs, such as CO_2 emissions, are treated as freely reducible without compromising desirable outputs. However, this is rarely the case in actual production settings, where emission reductions typically involve trade-offs, such as increased costs or reduced output [18]. To address this limitation, the recent literature has increasingly adopted the weak disposability assumption, which offers a more realistic representation of environmental constraints by recognizing that pollution abatement is not costless and often requires a sacrifice in economic performance.

Moreover, PEE is often calculated under the assumption of a homogeneous abatement factor across all ports, implicitly overlooking the heterogeneous carbon-reduction characteristics that exist in practice. Substantial disparities exist among ports in terms of operational scale, technological endowment, and environmental governance, which often lead to heterogeneous abatement capabilities. Consequently, disregarding this heterogeneity by applying a uniform abatement factor may lead to systematically biased efficiency scores [19].

To fill these aforementioned research gaps, this study proposes a novel evaluation framework to comprehensively evaluate port environmental performance. First, we employ an epsilon-based measure model with undesirable outputs (EBM-Undesirable) that combines radial and non-radial features, allowing for a more precise evaluation of PEE. Building on this foundation, the EBM-Undesirable model is further extended by integrating the weak disposability assumption and non-uniform abatement factors. This approach allows port operators to differentially adjust their production activity levels to curtail undesirable outputs, explicitly reflecting the heterogeneity in abatement capacity. On this basis, the dynamic characteristics and regional disparities of PEE are examined via the Global Malmquist index and the Dagum Gini coefficient. Drawing from this, the study answer encompasses the following: How does the overall trend in PEE unfold? What is the predominant driver affecting the TFP of ports? How has the spatial heterogeneity of PEE evolved?

This article's marginal contributions include: First, an EBM-Undesirable model was developed by incorporating weak disposability of undesired outputs to facilitate a precise evaluation of PEE. Second, heterogeneous carbon abatement capabilities across ports were accounted for in the efficiency measures, preventing the overestimation that can occur with a uniform abatement factor. Third, the Global Malmquist index and Gini coefficient of PEE are deeply decomposed, enabling the accurate identification of underlying drivers and the sources of regional disparity, underpinning targeted port development strategies.

The paper unfolds as follows. Section 2 presents the literature review. Section 3 proposes the methodology and data descriptions. Section 4 expounds the results and discussion. Section 5 summarizes the conclusions and proffers several recommendations.

2. Literature Review

2.1. DEA Application in Port Efficiency Evaluation

Port efficiency refers to the effective utilization of inputs to generate maximum possible outputs, or conversely, to achieve target outputs with minimal resource consumption [20]. The DEA approach is extensively utilized for quantifying port efficiency, with an emphasis on economic efficiency, environmental efficiency, and other applications [21].

In terms of economic efficiency evaluation, Tovar and Wall [22] implemented a DEA-Malmquist using meta-frontier analysis to calculate the efficiency, uncovering that port complexity and its scale are major determinants. Liu et al. [23] undertook benchmarking by assessing port efficiency of six typical Chinese Pilot Free Trade Zones (PFTZs), which revealed that the establishment of PFTZs positively influenced port efficiency. Agüero-Tobar et al. [24] utilized the DEA method for evaluating port operating performance in Chile and proposed a way for better understanding the economic effects of efficient operations. Moya et al. [25] employed the DEA meta-frontier approach to quantify port technical efficiency by adding an indicator of berth duration. Furthermore, Zhu et al. [14] explored the geographic disparities and changing tendency of inland waterway transportation efficiency. Kong et al. [26] established an integrated framework incorporating economic, environmental, and social elements for benchmarking port sustainable efficiency.

In terms of environmental efficiency evaluation, Chin and Low [27] incorporated environmental externalities into their port performance assessment framework and examined Asian port efficiency. In addition, Li et al. [28] and Wang et al. [29] respectively carried out studies on port environmental efficiency from the viewpoints of CO_2 and PM emissions. Castellano et al. [30] benchmarked the environmental efficiency of Italian ports, demonstrating that considering negative externalities and green policies is of utmost importance for assessing port competitiveness. Wang et al. [31] discovered that the green port efficiency is generally poor and unevenly developed, with most ports displaying a fluctuating growing trend within five years. In addition, Sun and Zhao [32] assessed the efficiency of China's port-region (PR) system, demonstrating that overall efficiency is rising, with management level being a crucial impetus affecting efficiency. Lin et al. [33] evaluated the port environmental efficiency of the Yangtze River Delta Pilot Free Trade Zone (YRD PFTZ) from both static and dynamic perspectives and found that these ports generally exhibit a high level of environmental efficiency.

Regarding the evaluation of other application fields in ports, a relatively limited number of existing studies are found. In this regard, Nguyen et al. [34] investigated market concentration in Southeast Asian container ports and observed an association with operational efficiency. Kong and Liu [35] gauged port-cities' sustainability by combining internal and external dimensions, emphasizing the implication of port-city interactions on sustainability. Additionally, Zhang et al. [36] undertook benchmarking the rail-water intermodal transport system efficiency. Jiang et al. [37] compared the efficiency and dynamic variations in the hinterland transportation system, concluding that road-rail transport shows exemplary efficiency. Zhang et al. [38] examined the low-carbon performance of the "ship-port" system and found that port system efficiency exhibited a sustained upward trend, whereas ship system efficiency experienced a volatile decline.

2.2. DEA Based on Weak Disposability (WD) Assumption

The conventional DEA model typically assumes strong disposability (SD) for undesired outputs in the production process, as posited by Färe et al. [39]. Under this axiom, undesired outputs, such as pollution, can be reduced independently and without cost, which neglects the potential impact of diminishing undesired outputs on desired outputs. However, this assumption may contradict economic rationality, especially when dealing with carbon emissions and other pollutants. In response to this issue, researchers progressively took a more realistic approach to managing undesirable outputs. Färe and Grosskopf [40] presented the concept of WD, which proposed that undesired outputs should be reduced in coordination with desired outputs rather than being disposed of independently and without cost. They argue that reducing undesirable outputs usually entails sacrificing some quantity of desired outputs or adding extra inputs. The traditional SD approach may cause estimation biases of production efficiency.

Early research on WD predominantly adopted a single abatement factor to reduce undesired outputs. While this approach is straightforward and intuitive, it disregards the heterogeneity across different firms in mitigating undesired outputs. Kuosmanen [41] put forward a method of multiple abatement factors that allows for a more accurate simulation of WD within non-parametric production models. Specifically, Kuosmanen's approach permits firms to utilize non-uniform abatement factors tailored to their specific characteristics, thereby decreasing undesired outputs more effectively. This approach offers a more realistic reflection of production and averts the problem of efficiency overestimation caused by adopting a uniform abatement factor.

In practical applications, Chang et al. [42] modified the constraint of the SBM model such that undesirable outputs increase proportionally with improvements in output slack, thus aligning with the WD assumption. They applied this adjusted model to assess global airlines' economic efficiency. Similarly, Jo and Chang [18] and Taleb et al. [43] benchmarked the port environmental efficiency by utilizing the SBM-DEA model with WD. It is worth noting that our previous study [44] also adopted the SBM-DEA framework with the weak disposability assumption to evaluate port environmental efficiency. Nevertheless, both studies relied on a uniform abatement factor to tackle the issue of WD within the DEA model framework. In contrast, Yu and See [45] utilized non-uniform abatement factors to compute the efficiency of airlines in diminishing undesired outputs such as CO_2 emissions, thereby obtaining unbiased efficiency estimates for airlines. However, there are scarce studies that have benchmarked the port's environmental efficiency considering WD with the non-uniform abatement factor.

A review of the existing literature (see Table 1) shows that most studies employed either radial or non-radial models to compute port efficiency. These models are limited in their ability to simultaneously integrate radial and non-radial dimensions, potentially leading to deviations in the evaluation results. Hence, emerging radial and non-radial merits are essential for constructing an integrated framework to assess port performance. Additionally, many prior studies have disregarded the implications of the disposability assumption on port efficiency evaluation, merely considering SD or WD with the unified abatement factor, ignoring the distinct abatement technology characteristics across disparate ports. This oversight typically gives rise to an overestimation of efficiency. Consequently, there is a clear need for further research that incorporates WD with a non-uniform abatement factor, permitting port operators to modulate their production activity levels in varying proportions to abate undesired outputs. This approach would enable a more refined and accurate evaluation of PEE.

Table 1. Summary of literature evaluating port efficiency with DEA model.

	Research Object	Method	Type	Inputs	Outputs
[10]	19 container ports	CCR, BCC	R	Terminal area, Quay length, Quay crane, Yard equipment, Maximum draft	Throughput
[11]	Kaohsiung Port	CCR, BCC	R	Gantry crane capacity, Gantry crane number, Berth length, Container yard area, Fixed costs, Variable costs	Total loading and unloading capacity
[12]	26 Spanish ports	DDF	R	Labor, Capital, Intermediate	Ships, Cargo traffic, Passenger traffic, CO_2 emissions
[13]	13 global ports	Dynamic NDEA	NR	Fleet capacity, Expenses, Employees	Lifting, Revenue
[14]	11 Chinese ports	SBM	NR	Employees, Motor ship, Urban fixed asset investment	NOx emission, Passenger volume, Cargo volume
[15]	10 Chinese ports	RAM	NR	Staff members, Annual cash investment, Production berths	Container throughput, Cargo throughput, Main business income, CO_2 emissions
[22]	26 Spanish ports	DEA-Malmquist	R	Labor, Intermediate consumption expenditures, Capital assets, Deposit surface area	Liquid bulk cargo, Solid bulk cargo, Container cargo, General non-container cargo, Passengers
[23]	6 Chinese ports	Super-SBM	NR	Berth number, Terminal length, Net asset, Cost of goods sold, Employees	Cargo throughput, Container throughput, Sales revenue, SO_2 emission, NO_x emission
[24]	12 Chilean ports	CCR, BCC	R	Maximum draft, Quay length, Berth number	TEUs transferred, Number of vessels
[25]	19 Mediterranean ports	CCR	R	Berth length, Yard area, Quay cranes number	Number of TEU handled per hour
[27]	13 East Asian ports	SBM	NR	The frequency of shipping services, Bilateral trade flows	Container capacity flows, Gaseous emissions
[28]	16 Chinese ports	SBM	NR	Labor, Fixed assets	Container throughput, Cargo throughput, CO_2 emission
[29]	11 Chinese ports	DDF	R	Terminal length, Berth quantity, Labor, Total assets	Container throughput, PM emission
[30]	24 Italian ports	CCR, BCC	R	Investments, Terminal area, Employees, Green Port Efforts (GPE)	Solid bulk, Liquid bulk, Containers, Environmental Quality Index (EQI)
[31]	18 Chinese ports	Cross-efficiency model	NR	Berth number, Terminal length, Employees, Total fixed assets	Cargo throughput, NOx emissions, SO_x emissions, Solid waste containers
[32]	22 Chinese ports	DDF	NR	Berth number, Terminal length, Employees, Barges	Cargo throughput
[36]	14 Chinese ports	NDEA	NR	Length of railways, Railway labor, Berth quantity, Port labor	Railway-port freight volumes, Cargo throughput, CO_2 emission

Note: R = radial, NR = non-radial.

3. Methodology and Data Descriptions

3.1. An EBM Model with Non-Uniform Abatement Factors

Conventional DEA models are typically either input-oriented or output-oriented. In these models, decision-makers can improve the efficiency of DMUs only by reducing inputs or increasing outputs but cannot optimize both inputs and outputs simultaneously. Furthermore, most of these models are radial DEA models, implying that decision-makers can only reduce inputs or increase outputs in equal proportions. Consequently, the traditional radial DEA approach struggles in tackling the slackness of inputs and outputs, together with overlooking undesired outputs [46]. To overcome these limitations, Tone [47] introduced the SBM model, incorporating slack variables within the objective function. This model helps avoid the deviation caused by differences in radial and oriented models. Despite this refinement, the efficiency values are underestimated due to the objective function of the SBM model, which aims to maximize input-output inefficiencies [17]. Subsequently, Tone and Tsutsui [48] proposed the EBM model, which is distinguished by its ability to merge radial proportions and non-radial slacks, thereby facilitating a more precise efficiency evaluation.

However, the basic EBM model disregards negative effects generated during the production process, making it challenging to accurately reflect the true efficiency of decision-making units (DMUs). Therefore, this study considers undesirable outputs incorporating WD. In practical applications, different DMUs may operate under varying returns to scale. To reflect this, this study adopts the variable returns to scale (VRS) assumption, which allows for the possibility that the proportionate change in outputs may not be constant relative to the change in inputs [49]. Under VRS assumption, an EBM-Undesirable model imposing weak disposability (EBM-WD model) is developed, applying non-uniform abatement factors across disparate ports. Suppose there are n DMUs, each input and output dimensions represented as $x \in R^m, y \in R^s, b \in R^q$, and corresponding matrices are: $X = [x_1, \ldots, x_n] \in R^{m\times n}$, $Y = [y_1, \ldots, y_n] \in R^{s\times n}$, and $B = [b_1, \ldots, b_n] \in R^{q\times n}$, where $X > 0, Y > 0$ and $B > 0$. The model is formulated as follows:

$$
\begin{aligned}
min\,\rho = &\frac{\theta - \varepsilon_x \sum_{i=1}^{m} \frac{w_i^- s_i^-}{x_{i0}}}{\varphi + \varepsilon_y \left(\sum_{r=1}^{s} \frac{w_r^{g+} s_r^{g+}}{y_{ro}} + \sum_{p=1}^{q} \frac{w_p^{b-} s_p^{b-}}{b_{po}} \right)} \\
s.t.\ &\sum_{j=1}^{n} X_j z_j + s^- = \theta x_0 \\
&\sum_{j=1}^{n} Y_j z_j \delta^j - s^{g+} = \varphi y_0 \\
&\sum_{j=1}^{n} B_j z_j \delta^j + s^{b-} = \varphi b_0 \\
&\sum_{j=1}^{n} z_j = 1 \\
&\theta \le 1, \varphi \ge 1 \\
&s^- \ge 0, s^{g+} \ge 0, s^{b-} \ge 0, z_j \ge 0, 0 \le \delta^j \le 1, j = 1, 2, \ldots, n
\end{aligned} \tag{1}
$$

where z represents intensity variable; ρ^* is the efficiency value; δ^j stands for the abatement factor, and $\delta^1 \ne \delta^2 \ne \ldots \ne \delta^n$ enables each DMU_j to be reduced by a distinct abatement factor; the constraints in model (1) are non-linear, and the intensity variables could be separated into two sections: $z_j = \lambda_j + \mu_j$, that is, $\mu_j = (1 - \delta^j)z_j$, $\lambda_j = \delta^j z_j \ldots$, the non-linear equations in model (1) can be converted into linear equations. By this setting, model (1) can be expressed as:

$$
\begin{aligned}
min\,\rho = &\frac{\theta - \varepsilon_x \sum_{i=1}^{m} \frac{w_i^- s_i^-}{x_{i0}}}{\varphi + \varepsilon_y \left(\sum_{r=1}^{s} \frac{w_r^{g+} s_r^{g+}}{y_{ro}} + \sum_{p=1}^{q} \frac{w_p^{b-} s_p^{b-}}{b_{po}} \right)} \\
s.t.\ &\sum_{j=1}^{n} X_j (\lambda_j + \mu_j) + s^- = \theta x_0 \\
&\sum_{j=1}^{n} Y_j \lambda_j - s^{g+} = \varphi y_0 \\
&\sum_{j=1}^{n} B_j \lambda_j + s^{b-} = \varphi b_0 \\
&\sum_{j=1}^{n} (\lambda_j + \mu_j) = 1 \\
&\theta \le 1, \varphi \ge 1 \\
&s^- \ge 0, s^{g+} \ge 0, s^{b-} \ge 0, \lambda_j \ge 0, \mu_j \ge 0, j = 1, 2, \ldots, n
\end{aligned} \tag{2}
$$

Here, μ_j signifies the segment of DMU_j's output that is mitigated through a decrease in activity level, while λ_j corresponds to the segment of DMU_j's output that remains active. s^-, s^{g+}, s^{b-} respectively denote the slack of inputs, desirable outputs and undesired outputs. $w_i{}^-, w_r{}^{g+}, w_p{}^{b-}$ respectively depict the weights of inputs, desirable outputs and undesired outputs, and meet $\sum_{i=1}^{m} w_i^- = 1$ and $\sum_{r=1}^{s} w_g^{g+} + \sum_{p=1}^{q} w_p^{b-} = 1$. ε_x and ε_y are crucial parameters that signify the non-radial component's significance in efficiency assessment, with feasible domain of [0, 1]. To prevent measurement inaccuracies, constraints on θ and φ are imposed, that is $\theta \le 1$, $\varphi \ge 1$. Parameters ε and w need to be determined beforehand. The projection values of multiple input and output metrics are derived via the SBM model with undesirable outputs, written as $\hat{x}_{ik} = x_{ik} - s_i^-, \hat{y}_{rk}^g = y_{rk}^g + s_r^{g+}, \hat{y}_{pk}^b = y_{pk}^b - s_p^{b-}$, and the projection value matrix is built.

Taking input indicators as example and referring the treatment methods of [48,50], the steps for obtaining parameters ε_x and w_i^- are outlined below:

Step 1: Projection values and substitutability interpretation.

The projection values of each input variable are obtained through the SBM model with undesirable outputs, denoted as $\hat{x}_k (k = 1, \ldots, m)$. The correlation between the projection values of two inputs $\hat{x}_i$ and $\hat{x}_j$ reflects their proportional relationship and indicates the degree of substitutability in the production process. A strong positive linear correlation suggests weak substitutability, implying a predominantly radial adjustment pattern; thus, ε should be close to 0. In contrast, a strong negative correlation implies strong substitutability and the adoption of non-radial adjustment, where ε should approach 1.

Step 2: Construct the diversity matrix R.

The diversity matrix $R = [R(\hat{x}_i, \hat{x}_j)]_{m \times m}$ reflects the pairwise linear relationships among the projected values of input indicators. For any two inputs $\hat{x}_i$ and $\hat{x}_j$, the corresponding element $R(\hat{x}_i, \hat{x}_j)$ is computed as the Pearson correlation coefficient between their projection vectors. See Equation (3) for details.

$$R(\hat{x}_i, \hat{x}_j) = \frac{\sum_{k=1}^{n} (\hat{x}_{ik} - \overline{x}_i)(\hat{x}_{jk} - \overline{x}_j)}{\sqrt{\sum_{k=1}^{n} (\hat{x}_{ik} - \overline{x}_i)^2} \sqrt{\sum_{k=1}^{n} (\hat{x}_{jk} - \overline{x}_j)^2}} \tag{3}$$

where $\hat{x}_i$ and $\hat{x}_j$ represent the sequences of projection values for i and j for the $\boldsymbol{k}$-th DMU, and $\overline{x}_i$, $\overline{x}_j$ are their respective means.

This definition ensures that R is a symmetric matrix with diagonal elements equal to 1, and effectively captures the similarity or substitutability among input indicators.

Step 3: Establish the affinity matrix S.

The new affinity matrix $S = [S(\hat{x}_i, \hat{x}_j)]_{m \times m}$ is constructed from the diversity matrix R via the specified formula $S(\hat{x}_i, \hat{x}_j) = 0.5 + 0.5R(\hat{x}_i, \hat{x}_j)$, which allows it to quantify both the strength and direction of correlation among the indicators.

Step 4: Compute the parameters ε_x and w_i^- from the affinity matrix S. The parameters ε_x is obtained as:

$$\varepsilon_x = \begin{cases} \frac{m-\rho_x}{m-1} & , \quad m > 1 \\ 0 & , \quad m = 1 \end{cases} \tag{4}$$

And the weights are normalized as:

$$w_i^- = \frac{w_{ix}}{\sum_{i=1}^{m} w_{ix}} \tag{5}$$

where ρ_x is the largest eigenvalue of affinity matrix S, $W_x = (w_{1x}, w_{2x}, \ldots, w_{mx})$ represents its corresponding eigenvector.

The methods for calculating parameters ε_y and w_r^{g+}, w_p^{b-} are consistent with the above steps.

3.2. The Global Malmquist Index

The DEA model typically provides a static analysis based on a time cross-section. When the evaluation unit involves panel data with multiple consecutive time points, the Malmquist Productivity Index (M index) is employed for dynamic analysis [51]. Nevertheless, the conventional M index can only compare productivity changes between adjacent periods and cannot facilitate intertemporal comparisons. Furthermore, it is prone to issues of infeasibility under VRS. To address these shortcomings, Pastor and Lovell [52] proposed the Global Malmquist Productivity Index (GM index). This index uses a joint frontier constructed by all periods, thus overcoming the defects of the traditional M index and possessing the advantage of intertemporal comparability. To incorporate undesirable outputs into the study, the GM index is further extended and utilized to calculate the TFP of the DMUs. Assume the study period is T; the production possibility set (PPS) in period t is expressed as:

$$P^t = \{ (x^t, y^t, b^t) : x^t \text{ can produce } (y^t, b^t) \} \tag{6}$$

The global PPS is constructed as the union of all period-specific PPSs and is formally defined as follows: $P^G = P^1 \cup P^2 \cup \ldots \cup P^T$.

The GM index via formula outlined below:

$$GM_{t,t+1}(x_{t+1}, y_{t+1}, b_{t+1}, x_t, y_t, b_t) = \frac{D^G(x_{t+1}, y_{t+1}, b_{t+1})}{D^G(x_t, y_t, b_t)} \tag{7}$$

where (x_t, y_t, b_t), $(x_{t+1}, y_{t+1}, b_{t+1})$ are the input and output direction vectors for period t and period $t+1$, respectively.$D^G(x, y, b)$ represents the global hybrid distance functions, while $D^t(x, y, b)$, $D^{t+1}(x, y, b)$ depict the distance functions in two periods. A value of $GM > 1$ suggests TFP growth over time; whereas $GM < 1$ indicates a degradation.

Additionally, this study decomposes the GM index into two components: GM efficiency change (GMEC) and GM technological change (GMTC).

$$\begin{aligned} & GM_{t,t+1}(x_{t+1}, y_{t+1}, b_{t+1}, x_t, y_t, b_t) \\ & = \frac{D^G(x_{t+1}, y_{t+1}, b_{t+1})}{D^G(x_t, y_t, b_t)} \\ & = \frac{D^t(x_{t+1}, y_{t+1}, b_{t+1})}{D^t(x_t, y_t, b_t)} \times \left[\frac{D^G(x_{t+1}, y_{t+1}, b_{t+1})}{D^{t+1}(x_{t+1}, y_{t+1}, b_{t+1})} \times \frac{D^t(x_t, y_t, b_t)}{D^G(x_t, y_t, b_t)} \right] \\ & = GMEC_{t,t+1} \times GMTC_{t,t+1} \end{aligned} \tag{8}$$

where $GMEC_{t,t+1}$ and $GMTC_{t,t+1}$ represent changes in technical efficiency and technological progress between period t and period $t+1$. Specifically, when $GMEC > 1$, it suggests technical efficiency catch-up, and vice versa; while $GMTC < 1$ demonstrates technological innovation, and vice versa.

3.3. Dagum Gini Coefficient

In examining regional disparities, many approaches such as the traditional Gini coefficient, variation coefficient, and Theil index, while possessing certain explanatory capabilities, have limitations in addressing the intersection and overlap of data distributions. To overcome the above drawbacks, Dagum [53] proposed the Dagum Gini coefficient, which divides the entire sample into distinct subgroups, and, respectively, examines the disparities within and between these subgroups, thereby providing a clearer understanding of the origins of these discrepancies. Furthermore, it adeptly resolves the issues of data intersection and overlap, as well as precisely revealing the relative difference features among different regions [54].

This study utilizes the Dagum Gini coefficient to explore PEE spatial disparities and sources, calculated as follows:

$$G = \frac{\sum_{j=1}^{k}\sum_{h=1}^{k}\sum_{i=1}^{n_j}\sum_{r=1}^{n_h}\left|y_{ji} - y_{hr}\right|}{2n^2\overline{y}} \tag{9}$$

where G signifies the overall Gini coefficient, the higher the value, more pronounced the general discrepancy. y_{ji} (y_{hr}) stands for environmental efficiency of port $i(r)$ in the region $j(h)$. $\overline{y}$ indicates the average value of all port efficiencies, n and k respectively denote the number of ports and region divisions. n_j (n_h) expresses the number of ports in region $j(h)$. It should be emphasized that before computing the Gini coefficient, sorting by the average port environmental efficiency in every region is required, outlined in Equation (10).

$$\overline{Y}_h \leq \ldots \leq \overline{Y}_j \leq \ldots \leq \overline{Y}_k \tag{10}$$

If the approach of calculating the overall Gini coefficient is solely applied to region j, the intraregional Gini coefficient G_{jj} of region j can be obtained.

$$G_{jj} = \frac{\sum_{i=1}^{n_j}\sum_{r=1}^{n_j}\left|y_{ji} - y_{jr}\right|}{2n_j^2\overline{y}_j} \tag{11}$$

Similarly, the interregional Gini coefficient G_{jh} can be derived.

$$G_{jh} = \frac{\sum_{i=1}^{n_j}\sum_{r=1}^{n_h}\left|y_{ji} - y_{hr}\right|}{n_j n_h(\overline{y}_j + \overline{y}_h)} \tag{12}$$

The overall Gini coefficient G can be further decomposed into the contribution of intraregional differences G_w, the contribution of interregional differences G_{nb}, and the contribution of the intensity of transvariation G_t, and it meets $G = G_w + G_{nb} + G_t$. The expression is as follows:

$$G_w = \sum_{j=1}^{k} G_{jj} p_j s_j \tag{13}$$

$$G_{nb} = \sum_{j=2}^{k}\sum_{h=1}^{j-1} G_{jh}(p_j s_h + p_h s_j) D_{jh} \tag{14}$$

$$G_t = \sum_{j=2}^{k}\sum_{h=1}^{j-1} G_{jh}(p_j s_h + p_h s_j)(1 - D_{jh}) \tag{15}$$

In Equation (14), $p_j = n_j/n, s_j = n_j\overline{y}_j/n\overline{y}$ $(j = 1, 2, \ldots, k)$, D_{jh} depicts the relative disparities of port efficiency between region j and region h, as given by Equation (16).

$$D_{jh} = \frac{d_{jh} - p_{jh}}{d_{jh} + p_{jh}} \tag{16}$$

$$d_{jh} = \int_0^{\infty} dF_j(y) \int_0^{y} (y - x) dF_h(x) \tag{17}$$

$$p_{jh} = \int_0^\infty dF_h(y) \int_0^y (y - x) dF_j(x) \tag{18}$$

where $F_j(\cdot)$ and $F_h(\cdot)$ respectively represent the cumulative distribution functions of region j, h. d_{jh} signifies the difference in the improvement of PEE across regions, while p_{jh} is hypervariable first-order moment, which are, respectively, the mathematical expectation of the sum of all sample values $y_{ji} - y_{hr} > 0$ and $y_{hr} - y_{ji} > 0$ in region j and h.

3.4. Data Collection and Descriptions

This study selects the data from 2013 to 2022 and takes 23 major ports in the five port groups along China's coastal ports as the research object, as shown in Table 2. Given the availability of data and based on relevant literature [55–57], this study selected input-output indicators, as detailed in Table 3, to calculate PEE from the perspective of port production and utilization.

Table 2. Port distribution within the port group.

	Port Group	Contains Port
1	Bohai Sea port group	Dalian Port, Yingkou Port, Qinhuangdao Port, Tianjin Port, Yantai Port, Weihai Port, Qingdao Port, Rizhao Port
2	Yangtze River Delta port group	Shanghai Port, Lianyungang Port, Ningbo-Zhoushan Port, Taizhou Port, Wenzhou Port
3	Southeast Coastal port group	Fuzhou Port, Xiamen Port
4	Pearl River Delta port group	Shantou Port, Shenzhen Port, Guangzhou Port, Zhuhai Port
5	Southwest Coastal port group	Zhanjiang Port, Beihai Port, Fangcheng Port, Haikou Port

Table 3. Port environmental efficiency evaluation index.

	Variable	Unit
Inputs	Quay length	m
	Berth number	pcs
	10,000 Ton Class Berth number	pcs
Desirable outputs	Cargo throughput	10,000 tons
	Container throughput	10,000 TEU
Undesirable outputs	CO_2 emission	10,000 tons

Since official yearbooks do not provide precise statistics on coastal port carbon emissions, indirect estimation methods are commonly used. While bottom-up approaches can provide detailed estimates at the vessel level, they require high-frequency operational data that are not consistently available for Chinese coastal ports [58]. By contrast, the top-down approach calculates emissions by multiplying fuel consumption by emission factors, offering a practical solution for regional studies [59]. Accordingly, and following prior studies [44,57,60], this paper adopts a top-down method to estimate port carbon emissions, as outlined in Equation (19).

$$C = SC \times HC \times \alpha \tag{19}$$

Here, C denotes the CO_2 emissions of the port; SC refers to the standardized coal-equivalent energy consumption per 10,000 tons of port cargo throughput, expressed in metric tons of standard coal. Annual SC values covering the period 2013–2022 were obtained from the Statistical Bulletin on the Development of China's Transportation Industry, as shown in Table 4 (drawing on prior research [44]). HC stands for the annual cargo throughput of a given port (in 10,000 tons); and α is the carbon emission coefficient, referring to the unit CO_2 emissions consumed per ton of standard coal in the port, taking the value of 2.4589. This value is derived from the theoretical carbon-to-CO_2 conversion process during the complete combustion of standard coal: 1 ton of standard coal contains approximately 0.67 tons of carbon, and full oxidation of carbon generates 3.67 tons of CO_2 per ton of carbon, resulting in 0.67·3.67 = 2.4589 tons of CO_2 per ton of standard coal.

Table 4. Reference coefficient of port coal-equivalent energy consumption per unit.

Period	2013	2014	2015	2016	2017	2018	2019	2020	2021	2022
SC	2.9	2.7	2.6	2.5	2.4	2.3	2.1	2.0	1.9	1.75

Table 5 presents the descriptive statistics of the indicators.

Table 5. Indicator data description Statistical analysis.

	Variable	Min	Max	Mean	SD
Total	Quay length	3949	126,921	29,714.89	25,957.33
	Berth number	15	1238	217.3	243.2
	10,000 Ton Class Berth number	7	224	64.41	46.84
	Cargo throughput of port	2078	126,134	29,260.64	24,637.02
	Container throughput of port	9	4730	853.47	1088.55
	CO_2 emission	14,817.82	613,273.5	164,683.9	137,513.4
Bohai Sea port group	Quay length	3949	48,211	26,177.74	12,195.15
	Berth number	15	257	128.4375	66.87071
	10,000 Ton Class Berth number	12	131	72.5375	30.82821
	Cargo throughput of port	3730	65,754	34,804.93	16,447.94
	Container throughput of port	39	2567	730.6	699.9425
	CO_2 emission	18,997.46	358,520.9	196,752	93,388.28
Yangtze River Delta port group	Quay length	11,361	126,921	50,856.74	44,364.47
	Berth number	53	1238	451.46	398.7322
	10,000 Ton Class Berth number	7	224	89.7	77.7196
	Cargo throughput of port	4901	126,134	41,503.86	39,115.48
	Container throughput of port	15	4730	1434.8	1640.272
Southeast Coastal port group	CO_2 emission	25,036.52	613,273.5	233,099.5	216,859
	Quay length	23,025	33,274	28,414.95	3054.073
	Berth number	158	207	180.1	12.2942
	10,000 Ton Class Berth number	48	81	68.85	9.852998
	Cargo throughput of port	12,759	30,164	20,158.35	4500.461
	Container throughput of port	198	1243	663.6	392.3723
	CO_2 emission	87,564.38	136,126.7	111,955.4	17,475.09
Pearl River Delta port group	Quay length	5013	56,055	27,764.78	16,000.29
	Berth number	34	621	235.925	187.8966
	10,000 Ton Class Berth number	11	96	49.1	26.95657
	Cargo throughput of port	3155	62,906	24,258.18	20,197.23
	Container throughput of port	88	3004	1236.175	1125.693
	CO_2 emission	16,291.44	336,395.2	136,400.3	112,017.4
Southwest Coastal port group	Quay length	4563	23,577	12,961.98	5504.1
	Berth number	32	177	102.3	44.48117
	10,000 Ton Class Berth number	11	51	29.625	12.04199
	Cargo throughput of port	2078	30,185	12,421.68	7857.261
	Container throughput of port	9	215	84.75	61.18142
	CO_2 emission	14,817.82	170,710.4	69,676.27	44,401.43

Note Data sources: National Bureau of Statistics, China Statistical Yearbook (2014–2023), China Port Yearbook (2014–2023).

4. Empirical Analysis

4.1. Comparisons of the PEE Under Different Models

To validate the applicability of the proposed EBM-WD model, this study conducted a comparative analysis of PEE with several widely recognized DEA models. These models were selected to represent different approaches to efficiency measurement: the radial BCC model, the non-radial SBM model, and the hybrid EBM model. The comparative results are summarized in Tables 6 and 7.

As shown in Tables 6 and 7, the BCC model yields the highest mean PEE score of 0.816, while the SBM model provides the lowest at 0.689. This finding is consistent with the established literature [61]. The BCC model, as a radial measure, tends to overestimate efficiency by projecting inefficient DMUs onto the production frontier without fully accounting for non-radial slacks in inputs and outputs. Conversely, the SBM model, which is entirely based on slacks, often underestimates efficiency by neglecting the proportional relationship between inputs and outputs. The EBM model, as a hybrid approach, correctly addresses the limitations of both radial and non-radial models. Its mean PEE score of 0.776 lies between those of the BCC and SBM models, demonstrating a more balanced and realistic efficiency measure.

Table 6. Comparisons of average PEE from 2013 to 2022 based on different models.

Model	2013	2014	2015	2016	2017	2018	2019	2020	2021	2022
BCC	0.809	0.827	0.834	0.811	0.858	0.853	0.846	0.773	0.785	0.761
SBM	0.652	0.704	0.726	0.673	0.742	0.799	0.705	0.578	0.660	0.648
EBM	0.744	0.836	0.821	0.750	0.807	0.827	0.752	0.703	0.778	0.740
EBM-WD	0.742	0.842	0.810	0.748	0.808	0.816	0.754	0.704	0.781	0.742

Table 7. Comparisons of average PEE between ports under different models.

Port	BCC	SBM	EBM	EBM-WD
Dalian	0.571	0.517	0.679	0.683
Yingkou	0.829	0.738	0.815	0.807
Qinhuangdao	0.814	0.226	0.655	0.652
Tianjin	0.798	0.819	0.829	0.830
Yantai	0.584	0.294	0.583	0.588
Weihai	0.734	0.657	0.534	0.533
Qingdao	1.000	1.000	1.000	1.000
Rizhao	1.000	1.000	1.000	1.000
Shanghai	1.000	1.000	1.000	1.000
Lianyungang	0.819	0.851	0.885	0.844
Ningbo-Zhoushan	1.000	1.000	1.000	1.000
Taizhou	1.000	1.000	1.000	1.000
Wenzhou	0.711	0.302	0.592	0.597
Fuzhou	0.487	0.377	0.545	0.549
Xiamen	0.557	0.534	0.514	0.514
Shantou	0.850	0.786	0.683	0.682
Shenzhen	1.000	1.000	1.000	1.000
Guangzhou	0.999	0.954	0.997	0.997
Zhuhai	0.682	0.373	0.596	0.597
Zhanjiang	0.920	0.615	0.906	0.910
Beihai	0.927	0.825	0.751	0.752
Fangcheng	0.542	0.100	0.381	0.389
Haikou	0.935	0.875	0.900	0.895
Mean	0.816	0.689	0.776	0.775

The mean PEE score of our proposed EBM-WD model (0.775) is very close to that of the conventional EBM model, and both fall within the range defined by the BCC and SBM models. This confirms that the EBM-WD model retains the hybrid strengths of the EBM framework while enhancing its environmental realism. Although the overall efficiency levels are similar, the results are not entirely identical: several ports experience shifts in their PEE rankings under the EBM-WD model, reflecting the differentiated impact of non-uniform abatement factors under the weak disposability assumption. These rank adjustments validate the methodological extension and underscore the necessity of incorporating non-uniform abatement factors to provide a more nuanced and realistic evaluation of port environmental efficiency.

4.2. Overall Analysis of Port Environmental Efficiency

This study utilizes the EBM-WD model to evaluate the environmental efficiency of China's coastal ports from 2013 to 2022. The results are provided in Table 8. As shown in Figure 1a, the overall PEE underwent a fluctuating increase throughout the research timeframe. A similar pattern was also observed in the study by Zhang et al. [38]. From 2013 to 2018, PEE increased gradually, with the mean value rising from 0.742 to 0.816, signifying a 9.973% enhancement. This growth can largely be ascribed to significant efforts in promoting smart ports, port integration, and prioritizing sustainable development [62,63].

Table 8. The PEE values of China's coastal ports from 2013 to 2022.

Port	2013	2014	2015	2016	2017	2018	2019	2020	2021	2022	Mean	SD
Dalian	0.611	0.948	0.687	0.637	1.000	0.658	0.506	0.420	1.000	0.358	0.683	0.220
Yingkou	0.910	0.969	0.855	0.868	1.000	0.944	0.655	0.615	0.596	0.661	0.807	0.150
Qinhuangdao	0.641	0.662	0.676	0.550	0.811	0.719	0.680	0.566	0.580	0.636	0.652	0.074
Tianjin	1.000	1.000	1.000	0.954	0.772	0.696	0.636	0.629	1.000	0.611	0.830	0.167
Yantai	0.525	0.733	0.757	0.603	0.483	0.666	0.545	0.501	0.513	0.553	0.588	0.093
Weihai	1.000	1.000	1.000	1.000	0.233	0.288	0.196	0.170	0.195	0.246	0.533	0.383
Qingdao	1.000	1.000	1.000	1.000	1.000	1.000	1.000	1.000	1.000	1.000	1.000	0.000
Rizhao	1.000	1.000	1.000	1.000	1.000	1.000	1.000	1.000	1.000	1.000	1.000	0.000
Shanghai	1.000	1.000	1.000	1.000	1.000	1.000	1.000	1.000	1.000	1.000	1.000	0.000
Lianyungang	0.797	1.000	0.644	0.646	1.000	1.000	1.000	0.708	0.645	1.000	0.844	0.162
Ningbo-Zhoushan	1.000	1.000	1.000	1.000	1.000	1.000	1.000	1.000	1.000	1.000	1.000	0.000
Taizhou	1.000	1.000	1.000	1.000	1.000	1.000	1.000	1.000	1.000	1.000	1.000	0.000
Wenzhou	0.591	0.586	0.625	0.568	0.571	0.561	0.592	0.603	0.645	0.627	0.597	0.026
Fuzhou	0.362	0.387	0.355	0.364	0.359	1.000	0.465	0.529	1.000	0.670	0.549	0.245
Xiamen	0.478	0.621	0.530	0.495	0.517	0.499	0.494	0.495	0.485	0.531	0.514	0.039
Shantou	0.376	0.477	0.471	0.478	0.772	0.659	0.590	1.000	1.000	1.000	0.682	0.233
Shenzhen	1.000	1.000	1.000	1.000	1.000	1.000	1.000	1.000	1.000	1.000	1.000	0.000
Guangzhou	0.972	1.000	1.000	1.000	1.000	1.000	1.000	1.000	1.000	1.000	0.997	0.009
Zhuhai	0.486	0.555	0.594	0.626	0.692	0.712	0.750	0.585	0.557	0.418	0.597	0.097
Zhanjiang	0.688	0.992	1.000	1.000	1.000	1.000	0.909	0.740	0.771	1.000	0.910	0.120
Beihai	0.238	1.000	1.000	0.364	1.000	1.000	1.000	0.536	0.755	0.630	0.752	0.280
Fangcheng	0.391	0.433	0.436	0.380	0.377	0.374	0.332	0.356	0.412	0.403	0.389	0.031
Haikou	1.000	1.000	1.000	0.680	1.000	1.000	1.000	0.737	0.814	0.721	0.895	0.132
Average	0.742	0.842	0.810	0.748	0.808	0.816	0.754	0.704	0.781	0.742	0.775	0.041

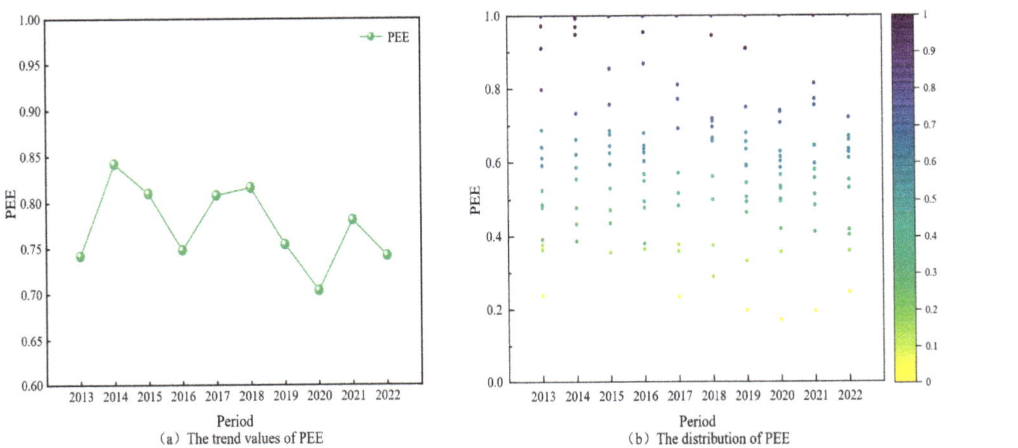

Figure 1. Trends of PEE from 2013 to 2022.

However, from 2018 to 2022, PEE declined, with the average value falling from 0.816 to 0.742, a decrease of 9.069%. This downturn was primarily caused by the COVID-19 pandemic, which disrupted port operations through reduced cargo volumes, temporary shutdowns, and labor shortages [64]. These disruptions constrained the effective utilization of inputs and led to a decline in port efficiency [65]. Notably, these effects are largely temporary, and port efficiency is expected to rebound as trade flows and port operations recover. Furthermore, Figure 1b depicts the distribution of PEE, highlighting a gradual rise in the number of ports at the effective frontier over the research period.

As depicted by Figure 2, the mean PEE across all ports stands at 0.775, reflecting a relatively high overall efficiency level. However, pronounced inter-port disparities are observed, which can be attributed to differences in management practices, technological adoption, and environmental policy implementation. The ports with the top efficiency scores comprise the ports of Qingdao, Shanghai, Taizhou, Rizhao, Ningbo-Zhoushan, and Shenzhen, all attaining an average PEE score of 1, which signifies DEA efficiency. Following closely are Guangzhou Port and Haikou Port, boasting PEE scores of 0.997 and 0.910, respectively, indicating considerable advances in low-carbon port construction. These ports benefit from advanced operational management systems, significant investment in low-carbon technologies, electrification of port equipment, and rigorous enforcement of local environmental regulations, collectively sustaining their high environmental performance.

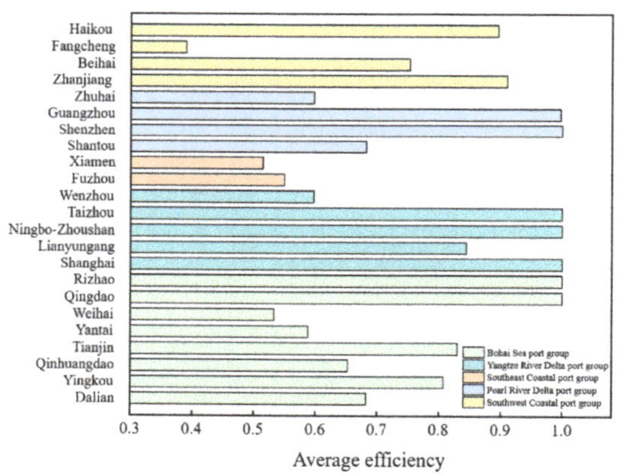

Figure 2. The average PEE from 2013 to 2022.

In contrast, Fangcheng Port, Xiamen Port, and Weihai Port showed unsatisfactory environmental performance, with average PEE values of 0.389, 0.514, and 0.533, respectively. To mitigate their environmental performance gaps, these ports should prioritize low-carbon transition strategies, encompassing the optimization of energy mixes and the implementation of targeted regulatory measures. Furthermore, it can be discerned from the last column of Table 8 that the standard deviations of PEE of Qingdao, Rizhao, Shanghai, Ningbo-Zhoushan, Taizhou, Shenzhen, and Guangzhou are relatively small during 2013–2022, illustrating that their environmental efficiency levels were relatively stable. On the contrary, the ports of Weihai, Beihai, Shantou, Fuzhou, and Dalian showed larger standard deviations in PEE, indicating relative volatility in their environmental efficiency performance.

Table 9 describes the PEE values of five major port groups from 2013 to 2022. During the research period, the Yangtze River Delta port group ranked first in seven years, consistently displaying a superior level. This high performance can be ascribed to the region's advantageous geographical location, superior infrastructure, robust hinterland economy, and efficient management systems. The Pearl River Delta port group experienced steady growth in environmental performance, driven by recent policy and rapid regional economic development, which boosted the environmental efficiency of Shenzhen Port and Guangzhou ports. In addition, the PEE of the Southwest and Southeast coastal port group showed a fluctuating upward tendency with significant volatility, indicating substantial potential for further development. Conversely, the Bohai Sea port group's PEE expressed a downward trend, owing mostly to improper resource allocation and overinvestment, resulting in poor overall environmental performance. These findings extend the research of Chen et al. [66].

Table 9. The PEE of five major group ports from 2013 to 2022.

Port Group	2012	2013	2014	2015	2016	2017	2018	2019	2020	2021	2022	Mean	SD
Bohai Sea port group	0.841	0.836	0.914	0.872	0.826	0.787	0.746	0.652	0.613	0.735	0.633	0.769	0.097
Yangtze River Delta port group	0.881	0.878	0.917	0.854	0.843	0.914	0.912	0.918	0.862	0.858	0.925	0.888	0.029
Southeast Coastal port group	0.411	0.420	0.504	0.443	0.430	0.438	0.750	0.479	0.512	0.742	0.601	0.521	0.118
Pearl River Delta port group	0.743	0.709	0.758	0.766	0.776	0.866	0.843	0.835	0.896	0.889	0.854	0.812	0.061
Southwest Coastal port group	0.565	0.579	0.856	0.859	0.606	0.844	0.844	0.810	0.592	0.688	0.689	0.721	0.117

Figure 3 depicts an evolutionary trend in the national kernel density curve, initially moving rightward before gradually shifting leftward. This dynamic pattern indicates that PEE experienced an initial rise and a subsequent decline over the research duration. The peak of the curve became higher and narrower, signifying a heightened concentration of PEE. This observation underscores an increasingly pronounced agglomeration effect among ports. Furthermore, the curve's distribution is right-skewed, implying that the majority of coastal ports demonstrate relatively high environmental efficiency. Notably, the primary peak shows a unimodal distribution, which suggests that the phenomenon of polarization is not notably evident.

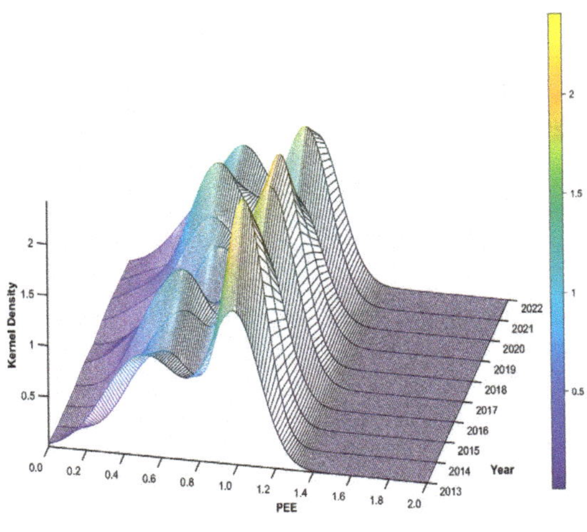

Figure 3. Kernel density of PEE during 2013–2022.

4.3. *Dynamic Characteristics Analysis of Port Environmental Efficiency*

To further examine the dynamic characteristics of PEE throughout the research period, the GM index was utilized. Table 10 lists the GM index and its decomposition items.

The findings demonstrate that the mean annual GM index for China's coastal ports during 2013–2022 is greater than 1, at 1.024, manifesting a slight upward tendency in

TFP of the coastal ports with an average yearly growth rate of 2.4%. Regarding the GM index decomposition, GMEC and GMTC increased by 4.2% and 3.7%, respectively. These findings confirm the joint enhancement of technological level and technical efficiency has contributed to the rise in port productivity, which is in line with Li et al. [28]. The improvement in technical efficiency is the predominant driving element behind TFP growth.

Table 10. Changes in the GM index and decomposition items from 2013 to 2022.

Period	GM	GMEC	GMTC
2013–2014	0.992	1.254	0.864
2014–2015	0.988	0.966	1.037
2015–2016	0.991	0.929	1.103
2016–2017	0.997	1.155	0.915
2017–2018	1.048	1.070	1.035
2018–2019	0.972	0.916	1.094
2019–2020	1.092	0.951	1.148
2020–2021	1.050	1.159	0.957
2021–2022	1.085	0.980	1.176
Mean	1.024	1.042	1.037

Overall, the TFP trend for China's coastal ports shows a pattern of initial decline followed by recovery. The TFP declined during 2013–2017 and 2018–2019 but rebounded between 2019 and 2022, indicating a positive trajectory in the recent development of coastal ports. As regards GMEC and GMTC, their overall fluctuations are pronounced, with alternating states of upward increases and downward decreases. However, there are few synchronized changes between the two, which suggests that the emphasis on technical efficiency and technological level varies during different periods of coastal port development. In agreement with Cui et al. [57], this result demonstrates that the effect of balancing the joint development of both is not ideal, leading to the overall port TFP not improving steadily.

As illustrated in Figure 4, 16 ports, accounting for 70% of the sample, exhibited a GM index surpassing 1, indicating an enhancement in their TFP. Shantou Port had the highest average GM value at 1.250, while Dalian Port had the lowest value at 0.954. Notably, there are pronounced spatial variations in TFP among coastal ports. The ports of Qingdao, Fuzhou, Shanghai, Beihai, Xiamen, and Ningbo-Zhoushan have demonstrated significant improvements in TFP, reflecting effective efforts in enhancing environmental performance through improved operational management and technological innovation. In contrast, ports such as Yingkou, Tianjin, and Shantou have experienced prolonged declines in TFP, suggesting persistent inefficiencies. To improve their environmental performance, these ports should consider optimizing governance mechanisms and accelerating the adoption of green technologies.

The decomposition results further reveal that for ports with increased TFP, such as Shanghai, Ningbo-Zhoushan, Guangzhou, Taizhou, and Shenzhen, GMTC has increased significantly more than GMEC, demonstrating that technological innovation serves as the primary impetus of TFP improvement. Conversely, for ports such as Shantou, Beihai, Fuzhou, Lianyungang, and Zhanjiang, the average growth rate of GMEC notably exceeds GMTC, suggesting that enhancements in technical efficiency predominate. For ports like Wenzhou, Fangcheng, Xiamen, Rizhao, and Qingdao, no marked distinction between GMEC and GMTC was observed, implying that both factors contribute equally to productivity fluctuations. Notably, ports exhibiting GMTC below the unity threshold, such as Qinhuangdao and Yingkou, demonstrate insufficient technological progress. These ports could benchmark against their high-performing counterparts to establish investment priorities and thereby enhance their environmental performance. The slight increase in GMTC for the ports of Zhuhai, Haikou, Weihai, and Tianjin was offset by a widespread decline in GMEC, thereby impeding productivity improvement.

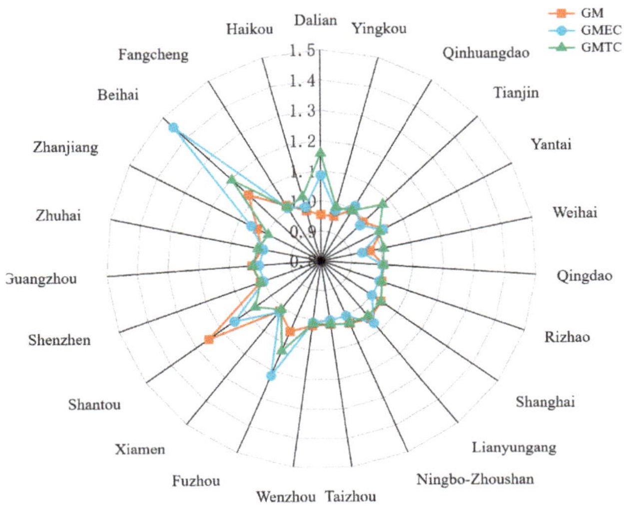

Figure 4. The average GM index and decomposition items radar diagram.

The average GM index for the five major port groups is displayed in Table 11. Across all port groups, with the exception of the Bohai Sea, the average GM index surpassed unity, indicating a positive trend in TFP from 2013 to 2022. This suggests relatively effective resource allocation and operational management, contributing to advancements in environmental performance and reduced emissions. Notably, the Pearl River Delta port group demonstrated the most significant advancement, with its TFP amplified by 7.2%, reflecting considerable gains in enhancing port environmental performance [67]. Conversely, the Bohai Sea port group registered a GM index of 0.987, signifying a marginal decline in TFP and an average yearly reduction of 1.3%. This indicates areas where environmental efficiency and emission reduction strategies may need further attention in this region.

Table 11. Mean values of GM index for five major port groups.

Port Group	2013/2014	2014/2015	2015/2016	2016/2017	2017/2018	2018/2019	2019/2020	2020/2021	2021/2022	Mean
Bohai Sea port group	0.967	1.001	1.012	0.877	1.123	0.857	0.984	1.020	1.043	0.987
Yangtze River Delta port group	0.956	0.951	1.029	1.075	0.991	1.018	0.941	1.100	1.203	1.029
Southeast Coastal port group	1.026	0.974	0.987	1.012	1.012	1.132	1.102	1.048	1.027	1.036
Pearl River Delta port group	1.039	1.008	1.000	1.058	0.955	1.059	1.516	0.967	1.043	1.072
Southwest Coastal port group	1.023	0.994	0.892	1.070	1.081	0.979	1.069	1.131	1.091	1.037

The average GM index and its decomposition items for the five port groups are illustrated in Figure 5. The primary inhibitory factor for the productivity growth in Bohai Sea port group is the decline in technical efficiency. Despite its abundant resources and infrastructure, this port group struggles to harness them effectively, thereby leading to reduced TFP. In contrast, technical efficiency has played a dominant role in driving the TFP growth of the Southwest and Southeast coastal port groups. Furthermore, the TFP growth observed in both the Yangtze River Delta and Pearl River Delta port groups can be attributed to the simultaneous improvement of both technical efficiency and technological progress. This indicates both port groups are both optimizing existing operations and embracing new technologies to boost PEE and cut emissions. In summary, these distinctions underscore the varying developmental stages and strategic priorities of each port group, suggesting that tailored approaches are necessary to further improve TFP and environmental performance within each region.

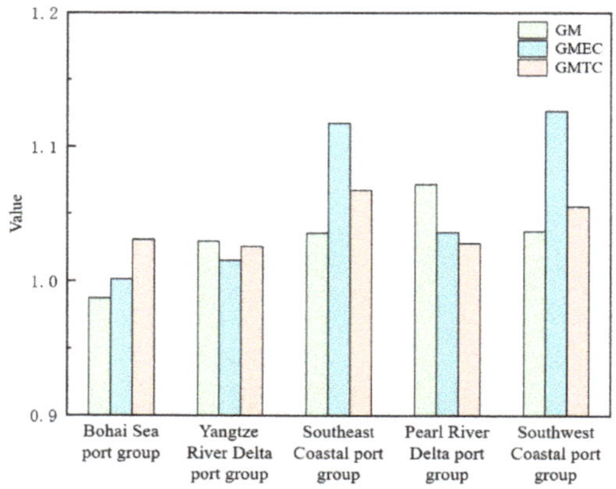

Figure 5. Average values of GM index and its decomposition for five port groups.

4.4. Regional Differences Analysis of Port Environmental Efficiency

To delve deeper into regional disparities in PEE, the Dagum Gini coefficient of PEE was calculated spanning 2013–2022. The findings are summarized in Table 12 and the Gini coefficient trend illustrated in Figure 6.

Table 12. Dagum Gini coefficient and its decomposition.

Period	Total	Intraregional	Interregional	Intensity of Transvariation	Subgroup				
					1	2	3	4	5
2013	0.194	0.035	0.102	0.057	0.120	0.093	0.069	0.208	0.279
2014	0.129	0.021	0.064	0.044	0.066	0.072	0.116	0.166	0.125
2015	0.145	0.025	0.058	0.062	0.084	0.104	0.098	0.162	0.123
2016	0.177	0.033	0.083	0.062	0.119	0.116	0.076	0.156	0.228
2017	0.159	0.032	0.068	0.059	0.175	0.075	0.090	0.083	0.138
2018	0.143	0.030	0.043	0.070	0.158	0.077	0.167	0.097	0.139
2019	0.182	0.034	0.096	0.051	0.204	0.071	0.015	0.111	0.162
2020	0.194	0.037	0.106	0.050	0.233	0.101	0.017	0.087	0.143
2021	0.165	0.036	0.052	0.077	0.207	0.099	0.174	0.093	0.111
2022	0.184	0.036	0.091	0.056	0.218	0.065	0.058	0.128	0.171
Mean	0.167	0.032	0.076	0.059	0.159	0.087	0.088	0.129	0.162

The mean Gini coefficient for the whole PEE was 0.167. The lowest value occurs in 2014, at 0.129, which suggests relatively balanced development in that year. The highest value of 0.194 appeared in 2013 and 2020, demonstrating a pronounced spatial imbalance in PEE during these two years. As indicated in Figure 6, the Gini coefficient for the overall environmental efficiency declines from 0.194 in 2013 to 0.184 in 2022, a decline of around 5.2%; the environmental efficiency gap is gradually narrowing. The trend can be characterized as fluctuating, with environmental efficiency gaps first narrowing and then widening across three distinct periods: 2013–2016, 2016–2020, and 2020–2022. Furthermore, the interregional Gini coefficient showed a similar evolution pattern, spanning 0.043 to 0.106. The Gini coefficient of intensity of transvariation is 0.057 in 2013 and then reaches above 0.070 in 2018 and 2021, separately, showing a certain degree of increase, and then drops to 0.056 at the end. In contrast, the intraregional Gini coefficient remained relatively stable, ranging from 0.020 to 0.040.

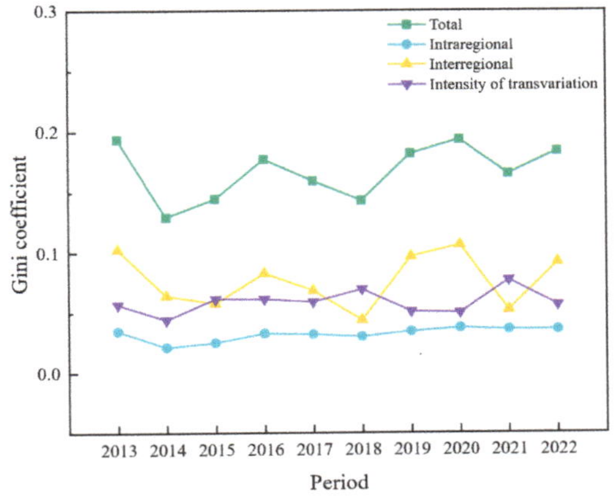

Figure 6. Trend of the Gini coefficient across 2013–2022.

The tendency of the intraregional Gini coefficient is represented in Figure 7. In terms of intraregional differences, the five major port groups have average Gini coefficients that span 0.085 to 0.165. The Yangtze River Delta port group had the smallest average intraregional Gini coefficient at 0.087, indicating minimal environmental efficiency disparities within the group. This balanced development could be attributed to the strong economic foundation, robust infrastructure, policy support, and technological cooperation, which have collectively fostered coordination and cooperation across ports. On the contrary, the Southwest Coastal port group had the highest average intraregional Gini coefficient at 0.162, reflecting significant disparities due to factors such as limited economic development, challenging geographical conditions, and insufficient investment. As depicted in Figure 7, the Bohai Sea port group witnessed a rising Gini coefficient from 2014 onward, highlighting a growing imbalance, while the Yangtze River Delta maintained relatively balanced development. The Pearl River Delta port group demonstrated a general decline in its Gini coefficients, suggesting a narrowing of intraregional environmental efficiency gaps. The same pattern is seen in the southwestern coastal port group. In addition, the Southeast Coastal port group maintained stable equilibrium levels, except for fluctuations in 2019 and 2021.

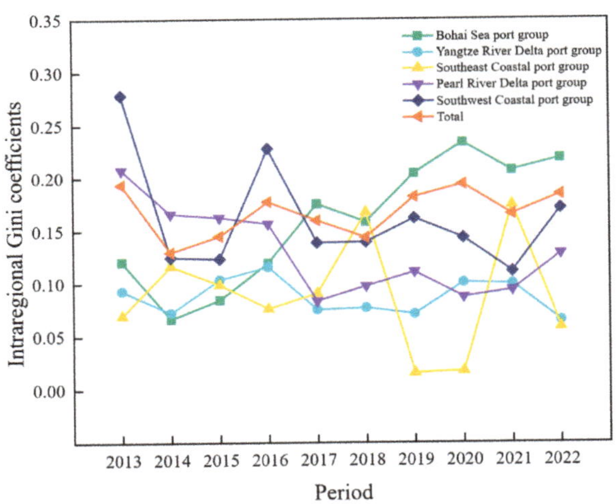

Figure 7. Trend of the intraregional Gini coefficient across 2013–2022.

Figure 8 visualizes the variations in the Gini coefficients among regions. Overall, the shaded areas in Figure 8 are generally decreasing, showing that the polarization of PEE among regions is diminishing during the observation period. It is noteworthy that the Yangtze River Delta and Pearl River Delta port groups have the smallest regional disparity (Gini coefficient = 0.122). This showcases similarities in economic development, port operations, resource allocation, and technological advancements. In contrast, the Southeast Coastal and Yangtze River Delta port groups showed the largest disparity, with a Gini coefficient of 0.274. This disparity may stem from variations in economic conditions, geographic features, investment levels, and policy support. The Gini coefficient among most port groups exhibit an alternating process of first falling and then rising, with an overall declining trend. This result demonstrates that the discrepancies among these port groups are gradually narrowing. In contrast, the Gini coefficient between the Bohai Sea port group and the Yangtze River Delta port group exhibits a slow rising trend, suggesting that the discrepancies between these two groups are widening.

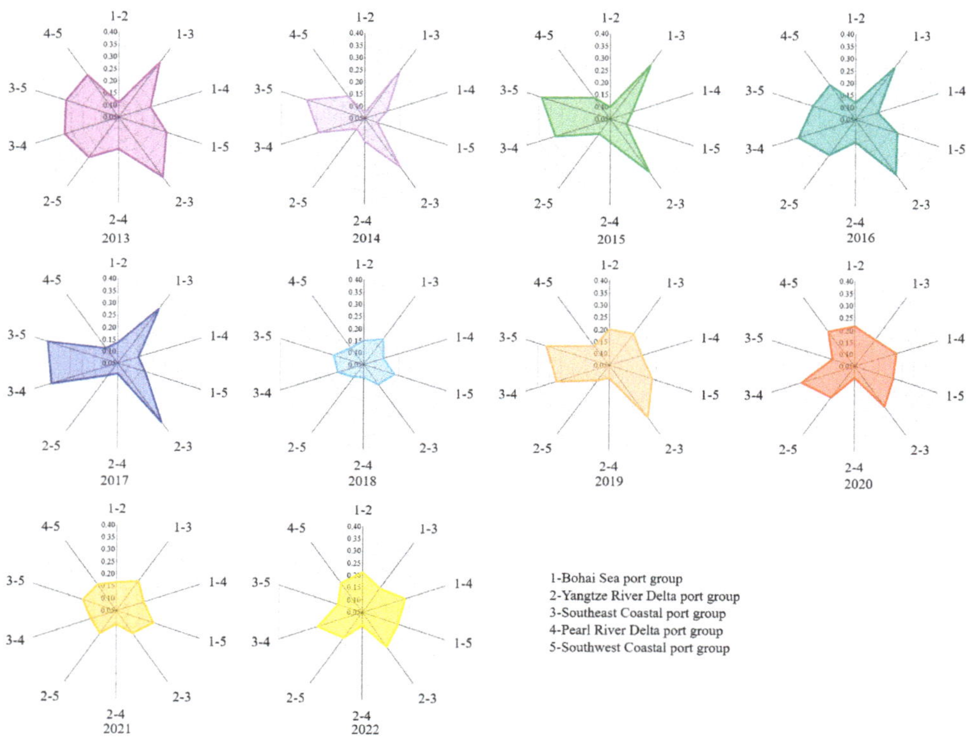

Figure 8. Changes in the interregional Gini coefficient.

Figure 9 illustrates the regional differentiation source and contribution rate. Interregional disparities accounted for the largest share at 45.1%, followed by the intensity of transvariation at 35.8%, and intraregional disparities at 19.1%. This demonstrates that the main sources of environmental efficiency discrepancies among China's coastal ports are interregional disparities and the intensity of transvariation, with interregional disparities being the dominant factor. Over time, the impact of interregional discrepancies has diminished, whereas the contribution from the intensity of transvariation has exhibited a slow volatile trend. This indicates that the contribution of the variation stemming from the overlapping interactions among various regions is progressively increasing. In addition, the contribution rate of intraregional disparities exhibits slight volatility and remains stable.

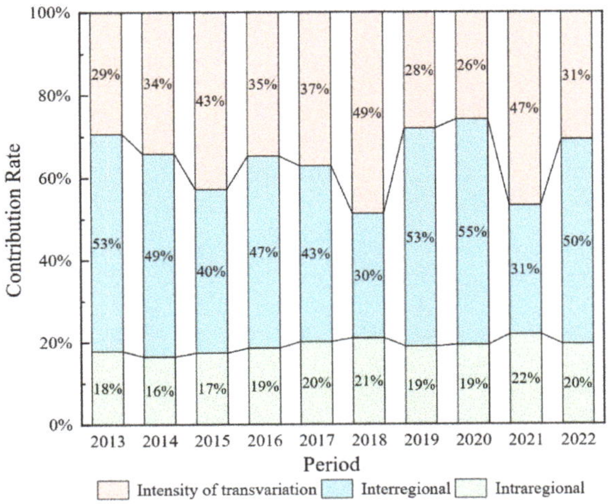

Figure 9. The contribution of Gini coefficient.

5. Conclusions and Policy Implications

5.1. Main Conclusions

Enhancing PEE is crucial for fostering sustainable port development; consequently, refining PEE assessment is imperative, particularly when considering the heterogeneous abatement capabilities of individual ports. This study proposes a modified EBM-Undesirable model that incorporates weak disposability and non-uniform abatement factors to more accurately evaluate PEE. The Global Malmquist index and the Dagum Gini coefficient were further employed to analyze the temporal dynamics and regional discrepancies of PEE. The main conclusions are summarized as follows.

First, PEE exhibited a fluctuating upward trend from 2013 to 2022. Moreover, while the general environmental efficiency level is relatively high, significant heterogeneity persists among individual ports and across port groups. Regionally, the Yangtze River Delta and Pearl River Delta port groups demonstrate considerable environmental performance and green transformation capacity, whereas the Bohai Sea port group experienced a decline.

Second, the dynamic analysis indicates a moderate overall TFP growth, primarily driven by improvements in technical efficiency. However, there remains untapped potential in advancing technological levels. The sources of TFP improvement vary across port groups; some have benefited from technical efficiency gains or technological innovation, while others have experienced decline due to resource misallocation.

Third, the regional disparity analysis uncovers an imbalance in the spatial distribution of PEE. In general, regional development within and between port groups still exhibits discrepancies, albeit the overall gap is gradually narrowing. Interregional discrepancies and the density of transvariation are the predominant factors contributing to these disparities, while intraregional discrepancies play a comparatively minor role.

5.2. Policy Implications

Drawing on the discoveries, several key policy recommendations are proposed to facilitate the high-quality, sustainable development of the port sector. It can proffer valuable insights for policymakers in formulating sustainable port development strategies.

Firstly, the allocation of resources should be optimized. Ports with relatively low efficiency should enhance managerial capabilities and operational practices through targeted policy support and increased capital investment. Resource allocation should not only aim at expanding cargo throughput but also prioritize the implementation of energy-saving and emission-reduction initiatives.

Secondly, differentiated development strategies should be implemented. Ports should move beyond a one-size-fits-all approach and formulate strategies tailored to their specific characteristics, including location, development stage, and operational efficiency. At the port-group level, region-specific measures are recommended: the Bohai Sea group should prioritize improvements in operational management and technical efficiency; the Yangtze River Delta and Pearl River Delta groups should emphasize technological innovation and smart port transformation; and the Southwest and Southeast coastal groups should enhance regional coordination and infrastructure connectivity.

Thirdly, interregional cooperation and integration should be strengthened. The fragmented development of coastal ports has led to redundant competition and inefficient resource use. Establishing interregional cooperation mechanisms would facilitate the sharing of resources, technologies, and managerial expertise. Coordinated planning and policy alignment can improve the overall competitiveness of the port system and promote regional integration.

Lastly, the construction of smart ports should be continuously advanced. Advancing smart port construction is essential for future competitiveness. Ports should accelerate the development of intelligent infrastructure, such as big data platforms, IoT applications, and automated terminals, while also embedding low-carbon technologies, including shore power facilities, renewable energy integration, and intelligent energy management.

5.3. Research Limitations and Further Expansions

While this study provides a comprehensive evaluation of port environmental efficiency and spatio-temporal characteristics, several limitations should be acknowledged. First, the scope of this study is primarily confined to coastal ports. Future research should expand the analysis to inland ports and adopt a national-scale perspective. Second, owing to data availability and measurement constraints, certain input metrics, such as labor and assets, were not incorporated. Moreover, port CO_2 emission data was estimated indirectly from cargo throughput and standardized coal coefficients. Although this approach is widely used, it may not fully reflect port-specific energy structure. Future studies should address this limitation by pursuing collaborative research with port authorities or government agencies to gain access to more detailed, disaggregated data. This will enable a more nuanced and accurate assessment of port environmental performance.

Author Contributions: Conceptualization, J.Z. and G.G.; methodology, J.Z.; formal analysis, J.Z.; software, J.Z.; resources, G.G.; data curation, J.Z.; writing—original draft preparation, J.Z.; writing—review and editing, G.G.; funding acquisition, G.G. All authors have read and agreed to the published version of the manuscript.

Funding: This research was funded by the Social Science Fund of Liaoning Province (Grant Number L23CGL017).

Data Availability Statement: The raw data supporting the conclusions of this article will be made available by the authors on request.

Acknowledgments: The authors sincerely appreciate the editor and anonymous reviewers for their valuable feedback on the paper.

Conflicts of Interest: The authors declare no conflicts of interest.

Abbreviations

The following abbreviations are used in this manuscript:

PEE	Port Environmental Efficiency
DEA	Data Envelopment Analysis
SFA	Stochastic Frontier Analysis
DMU	Decision-Making Unit
DDF	Directional Distance Function
NDEA	Network DEA
SBM	Slack Based Measure
RAM	Range Adjusted Measure
EBM	Epsilon-based Measure
SD	Strong Disposability
WD	Weak Disposability
TFP	Total Factor Productivity
MI	Malmquist Productivity Index
GMI	Global Malmquist Productivity Index
PPS	Production Possibility Set
EC	Efficiency Change
TC	Technological Change

References

1. Chen, X.; Ma, F.; Wu, Y.; Han, B.; Luo, L.; Biancardo, S.A. MFMDepth: MetaFormer-Based Monocular Metric Depth Estimation for Distance Measurement in Ports. *Comput. Ind. Eng.* **2025**, *207*, 111325. [CrossRef]
2. Xiao, G.; Wang, Y.; Wu, R.; Li, J.; Cai, Z. Sustainable Maritime Transport: A Review of Intelligent Shipping Technology and Green Port Construction Applications. *J. Mar. Sci. Eng.* **2024**, *12*, 1728. [CrossRef]
3. Xu, J.; Chen, J.; Li, J.; Shi, J.; Jiang, H.; Mei, Q.; Liu, Y.; Liu, J. Transitioning to Clean Energy in Global Shipping: Country-Based Differences and Industrial Chain Collaboration. *Ocean Coast. Manag.* **2025**, *269*, 107752. [CrossRef]
4. Buonomano, A.; del Papa, G.; Giuzio, G.F.; Palombo, A.; Russo, G. Future Pathways for Decarbonization and Energy Efficiency of Ports: Modelling and Optimization as Sustainable Energy Hubs. *J. Clean. Prod.* **2023**, *420*, 138389. [CrossRef]
5. Cammin, P.; Brüssau, K.; Voß, S. Classifying Maritime Port Emissions Reporting. *Marit. Transp. Res.* **2022**, *3*, 100066. [CrossRef]
6. Zhang, J.; Yang, D.; Luo, M. Port Efficiency Types and Perspectives: A Literature Review. *Transp. Policy* **2024**, *156*, 13–24. [CrossRef]
7. Durán, C.; Derpich, I.; Moreno, F.; Karbassi Yazdi, A.; Tan, Y. Modelling Sustainable Port Operations: Balancing Inputs and Outputs with the Cobb–Douglas Function. *J. Mar. Sci. Eng.* **2024**, *12*, 2285. [CrossRef]
8. Sun, J.; Yuan, Y.; Yang, R.; Ji, X.; Wu, J. Performance Evaluation of Chinese Port Enterprises under Significant Environmental Concerns: An Extended DEA-Based Analysis. *Transp. Policy* **2017**, *60*, 75–86. [CrossRef]
9. Pérez-López, G.; Prior, D.; Zafra-Gómez, J.L. Temporal Scale Efficiency in DEA Panel Data Estimations. An Application to the Solid Waste Disposal Service in Spain. *Omega* **2018**, *76*, 18–27. [CrossRef]
10. Almawsheki, E.S.; Shah, M.Z. Technical Efficiency Analysis of Container Terminals in the Middle Eastern Region. *Asian J. Shipp. Logist.* **2015**, *31*, 477–486. [CrossRef]
11. Hsu, W.K.K.; Huang, S.H.S.; Huynh, N.T. An Assessment of Operating Efficiency for Container Terminals in a Port—An Empirical Study in Kaohsiung Port Using Data Envelopment Analysis. *Res. Transp. Bus. Manag.* **2023**, *46*, 100823. [CrossRef]
12. Tovar, B.; Wall, A. Environmental Efficiency for a Cross-Section of Spanish Port Authorities. *Transp. Res. Part D Transp. Environ.* **2019**, *75*, 170–178. [CrossRef]
13. Chao, S.L.; Yu, M.M.; Hsieh, W.F. Evaluating the Efficiency of Major Container Shipping Companies: A Framework of Dynamic Network DEA with Shared Inputs. *Transp. Res. Part A Policy Pract.* **2018**, *117*, 44–57. [CrossRef]
14. Zhu, J.; Zhao, Y.; Yang, Q.; Jiang, J. Analysis of Regional Differences and Evolution Features for Waterway Transport Efficiency in the Yangtze River Economic Belt Considering Undesired Outputs. *Ocean Coast. Manag.* **2024**, *253*, 107122. [CrossRef]
15. Cui, Q. Environmental Efficiency Measures for Ports: An Application of RAM-Tobit-RAM with Undesirable Outputs. *Marit. Policy Manag.* **2017**, *44*, 551–564. [CrossRef]
16. Yang, L.; Wang, K.-L.; Geng, J.-C. China's Regional Ecological Energy Efficiency and Energy Saving and Pollution Abatement Potentials: An Empirical Analysis Using Epsilon-Based Measure Model. *J. Clean. Prod.* **2018**, *194*, 300–308. [CrossRef]

17. Wu, P.; Wang, Y.; Chiu, Y.; Li, Y.; Lin, T.-Y. Production Efficiency and Geographical Location of Chinese Coal Enterprises—Undesirable EBM DEA. *Resour. Policy* **2019**, *64*, 101527. [CrossRef]
18. Jo, A.H.; Chang, Y.T. Environmental Efficiency of Seaports Correcting Estimation Biases: Weak Disposability and Bootstrapping Method. *Transp. Res. Part D Transp. Environ.* **2023**, *122*, 103884. [CrossRef]
19. Piao, S.; Li, J.; Ting, C. Assessing Regional Environmental Efficiency in China with Distinguishing Weak and Strong Disposability of Undesirable Outputs. *J. Clean. Prod.* **2019**, *227*, 748–759. [CrossRef]
20. Fei, H.; Shi, H.; Pan, X. A Systematic Review of Green Port Evaluation: Methods, Subjects, and Indicators. *J. Mar. Sci. Eng.* **2025**, *13*, 604. [CrossRef]
21. Krmac, E.; Mansouri Kaleibar, M. A Comprehensive Review of Data Envelopment Analysis (DEA) Methodology in Port Efficiency Evaluation. *Marit. Econ. Logist.* **2023**, *25*, 817–881. [CrossRef]
22. Tovar, B.; Wall, A. Are Larger and More Complex Port More Productive? An Analysis of Spanish Port Authorities. *Transp. Res. Part A Policy Pract.* **2019**, *121*, 265–276. [CrossRef]
23. Liu, J.; Wang, X.; Guo, J. Port Efficiency and Its Influencing Factors in the Context of Pilot Free Trade Zones. *Transp. Policy* **2021**, *105*, 67–79. [CrossRef]
24. Agüero-Tobar, M.A.; González-Araya, M.C.; González-Ramírez, R.G. Assessment of Maritime Operations Efficiency and Its Economic Impact Based on Data Envelopment Analysis: A Case Study of Chilean Ports. *Res. Transp. Bus. Manag.* **2023**, *46*, 100821. [CrossRef]
25. Martínez-Moya, J.; Mestre-Alcover, A.; Sala-Garrido, R.; Furió-Pruñonosa, S. Are Transhipment Ports More Efficient in the Mediterranean Sea? Analysing the Role of Time at Ports Using DEA Metafrontier Approach. *J. Transp. Geogr.* **2024**, *116*, 103866. [CrossRef]
26. Kong, Y.; Tian, X.; Sun, J.; Zhou, H. Charting Sustainable Vistas: Analysis of Internal and External Sustainability Performance of Chinese Ports. *Transp. Policy* **2025**, *161*, 31–43. [CrossRef]
27. Chin, A.T.H.; Low, J.M.W. Port Performance in Asia: Does Production Efficiency Imply Environmental Efficiency? *Transp. Res. Part D Transp. Environ.* **2010**, *15*, 483–488. [CrossRef]
28. Li, Y.; Li, J.; Gong, Y.; Wei, F.; Huang, Q. CO_2 Emission Performance Evaluation of Chinese Port Enterprises: A Modified Meta-Frontier Non-Radial Directional Distance Function Approach. *Transp. Res. Part D Transp. Environ.* **2020**, *89*, 102605. [CrossRef]
29. Wang, Z.; Wu, X.; Guo, J.; Wei, G.; Dooling, T.A. Efficiency Evaluation and PM Emission Reallocation of China Ports Based on Improved DEA Models. *Transp. Res. Part D Transp. Environ.* **2020**, *82*, 102317. [CrossRef]
30. Castellano, R.; Ferretti, M.; Musella, G.; Risitano, M. Evaluating the Economic and Environmental Efficiency of Ports: Evidence from Italy. *J. Clean. Prod.* **2020**, *271*, 122560. [CrossRef]
31. Wang, L.; Zhou, Z.; Yang, Y.; Wu, J. Green Efficiency Evaluation and Improvement of Chinese Ports: A Cross-Efficiency Model. *Transp. Res. Part D Transp. Environ.* **2020**, *88*, 102590. [CrossRef]
32. Sun, J.; Zhao, R. Analysis of Environmental Performance and Interactivity of Ports and Regions. *Ocean Coast. Manag.* **2023**, *239*, 106602. [CrossRef]
33. Lin, X.; Jing, X.; Cheng, F.; Wang, M. Exploration of Port Environmental Efficiency Measurement and Influential Factors in the Yangtze River Delta Pilot Free Trade Zone. *Mar. Pollut. Bull.* **2024**, *206*, 116766. [CrossRef]
34. Nguyen, P.N.; Woo, S.H.; Beresford, A.; Pettit, S. Competition, Market Concentration, and Relative Efficiency of Major Container Ports in Southeast Asia. *J. Transp. Geogr.* **2020**, *83*, 102653. [CrossRef]
35. Kong, Y.; Liu, J. Sustainable Port Cities with Coupling Coordination and Environmental Efficiency. *Ocean Coast. Manag.* **2021**, *205*, 105534. [CrossRef]
36. Zhang, W.; Wu, X.; Guo, J. CO_2 Emission Efficiency Analysis of Rail-Water Intermodal Transport: A Novel Network DEA Model. *J. Mar. Sci. Eng.* **2022**, *10*, 1200. [CrossRef]
37. Jiang, X.; Zhan, Z.; Ren, W.; Tao, X.; Zheng, S.; Luo, M. Balancing Economic and Environmental Strategies in Regional Hinterland Transport: A Dynamic Network Cross Efficiency Analysis. *Transp. Policy* **2025**, *162*, 128–154. [CrossRef]
38. Zhang, W.; Chen, W.; Wu, X.; Chen, J.; Zhang, Z. Evaluation and Spatiotemporal Analysis of Low-Carbon Efficiency of the "Ship-Port" System. *Marit. Policy Manag.* **2025**, 1–33. [CrossRef]
39. Faere, R.; Grosskopf, S.; Lovell, C.K.; Pasurka, C. Multilateral Productivity Comparisons When Some Outputs Are Undesirable: A Nonparametric Approach. *Rev. Econ. Stat.* **1989**, *71*, 90–98. [CrossRef]
40. Färe, R.; Grosskopf, S. Nonparametric Productivity Analysis with Undesirable Outputs: Comment. *Am. J. Agric. Econ.* **2003**, *85*, 1070–1074. [CrossRef]
41. Kuosmanen, T. Weak Disposability in Nonparametric Production Analysis with Undesirable Outputs. *Am. J. Agric. Econ.* **2005**, *87*, 1077–1082. [CrossRef]
42. Chang, Y.T.; Park, H.; Jeong, J.; Lee, J. Evaluating Economic and Environmental Efficiency of Global Airlines: A SBM-DEA Approach. *Transp. Res. Part D Transp. Environ.* **2014**, *27*, 46–50. [CrossRef]

43. Taleb, M.; Khalid, R.; Emrouznejad, A.; Ramli, R. Environmental Efficiency under Weak Disposability: An Improved Super Efficiency Data Envelopment Analysis Model with Application for Assessment of Port Operations Considering NetZero. *Environ. Dev. Sustain.* **2023**, *25*, 6627–6656. [CrossRef]
44. Gu, G.; Zhang, J.; Pan, X. A Meta-Frontier Approach to Evaluating the Environmental Efficiency of Coastal Ports: Implications for Port Sustainability. *J. Mar. Sci. Eng.* **2025**, *13*, 1272. [CrossRef]
45. Yu, M.M.; See, K.F. Evaluating the Efficiency of Global Airlines: A New Weighted SBM-NDEA Approach with Non-Uniform Abatement Factor. *Res. Transp. Bus. Manag.* **2023**, *46*, 100860. [CrossRef]
46. Wang, J.; Wang, S.; Li, S.; Cai, Q.; Gao, S. Evaluating the Energy-Environment Efficiency and Its Determinants in Guangdong Using a Slack-Based Measure with Environmental Undesirable Outputs and Panel Data Model. *Sci. Total Environ.* **2019**, *663*, 878–888. [CrossRef] [PubMed]
47. Tone, K. A Slacks-Based Measure of Efficiency in Data Envelopment Analysis. *Eur. J. Oper. Res.* **2001**, *130*, 498–509. [CrossRef]
48. Tone, K.; Tsutsui, M. An Epsilon-Based Measure of Efficiency in DEA—A Third Pole of Technical Efficiency. *Eur. J. Oper. Res.* **2010**, *207*, 1554–1563. [CrossRef]
49. Janmontree, J.; Zadek, H.; Ransikarbum, K. Analyzing Solar Location for Green Hydrogen Using Multi-Criteria Decision Analysis. *Renew. Sustain. Energy Rev.* **2025**, *209*, 115102. [CrossRef]
50. Wu, D.; Wang, Y.; Qian, W. Efficiency Evaluation and Dynamic Evolution of China's Regional Green Economy: A Method Based on the Super-PEBM Model and DEA Window Analysis. *J. Clean. Prod.* **2020**, *264*, 121630. [CrossRef]
51. Färe, R.; Grosskopf, S.; Roos, P. Productivity and Quality Changes in Swedish Pharmacies. *Int. J. Prod. Econ.* **1995**, *39*, 137–144. [CrossRef]
52. Pastor, J.T.; Lovell, C.K. A Global Malmquist Productivity Index. *Econ. Lett.* **2005**, *88*, 266–271. [CrossRef]
53. Dagum, C. A New Approach to the Decomposition of the Gini Income Inequality Ratio. *Empir. Econ.* **1997**, *22*, 515–531. [CrossRef]
54. Gao, K.; Yuan, Y. Spatiotemporal Pattern Assessment of China's Industrial Green Productivity and Its Spatial Drivers: Evidence from City-Level Data over 2000–2017. *Appl. Energy* **2022**, *307*, 118248. [CrossRef]
55. Li, X.; Li, F.; Zhao, N.; Zhu, Q. Measuring Environmental Sustainability Performance of Freight Transportation Seaports in China: A Data Envelopment Analysis Approach Based on the Closest Targets. *Expert Syst.* **2020**, *37*, e12334. [CrossRef]
56. Huang, T.; Chen, Z.; Wang, S.; Jiang, D. Efficiency Evaluation of Key Ports along the 21st-Century Maritime Silk Road Based on the DEA–SCOR Model. *Marit. Policy Manag.* **2021**, *48*, 378–390. [CrossRef]
57. Cui, L.; Chen, L.; Yang, X. Evaluation and Analysis of Green Efficiency of China's Coastal Ports under the "Double Carbon" Goal: Two Improved DEA Models with CO_2 Emissions. *Environ. Dev. Sustain.* **2023**, *26*, 29099–29128. [CrossRef]
58. Wang, L.; Li, Y. Estimation Methods and Reduction Strategies of Port Carbon Emissions—What Literatures Say? *Mar. Pollut. Bull.* **2023**, *195*, 115451. [CrossRef]
59. Xu, X.; Liu, X.; Feng, L.; Yap, W.Y.; Feng, H. Emission Estimation and Spatiotemporal Distribution of Passenger Ships Using Multi-Source Data: A Case from Zhoushan (China). *J. Mar. Sci. Eng.* **2025**, *13*, 168. [CrossRef]
60. Kuang, H.; Zhu, J.; Bai, Z. Study on the Interaction between Green Competitiveness of Coastal Ports and Hinterland Economy. *Sustainability* **2023**, *15*, 1364. [CrossRef]
61. Mo, R.; Huang, H.; Zhang, J.; Liu, Y.; Zhao, X. Green Economic Efficiency and Productivity for Sustainable Development in China: A Ray Epsilon-Based Measure Model Analysis. *Environ. Sci. Policy* **2024**, *160*, 103860. [CrossRef]
62. Zhou, Y.; Li, Z.; Duan, W.; Deng, Z. The Impact of Provincial Port Integration on Port Efficiency: Empirical Evidence from China's Coastal Provinces. *J. Transp. Geogr.* **2023**, *108*, 103574. [CrossRef]
63. Zhang, C.; Hu, H.; Yang, Y.; Wang, N. Optimizing the Portfolio of Smart Technologies to Maximize Port Carbon Emission Efficiency: A Configurational Theory Approach. *Marit. Policy Manag.* **2025**, 1–39. [CrossRef]
64. Nguyen, P.N.; Kim, H. The Effects of the COVID-19 Pandemic on Connectivity, Operational Efficiency, and Resilience of Major Container Ports in Southeast Asia. *J. Transp. Geogr.* **2024**, *116*, 103835. [CrossRef]
65. Xu, L.; Yang, S.; Chen, J.; Shi, J. The Effect of COVID-19 Pandemic on Port Performance: Evidence from China. *Ocean Coast. Manag.* **2021**, *209*, 105660. [CrossRef]
66. Chen, J.R.; Choi, J.W.; Seo, Y.J. Environmental Efficiency Assessment of Coastal Ports in China: Implications for Sustainable Port Management. *Mar. Pollut. Bull.* **2025**, *211*, 117436. [CrossRef] [PubMed]
67. Liu, P.; Pan, Q.; Zhu, B. Dynamic Evolution of Green Total Factor Productivity Growth of China's Coastal Ports. *Marit. Policy Manag.* **2024**, *51*, 1535–1557. [CrossRef]

Journal of
Marine Science and Engineering

MDPI

Article

Two-Dimensional Thompson Sampling for Joint Beam and Power Control for Uplink Maritime Communications

Kyeong Jea Lee [1], Joo-Hyun Jo [2], Sungyoon Cho [3], Ki-Won Kwon [3] and DongKu Kim [1,*]

[1] School of Electrical and Electronic Engineering, Yonsei University, Seoul 03722, Republic of Korea; kjlee92@yonsei.ac.kr
[2] 6G Connected Mobility Standard TP, LG Electronics, Seoul 08826, Republic of Korea; joohyun.jo@lge.com
[3] Smart Network Research Center, Korea Electronics Technology Institute (KETI), Seoul 02456, Republic of Korea; sycho@keti.re.kr (S.C.); kwonkw@keti.re.kr (K.-W.K.)
* Correspondence: dkkim@yonsei.ac.kr

Abstract

In a cellular maritime communication system, ocean buoys are essential to enable environmental monitoring, offshore platform management, and disaster response. Therefore, energy-efficient transmission from the buoys is a key requirement to prolong their operational time. A fixed uplink beamforming can be considered to save energy by leveraging its beam gain while managing the target link reliability. However, the dynamic condition of ocean waves causes buoys' random orientation, leading to frequent misalignment of their predefined beam direction aimed at the base station, which degrades both the link reliability and energy efficiency. To address this challenge, we propose a wave-adaptive beamforming framework to satisfy data-rate demands within limited power budgets. This strategy targets scenarios where sea state information is unavailable, such as in network-assisted systems. We propose a Two-Dimensional Thompson Sampling (2DTS) scheme that jointly selects beamwidth and transmit power to satisfy the target-rate constraint with minimal power consumption and thus achieve maximal energy efficiency. This adaptive learning approach effectively balances exploration and exploitation, enabling efficient operation in uncertain and changing sea conditions. In simulation, under a moderate sea state, 2DTS achieves an energy efficiency of 1.26×10^4 bps/Hz/J at round 600, which is 73.7% of the ideal (1.71×10^4), and yield gains of 96.9% and 447.8% over exploration-based TS and conventional TS, respectively. Under a harsh sea state, 2DTS attains 3.09×10^4 bps/Hz/J (85.6% of the ideal 3.61×10^4), outperforming the exploration-based and conventional TS by 83.9% and 113.1%, respectively. The simulation results demonstrate that the strategy enhances energy efficiency, confirming its practicality for maritime communication systems constrained by limited power budgets.

Keywords: maritime communication; 2DTS; beam management; energy efficiency

1. Introduction

The growing demand for advanced maritime applications, such as surveillance, environmental monitoring, offshore platform management, and disaster response services, requires buoy-based ocean communication networks [1,2]. In particular, lighthouses and maritime buoys now play a critical role, going beyond simple signaling to transmitting real-time environmental and situational data including wave height, surrounding visibility, and environmental video feeds. These high data-rate requirements necessitate more efficient wireless transmission strategies than the low-rate, omnidirectional links traditionally used for basic telemetry (e.g., wave height, air temperature).

Unlike terrestrial networks, maritime communication systems must deal with a highly dynamic ocean environment. Waves, wind and currents induce buoy motions that tilt, heave and rotate, increasing the probability of outage and energy consumption [3]. Although these effects are governed by the sea state, real-time sea state information is often unavailable. As a result, static power allocation schemes often result in inefficiencies in spectral and energy resources [4]. To address this, recent studies have explored adaptive techniques such as beamforming and multi-hop relaying to strengthen maritime links [5,6]. While directional beamforming can focus energy toward the intended receiver [7], most existing methods presume prior knowledge of sea conditions at the buoys, limiting their utility in the use cases when the sea conditions or channel state information are not available [8,9]. Buoys generally possess only partial and delayed knowledge of the prevailing sea state for uplink beamforming, whereas the base station (BS) can infer sea-state dynamics more reliably from satellite-based, wide-area forecasts. Moreover, it is noted that the instantaneous beamforming that adapts to the changing orientation of the buoy requires channel state information feedback from the base station, where cellular maritime communication such as Long-Term Evolution for Maritime (LTE-M) lacks a Multiple-Input Multiple-Output (MIMO) framework.

To address this challenge with respect to the uncertainty of the sea state and its related feedback channel unavailability in the current marine cellular system, energy-efficient uplink transmission becomes a crucial design goal, since maritime buoys operate under limited battery capacity and must maintain reliable connectivity under harsh sea-state variations. Recent studies have emphasized the importance of optimizing power efficiency and robustness in underwater and backscatter communication scenarios, such as in the Anti-Jamming Colonel Blotto Game for Underwater Acoustic Backscatter Communication [10]. Motivated by these challenges, we aim to minimize power consumption while satisfying the target rate constraint in uncertain sea-state environments. Therefore, we propose an adaptive beam and power control method based on Thompson Sampling (TS)-based reinforcement learning [11] without knowledge on the sea state directly from the base station. Because the base station cannot have knowledge on the sea state directly, we exploit the feedback information to implicitly infer the current sea state. To optimize beamwidth and power allocation under uncertain conditions, we employ the Thompson Sampling (TS) framework, a Bayesian algorithm widely used in multi-armed bandit (MAB) problems [12–14]. TS is known to efficiently balance exploration and exploitation by sampling from the posterior belief distributions of each action. In particular, the posterior is updated based on binary feedback success or failure, while allowing the algorithm to gradually converge to the most promising actions. TS has been successfully applied in areas such as online advertising, recommendation systems, and clinical trials. In communication systems, it has been used for channel selection, interference management, and power control, including underwater acoustic communications and dynamic spectrum access [14,15].

Unfortunately, most of the above-mentioned applications focus on maximizing performance metrics like throughput or reliability, with limited attention paid to energy efficiency [16]. Unlike these existing applications, in our scenario, beamforming performance is highly sensitive to sea-state conditions, where dynamic wave motion causes variations in signal alignment and channel quality. Alongside strict power constraints at buoys, this creates a multi-dimensional optimization problem involving both beamwidth and transmit power, which makes standard TS not suitable for our problem. This limitation arises because conventional Thompson Sampling (TS) is designed to maximize the expected reward of a single action rather than to handle constrained or multi-dimensional objectives. If the reward is defined as the achieved rate, TS tends to converge toward high-power actions, since higher power monotonically increases the rate. Even when the reward is represented as a binary indicator of meeting the target rate, the algorithm still favors the high power that maximizes success probability, without accounting for power minimization. By contrast, our problem follows an objective: minimize transmit power subject to satisfying a predefined target rate,which cannot be directly encoded in the standard TS formulation. Moreover, unlike traditional TS that optimizes a single action, our setting requires joint exploration and exploitation over two coupled decision variables, beamwidth and transmit power, which form a two-dimensional action space. To address this challenge, we develop a Two-Dimensional Thompson Sampling (2DTS) approach that maintains beta posteriors for each beamwidth–power pair and applies a dominance-aware selection rule. Among all feasible pairs whose sampled success probabilities exceed the target reliability, 2DTS selects the configuration with the lowest power, thereby ensuring energy-efficient yet reliable transmission under uncertain sea-state conditions. The goal of 2DTS is to jointly select the optimal beamwidth and the minimum transmission power required to meet a predefined target data rate. This is achieved by updating the shape parameters α and β of the posterior beta distribution for each beamwidth and power pair selection actions, using binary feedback from the base station indicating whether the target rate was successfully achieved. This approach enables adaptive learning that accounts for underlying sea-state variations while minimizing power consumption. By integrating 2DTS into our framework, we provide an efficient solution that balances exploration and exploitation, ensuring reliable and energy-efficient communication in dynamic and uncertain maritime environments.

The paper is organized as follows. Section 2 models wave-induced fluctuations in the uplink transmit beam gain by incorporating buoy motion and sea-surface dynamics. Section 3 presents the Two-Dimensional Thompson Sampling (2DTS) algorithm, which jointly selects beamwidth and transmit power to meet the target-rate constraint with minimal power, thereby maximizing energy efficiency. Section 4 reports simulation results that validate the modeling and quantify the gains of the proposed approach. Finally, Section 5 concludes the paper.

2. System Model

As illustrated in Figure 1, we consider an uplink maritime communication scenario where a buoy equipped with N transmit antennas communicates with a single-antenna base station (BS). The buoy's location and orientation are assumed to be known, so the buoy employs predefined fixed transmit beamforming aiming at the base station without dynamic adjustments. The buoy transmits the scalar information symbol $x \sim \mathcal{CN}(0, P)$ to the BS through a linear beamforming vector $\mathbf{f} \in \mathbb{C}^{N\times 1}$. Accordingly, the received signal, denoted as y at the BS, is represented as

$$y = \mathbf{h}\mathbf{f}x + n, \tag{1}$$

where $\mathbf{h} \in \mathbb{C}^{1\times N}$ is the channel between the buoy and BS, and $n \in \mathbb{C}$ follows an additive white Gaussian noise (AWGN) distribution [17]. For simplicity, $\mathbf{f}$ is chosen from a predefined vertical beam vector set $\mathcal{F} = \{\mathbf{f}_1, \cdots, \mathbf{f}_S\}$, where they are the arrays steering to BS with different vertical beamwidths. It is assumed that the uplink beam vectors are designed to be horizontally isotropic and vertically directional because the buoy can frequently be rotating due to the ocean current. $\mathbf{f}_1$ indicates the beamforming vector with beamwidth $\theta_1 = \frac{\pi}{4}$, and subsequent beamwidths are progressively halved, given by

$$\theta_s = \frac{\pi}{2^{s+1}}, \quad s = 1,2,\ldots,5, \tag{2}$$

corresponding to $\frac{\pi}{4}, \frac{\pi}{8}, \frac{\pi}{16}, \frac{\pi}{32}$, and $\frac{\pi}{64}$ [18]. For modeling the channel $\mathbf{h}$, following [19], we assume

$$\mathbf{h} = \sqrt{\frac{K}{K+1}}\mathbf{h}_{\mathsf{LoS}} + \sqrt{\frac{1}{K+1}}\mathbf{h}_{\mathsf{NLoS}}, \tag{3}$$

where $\mathbf{h}_{\mathsf{LoS}}$ and $\mathbf{h}_{\mathsf{NLoS}}$ represent the Line-of-Sight (LoS) and the Non Line-of-Sight (NLoS) components of the channel, respectively. For the NLoS component, we assume that each element in $\mathbf{h}_{\mathsf{NLoS}}$ is drawn from $\mathcal{CN}(0,1)$. This part captures the scattered multipath components. The Rician factor K represents the ratio of the power in the LoS component to the power in the NLoS component. The LoS component, $\mathbf{h}_{\mathsf{LoS}}$, reflects the deterministic nature of the direct transmission path, which is particularly dominant in the maritime environment due to the lack of obstacles between the BS and the buoy [20]. Although the buoy is placed at known locations, the sea surface moves dynamically and causes buoy orientation changes, resulting in misalignment of the direction of the transmit array of the buoy and degradation of the gain at BS [21].

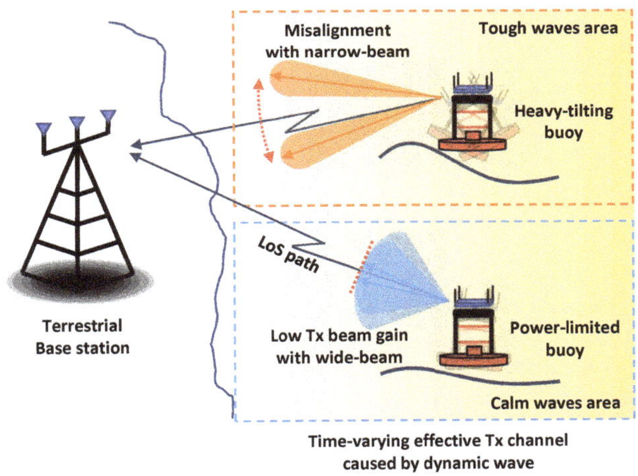

Figure 1. System model of the maritime buoy to base station uplink.

Statistical Modeling of Ocean Beam Channel

Incorporating the dynamic ocean environment into the channel model, the modified channel is expressed as

$$\mathbf{h}_{\mathsf{dyn}} = \sqrt{\frac{K}{K+1}}\hat{\mathbf{h}}_{\mathsf{LoS}} + \sqrt{\frac{1}{K+1}}\mathbf{h}_{\mathsf{NLoS}}, \tag{4}$$

where

$$\hat{\mathbf{h}}_{\mathsf{LoS}}[i] = e^{j\pi(i-1)\sin(\Theta_{\mathrm{AoD}}+\Omega_0)} \tag{5}$$

where Θ_{AoD} is the angle of departure at the buoy and Ω_0 is derived from the buoy's angular displacement caused by ocean wave dynamics modeled via the ISSC spectrum [3]. To account for the impact of the beamwidth on the overall channel performance, the effective channel gain is defined as the projection of the channel vector onto the beamforming vector. Specifically, the effective channel gain H_G is expressed as follows:

$$H_G = ||\mathbf{h}_{\mathsf{dyn}}\mathbf{f}||^2, \tag{6}$$

where H_G denotes the effective channel gain after considering the effects of the dynamic ocean environment, and $\mathbf{f}$ represents the beamforming vector corresponding to a specific beamwidth [22]. In high sea states, large wave heights induce substantial angular variations in the buoy, thereby increasing the probability of steering misalignment, particularly when employing narrower transmit beams. While a narrower beam can provide larger array gain, its susceptibility to misalignment may severely degrade system performance under adverse conditions. Consequently, the buoy must adapt its beam vector to preserve robustness against wave-induced motion in the current sea state. The optimal beamwidths and the corresponding minimum transmit powers required to meet the target across different sea states are summarized in Table 1.

Table 1. Optimal beamwidth and minimum power required per sea state.

Sea State	Optimal Beamwidth (°)	Minimum Power (W)
1	2.81	0.0013
2	5.62	0.0031
3	11.25	0.0106
4	11.25	0.0304
5	22.50	0.0416
6	22.50	0.0563
7	22.50	0.0779
8	45.00	0.1373
9	45.00	0.1625

Note: Values are computed for a target rate of 6 bps/Hz.

3. Joint Beamwidth and Power Selection Using 2DTS

This section introduces a joint beamwidth and power selection strategy for maritime communication using 2DTS. In scenarios where the sea-state information is known at the buoy, effective adaptation is crucial. Unlike the Upper Confidence Bound (UCB) algorithm, which struggles under dynamic conditions due to its reliance on confidence intervals, TS achieves a better balance between exploration and exploitation through probabilistic sampling. The proposed approach extends standard TS by jointly selecting the best beamwidth and minimum transmit power, enabling network-assisted real-time adaptation to varying maritime environments. Simulation shows that 2DTS outperforms standard TS, leading to lower power consumption, higher reliability, and robust operation under dynamic and uncertain sea states.

3.1. Standard Thompson Sampling

Thompson Sampling (TS) is a popular Bayesian decision-making algorithm tailored for uncertain environments, particularly effective in MAB problems. The fundamental objective in a MAB setting is to maximize cumulative rewards by choosing among K possible actions (or arms), each with an unknown reward distribution. TS accomplishes this by sampling from posterior distributions of the reward means and updating these distributions as new data (rewards) are observed. Over time, it allocates resources to actions believed to yield higher returns while still periodically exploring other actions to gather new information. Below, we provide an expanded description of TS, emphasizing the Bayesian structure and adaptive update rules.

3.1.1. Bayesian Multi-Armed Bandit Formulation

Consider a MAB with K arms, where each arm a_k is associated with an unknown reward distribution. Let μ_k denote the (unknown) mean reward of arm k. The agent, or decision-maker, seeks to maximize the cumulative reward obtained by sequentially selecting arms over T rounds. At each round t, the agent selects an arm a_{k^*}, observes a reward r_{k^*} drawn from the distribution with mean μ_{k^*}, and subsequently updates its posterior belief about μ_{k^*} based on the observed reward. The central challenge lies in balancing exploration of insufficiently understood arms with exploitation of arms already identified as yielding high rewards.

3.1.2. Bayesian Perspective and Thompson Sampling Logic

From a Bayesian standpoint, each mean reward μ_k is modeled with a prior distribution that reflects initial uncertainty. As rewards are observed, Bayes' rule is applied to update this belief, yielding a posterior distribution $P(\mu_k \mid \text{data})$. Thompson Sampling exploits this posterior by sampling a candidate value of μ_k for each arm and selecting the arm with the highest sampled value. This naturally balances exploration and exploitation, as arms with uncertain but potentially high means can still be chosen. The procedure can be described step by step as follows:

(1) Initialization.

Define a prior distribution $P(\mu_k)$ for each arm $k \in \{1, \ldots, K\}$. The choice of prior depends on the reward type:

- For Bernoulli rewards (i.e., rewards are either 0 or 1), a beta distribution is typically selected:

$$\mu_k \sim \text{Beta}(\alpha_k, \beta_k), \tag{7}$$

  where α_k and β_k are positive hyperparameters. This is because the beta distribution is the conjugate prior for the Bernoulli likelihood, ensuring a simple form of posterior updates.
- For Gaussian rewards, a normal prior is often chosen:

$$\mu_k \sim \mathcal{N}(\mu_{0,k}, \sigma_{0,k}^2), \tag{8}$$

  where $\mu_{0,k}$ is the initial mean, and $\sigma_{0,k}^2$ is the initial variance. The normal distribution is a conjugate prior for the Gaussian likelihood, which keeps posterior updates tractable.

These priors capture the agent's initial beliefs about the reward means before any data is observed.

(2) Sampling.

In each round t, for each arm a_k, sample a reward estimate $\hat{\mu}_k$ from the posterior distribution $P(\mu_k|\text{data})$. This posterior reflects the agent's updated knowledge after previous selections. Concretely,

- Bernoulli rewards (Beta prior):

$$\hat{\mu}_k \sim \text{Beta}(\alpha_k + S_k,\ \beta_k + F_k), \tag{9}$$

  where S_k is the total number of observed "successes" (reward = 1) and F_k the total number of observed "failures" (reward = 0) for arm k. Each time arm k is played, the parameters α_k, β_k are updated accordingly.
- Gaussian rewards (Normal prior):

$$\hat{\mu}_k \sim \mathcal{N}\left(\frac{\mu_{0,k}/\sigma_{0,k}^2 + \sum r_k/\sigma^2}{1/\sigma_{0,k}^2 + n_k/\sigma^2},\ \frac{1}{1/\sigma_{0,k}^2 + n_k/\sigma^2}\right). \tag{10}$$

Here, $\sum r_k$ is the sum of all rewards observed from arm k, n_k is the number of times arm k has been selected, and σ^2 is the known variance of the underlying reward distribution.

By drawing a sample $\hat{\mu}_k$ from each posterior, TS naturally incorporates both the estimated mean and associated uncertainty.

(3) Action Selection.

After sampling one candidate $\hat{\mu}_k$ for each arm k, the algorithm chooses the arm

$$a_{k^*} = \arg\max_k \hat{\mu}_k. \tag{11}$$

Thus, the selected arm is the one that this particular sample indicates has the largest mean reward. Over many rounds, arms with higher posterior means or lower uncertainty will be chosen more often, but less-known arms still have a chance to be sampled if their posterior draws occasionally exceed those of well-explored arms.

(4) Update.

The chosen arm a_{k^*} is played, and the resulting reward r_{k^*} is observed. The posterior for μ_{k^*} is then updated according to Bayesian rules:

- Bernoulli rewards: If $r_{k^*} \in \{0,1\}$, update

$$\alpha_{k^*} \leftarrow \alpha_{k^*} + r_{k^*}, \quad \beta_{k^*} \leftarrow \beta_{k^*} + (1 - r_{k^*}). \tag{12}$$

- Gaussian rewards: Update the mean and variance parameters of the normal posterior using standard Gaussian conjugate updating formulas, incorporating the new data point r_{k^*}.

This feedback loop is repeated in each subsequent round, enabling the algorithm to progressively refine its understanding of each arm's reward potential.

3.2. Problem Formulation

In this paper, the objective is to select the beamwidth and the transmit power level so that the achieved rate meets or exceeds a predefined threshold while minimizing the power used. To achieve this, we utilized 2DTS to select the best beamwidth and minimum power level.

Let θ_i denote the available beamwidth options, where $i \in 1, 2, \ldots, N$, and P_j denote the available power levels, where $j \in 1, 2, \ldots, M$. The objective is to maximize the cumulative number of successful transmissions over multiple trials while minimizing energy consumption.

3.3. Thompson Sampling with Joint Beamwidth and Power Selection

We apply 2DTS to jointly select the beamwidth and the power level in each trial to find the beamwidth and the minimum power that meet a given rate threshold r. The algorithm operates as follows:

- Initialize the parameters α and β for the beta distribution for each beamwidth i and each power level j, with $\alpha_i = 1$, $\beta_i = 1$, $\alpha_{i,j} = 1$, and $\beta_{i,j} = 1$ for all i and j.
- For each trial t
  - Sample θ_i from $\text{Beta}(\alpha_i, \beta_i)$ for each beamwidth i.
  - Select the beamwidth b_t with the highest sampled value:

$$b_t = \arg\max_i \theta_i \tag{13}$$

– For the selected beamwidth b_t, establish the power value corresponding to the maximum sampled probability from the beta distribution $P_{i,j}$ from $\text{Beta}(\alpha_{i,j}, \beta_{i,j})$ for each power level j, then select the power level p_t with the highest sampled value:

$$p_t = \arg\max_j P_{i,j} \tag{14}$$

The algorithm selects the power level p_t, aiming to minimize power usage by choosing the lowest possible power level that still satisfies the target threshold r. Here, R_t denotes the instantaneous channel capacity, while the threshold r is set strictly below this maximum capacity to represent the actual transmission rate. Thus, the algorithm checks whether the achievable rate R_t exceeds r, treating r as the success criterion for transmission.

– Calculate the SNR for the selected beamwidth and power level:

$$\text{SNR}_t = \frac{p_t H_G}{N}, \tag{15}$$

where H_G is the effective channel gain associated with beamwidth b_t and N is the noise power. Using this SNR, compute the rate R_t as follows:

$$R_t = \log_2(1 + \text{SNR}_t), \tag{16}$$

– Determine the reward r_t based on whether the achievable rate R_t meets or exceeds the target threshold r:

$$r_t = \begin{cases} 1, & \text{if } R_t \geq r \\ 0, & \text{otherwise} \end{cases} \tag{17}$$

– Update the beta distribution parameters for the selected beamwidth and power level based on the observed reward:

  * For the selected beamwidth b_t

$$\alpha_{b_t} = \alpha_{b_t} + r_t, \tag{18}$$

$$\beta_{b_t} = \beta_{b_t} + (1 - r_t), \tag{19}$$

  * For the selected power level p_t of the selected beamwidth b_t

    · If $r_t = 1$, this means success (the rate threshold is met or exceeded); increase α for the selected power level and the level immediately below it (if it exists). This update encourages the selection of lower power levels that can meet the threshold, improving power efficiency:

$$\alpha_{b_t,p_t} = \alpha_{b_t,p_t} + r_t, \tag{20}$$

$$\alpha_{b_t,p_{t-1}} = \alpha_{b_t,p_{t-1}} + r_t \quad \text{if } p_{t-1}\text{exists}, \tag{21}$$

    · If $r_t = 0$, this means fail (the rate threshold is not met); increase β for the selected power level and all power levels below it. This penalizes lower power levels that fail to meet the threshold, encouraging the selection of higher power levels when necessary:

$$\beta_{b_t,1:p_t} = \beta_{b_t,1:p_t} + (1 - r_t), \tag{22}$$

In this approach, 2DTS iteratively updates its estimates of the success probabilities for each power level, allowing it to identify the minimum power level that achieves the required rate with high confidence. Adjusting α and β for each beamwidth and power level based on the observed rate, the algorithm efficiently balances exploration and exploitation, focusing on power levels that maximize energy efficiency while meeting the communication threshold Algorithm 1.

Algorithm 1: 2DTS for Beamwidth and Power selection

Require: Threshold r, Noise power N_0, Beamwidth options θ, Power levels P, Rate R .

Initialize $\alpha_i = 1, \beta_i = 1$ for all $i \in b$, and $\alpha_{i,j} = 1, \beta_{i,j} = 1$ for all $i \in \theta, j \in P$.
for $t = 1$ to T **do**
    Sample $\theta_i \sim \text{Beta}(\alpha_i, \beta_i)$ for each $i \in \theta$.
    Select beamwidth $b_t = \arg\max_i \theta_i$.
    For the selected b_t, sample $P_{i,j} \sim \text{Beta}(\alpha_{i,j}, \beta_{i,j})$ for each $j \in P$.
    Select power level $p_t = \arg\max_j P_{i,j}$.
    Compute SNR: $\frac{p_t H_g}{N}$.
    Compute rate: $R_t = \log_2(1 + \text{SNR}_t)$.
    Compute reward $r_t = \begin{cases} 1, & \text{if } R_t \geq r \\ 0, & \text{otherwise} \end{cases}$.
    if $r_t = 1$ **then**
        Update $\alpha_{b_t} \leftarrow \alpha_{b_t} + r_t$.
        Update $\alpha_{b_t,p_t} \leftarrow \alpha_{b_t,p_t} + r_t$.
        if p_{t-1} exists **then**
            Update $\alpha_{b_t,p_{t-1}} \leftarrow \alpha_{b_t,p_{t-1}} + r_t$.
        end if
    else
        Update $\beta_{b_t} \leftarrow \beta_{b_t} + r_t$.
        Update $\beta_{b_t,1:p_t} \leftarrow \beta_{b_t,1:p_t} + r_t$.
    end if
end for

3.4. Energy Efficiency

We confirmed the proposed 2DTS joint beamwidth–power selection over T trials. At each trial t, 2DTS samples a beam and a power. The cumulative number of successful transmissions at trial t is

$$R_T = \sum_{t=1}^{T} r_t. \tag{23}$$

where $r_t \in \{0, 1\}$ denotes the binary reward at round t. The cumulative power used over T trials is

$$P_{\text{total}} = \sum_{t=1}^{T} p_t, \qquad 0 < p_t \leq P_{\max}, \ b_t \in \theta, \ p_t \in P. \tag{24}$$

Accordingly, the overall energy efficiency after T trials is

$$\text{EE}_T = \frac{R_t}{P_{\text{total}}} = \frac{\sum_{t=1}^{T} r_t}{\sum_{t=1}^{T} p_t}. \tag{25}$$

Equivalently, 2DTS seeks to maximize EE_T over the following sequence:

$$\begin{aligned} \underset{\{(b_t,p_t)\}_{t=1}^{T}}{\text{maximize}} \quad & \mathrm{EE}_T = \frac{R_t}{P_{\text{total}}} \\ \text{subject to} \quad & r_t = \begin{cases} 1, & \log_2(1+\mathrm{SNR}_t) \geq r, \\ 0, & \text{otherwise}, \end{cases} \\ & 0 < p_t \leq p_{\max}, \quad b_t \in \theta, \quad p_t \in P, \quad t = 1, \ldots, T. \end{aligned}$$

The results indicate that the algorithm successfully converges to the optimal beamwidth and power level, achieving reliable communication with minimum energy consumption.

4. Simulation Results

In this section, we present simulation results that quantitatively demonstrate the effectiveness of the proposed 2DTS algorithm in selecting the optimal beamwidth and transmit power under varying sea states. Performance is measured in terms of cumulative energy efficiency (r/p) in iterative rounds. Here, r represents the achieved data rate (in bps/Hz) obtained from the simulation, rather than the binary success indicator used within the Thompson Sampling process. Accordingly, the energy efficiency (EE) metric has the physical unit of bits/Joule/Hz, consistent with the standard definition of spectral energy efficiency. For clarity, all relevant figures are labeled as "Energy Efficiency (bps/Hz/J)". In buoy-based maritime uplinks, energy efficiency (EE) is a primary design objective due to strict power budgets. Throughout this section, we use $\mathrm{EE} = r/P_t$, where r denotes the achieved rate in bps/Hz (i.e., spectral efficiency, SE). Hence, EE implicitly captures SE in its numerator while simultaneously reflecting the rate–power trade-off that is central to battery-constrained buoy operation. In our target-rate setting, methods that satisfy the reliability constraint yield comparable SE at the threshold, so the observed EE gains of 2DTS primarily arise from using lower transmit power to meet the same target. Qualitatively, therefore, SE trends are consistent with the reported EE improvements. Figures 2 and 3 show the energy efficiency achieved by four schemes: (i) an ideal case that assumes perfect prior knowledge of the optimal beamwidth and power at each round and (ii) the proposed 2DTS algorithm. (iii) The "modified TS with a dedicated exploration phase" (Exploration TS) follows an explore–then–commit protocol over the joint action space of beamwidth and power. During the first T_{exp} rounds, actions are sampled uniformly (no exploitation) and beta posteriors are updated from binary success/failure. At $t = T_{\text{exp}}+1$, the algorithm commits to a single pair: among arms whose posterior-mean success probability meets the target reliability, it picks the minimum-power configuration; if none is feasible, it selects the arm with the largest posterior mean and uses it for the remaining rounds and (iv) a conventional TS that independently selects beamwidth and power using standard TS. As shown in Figure 2, under a moderate sea state (sea state 4), the proposed 2DTS achieves an energy efficiency of approximately 1.26×10^4 (bps/Hz/J) at round 600, which is 73.5% of the ideal case performance (1.71×10^4 (bps/Hz/J)). This result clearly outperforms both the exploration-based variant (0.64×10^4) and the conventional TS (0.23×10^4), indicating relative gains of 96.9% and 447.8% over the exploration-based and conventional TS baselines, respectively. As shown in Figure 3, for a harsh sea state (sea state 9), the performance gap narrows further. In round 600, 2DTS achieves 3.09×10^4 (bps/Hz/J), corresponding to 85.6% of the ideal case (3.61×10^4), while exploration TS and conventional TS reach 1.68×10^4 and 1.45×10^4, respectively. This translates to a performance gain of 84.0% over conventional TS, again highlighting the robustness of the 2DTS framework in dynamically fluctuating maritime environments. The superior performance of 2DTS arises from its joint optimization structure.

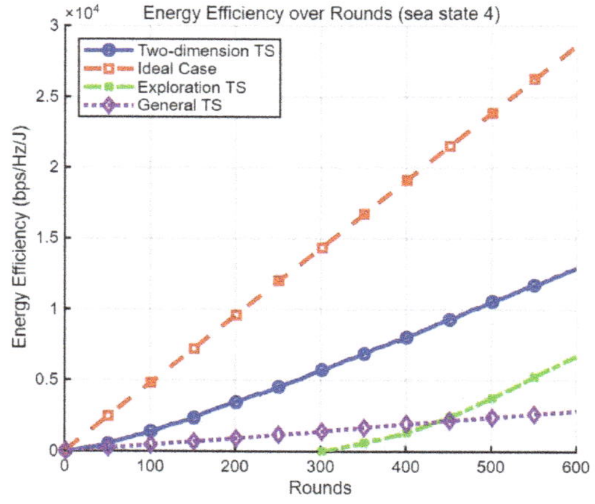

Figure 2. Energy efficiency over rounds (sea state 4).

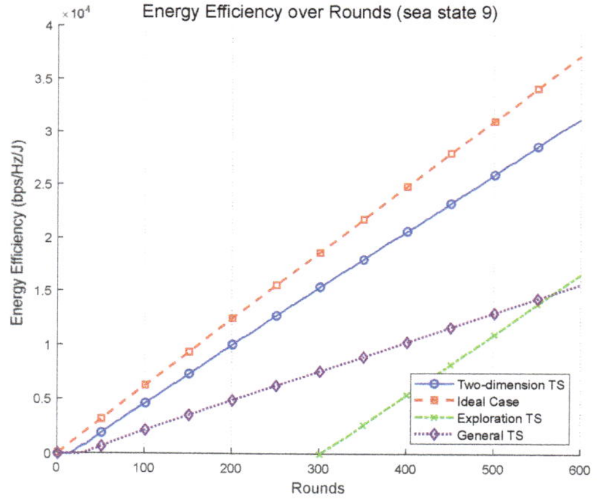

Figure 3. Energy efficiency over rounds (sea state 9).

5. Discussion

5.1. Key Findings

Across moderate (Sea State 4) and harsh (Sea State 9) sea states, the proposed 2DTS consistently achieves high spectral energy efficiency (EE) while meeting the target-rate constraint. At round 600, 2DTS attains 1.26×10^4 bps/Hz/J in sea state 4 (73.5% of the ideal) and 3.09×10^4 bps/Hz/J in sea state 9 (85.6% of the ideal), outperforming exploration–then–commit TS and conventional TS by large margins. These results indicate that feasibility-first selection combined with minimum-power control is effective under sea-state uncertainty.

5.2. Joint vs. Independent Selection

Unlike conventional TS that treats beamwidth and power independently, 2DTS jointly updates and selects (beamwidth, power) pairs. This joint treatment captures the coupling between directional gain (beamwidth) and energy usage (power), reducing both misalignment-induced losses and unnecessary transmit power. The observed gains over conventional TS in Figures 2 and 3 are consistent with this mechanism.

5.3. *Effect of Explore–Then–Commit*

The explore–then–commit baseline improves naive independent selection by dedicating an initial exploration budget; however, its one-shot commitment is vulnerable to subsequent sea-state shifts. In contrast, 2DTS continuously re-balances exploration and exploitation, which leads to higher cumulative EE across rounds.

5.4. *Sensitivity and Limitations*

Performance depends on the reliability threshold, the discretization of the (beamwidth, power) action grid, and the duration of the exploratory budget in the explore–then–commit baseline (T_{exp}). While our setup requires only binary success/failure feedback (compatible with LTE-M systems lacking full MIMO/Channel State Information), extremely severe sea states or overly coarse action grids may slow convergence. Incorporating coarse contextual signals (e.g., forecasted sea state) or adaptive action refinement could further improve sample efficiency.

5.5. *Practical Implications and Future Work*

The results suggest that lightweight, feedback-driven learning can provide reliable and energy-efficient operation for buoy-based uplinks without explicit sea-state information. Future work includes context-aware 2DTS (e.g., integrating satellite-based forecasts as side information), multi-buoy cooperation, and relay-assisted links.

Summary and Novelty

Unlike standard TS, which treats each parameter selection problem independently, our 2DTS algorithm simultaneously selects the optimal beamwidth–power pair through a joint posterior update that captures both directional-gain dynamics (via beamwidth) and power–efficiency trade-offs. This design is tailored to maritime systems where beam misalignment and power constraints are tightly coupled due to wave-induced platform motion. To the best of our knowledge, no existing TS-based algorithm directly addresses the joint selection of beamwidth and power levels under unknown and dynamic sea-state conditions. Although advanced variants of TS (e.g., contextual or neural TS) have been proposed in other domains, they are generally ill-suited to the maritime scenario considered here. In our setting, the decision space is discrete and multidimensional and is tightly coupled with physical constraints such as misalignment and energy efficiency. Therefore, our approach constitutes a novel TS formulation tailored to this setting, offering a practical solution when sea-state information is unavailable. In summary, 2DTS achieves up to 85.6% of the ideal energy efficiency across varying sea states while significantly outperforming baseline methods, demonstrating its suitability for energy-constrained maritime communication environments.

6. Conclusions

In this work, we proposed an adaptive approach for buoy-embedded maritime communication systems, focusing on the selection of beamforming and power control under dynamic sea conditions. For unknown sea states, we developed a novel 2DTS algorithm that simultaneously adjusts beamwidth and transmission power, achieving robust communication even without direct sea-state information. This adaptive framework balances exploration and exploitation to optimize communication parameters over time, closely approximating ideal performance and ensuring energy-efficient operation. Our simulation results validate the effectiveness of this method, demonstrating significant improvements in energy efficiency and reliability in dynamic maritime environments. Our approach advances the field of maritime communications by providing a robust, energy-efficient strategy for dynamic and unpredictable sea conditions.

Author Contributions: Conceptualization, K.J.L. and D.K.; methodology, K.J.L.; software, K.J.L.; validation, J.-H.J. and D.K.; writing—original draft, K.J.L.; writing—review and editing, J.-H.J., S.C., K.-W.K., and D.K.; supervision, D.K. All authors have read and agreed to the published version of the manuscript.

Funding: This research was supported by the Korea Institute of Marine Science and Technology Promotion (KIMST), funded by the Ministry of Oceans and Fisheries, under grant number RS-2021-KS211516.

Data Availability Statement: Data supporting the findings of this study are available from the corresponding author upon reasonable request.

Acknowledgments: The authors also appreciate all collaborators for valuable discussions and data sharing.

Conflicts of Interest: The authors declare no conflicts of interest.

References

1. Palma, D. Enabling the Maritime Internet of Things: CoAP and 6LoWPAN Performance Over VHF Links. *IEEE Internet Things J.* **2018**, *5*, 5205–5212. [CrossRef]
2. Yang, T.; Zheng, Z.; Liang, H.; Deng, R.; Cheng, N.; Shen, X. Green Energy and Content-Aware Data Transmissions in Maritime Wireless Communication Networks. *IEEE Trans. Intell. Transp. Syst.* **2015**, *16*, 751–762. [CrossRef]
3. Huo, Y.; Dong, X.; Beatty, S. Cellular Communications in Ocean Waves for Maritime Internet of Things. *IEEE Internet Things J.* **2020**, *7*, 9965–9979. [CrossRef]
4. Zhou, Z.; Ge, N.; Wang, Z. Two-Timescale Beam Selection and Power Allocation for Maritime Offshore Communications. *IEEE Commun. Lett.* **2021**, *25*, 3060–3064. [CrossRef]
5. Guan, S.; Wang, J.; Jiang, C.; Duan, R.; Ren, Y.; Quek, T.Q.S. MagicNet: The Maritime Giant Cellular Network. *IEEE Commun. Mag.* **2021**, *59*, 117–123. [CrossRef]
6. Kim, H.J.; Tiwari, S.V.; Chung, Y.H. Multi-hop relay-based maritime visible light communication. *Chin. Opt. Lett.* **2016**, *14*, 050607. [CrossRef]
7. Wang, W.; Gill, E.W. Evaluation of Beamforming and Direction Finding for a Phased Array HF Ocean Current Radar. *J. Atmos. Ocean. Technol.* **2016**, *33*, 2599–2613. [CrossRef]
8. Romdhane, I.; Kaddoum, G. A Reinforcement-Learning-Based Beam Adaptation for Underwater Optical Wireless Communications. *IEEE Internet Things J.* **2022**, *9*, 20270–20281. [CrossRef]
9. Jo, S.W.; Shim, W.S. LTE-Maritime: High-Speed Maritime Wireless Communication Based on LTE Technology. *IEEE Access* **2019**, *7*, 53172–53181. [CrossRef]
10. Zhang, L.; Wang, Z.; Zhang, H.; Min, M.; Wang, C.; Niyato, D.; Han, Z. Anti-Jamming Colonel Blotto Game for Underwater Acoustic Backscatter Communication. *IEEE Trans. Veh. Technol.* **2024**, *73*, 10181–10195. [CrossRef]
11. Ibrahim, S.; Mostafa, M.; Jnadi, A.; Salloum, H.; Osinenko, P. Comprehensive Overview of Reward Engineering and Shaping in Advancing Reinforcement Learning Applications. *IEEE Access* **2024**, *12*, 175473–175500. [CrossRef]
12. Russo, D.J.; Roy, B.V.; Kazerouni, A.; Osband, I.; Wen, Z. A Tutorial on Thompson Sampling. *Found. Trends Mach. Learn.* **2018**, *11*, 1–96. [CrossRef]
13. Chapelle, O.; Li, L. An Empirical Evaluation of Thompson Sampling. In *Proceedings of the Advances in Neural Information Processing Systems*; Shawe-Taylor, J., Zemel, R., Bartlett, P., Pereira, F., Weinberger, K., Eds.; Curran Associates, Inc.: San Francisco, CA, USA, 2011; Volume 24. Available online: https://proceedings.neurips.cc/paper_files/paper/2011/file/e53a0a2978c28872a4505bdb51db06dc-Paper.pdf (accessed on 12 December 2011).
14. Deng, W.; Kamiya, S.; Yamamoto, K.; Nishio, T.; Morikura, M. Thompson Sampling-Based Channel Selection Through Density Estimation Aided by Stochastic Geometry. *IEEE Access* **2020**, *8*, 14841–14850. [CrossRef]
15. Tong, J.; Fu, L.; Wang, Y.; Han, Z. Model-Based Thompson Sampling for Frequency and Rate Selection in Underwater Acoustic Communications. *IEEE Trans. Wireless Commun.* **2023**, *22*, 6946–6961. [CrossRef]

16. Komiyama, J.; Honda, J.; Nakagawa, H. Optimal Regret Analysis of Thompson Sampling in Stochastic Multi-armed Bandit Problem with Multiple Plays. In Proceedings of the 32nd International Conference on Machine Learning, Lille, France, 7–9 July 2015; Volume 37, pp. 1152–1161. [CrossRef]
17. Bai, L.; Han, R.; Liu, J.; Choi, J.; Zhang, W. Random Access and Detection Performance of Internet of Things for Smart Ocean. *IEEE Internet Things J.* **2020**, *7*, 9858–9869. [CrossRef]
18. Goodwin, M.; Elko, G. Constant beamwidth beamforming. In Proceedings of the 1993 IEEE International Conference on Acoustics, Speech, and Signal Processing, Minneapolis, MN, USA, 27–30 April 1993; Volume 1, pp. 169–172. [CrossRef]
19. Duan, R.; Wang, J.; Zhang, H.; Ren, Y.; Hanzo, L. Joint Multicast Beamforming and Relay Design for Maritime Communication Systems. *IEEE Trans. Green Commun. Netw.* **2020**, *4*, 139–151. [CrossRef]
20. Wang, J.; Zhou, H.; Li, Y.; Sun, Q.; Wu, Y.; Jin, S.; Quek, T.Q.S.; Xu, C. Wireless Channel Models for Maritime Communications. *IEEE Access* **2018**, *6*, 68070–68088. [CrossRef]
21. Yau, K.L.A.; Syed, A.R.; Hashim, W.; Qadir, J.; Wu, C.; Hassan, N. Maritime Networking: Bringing Internet to the Sea. *IEEE Access* **2019**, *7*, 48236–48255. [CrossRef]
22. Love, D.; Heath, R. Grassmannian beamforming on correlated MIMO channels. In Proceedings of the IEEE Global Telecommunications Conference, GLOBECOM '04, Dallas, TX, USA, 29 November–3 December 2004; Volume 1, pp. 106–110. [CrossRef]

Journal of *Marine Science and Engineering*

MDPI

Article

Integrated Scheduling of Handling and Spraying Operations in Smart Coal Ports: A MAPPO-Driven Adaptive Micro-Evolutionary Algorithm Framework

Yidi Wu, Shiwei He *, Haozhou Tang, Zeyu Long and Aibing Xiang

School of Traffic and Transportation, Beijing Jiaotong University, Beijing 100044, China
* Correspondence: shwhe@bjtu.edu.cn

Abstract

This study explores the integrated scheduling optimization of coal port operations, addressing the dual challenges of handling efficiency and resource conservation by coordinating equipment scheduling with stockyard spraying operations. Through a systematic analysis of operational processes in coal ports, a mixed-integer linear programming (MILP) model is developed to achieve global optimization while explicitly quantifying water and electricity consumption in spraying operations. To address this complex problem, we propose a novel hybrid algorithm that integrates a micro-evolutionary algorithm (MEA) framework with multi-agent proximal policy optimization (MAPPO), enabling adaptive decision-making for large-scale real-time scheduling. Three specialized agents for crossover, mutation, and neighborhood search achieve collaborative optimization by observing population features as states, selecting evolutionary operators as actions, and receiving composite rewards based on both population improvement and individual contributions. This strategy facilitates adaptive operator selection and optimal evolutionary direction derivation, collectively guiding population evolution toward high-quality solutions. Extensive experiments on ten scaled instances of a real-world coal port confirm the proposed algorithm's superior performance. Compared with four other standard algorithms, it consistently yields higher hypervolume (HV) values and lower inverted generational distance (IGD) metrics, which collectively demonstrate stronger convergence capability and higher solution quality.

Keywords: smart port; integrated scheduling; resource conservation; micro-evolutionary algorithm; deep reinforcement learning

1. Introduction

In recent years, the global energy trade has been highly dynamic, with coal resources, a key pillar of the global energy system, experiencing unprecedented levels of demand [1]. As a critical node in the coal supply chain and a key hub in rail-water intermodal transportation, coal ports primarily handle the transfer of coal between trains and ships. Against the backdrop of rising coal demand and the gradual improvement of integrated transportation systems, port operational efficiency directly impacts the transportation efficiency and costs of coal within the multimodal framework. The workflow at coal ports connected to heavy-haul railway stations can be described as follows: trains arrive for unloading, coal is transferred through unloading, conveying, and storage equipment to the designated stockyard, and upon ship berthing, the reclaimer and ship loader transfer the coal onto vessels, which then depart once loading is completed. The formulation and execution of overall port operation plans involve collaboration among multiple departments. Traditionally, ports have relied heavily on manual expertise for equipment control and scheduling, where limited resources, complex constraints, and high operational demands make the process highly challenging.

With the deepening integration of automation and intelligent technologies, the construction of smart ports, combining intelligent scheduling and equipment management, has become a key direction for modern port transformation. Simultaneously, the development of sustainable port ecosystems has garnered significant attention from port operators, necessitating further advancements in environmental management at coal ports [2]. The stockyard, as the central area for coal transfer and storage, plays a vital role in ensuring efficient unloading and loading operations. However, dust emissions generated during coal handling operations in stockyards remain a major challenge for port environmental management. Precise spraying operations serve as an effective technical approach for suppressing dust, as they control the moisture content of coal to minimize dust generation. However, current spraying operations consume substantial amounts of water and electricity, creating conflict between environmental protection and resource conservation. Therefore, optimizing intelligent spraying systems in stockyards while effectively balancing resource consumption is critical for promoting green development in coal ports.

Existing research on bulk port operations has primarily focused on operational procedures or specific equipment [3]. Such isolated optimization of equipment scheduling or stockyard allocation leads to singular objectives and deficient systemic coordination. Additionally, current studies have not comprehensively addressed resource consumption in coal ports operations, especially in relation to resource metrics within scheduling models [4]. These limitations make existing optimization solutions inadequate for meeting the operational demands of smart ports in terms of efficiency and resource conservation. Modern port operations encounter the challenge of integrating green scheduling within an intelligent equipment management framework, requiring the balanced coordination of equipment scheduling and operational line allocation while simultaneously considering dust suppression and resource consumption in spraying operations. Therefore, the integrated scheduling model for coal ports handling and spraying operations proposed in this study is essential for achieving the dual objectives of operational efficiency and resource conservation, supporting the port's green and intelligent transformation.

Specifically, the main contributions of this paper are as follows:

(1) This study achieves a comprehensive analysis of coal ports' operational processes, integrating train splitting, handling equipment scheduling, stockyard allocation, and dust suppression spraying operations. Focusing on the resource consumption characteristics of spraying operations, it quantifies water and electricity consumption by combining environmental factors with intrinsic coal properties like moisture content and coal type and incorporates these quantitative results into the optimization objectives. Specifically, the established mixed-integer linear programming model takes minimizing the total operating time of trains and ships and reducing resource consumption costs as dual goals, breaking through the limitations of traditional segmented optimization that separates equipment scheduling from spraying operations and realizing the overall optimization of efficiency and resource conservation.

(2) A novel hybrid optimization algorithm is developed, which combines the global search capability of the micro-evolutionary algorithm (MEA) framework with the adaptive decision-making ability of multi-agent proximal policy optimization (MAPPO). The algorithm adopts a two-layer chromosome encoding structure to capture key decision variables such as unloading and loading line selection, ass well as operation sequence, ensuring comprehensive coverage of the solution space while

reducing encoding complexity. The dominance recording matrix (DRM) dynamically extracts and records dominant gene structures from elite individuals, providing directional guidance for evolution operations. Three specialized agents for crossover, mutation, and neighborhood search perceive population states and environmental feedback, collaboratively adjust evolutionary operators in real time, and balance exploration and exploitation through adaptive strategy adjustment, enabling an efficient solution of large-scale multi-objective scheduling problems.

(3) The effectiveness and applicability of the proposed algorithm are verified through in-depth case studies and multi-scale comparative experiments. In the case study of a major Chinese coal port, two typical solutions on the Pareto front, namely the efficiency-maximizing solution and the optimal compromise solution, are analyzed in detail, and the trade-off relationship between their operational time and resource consumption is quantified, which reflects the algorithm's ability to balance dual objectives. Moreover, comparative experiments with four benchmark algorithms on ten different scale instances show that the proposed algorithm has obvious advantages in hypervolume (HV) and inverted generational distance (IGD), which confirms its strong convergence and high solution quality in different operation scales, and fully demonstrates its practical application value.

The remainder of this paper is organized as follows: Section 2 presents a review of the relevant literature. Section 3 describes the studied problem, followed by the mathematical model. Section 4 presents the hybrid algorithm framework and technical details. Section 5 conducts numerical experiments. Section 6 concludes the paper.

2. Literature Review

The optimization of port loading and unloading operations is a core issue in port operations management and has long been a subject of extensive academic inquiry, generating a substantial body of research in this field [5]. This paper reviews prior research from three perspectives: intelligent scheduling, green strategies, and reinforcement learning in port operations.

2.1. Intelligent Scheduling in Port Operations

In the field of port intelligent scheduling research, container terminals have been a focus of early studies due to their high operational standardization and clear equipment coordination demands. These studies have now established foundational methodologies for optimizing equipment scheduling and resource allocation. Golias et al. [6] introduced a robust berth scheduling method that minimizes total service time under uncertainty. Gambardella et al. [7] proposed a hierarchical optimization framework for intermodal terminals, combining resource allocation with discrete-event simulation to reduce crane conflicts. Chen et al. [8] formulated container terminal operations as a hybrid flow shop scheduling problem and proposed a tabu search algorithm, achieving improvements in the makespan for ship handling tasks. The integration of hybrid algorithms has become a key trend in addressing complex port scheduling challenges [9]. Zeng and Yang [10] integrated genetic algorithms with neural network surrogate models for container loading operations, reducing simulation runtime while maintaining solution accuracy.

As the automation level of container terminals continues to improve, automated guided vehicles (AGVs) serve as key equipment connecting various operational links. They have gradually emerged as a focal point in research on automated container terminals, and their scheduling and path planning directly impact port operational efficiency and cost control [11–13]. Chen et al. [14] proposed an artificial potential field and twin delayed deep deterministic policy gradient model for AGV path optimization, reducing path lengths while improving safety. Liang et al. [15] proposed a three-stage scheduling algorithm integrating genetic algorithms and simulation modeling to reduce the AGV locking situation. Recent research on sea-rail automated container terminals has demonstrated that integrated scheduling of port and railway operations is critical for improving logistics efficiency [16]. Chen et al. [17] demonstrated through MILP models that synchronous operations reduced crane task completion time. Munuzuri et al. [18] applied heuristic method to optimize integrated scheduling of vessel, crane and train operations.

Beyond container terminals, research on intelligent scheduling for dry bulk ports has also achieved several notable advances. De Andrade and Menezes [19] proposed a column generation-based heuristic that effectively solved the complex integrated planning problem of yard and berth allocation, showing superior performance over commercial solvers for large terminals. For iron ore port, Tang et al. [20] optimized storage allocation considering ore-mixing requirements. Ouhaman et al. [21] presented a practical heuristic method specifically designed for export bulk terminals, enabling efficient scenario testing and yard planning. Van Vianen et al. [22] proposed a simulation approach to optimize stacker-reclaimer scheduling, effectively reducing port time of trains while preserving ship service level. Xin et al. [23] formulated a predictive control strategy based on a mixed logical dynamical model to maximize profit in dry bulk terminal storage allocation. Bouzekri et al. [24] developed an integer programming model for the integrated laycan and berth allocation problem in bulk ports, incorporating practical constraints including conveyor routing, ship stability, and maintenance activities, and validated its computational efficiency through case studies from the port of Jorf Lasfar in Morocco. Babu et al. [25] employed two heuristic greedy construct algorithms to minimize ship delays through the simultaneous optimization of ship, stockyard, and train scheduling, validating the model's efficacy using operational data from a port on India's east coast.

To summarize, existing studies on container terminal intelligent scheduling have covered berth allocation, crane coordination, and AGV collaborative optimization, forming relatively mature technical frameworks especially in hybrid algorithm integration and multi-equipment conflict avoidance. These achievements are mostly tailored to the standardized operations of container terminals though, and their applicability to bulk terminals with more complex material flow and diverse storage requirements remains limited. Similarly, existing research on dry bulk ports has primarily focused on segmental optimization such as yard-berth coordination or stacker-reclaimer scheduling for general bulk cargo. Coal ports however have unique operational characteristics including heavy-haul train splitting, multi-type coal storage constraints and tight coupling between unloading-loading processes and stockyard operations. The complexity of multi-equipment, multi-process coordinated scheduling in coal terminals, especially the integration of handling operations with environmental protection processes, has not been sufficiently explored.

2.2. Green Strategies in Port Operations

Extant literature on green port operations demonstrates that achieving sustainable development requires addressing both conventional equipment coordination challenges and emerging environmental considerations [26]. Zheng et al. [27] proposed a genetic algorithm integrating quay crane and yard truck scheduling, achieving carbon emission reductions through three-dimensional chromosome representation and heuristic mutation operators. Peng et al. [28] quantified through simulation that strategic yard crane deployment could reduce emissions, though with diminishing returns beyond optimal equipment numbers. Niu et al. [29] introduced a Monte Carlo tree search algorithm for U-shaped terminals that achieved energy savings while resolving equipment conflicts. For specialized terminal layouts, Peng et al. [30] developed an improved multi-objective particle swarm optimization algorithm, yielding 11.72% energy efficiency improvements through conflict-free spatiotemporal path planning.

Several studies have specifically targeted energy consumption optimization in handling equipment. He et al. [31] proposed a hybrid algorithm for yard crane scheduling formulated as a vehicle routing problem with soft time windows, successfully balancing efficiency and energy consumption. Lu et al. [32] applied particle swarm optimization-genetic algorithm to reduce carbon emissions by 15.3% in container terminals.

In bulk cargo port operations, Jiang et al. [33] proposed a multi-objective genetic algorithm for the integrated scheduling of channels, berths, and yards, demonstrating emission reductions while maintaining throughput. Notably, Wang et al. [34] made significant progress by incorporating spraying operation optimization into stockyard allocation models, achieving 30% reductions in water consumption.

Existing green port research mainly focuses on container terminals, with optimization centered on equipment energy efficiency and layout-based emission reduction. These studies provide effective paths for container terminal carbon reduction, but their focus on resource conservation often stays at overall energy reduction. For bulk ports, while some studies explore emission reduction in core handling links, few have quantitatively modeled the resource consumption of key environmental protection processes or established dynamic correlations between such consumption and real-time operational conditions. This gap limits green scheduling precision for coal ports, where spraying is critical for dust suppression and incurs non-negligible water and electricity costs. Thus, intelligent spraying and its integration into overall port scheduling still need in-depth research.

2.3. Reinforcement Learning in Port Operations

Reinforcement learning (RL) has been extensively applied to the optimization of port operations scheduling, particularly in complex decision-making scenarios such as equipment dispatching, path planning, and energy consumption management. By interacting with dynamic environments, RL effectively handles scheduling uncertainties, improves efficiency, and promotes energy sustainability, emerging as a key method for intelligent port optimization [35]. Yue and Fan [36] combined Dijkstra's algorithm with Q-learning for dynamic AGV path planning, reducing the transportation cost of AGVs under the constraint of laytime. Similarly, Zeng et al. [37] introduced a Q-learning and simulation integration method to minimize container terminal operation time.

The advancement of deep reinforcement learning (DRL) methods has enhanced the precision and efficiency of port scheduling operations [38]. Che et al. [39] proposed a MAPPO framework with graph neural networks, demonstrating superior performance in handling charging station constraints compared to traditional heuristics. Tang et al. [40] developed a Double Deep Q-Network (DDQN) framework for U-shaped automated terminals, optimizing twin cantilever rail crane operations to simultaneously improve efficiency and energy savings. Zhu et al. [41] introduced multi-attention DRL for unmanned vessel scheduling, achieving costs reductions while ensuring conflict-free paths. At the strategic level, Li et al. [42] demonstrated that Dueling Double DQN could enhance multi-terminal berth allocation, reducing total dwelling time by 3.7% compared to conventional algorithms. Bulk cargo terminals have also benefited from specialized DRL implementations. Li et al. [43] adapted DDQN with an improved ε-greedy strategy for conveyor path scheduling, achieving faster loading processes in coal terminals. Ai et al. [44] employed Softmax-based Dueling Double DQN, which reduced berth allocation costs in bulk ports while eliminating manual scheduling errors.

3. Problem Description and Model Formulation

Focusing on the intelligent scheduling of port equipment and the optimization of operational resources, this section analyzes the coal handling process and investigates the influencing factors of stockyard spraying operations, along with their interaction with scheduling tasks. An integrated scheduling model is developed to balance coal ports handling efficiency and resource consumption.

3.1. Problem Description

The study examines a coal rail-water intermodal transportation port in China, analyzing equipment scheduling and the end-to-end coal handling process from unloading to ship loading. The operational workflow and stockyard layout are illustrated in Figure 1. The coal transfer operations involve multiple procedures, including train arrival and splitting, unloading, storage, and ship loading. These processes encompass train splitting plans, berth allocation plans, unloading-stacking operation plans, stockyard spraying plans and reclaiming-loading operation plans, all interconnected through spatiotemporal resource constraints to form a complex dynamic decision-making problem.

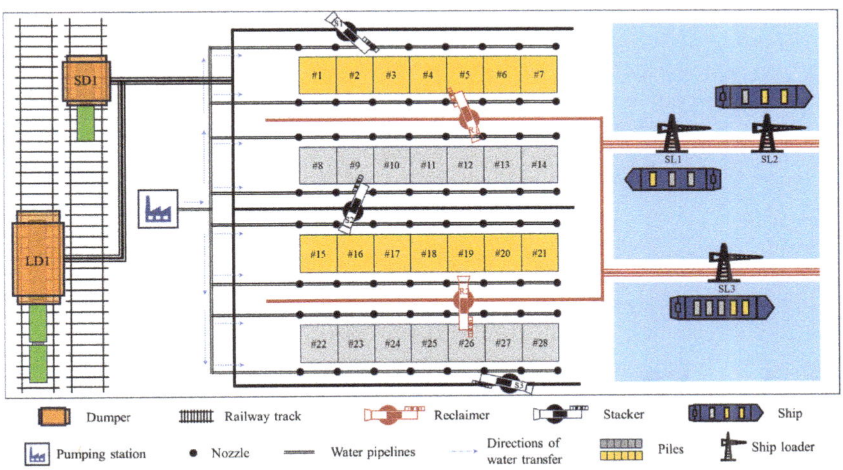

Figure 1. Layout of coal port.

Upon arrival at the port, heavy-haul trains undergo technical inspection before being split into unit trains (5 kt or 10 kt) to match the capacity of the unloading systems. Notably, each inbound train carries coal of a single uniform type. Different train formations correspond to different splitting plans, as detailed in Table 1. Ports typically deploy two capacity classes of dumpers: large dumpers (LD) for 10 kt unit trains and small dumpers (SD) for 5 kt unit trains. The varying unloading efficiencies of these systems mean unbalanced train splitting may lead to idle equipment. Therefore, refined train splitting plans and unloading sequences are required to balance the workload between the two dumper systems, making rational splitting and unloading plans essential for efficient coal transfer.

Table 1. Formation and splitting plan of heavy-haul train.

Heavy-Haul Train	Splitting Plan
5000 t	① 5 kt
10,000 t	① 10 kt ② 5 kt + 5 kt
15,000 t	① 10 kt + 5 kt ② 5 kt + 5 kt + 5 kt
20,000 t	① 10 kt + 10 kt ② 10 kt + 5 kt + 5 kt ③ 5 kt + 5 kt + 5 kt + 5 kt

Following splitting operations, dispatchers formulate coal unloading plans based on current production conditions, specifying the dumper and sequence for unloading operations along with candidate stockyard piles, generating the unloading-stacking operation plan. To facilitate subsequent model construction, this paper defines the complete set of equipment and their interconnections involved in this process as an unloading operational line, illustrated in Figure 1 as: SD1-stacker (S1)-pile (#1). During unloading operations, several critical constraints must be considered, including coal type matching, pile capacity limits, and stacker capability and accessibility. Coal type matching means the coal type of the inbound train must match that of the target pile, as each pile exclusively stores one coal type. Stacker capability and accessibility reflect that stackers are restricted to operating within corresponding piles by physical constraints. These constraints collectively influence the scheduling decisions for unloading operational line equipment selection.

The coal is transferred to the stockyard following the completion of unloading operations. Under stockyard operations, the fundamental challenge lies in maintaining adequate coal inventory for loading demands while optimizing handling schedules and piles allocation strategies to minimize spraying resource consumption in stockyard and reduce environmental impacts during unfavorable operating conditions. The spraying system in stockyard consists of three components: a pumping station, an array of high-pressure nozzles positioned at each pile's four corners, and a pipeline network connecting them. When spraying is activated for a target pile, the pumping station extracts water and delivers it through the pipeline network to the designated nozzles, which convert the water into mist and spray it onto the pile surface to increase the moisture content of the surface layer. The amount of water consumed in spraying operations on piles depends on two key variables: the current amount of stored coal in the pile and a dynamically adjusted spraying coefficient. This coefficient varies in response to external environmental factors, such as ambient temperature, relative humidity, and wind speed, as well as intrinsic coal properties, including existing moisture content and coal type. Through real-time monitoring of these parameters, the port's intelligent spraying system dynamically calibrates operational parameters. The water consumption calculation per nozzle operation, as shown in Formula (1):

$$\sigma_{dt} G_t \varphi_{dt} = \theta_{hdt} B, \forall \mathrm{h} \in H, d \in D_h^3, t \in T_2 \quad (1)$$

In the Formula, σ_{dt} represents the amount of stored coal in pile d at time t. The coefficient of external environmental factors affecting the stockyard's water consumption at time t is denoted by G_t, while φ_{dt} indicates the coefficient of intrinsic coal properties influencing the spraying requirement for pile d at time t. The variable θ_{hdt} stands for the water consumption by nozzle h for pile d at time t, and B refers to the number of nozzles allocated to each pile.

To meet spraying operation requirements, the pumping station consumes electricity to convert it into mechanical energy for water delivery. During operations, the electricity consumption of the pumping station is proportional to the water delivery distance. Therefore, the stockyard allocation strategy has a direct impact on resource consumption throughout all operational procedures, necessitating a dynamic balance between spraying resource consumption and loading efficiency to achieve optimal overall operational costs.

During the ship loading procedure, ship information and coal loading requirements, including maximum load capacity, coal type and quantity per cabin, and loading sequence specifications, are provided prior to arrival. Based on these requirements and current stockyard conditions, dispatchers develop loading plans for each cabin, generating reclaiming-loading operation plans. This paper defines the complete set of equipment and their interconnections involved in this process as a loading operational line, illustrated in Figure 1 as: pile (#7)-reclaimer (R1)-ship loader (SL1). Loading tasks must follow specified sequences while adhering to equipment capability and accessibility constraints for both reclaimers and ship loaders. Through this process, the coal is ultimately loaded onto the ship, marking the completion of the entire operational procedure.

The systematic analysis of the integrated coal ports handling and spraying operations clearly demonstrates the multidimensional constraints and complex interactions characteristics of the operational processes. To address these challenges, this study establishes two primary objectives: first, minimizing total dwell and operating time of trains and ships to enhance throughput capacity and operational efficiency; second, minimizing resource consumption, including water and electricity costs in stockyard spraying operations as well as electricity costs in unloading and loading operational lines. The second objective aims to reduce operational costs while supporting green and sustainable port development.

In summary, the integrated scheduling problem for coal ports handling and spraying operations studied in this paper can be summarized as follows: the research provides information such as train and ship arrival schedules, equipment layouts and capacities, stockyard configurations, and storage limits, with optimization objectives focusing on minimizing both operating time costs and resource consumption costs to produce scheduling solutions that align with smart and green port requirements.

3.2. Model Formulation

A MILP model is developed based on the above problem description, incorporating the following foundational assumptions:

(1) Each piece of equipment serves only one operational line at any given time without interruption.
(2) Equipment productivity remains constant, with synchronized start and stop times for all equipment within the same operational line.
(3) Each pile is dedicated to storing a single coal type exclusively.
(4) Handling duration is determined by both operation volume and pile location.
(5) Every berth is equipped with its corresponding ship loader.

The model employs the sets and parameters detailed in Table 2, with variables specified in Table 3.

Table 2. Definitions of sets, indices, and parameters.

Notation	Definition
I	Set of inbound trains, indexed by $i, i' \in I$.
J_i	Set of splitting plans for inbound train i, indexed by $j \in J_i$.
M_i	Set of all potential unit trains for inbound train i, indexed by $m, m' \in M_i$.

Table 2. *Cont.*

Notation	Definition
M_{ij}	Set of all potential unit trains for inbound train i in splitting plan j, indexed by $m \in M_{ij}$.
L_{im}	Set of candidate unloading operational lines for unit train m of inbound train i, indexed by $l \in L_{im}$.
F_{im}	Set of candidate dumpers for unit train m of inbound train i, indexed by $f \in F_{im}$.
L_f	Set of all unloading operational lines corresponding to dumper f, indexed by $l \in L_f$.
K_{im}	Set of candidate stackers for unit train m of inbound train i, indexed by $k \in K_{im}$.
L_k	Set of all unloading operational lines corresponding to stacker k, indexed by $l \in L_k$.
D^1_{im}	Set of candidate piles for unit train m of inbound train i, indexed by $d \in D^1_{im}$.
L_d	Set of all unloading operational lines corresponding to pile d, indexed by $l \in L_d$.
U	Set of arriving ships, indexed by $u, u' \in U$.
V_u	Set of candidate berths for ship u, indexed by $v \in V_u$.
W_u	Set of loading operations for ship u, indexed by $w, w' \in W_u = \{1, 2, \ldots, w^0_u\}$.
P_{uw}	Set of candidate loading operational lines for loading operation w of ship u, indexed by $p \in P_{uw}$.
P_d	Set of all loading operational lines corresponding to pile d, indexed by $p \in P_d$.
D^2_{uw}	Set of candidate piles for loading operation w of ship u, indexed by $d \in D^2_{uw}$.
Q_{uw}	Set of candidate reclaimers for loading operation w of ship u, indexed by $q \in Q_{uw}$.
P_q	Set of all loading operational lines corresponding to reclaimer q, indexed by $q \in P_q$.
S	Set of ship loaders, indexed by $s \in S$.
H	Set of nozzles, indexed by $h \in H$.
D^3_h	Set of piles corresponding to nozzle h, indexed by $d \in D^3_h$.
T_1	Set of time within the scheduling period, indexed by $t \in T_1 = \{1, 2, \ldots, \lvert T_1 \rvert\}$.
T_2	Set of spraying operation time within the scheduling period, indexed by $t \in T_2$.
a_{im}	Formation type of unit train m of inbound train i, $a_{im} \in \{0.5, 1\}$.
n_i	Formation type of heavy-haul train i, $n_i \in \{0.5, 1, 1.5, 2\}$.
s_p	Ship loader corresponding to loading operational line p.
s_v	Ship loader corresponding to berth p.
t^0_{im}	Earliest unloading start time for unit train m of inbound train i.
t_{iml}	Unloading-stacking time for unit train m of inbound train i on unloading operational line l.
t^0_u	Arrival time for ship u.
t_{uwp}	Reclaiming-loading time for loading operation w of ship u on loading operational line p.
w^0_u	The last loading operation of ship u.
A^0_d	Initial coal amount in pile d.
A^1_d	Maximum storage capacity of pile d.
A_{im}	Unloading coal volume for unit train m of inbound train i.
A_u	Deadweight tonnage of ship u.
A_{uw}	Loading coal volume for loading operation w of ship u.
B	Number of nozzles corresponding to each pile.

Table 2. *Cont.*

Notation	Definition
G_t	Spraying coefficient per unit coal at time t.
φ_{dt}	Intrinsic coal characteristic coefficient of pile d at time t.
ξ_h	Electricity consumption coefficient per unit water for nozzle h.
κ^1	Unit electricity cost.
κ^2	Unit water cost.
κ^3_{iml}	Cost for unit train m of inbound train i on unloading operational line l.
κ^4_{uwp}	Cost for loading operation w of ship u on loading operational line p.
N	A large enough positive number.

Table 3. Definitions of variables.

Variable	Definition
δ_{ij}	Decision binary variable equals 1 if inbound train i selects splitting plan j; 0 otherwise.
c_{im}	Decision binary variable equals 1 if inbound train i splits into unit train m; 0 otherwise.
ϕ_{iml}	Decision binary variable equals 1 if unit train m of inbound train i selects unloading operational line l; 0 otherwise.
x_{imf}	Binary variable equals 1 if unit train m of inbound train i selects dumper f for unloading; 0 otherwise.
$\varepsilon^1_{imi'm'f}$	Binary variable equals 1 if unit train m of inbound train i unloads before unit train m' of inbound train i' at dumper f; 0 otherwise.
t^1_{im}	Start time of unloading-stacking operation for unit train m of inbound train i.
y_{imk}	Binary variable equals 1 if coal from unit train m of inbound train i selects stacker k; 0 otherwise.
$\varepsilon^2_{imi'm'k}$	Binary variable equals 1 if coal from unit train m of inbound train i is stacked before coal from unit train m' of inbound train i' at stacker k; 0 otherwise.
z_{imd}	Binary variable equals 1 if coal from unit train m of inbound train i is stored in pile d; 0 otherwise.
$\varepsilon^3_{imi'm'd}$	Binary variable equals 1 if coal from unit train m of inbound train i is stacked before coal from unit train m' of inbound train i' in pile d; 0 otherwise.
μ_{uv}	Decision binary variable equals 1 if ship u selects berth v; 0 otherwise.
η_{uwp}	Decision binary variable equals 1 if loading operation w of ship u selects loading operational line p; 0 otherwise.
t^2_{uw}	Start time of reclaiming-loading operation for loading operation w of ship u.
g_{uwd}	Binary variable equals 1 if loading operation w of ship u reclaims from pile d; 0 otherwise.
$\lambda^1_{uwu'w'd}$	Binary variable equals 1 if loading operation w of ship u before loading operation w' of ship u' from pile d; 0 otherwise.
e_{uwq}	Binary variable equals 1 if loading operation w of ship u selects reclaimer q; 0 otherwise.

Table 3. *Cont.*

Variable	Definition
$\lambda^2_{uwu'w'q}$	Binary variable equals 1 if loading operation w of ship u before loading operation w' of ship u' at reclaimer q; 0 otherwise.
b_{uws}	Binary variable equals 1 if loading operation w of ship u selects ship loader s; 0 otherwise.
$\lambda^3_{uu'v}$	Binary variable equals 1 if ship u arrives before ship u' at berth v; 0 otherwise.
o^1_{imdt}	Binary variable equals 1 if unit train m of inbound train i stacks coal in pile d at time t; 0 otherwise.
o^2_{uwdt}	Binary variable equals 1 if loading operation w of ship u reclaims from pile d at time t; 0 otherwise.
λ^4_{imuwd}	Binary variable equals 1 if unit train m of inbound train i operates on pile d before loading operation w of ship u; 0 otherwise.
σ_{dt}	Amount of stored coal in pile d at time t.
θ_{hdt}	The amount of water consumed by nozzle h in pile d at time t.
θ_{ht}	The amount of water consumed by nozzle h at time t.

As mentioned in Section 3.1, our objective function consists of two parts. The first part represents the operating time for trains and ships.

$$\min \sum_{i\in I,m\in M_i} a_{im}(t^1_{im} - c_{im}t^0_{im}) + \sum_{i\in I,m\in M_i,l\in L_{im}} a_{im}\phi_{iml}t_{iml} + \sum_{u\in U} A_u(t^2_{uw^0_u} - t^0_u) + \sum_{u\in U,p\in P_{uw^0_u}} A_u\eta_{uw^0_u p}t_{uw^0_u p} \tag{2}$$

The second part accounts for the resource consumption costs.

$$\min \sum_{h\in H,t\in T_2} (\kappa^1\xi_h + \kappa^2)\theta_{ht} + \sum_{i\in I,m\in M_i,l\in L_{im}} \kappa^3_{iml}\phi_{iml} + \sum_{u\in U,w\in W_u,p\in P_{uw}} \kappa^4_{uwp}\eta_{uwp} \tag{3}$$

There are three fundamental categories of constraints.

(1) Train splitting and unloading-stacking constraints

$$\sum_{j\in J_i} \delta_{ij} = 1, \forall i \in I \tag{4}$$

$$\delta_{ij} \le c_{im}, \forall i \in I, j \in J_i, m \in M_{ij} \tag{5}$$

$$\sum_{m\in M_i} c_{im}a_{im} = n_i, \forall i \in I \tag{6}$$

$$\sum_{l\in L_{im}} \phi_{iml} = c_{im}, \forall i \in I, m \in M_i \tag{7}$$

$$t^1_{im} \ge t^0_{im}c_{im}, \forall i \in I, m \in M_i \tag{8}$$

$$t^1_{im} \le Nc_{im}, \forall i \in I, m \in M_i \tag{9}$$

$$x_{imf} = \sum_{l\in L_f} \phi_{iml}, \forall i \in I, m \in M_i, f \in F_{im} \tag{10}$$

$$\varepsilon^1_{imi'm'f} + \varepsilon^1_{i'm'imf} \geq x_{imf} + x_{i'm'f} - 1, \forall i \in I, i' \in I, m \in M_i, m' \in M_{i'}, f \in F_{im} \cap F_{i'm'}, i \neq i' \vee m \neq m' \quad (11)$$

$$\begin{aligned} & t^1_{im} + \sum_{l \in L_{im}} t_{iml}\phi_{iml} - t^1_{i'm'} \leq N(3 - \varepsilon^1_{imi'm'f} - x_{imf} - x_{i'm'f}), \\ & \forall i \in I, i' \in I, m \in M_i, m' \in M_{i'}, f \in F_{im} \cap F_{i'm'}, i \neq i' \vee m \neq m' \end{aligned} \quad (12)$$

$$y_{imk} = \sum_{l \in L_k} \phi_{iml}, \forall i \in I, m \in M_i, k \in K_{im} \quad (13)$$

$$\varepsilon^2_{imi'm'k} + \varepsilon^2_{i'm'imk} \geq y_{imk} + y_{i'm'k} - 1, \forall i \in I, i' \in I, m \in M_i, m' \in M_{i'}, k \in K_{im} \cap K_{i'm'}, i \neq i' \vee m \neq m' \quad (14)$$

$$\begin{aligned} & t^1_{im} + \sum_{l \in L_{im}} t_{iml}\phi_{iml} - t^1_{i'm'} \leq N(3 - \varepsilon^2_{imi'm'k} - y_{imk} - y_{i'm'k}), \\ & \forall i \in I, i' \in I, m \in M_i, m' \in M_{i'}, k \in K_{im} \cap K_{i'm'}, i \neq i' \vee m \neq m' \end{aligned} \quad (15)$$

$$z_{imd} = \sum_{l \in L_d} \phi_{iml}, \forall i \in I, m \in M_i, d \in D^1_{im} \quad (16)$$

$$\varepsilon^3_{imi'm'd} + \varepsilon^3_{i'm'imd} \geq z_{imd} + z_{i'm'd} - 1, \forall i \in I, i' \in I, m \in M_i, m' \in M_{i'}, d \in D^1_{im} \cap D^1_{i'm'}, i \neq i' \vee m \neq m' \quad (17)$$

$$\begin{aligned} & t^1_{im} + \sum_{l \in L_{im}} t_{iml}\phi_{iml} - t^1_{i'm'} \leq N(3 - \varepsilon^3_{imi'm'd} - z_{imd} - z_{i'm'd}), \\ & \forall i \in I, i' \in I, m \in M_i, m' \in M_{i'}, d \in D^1_{im} \cap D^1_{i'm'}, i \neq i' \vee m \neq m' \end{aligned} \quad (18)$$

Formula (4) requires each inbound train to follow a single splitting plan. Formulas (5) and (6) guarantee that the formation of unit trains exactly corresponds to the formations of the respective inbound trains, while Formula (7) assigns each unit train to a specific unloading line. Formulas (8) and (9) specify the earliest possible start times for unloading operations. Formula (10) ensures the association between dumpers and unloading lines. Formulas (11) and (12) prevent scheduling conflicts between dumpers. Formula (13) aligns stackers with their assigned operational lines. Formulas (14) and (15) coordinate the scheduling of stackers to avoid conflicts. Formula (16) defines the mapping between piles and unloading lines, while Formulas (17) and (18) ensure non-overlapping stacking sequences at the same piles.

(2) Reclaiming-loading constraints

$$\sum_{v \in V_u} \mu_{uv} = 1, \forall u \in U \quad (19)$$

$$\sum_{p \in P_{uw}} \eta_{uwp} = 1, \forall u \in U, w \in W_u \quad (20)$$

$$t^2_{uw} \geq t^0_u, \forall u \in U, w = 1 \quad (21)$$

$$t^2_{uw} + t_{uwp}\eta_{uwp} \leq t^2_{uw+1}, \forall u \in U, w \in W_u \backslash \left\{w^0_u\right\}, p \in P_{uw} \quad (22)$$

$$g_{uwd} = \sum_{p \in P_d} \eta_{uwp}, \forall u \in U, w \in W_u, d \in D^2_{uw} \quad (23)$$

$$\lambda^1_{uwu'w'd} + \lambda^1_{u'w'uwd} \geq g_{uwd} + g_{u'w'd} - 1, \forall u \in U, u' \in U, w \in W_u, w' \in W_{u'}, d \in D^2_{uw} \cap D^2_{u'w'}, u \neq u' \vee w \neq w' \quad (24)$$

$$t_{uw}^2 + \sum_{p \in P_{uw}} t_{uwp}\eta_{uwp} - t_{u'w'}^2 \leq N(3 - \lambda_{uwu'w'd}^1 - g_{uwd} - g_{u'w'd}),$$
$$\forall u \in U, u' \in U, w \in W_u, w' \in W_{u'}, d \in D_{uw}^2 \cap D_{u'w'}^2, u \neq u' \vee w \neq w' \tag{25}$$

$$e_{uwq} = \sum_{p \in P_q} \eta_{uwp}, \forall u \in U, w \in W_u, q \in Q_{uw} \tag{26}$$

$$\lambda_{uwu'w'q}^2 + \lambda_{u'w'uwq}^2 \geq e_{uwq} + e_{u'w'q} - 1, \forall u \in U, u' \in U, w \in W_u, w' \in W_{u'}, q \in Q_{uw} \cap Q_{u'w'}, u \neq u' \vee w \neq w' \tag{27}$$

$$t_{uw}^2 + \sum_{p \in P_{uw}} t_{uwp}\eta_{uwp} - t_{u'w'}^2 \leq N(3 - \lambda_{uwu'w'q}^2 - e_{uwq} - e_{u'w'q}),$$
$$\forall u \in U, u' \in U, w \in W_u, w' \in W_{u'}, q \in Q_{uw} \cap Q_{u'w'}, u \neq u' \vee w \neq w' \tag{28}$$

$$b_{uws} \geq \eta_{uwp}, \forall u \in U, w \in W_u, p \in P_{uw}, s = s_p \tag{29}$$

$$b_{uws} = \mu_{uv}, \forall u \in U, v \in V_u, w \in W_u, s = s_v \tag{30}$$

$$\lambda_{uu'v}^3 + \lambda_{u'uv}^3 \geq \mu_{uv} + \mu_{u'v} - 1, \forall u \in U, u' \in U, v \in V_u \cap V_{u'}, u \neq u' \tag{31}$$

$$t_{uw}^2 + \sum_{p \in P_{uw}} t_{uwp}\eta_{uwp} - t_{u'w'}^2 \leq N(3 - \lambda_{uu'v}^3 - \mu_{uv} - \mu_{u'v}), \forall u \in U, u' \in U, v \in V_u \cap V_{u'}, u \neq u', w = w_u^0, w' = 1 \tag{32}$$

$$t_{im}^1 + \sum_{l \in L_{im}} t_{iml}\phi_{iml} - t_{uw}^2 \leq N(3 - \lambda_{imuwd}^4 - z_{imd} - g_{uwd}), \forall i \in I, m \in M_i, u \in U, w \in W_u, d \in D_{im}^1 \cap D_{uw}^2 \tag{33}$$

$$t_{uw}^2 + \sum_{p \in P_{uw}} t_{uwp}\eta_{uwp} - t_{im}^1 \leq N(2 + \lambda_{imuwd}^4 - z_{imd} - g_{uwd}), \forall i \in I, m \in M_i, u \in U, w \in W_u, d \in D_{im}^1 \cap D_{uw}^2 \tag{34}$$

Formula (19) assigns each ship to a unique berth, while Formula (20) allocates loading operations to specific loading operational lines. Formulas (21) and (22) maintain a sequential order of loading operations for each ship. Formula (23) ensures that pile selections align with the corresponding loading operational lines. Formulas (24) and (25) prevent conflicts in pile reclaiming, and Formula (26) ensures that reclaimer selections are consistent with the loading operational lines. Formulas (27) and (28) avoid conflicts on the same reclaimer by sequential execution of reclaiming operations. Formulas (29) and (30) guarantee consistency among the selected ship loader, its associated loading operational line, and the assigned berth. Formulas (31) and (32) regulate the sequence of berth occupancy, while Formulas (33) and (34) prohibit simultaneous stacking and reclaiming operations at any pile.

(3) Spraying operation constraints

$$\sum_{t \in T_1} o_{imdt}^1 = z_{imd}, \forall i \in I, m \in M_i, d \in D_{im}^1 \tag{35}$$

$$\sum_{t \in T_1, d \in D_{im}^1} o_{imdt}^1 t = t_{im}^1, \forall i \in I, m \in M_i \tag{36}$$

$$\sum_{t \in T_1} o_{uwdt}^2 = g_{uwd}, \forall u \in U, w \in W_u, d \in D_{uw}^2 \tag{37}$$

$$\sum_{t \in T_1, d \in D_{uw}^2} o_{uwdt}^2 t = t_{uw}^2, \forall u \in U, w \in W_u \tag{38}$$

$$\sigma_{dt} = A_d^0, \forall d \in D, t = 1 \tag{39}$$

$$\sigma_{dt} = \sigma_{dt-1} + \sum_{i \in I, m \in M_i} o_{imdt}^1 A_{im} - \sum_{u \in U, w \in W_u} o_{uwdt}^2 A_{uw}, \forall d \in D, t \in T_1/\{1\} \tag{40}$$

$$\sigma_{dt} \geq 0, \forall d \in D, t \in T_1 \tag{41}$$

$$\sigma_{dt} \leq A_d^1, \forall d \in D, t \in T_1 \tag{42}$$

$$\theta_{ht} = \sum_{d \in D_h} \theta_{hdt}, \forall h \in H, t \in T_2 \tag{43}$$

$$\begin{gathered}\delta_{ij}, c_{im}, \phi_{iml}, x_{imf}, \varepsilon_{imi'm'f}^1, y_{imk}, \varepsilon_{imi'm'k}^2, z_{imd}, \varepsilon_{imi'm'd}^3, \mu_{uv}, \eta_{uwp}, g_{uwd}, \\ \lambda_{uwu'w'd}^1, e_{uwq}, \lambda_{uwu'w'q}^2, b_{uws}, \lambda_{uu'v}^3, o_{imdt}^1, o_{uwdt}^2, \lambda_{imuwd}^4 \in \{0,1\}\end{gathered} \tag{44}$$

Formulas (35) and (36) and (37) and (38) define the conversion constraints between decision variables and temporal variables for stacking and reclaiming operations at the piles, respectively. Formulas (39) and (40) dynamically track changes in coal inventory. Formulas (41) and (42) ensure the capacity limit of piles, while Formulas (1) and (43) calculate water spraying volumes of each nozzle by coal quantities with spraying coefficients. Formula (44) ensures that the decision variables are binary.

4. Algorithm Design

The integrated scheduling of coal ports operations, involving equipment coordination and stockyard allocation with spraying operations, presents an NP-hard problem that balances operating time costs and resource consumption costs under spatiotemporal constraints. As the research horizon expands and the number of inbound trains and ships increases, the complexity of global scheduling decisions increases significantly. Extensive research has demonstrated the advantage of heuristic algorithms in solving large-scale combinatorial optimization problems, significantly accelerating the discovery of feasible solutions [45]. Concurrently, DRL has seen increasingly profound and widespread application in scheduling domains, particularly for sequential decision-making problems in recent years.

Therefore, this paper proposes a scheduling optimization algorithm based on the MEA framework, incorporating a MAPPO-driven adaptive evolutionary strategy. This algorithm deeply integrates the global search capability of the MEA with the dynamic decision-making capability of MAPPO. On the one hand, dominant gene structures serve to guide the search direction. On the other hand, specialized agents, specifically a crossover agent, a mutation agent, and a neighborhood search agent, are designed. These agents dynamically and collaboratively adapt evolutionary strategies, such as operator selection, in response to the current state of the population and environmental reward feedback. This adaptive mechanism enables precise optimization of the population's evolutionary direction, facilitating a dynamic balance between handling efficiency and resource consumption.

4.1. Encoding and Gene Structure

Encoding design is a critical component of the algorithm, requiring a comprehensive representation of all decision variables and the solution space within the port scheduling problem. Interdependencies exist among these decision variables. For example, the splitting plan determines subsequent selection of unloading lines, while berth allocation influences loading lines selection. To reduce encoding complexity, we employ a hierarchical encoding strategy. The lower layer constructs a complete scheduling code using a two-segment chromosome structure: operational line assignment and operational priority sequencing. The upper layer utilizes a gene structure to record the splitting plan for each train and the berth allocation for each ship. A schematic diagram of this dual-layer chromosome and gene structure encoding is presented in Figure 2.

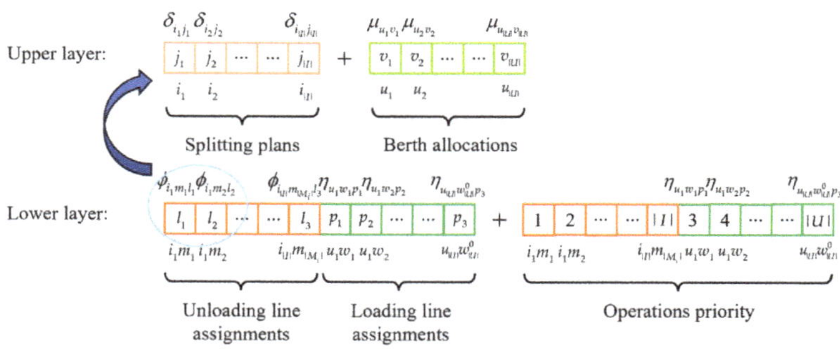

Figure 2. Schematic diagram of the dual-layer encoding.

The first segment of the chromosome encodes the selection of unloading lines and loading lines. Each unit train generated from the splitting of inbound trains is assigned a specific genetic locus. The value at this locus represents the selected unloading line. During chromosome generation, unloading line assignments must ensure the uniqueness of the train splitting plan and that the coal type corresponds to the designated piles. Correspondingly, each ship's loading operation occupies a genetic locus where its value denotes the chosen loading line, subject to berth allocation constraints and equipment accessibility. The second chromosome segment handles the operational sequence encoding. This study combines the priority sequence with the earliest available scheduling principle for decoding. Specifically, in each scheduling iteration, operations are selected based on their encoded priority sequence. The start time of an operation is determined by the completion time of its immediately preceding operation and the earliest available time of the assigned operational line, while simultaneously ensuring that no equipment usage conflicts occur and that pile capacity constraints are satisfied. By repeating this process, a complete feasible schedule is constructed. During initial population generation, values are randomly selected from the set of candidate operational lines for each genetic locus, and operational sequences are randomly generated, ensuring population diversity.

Splitting plans and berth allocations are not directly encoded in the lower-layer chromosome structure, they are represented as part of the upper-layer gene information. To ensure an effective mapping between the gene structure and the chromosome, the splitting plan gene structure assigns one locus for each inbound train, where the locus value indicates its corresponding splitting strategy. Similarly, the berth allocation gene structure assigns one locus for each ship, with the locus value indicating the allocated berth. The values of these upper-layer genes are indirectly determined by analyzing the unloading and loading line selections encoded in the chromosome. This dual-layer chromosome encoding and gene structure strategy comprehensively captures the interdependencies among decision variables and ensures full exploration of the solution space.

4.2. Dominance Recording Matrix

The DRM serves as the core mechanism that guides the high-quality evolution of the population. It dynamically records the distribution probabilities of gene structures among elite individuals in the population, capturing the values of dominant gene structures. This captured information subsequently guides the population toward evolving into more optimal regions of the solution space.

The DRM is constructed by extracting and statistically analyzing the splitting plan and berth allocation gene structures from the top $\mu\%$ of elite individuals located on the Pareto front. As depicted in Figure 3, the matrix element value represents the frequency of occurrence of the corresponding decision option within the elite individuals. By analyzing these frequency distributions, the DRM identifies the decisions that currently dominate the population. Genes associated with decisions exhibiting a frequency exceeding 80% are defined as high-frequency dominant gene structures. To prevent premature convergence of the population, the DRM undergoes periodic updates every few generations, ensuring its dynamic adaptability to the evolving search environment.

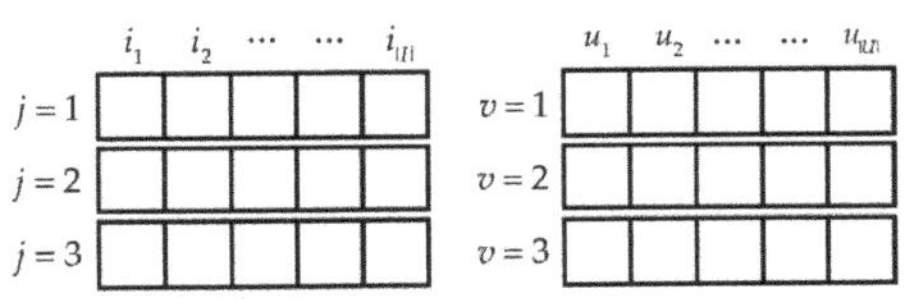

Figure 3. Illustration of dominance recording matrix.

4.3. MAPPO-Based Adaptive Evolutionary Strategy

To enhance algorithm performance and overcome the limitations of fixed parameter settings and random strategies inherent in traditional evolutionary operations, this study introduces an adaptive evolutionary strategy framework based on MAPPO. Within this MAPPO-driven framework, agents perceive the population state and receive environmental feedback, enabling them to dynamically learn and optimize their respective evolutionary strategies. These optimized strategies subsequently guide the evolutionary direction of the population. This adaptive strategy adjustment mechanism enables the algorithm to effectively drive the population toward rapid convergence to high-quality solutions while simultaneously maintaining population diversity.

4.3.1. State Representation

The construction of the state space must balance the global evolutionary characteristics of the population with the local correlations of individual gene structures. Therefore, we primarily extract key features relevant to each agent's responsibilities from dimensions such as population evolution dynamics and the distribution of dominant gene structures, forming their local observation state. Simultaneously, representative indicators reflecting the population's global characteristics are incorporated as public features. This results in a complete state vector for each agent by combining local key features with public features.

Public features need to reflect the current population quality and evolutionary state. Consequently, we select the average values all individuals on two objectives, denoted as $\overline{F}_1$, $\overline{F}_2$ and their variances $Var(F_1)$, $Var(F_2)$, along with the proportion of Pareto front solutions within the population, denoted as γ_{pa}, as public population features. Their calculations are as follows:

$$\overline{F}_k = \frac{1}{N}\sum_{n=1}^{N} f_k^n \quad k = 1,2 \tag{45}$$

$$Var(F_k) = \frac{1}{N}\sum_{n=1}^{N}(f_k^n - \overline{F}_k)^2 \quad k = 1,2 \tag{46}$$

$$\gamma_{pa} = \frac{N_{pa}}{N} \tag{47}$$

where f_k^n is the value of the $n-$th individual for objective k (operating time: $k = 1$; resource consumption: $k = 2$), N is the population size, and N_{pa} is the count of Pareto front solutions. The public state vector at generation t is $o_s^t = [\overline{F}_1, \overline{F}_2, Var(F_1), Var(F_2), \gamma_{pa}]$.

The state observation for the crossover agent needs to reflect the current structural diversity and genetic potential of the population. The diversity of the population's gene structure is quantified by the average entropy value per locus:

$$E_{avg} = \frac{1}{L}\sum_{i=1}^{L}(-\sum_{j=1}^{K_i} p_{ij}\log p_{ij}) \tag{48}$$

where L is the chromosome length, K_i is the number of possible values at the $i-$th locus, and p_{ij} is the frequency of value j occurring at that locus. Secondly, we introduce the proportion of dominant gene structures η_{high}:

$$\begin{gathered}\eta_{high} = \tfrac{1}{L_1}\sum_{i=1}^{L_1} DRM_{high}(i) \\ DRM_{high}(i) = \begin{cases} 1 \text{ if } \max(\ DRM[i]\ ) > 0.8 \\ \\ 0 \text{ otherwise} \end{cases}\end{gathered} \tag{49}$$

where L_1 is the length of the gene structures encoding, and $DRM_{high}(i)$ is an indicator function equaling 1 if the $i-$th locus is a high-frequency dominant gene structure. The features extracted above are concatenated with the public features to form the crossover agent's state vector: $o_{cx}^t = [o_s^t, E_{avg}, \eta_{high}]$.

The mutation agent is responsible for seeking a balance between exploration and exploitation, preventing premature convergence. It focuses on the dynamic changes in the Pareto front and the effects of mutation. For this purpose, we introduce the average objective value change in solutions on the Pareto front $\Delta\overline{F}_k^{pa}$, which measures the improvement rate of high-quality solutions during population evolution. It is calculated as follows:

$$\Delta\overline{F}_k^{pa} = \frac{1}{N_t^{pa}}\sum_{n=1}^{N_t^{pa}} f_k^n - \frac{1}{N_{t-1}^{pa}}\sum_{n=1}^{N_{t-1}^{pa}} f_k'^n k = 1,2 \tag{50}$$

where N_t^{pa} and N_{t-1}^{pa} are the number of individuals on the Pareto front in the current generation t and the previous generation $t-1$, respectively. Therefore, the mutation agent's state vector is $o_{mu}^t = [o_s^t, \Delta\overline{F}_1^{pa}, \Delta\overline{F}_2^{pa}]$.

The neighborhood search agent, however, needs to focus on scheduling operation characteristics. The features it extracts include the average waiting time of trains and ships $\overline{t}_{wait}^{pa}$ and the average idle time of equipment $\overline{t}_{idle}^{pa}$ among solutions on the Pareto front, resulting in its state vector is $o_{ns}^t = [o_s^t, \overline{t}_{wait}^{pa}, \overline{t}_{idle}^{pa}]$. The state vectors of each agent serve as input to their respective actor networks, enabling them to make rational action selections.

4.3.2. Action Space

The design of the action space aims to provide the three agents with executable instruction sets for dynamically and precisely controlling key parameters and strategies during the population evolution process. By integrating optimization capabilities of reinforcement learning into traditional evolutionary operations, it drives the population towards efficient convergence to the Pareto optimal front.

The crossover agent's actions a_{cx}^{t} are primarily responsible for strategically guiding the crossover operation at a global level, ensuring the generation of high-quality offspring while promoting effective exploration of the solution space. Its actions encompass parent selection, the choice of crossover operator, and the setting of the crossover probability. The parent selection mechanism includes two strategies: selecting elite individuals from the top 30% based on non-dominated rank or utilizing tournament selection. The choice of crossover operator determines the diversity and quality of offspring gene combinations. Guided by statistical information from the DRM, the crossover operator preferentially preserves dominant parental gene structures, thereby biasing offspring toward proven high-quality regions. Therefore, two crossover operators are proposed based on the dominant gene structure: a combined crossover operator and a uniform crossover operator. The former compares and retains the better gene structures from both parents to form a new individual. The latter only preserves the high-frequency dominant gene structures from the parents, while the remaining parts are determined through uniform crossover. To balance exploration and exploitation capabilities, the crossover probability should also be dynamically adjusted based on the population state.

The mutation agent's actions a_{mu}^{t} focus on determining the mutation location, mutation operator type, and mutation probability to maintain population diversity and promote local optimization. The mutation point selection strategy differentially processes individuals: for elite individuals, target loci with low-frequency dominance; for inferior-front individuals, preferentially select loci carrying high-frequency dominant genes. The mutation operators should embody strategies of local adjustment and structural perturbation, including reassigning operational lines at specific locus, such as swapping operational lines accessing proximal versus distal stockyard piles, and readjusting gene structures such as optimizing splitting plans or berth allocations. The mutation probability is adaptively selected from a predefined set based on the population's convergence state.

The neighborhood search agent's actions a_{ns}^{t} are designed for conducting refined local search on the sequence encoding to enhance the quality of solutions on the Pareto front. Its actions include selecting individuals, determining the neighborhood structure, and regulating the search depth. Individuals subjected to neighborhood search are selected from those on the current Pareto front. The neighborhood operators comprise three types: swap neighborhood, which involves exchanging the sequence of operations associated with the same equipment or the same pile; insertion neighborhood and reversal neighborhood, which perform local adjustments on the operation sequence corresponding to loci identified as high-frequency and low-frequency, respectively. The search depth is dynamically controlled by adapting the iteration count of neighborhood operations in response to Pareto front characteristics and local search effectiveness.

The action space for each agent is composed of the combinations of decisions across its different dimensions. Therefore, at each timestep, each agent outputs a composite strategy to guide the evolution of the current population.

4.3.3. Reward Function

The core purpose of the reward function is to quantify improvements in the Pareto front during population evolution and to effectively guide agents in adjusting their strategies. To precisely evaluate the differentiated contributions of the three agents during evolution, this paper designs a composite reward function. The hypervolume change ΔHV_t is utilized as the shared reward r_s. Simultaneously, the proportion of offspring generated by each agent itself that enter the current Pareto front is assigned as its local additional reward. The reward for each agent is calculated as follows:

$$r_s^t = \Delta HV_t = HV_t - HV_{t-1} \tag{51}$$

$$r_z^t = r_s^t + \frac{N_z}{N_{pa}} z \in \{cx, mu, ns\} \tag{52}$$

where z denotes the agent index, N_z is the number of offspring generated by agent z that enter the current Pareto front. This reward design provides feedback on the impact of each agent's actions on solution quality enhancement, enabling them to adjust their policies accordingly.

4.4. Algorithm Framework

The proposed algorithm employs a MEA as the external optimization framework, which internally integrates three specialized agents with advantage actor-critic (A2C) networks for crossover, mutation and neighborhood search operations, achieving collaborative evolutionary optimization. The strategy updates are implemented using PPO mechanism. The overall framework of the algorithm is illustrated in Figure 4.

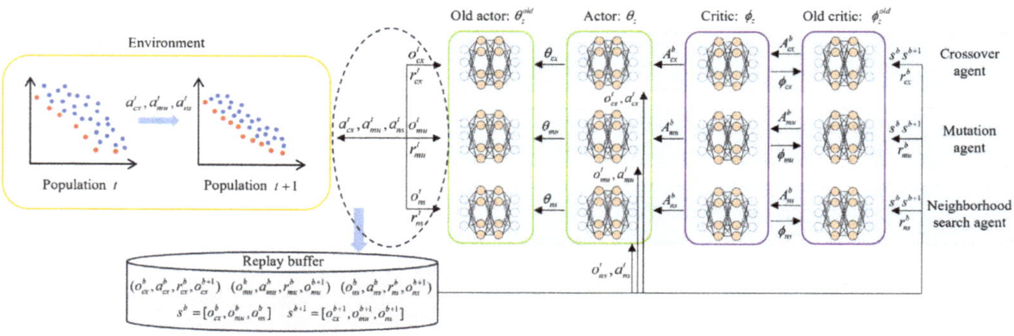

Figure 4. Overall framework of the algorithm.

First, within the current generation population, each of the three agents, based on its respective observation o_z^t and its old actor network $\pi_{\theta_z^{old}} = (a_z^t | o_z^t)$ (where θ_z^{old} represents the parameters of the old actor network), executes its corresponding evolutionary operation to generate its own offspring. The offspring from all agents are merged to form a new population. This new population undergoes non-dominated sorting to update the Pareto front solution set. The individual rewards r_z^t are then calculated for each agent, with experience tuples $(o_z^t, a_z^t, r_z^t, o_z^{t+1})$ stored in the replay buffer. The DRM is periodically updated by statistically analyzing gene structures of elite Pareto solutions.

When the number of tuples in the replay buffer reaches the threshold N, the network training process begins. After sampling a minibatch of data $|B|$, the critic network parameters ϕ_z are updated. The state for each agent s^b is formed by combining their respective local states into a global state representation. By sharing this global state information, the policies of the individual agents are collaboratively optimized. Specifically, the critic loss function is minimized:

$$L(\phi_z) = \frac{1}{2|B|} \sum_{b=1}^{|B|} \left[v_{\phi_z}(s^b) - (A_z^b + v_{\phi_z^{old}}(s^b)) \right]^2 \tag{53}$$

The advantage function A_z^b is computed using the old critic network ϕ_z^{old}. The critic network parameters are updated via stochastic gradient descent with learning rate α_ϕ: $\phi_z \leftarrow \phi_z - \alpha_\phi \nabla_{\phi_z} L(\phi_z)$. Secondly, the actor loss function is computed:

$$J(\theta_z) = -\frac{1}{|B|}\sum_{b=1}^{|B|}\left[\min\left(r_b(\theta_z)A_z^b, clip(r_b(\theta_z), 1-\varepsilon, 1+\varepsilon)A_z^b\right)\right] \tag{54}$$

$$r_b(\theta_z) = \frac{\pi_{\theta_z}\left(a_z^b \middle| o_z^b\right)}{\pi_{\theta_z^{old}}\left(a_z^b \middle| o_z^b\right)} \tag{55}$$

where $r_b(\theta_z)$ is the importance sampling ratio and ε is the clipping range. The actor network parameters are then updated via stochastic gradient descent with learning rate α_θ: $\theta_z \leftarrow \theta_z - \alpha_\theta \nabla_{\theta_z} L(\theta_z)$. Finally, the trained parameters ϕ_z and θ_z are assigned to ϕ_z^{old} and θ_z^{old}, respectively, and the replay buffer is cleared. This iterative process of evolution and training continues until the maximum predefined number of generations is reached, at which point the algorithm terminates.

5. Case Study

5.1. Case Introduction

This section presents a case to verify the effectiveness of the proposed algorithm, based on the actual layout and operational data of coal rail-water intermodal ports in northern China, as illustrated in Figure 5. For the unloading operation, the large dumper and small dumper can unload 8400 t and 4200 t of coal per operation, taking 80 min and 60 min, respectively, to unload to the nearest pile. For the loading operation, the ship loader takes 90 min to load coal from the nearest pile to the ship. The maximum storage capacity of each pile is limited to 30,000 t. Berths B3 and B6 have a maximum capacity of 35,000 deadweight tons, while others accommodate up to 50,000 deadweight tons.

The case study period was selected from 10:00 to 22:00 on a specific day. During this period, 15 heavy-haul trains arrived and were numbered in the order of their arrival. The arrival times, formulation, and the coal type carried by each train are summarized in Table 4. In addition, loading operations for six ships had to be completed, with the arrival time and loading requirements of each ship specified in Table 5.

Table 4. Train data.

No.	Time	Formulation	Coal Type	No.	Time	Formulation	Coal Type	No.	Time	Formulation	Coal Type
TR1	10:12	10 kt	3	TR6	13:52	10 kt	4	TR11	15:13	5 kt	1
TR2	11:27	20 kt	4	TR7	14:10	20 kt	1	TR12	16:08	10 kt	4
TR3	12:45	5 kt	2	TR8	14:19	20 kt	3	TR13	18:04	20 kt	5
TR4	13:01	20 kt	2	TR9	14:33	15 kt	3	TR14	18:37	15 kt	5
TR5	13:24	15 kt	5	TR10	14:50	20 kt	1	TR15	19:11	10 kt	2

The algorithm parameters were rigorously configured through empirical validation: the population size was set to 500, the number of iteration episodes was set to 10,000, the network learning rate was set to 0.0001, and the clipping parameter was set to 0.15. All experiments were executed on an Intel Core i7-11800H 2.3 GHz CPU with 16 GB of RAM (Intel Corporation, Santa Clara, CA, USA).

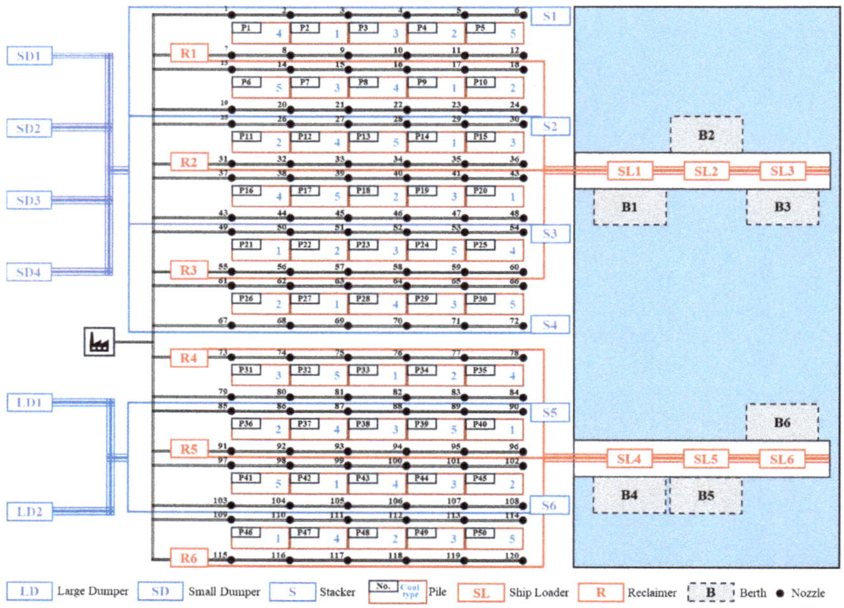

Figure 5. Port layout. This figure shows the layout of the port's loading and unloading equipment and the configuration of the stockyard. There are two stockyards in the port, one is in the shape of 6 rows and 5 columns, and the other is in the shape of 4 rows and 5 columns. The number in the upper left corner of each pile is its number, and the number in the middle indicates the type of coal that can be stored in it. There is a nozzle at each of the four vertices of each pile.

Table 5. Ship data.

No.	Time	Loading Operation
SP1	10:06	#1: 2/8500 t, #2: 1/8800 t, #3: 4/9400 t, #4: 1/8900 t, #5: 3/8500 t
SP2	10:20	#1: 3/8300 t, #2: 5/8500 t, #3: 3/9200 t, #4: 2/8700 t, #5: 4/8400 t
SP3	10:22	#1: 3/9000 t, #2: 4/9700 t, #3: 3/9200 t
SP4	10:27	#1: 4/9500 t, #2: 5/9800 t, #3: 1/8700 t, #4: 5/9200 t, #5: 3/9300 t
SP5	10:43	#1: 4/9000 t, #2: 4/9500 t, #3: 1/8900 t
SP6	11:17	#1: 2/9200 t, #2: 2/8800 t, #3: 1/9500 t, #4: 5/9000 t, #5: 3/8500 t

For instance, "#1:2/8500 t" indicates that the first loading operation of SP1 requires 8500 tons of coal type 2.

5.2. Results Analysis

5.2.1. Computational Results

By employing the proposed algorithm, we obtained the Pareto front solutions, as illustrated in Figure 6.

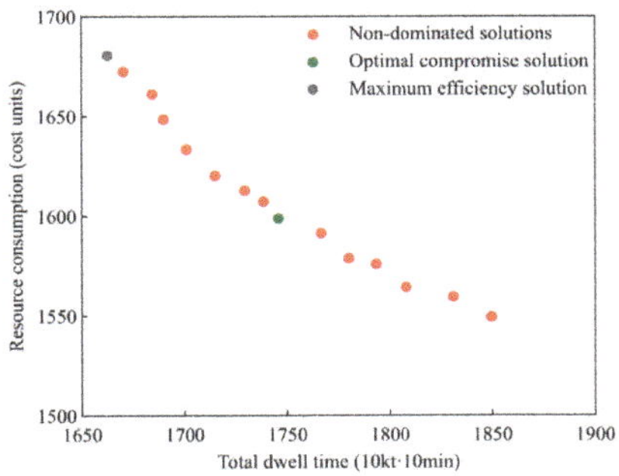

Figure 6. Pareto front. The horizontal axis represents the total time of loading and unloading operations, and the vertical axis represents the total unit cost of resource consumption.

The Pareto front provides a quantitative decision-making framework for navigating the trade-off between efficiency and resource conservation in port operations. As visualized in Figure 6, the Pareto front shows a distinct negative correlation: a reduction in total operational time is generally accompanied by an increase in resource consumption, embodying the inherent balance between the two objectives.

To screen out the optimal compromise solution (OC solution) that balances these dual goals from the Pareto front, we require an equitable evaluation basis. Given the significant scale disparities between the two objectives, this basis can be established through normalization. Therefore, we adopt min-max normalization, which maps both objectives to the unified interval [0, 1], to ensure their equitable contribution to the OC solution identification. The min-max normalization formula applied is:

$$\overline{f}_k^n = \frac{f_k^n - f_{k,\min}}{f_{k,\max} - f_{k,\min}} \tag{56}$$

where $\overline{f}_k^n$ is the normalized value of the $n-$ th individual for objective k, and $f_{k,\min}, f_{k,\max}$ are the minimum and maximum value of the objective k across all non-dominated solutions. After normalization, we calculate the Euclidean distance from each solution to the ideal point (0, 0) in the normalized objective space. The green node in Figure 6, which exhibits the minimum Euclidean distance, is selected as the OC solution. Subsequently, we select the operation plan that maximizes the loading and unloading efficiency from the Pareto front, which is the gray node in Figure 6 (ME solution).

The ME solution requires 25.75 Mt·min for train unloading operations and 140.50 Mt·min for ship loading operations, with total resource consumption reaching 1680.49 cost units. Comparatively, the OC solution demonstrates balanced scheduling performance: train unloading requires 30.85 Mt·min, ship operations require 143.72 Mt·min, while resource costs are reduced to 1598.71 cost units. This represents a strategic trade-off, in which a 4.87% reduction in resource consumption results in a 5.00% increase in operational time, thereby improving port sustainability. The following comparative analysis will examine the two solutions in terms of equipment scheduling for loading and unloading operations, stockyard allocation, and resource utilization.

(1) Analysis of unloading operations

The unloading-stacking operational plans for both the ME solution and OC solution are presented in Figure 7. The two solutions effectively balance the workloads between the two stockyards at the port, ensuring that neither of them is overloaded. By distributing unloading operations between the two stockyards based on the equipment capacity, they avoid bottlenecks in dumper and stacker usage. This is a prerequisite for generating an efficient and reasonable operation plan. The unloading operation lines and sequence should also be arranged reasonably to ensure non-conflicting equipment use in each process, minimize waiting times, and reduce train dwell time at the station.

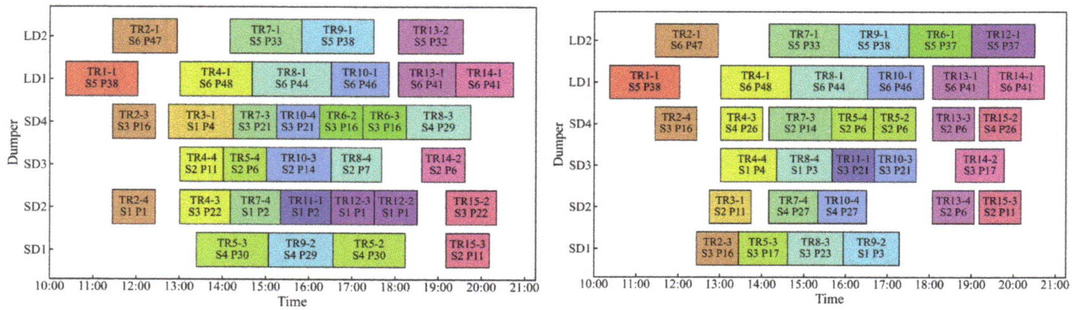

(**a**) (**b**)

Figure 7. Gantt chart of unloading-stacking operational plans. Among them, (**a**) is the ME solution, and (**b**) is the OC solution. The labels within each rectangle denote the unloading and stacking plans for unit trains. For instance, train TR2 of ME solution is split into TR2-1 (10 kt unit train), TR2-3 (5 kt unit train) and TR2-4 (5 kt unit train); TR2-1 completes unloading at LD2, after which the coal is transferred and stacked at P47 via S6.

As shown in the comparison between Figure 7a,b, the analysis reveals that both solutions have adjusted the unloading sequence of the train to prioritize turnover efficiency of different types of coal. During the research period, coal type 1 and coal type 3 are relatively scarce, while the reserve of coal type 5 is relatively abundant. To ensure that the loading operations are completed on time, the unloading operations of the train carrying the abundant coal type will be moderately postponed, and the unloading operation of the scarce coal type will be given priority. For example, the unloading operation of TR5, which carries coal type 5, is postponed to prioritize the unloading operations of TR7 carrying coal type 1 and TR8 carrying coal type 3.

During the unloading-stacking process, the resource consumption of the spraying operation is jointly affected by both environmental conditions during coal handling and the position of the pile. Severe external environmental factors can substantially increase the amount of dust suppression spray required per unit of coal. For the ME solution, it prioritizes matching the train arrival with the unloading equipment capacity to maintain the operation continuity, which may cause higher resource consumption. In contrast, the OC solution schedules unloading operations to reduce the duration of coal storage under unfavorable external environmental conditions by selecting appropriate operation times, thereby lowering overall resource consumption.

In terms of the allocation of piles, the ME solution prioritizes operational efficiency over the impact of pile location on resource consumption. As a result, piles located near the loading-side are preferentially selected, such as P30 is used for unloading coal from TR5-2 and TR5-3 to streamline loading operations. However, supplying water to these distant piles requires extra electrical energy. Additionally, coal type 5 has a high inventory level and slow turnover rate, leading to extended storage duration and increased resource consumption. The OC solution, by comparison, carefully considers pile locations and inventory levels. Coal types with abundant inventory are allocated to piles near the unloading-side close to the pumping station in order to take advantage of the lower resource consumption of these piles. For instance, the coal of the unit train TR5-2 and TR5-4 is unloaded at P6.

(2) Analysis of loading operations

The reclaiming-loading operational plans for both the ME solution and OC solution are presented in Figure 8. The results of the two plans show that the berth allocation aligns with stockyard capacity, ensuring that the loading operation at each berth is compatible with the coal supply capacity of the corresponding stockyard, thus preventing bottlenecks caused by mismatched loads.

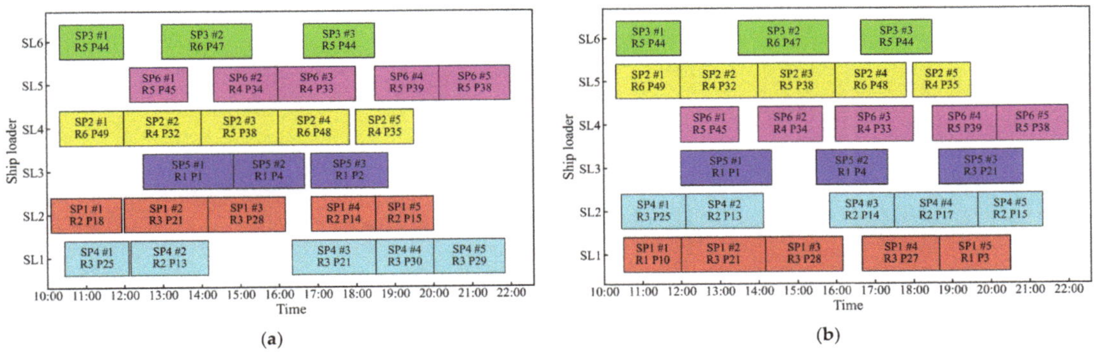

Figure 8. Gantt chart of reclaiming-loading operational plans. Among them, (**a**) is the ME solution, and (**b**) is the OC solution. The labels within each rectangle denote the reclaiming and loading plans for ships. For instance, ship SP3 of ME solution has been designated for loading operations with SL6 and the coal required for the first loading operation is reclaimed at P44 via R5.

The loading operations are restricted by the allocation of the piles and the initial coal inventory level of each pile. The longer the coal is stored in piles, the greater the negative impact on operational efficiency and resource consumption. The ME solution prioritizes seamless coordination between reclaimers and ship loaders. Therefore, except for the necessary time to wait for the replenishment of the scarce coal types, the loading operations at all berths are arranged continuously and efficiently. This focus on efficiency is evident in its pile selection, where loading-side piles are preferred to reduce loading time. For example, SP1's fifth loading operation reclaims coal from P15 instead of P3 to accelerate the process. While the OC solution maintains reasonable loading efficiency, it integrates resource consumption into its decision-making. It adjusts pile selection to balance speed and resource usage, such as choosing P3, which is located in the middle of stockyard, for SP1's fifth loading operation. Though this extends loading time slightly, it reduces the consumption of spraying resources.

Both solutions prioritize piles with sufficient inventory to avoid delays. The ME solution reduces reclaimers' and ship loaders' idle time through precise scheduling. While the OC solution avoids excessive delays and allows minor operational gaps to utilize low-consumption piles, ensuring conflict-free equipment use and alignment with long-term resource goals.

(3) Analysis of stockyard allocation and resource consumption

Based on the changes in the coal inventory of piles, 10 representative piles are selected from the ME solution and the OC solution, and the changes in their coal inventory over time are plotted, as shown in Figure 9. In addition, to visually display the resources, especially water and electricity, consumed by the spraying operation in each pile under the two plans, four heat maps are shown in Figure 10.

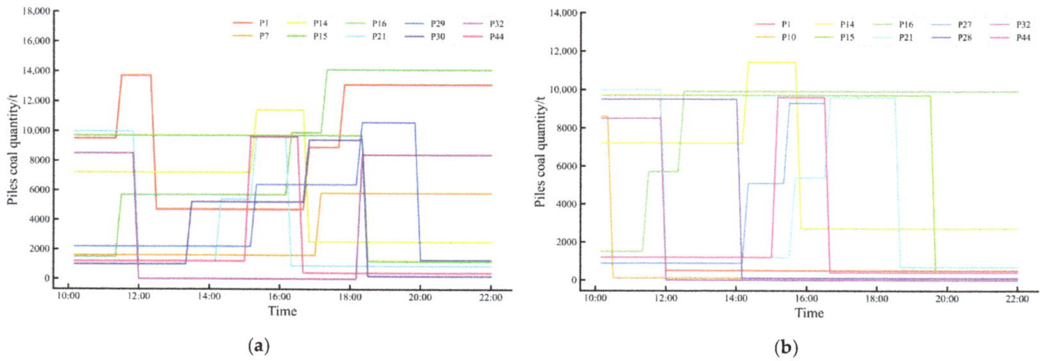

Figure 9. Curve graph depicting changes in coal volume at piles. Specifically, (**a**) is the ME solution, and (**b**) is the OC solution.

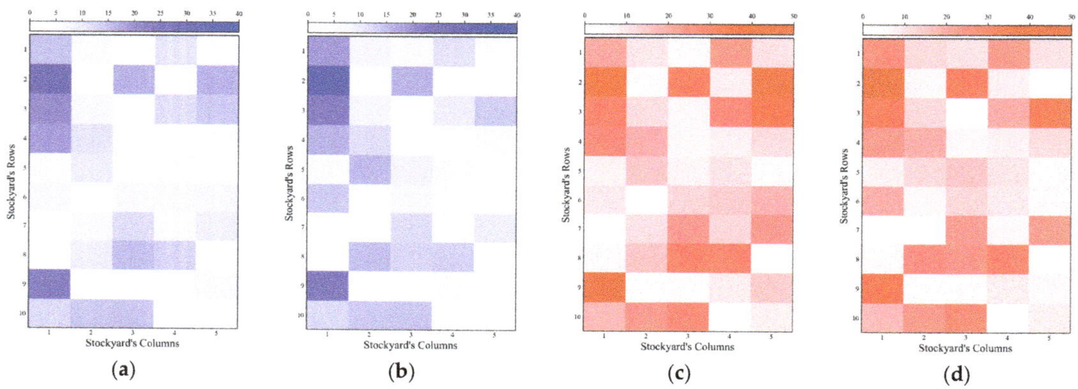

Figure 10. Heat maps illustrating resource consumption of spraying operations. Specifically, (**a**) and (**b**) depict the water resource consumption associated with the ME solution and the OC solution. Similarly, (**c**,**d**) represent the electricity consumption for the ME solution and the OC solution.

Under the ME solution, Figure 9a reveals pronounced inventory fluctuations across key piles such as P14 and P44, indicating concentrated stacking and reclaiming operations at shared locations. This operational clustering enhances equipment coordination but elevates average pile inventory levels, particularly in loading-side piles distant from pumping stations. Consequently, extended coal exposure duration intensifies spraying demands during peak operational periods, as reflected in the deeper hues across the ME solution's heat map in Figure 10a,c. While this strategy boosts short-term port throughput, it heavily relies on high resource consumption areas, which may raise long-term resource costs.

As shown in Figure 9b, the OC solution adopts a more balanced allocation strategy, with average inventory adjustments at each pile, keeping most inventory levels low by the end of the period. This strategy reduces spraying intensity by avoiding prolonged coal storage in intensive zones, especially in piles with high spraying demand near loading berths. The corresponding heatmap in Figure 10b,d visually confirms this advantage through uniformly lighter resource consumption patterns, especially in central piles accessible to both unloading and loading zones.

The fundamental divergence in stockyard allocation strategies manifests through distinct operational plans. Heat maps in Figure 10 provide empirical validation of these strategic foundations. The ME solution prioritizes handling immediacy by concentrating activities on loading-side piles, intentionally accepting elevated resource costs to minimize operational time. This plan prefers loading-side locations despite extended water transmission distances. However, the OC solution implements spatial equilibrium coordination by optimizing coal handling flows across two operational dimensions: the unloading-stacking path and the reclaiming-loading path. This integrated scheduling dynamically leverages applicable piles to balance equipment deployment efficiency. Through strategic sequencing of stacker and reclaimer operations, the solution minimizes localized resource pressure while maintaining throughput efficiency.

In summary, the two obtained solutions fully satisfy the dispatching requirements under various conditions in actual operational organization, and port dispatchers can select the most appropriate operational plan based on real-time demands.

5.2.2. Algorithm Performance

To verify the effectiveness and applicability of the proposed algorithm, test instances of varying scales are generated based on the number of arriving trains and ship operations. Comparative analyses are conducted against four benchmark algorithms: NSGA-II, IDQN, MAA2C, and MAPPO. An approximate Pareto optimal front is constructed by merging and filtering the non-dominated solution sets generated by each algorithm. HV and IGD are selected as evaluation metrics to assess solution set quality, with the calculated results presented in Table 6.

Table 6. Performance comparison of different algorithms.

Istance (Train, Ship)	HV					IGD				
	MEA-MAPPO	NSGA-II	IDQN	MA-A2C	MA-PPO	MEA-MAPPO	NSGA-II	IDQN	MA-A2C	MA-PPO
I1 (15, 6)	1.0573	0.8768	0.8302	0.9126	0.9053	0.0365	0.1493	0.2064	0.0892	0.0921
I2 (16, 6)	1.1483	0.7639	0.8252	0.8583	0.9205	0.0339	0.2904	0.2128	0.1976	0.1082
I3 (17, 7)	1.0885	0.8022	0.6816	0.7475	0.7794	0.0952	0.1866	0.3271	0.2793	0.2365
I4 (18, 7)	1.2062	0.9303	0.6573	0.8434	0.7252	0.0263	0.1437	0.4065	0.2070	0.3527
I5 (19, 7)	1.1704	0.8145	0.8957	0.8608	0.9032	0.0176	0.2672	0.1973	0.2164	0.1793
I6 (20, 8)	1.0948	0.6996	0.8489	0.7313	0.8177	0.0383	0.3974	0.2065	0.3325	0.2204
I7 (21, 8)	1.1501	0.6620	0.9037	0.7158	0.8640	0.0409	0.4083	0.1818	0.3578	0.2419
I8 (23, 9)	1.0949	0.7413	0.6801	0.9195	0.7518	0.0336	0.3056	0.3549	0.1691	0.2878
I9 (24, 9)	1.2620	0.8174	0.7342	0.7810	0.8393	0.0127	0.2722	0.3780	0.3309	0.2442
I10 (26, 10)	1.1363	0.7221	0.6128	0.8290	0.8424	0.0231	0.3874	0.4669	0.2547	0.2165

The results demonstrate that the proposed algorithm achieves optimal overall performance across all test instances. The HV metric directly reflects the combined performance of solution set convergence and diversity in the objective space. The HV values of the proposed algorithm are significantly superior to those of NSGA-II and IDQN. This advantage originates from the design of the hybrid framework, which integrates the multi-agent dynamic policy optimization concept into the adaptive operator selection process within the MEA. Agents can adaptively adjust their evolutionary strategies based on their respective observed state information. This mechanism transforms conventional random search during evolutionary operations into an optimized strategy guided by proven superior genetic information, effectively accelerating training efficiency and convergence speed. Furthermore, the proposed algorithm utilizes the hypervolume change per generation as a shared reward. By feeding back to the agents in a combined form incorporating global optimization information and local rewards, it enhances the quality of the agents' collaborative strategies for guiding population evolution. Consequently, the search efficiency and convergence precision within the solution space are optimized.

The IGD metric measures the proximity and coverage of the solution set relative to the true Pareto front. The proposed algorithm also demonstrates superior performance in terms of IGD. Its core advantage lies in the structured evolutionary mechanism guided by the DRM, which provides agents with structured prior knowledge for decision-making. In contrast, the MAA2C and MAPPO frameworks lack the integration of micro evolution principles. Their insufficient global search capability and difficulty in effectively balancing exploration and exploitation during training lead to uneven distribution of solution sets. Regarding the proposed algorithm, the DRM is periodically updated by extracting dominant gene structures from elite individuals, ensuring dynamic adaptation to population evolution trends. When agents execute decisions, the guidance based on the DRM ensures that historically validated efficient decision patterns are directionally transmitted to offspring, while retaining a degree of randomness in operator execution to avoid premature convergence.

The advantages of the proposed algorithm in solution quality and performance become more pronounced as the problem scale increases. This is attributed to the collaborative mechanism among the agents. In terms of state observation, agents dynamically adjust evolutionary operators based on real-time population states and Pareto front changes. Moreover, during the training process, agents share observed state information to construct a global state representation, enhancing the accuracy of the critic network in evaluating strategy value.

The action space and operator selection of each agent are designed with distinct focuses, thereby maintaining the diversity and convergence of the population. Specifically, during crossover operations, agents combine dominant gene structures, making new individuals more likely to inherit characteristics of high-quality solutions. For mutation operations, the mechanism selectively adjusts gene structures based on DRM, thereby effectively preventing the algorithm from converging to local optima. The neighborhood search agent performs targeted optimization on sparsely distributed regions of the Pareto front. The various operator combination strategies under adaptive selection can effectively balance exploration and exploitation, thereby improving the efficiency and quality of population evolution. With regard to reward assignment for each agent, while focusing on the improvement of HV, the local reward is determined by the proportion of each agent's contribution to the Pareto front. Through such division of labor, agents collaboratively optimize evolutionary strategies, effectively driving the population towards global optimality and ensuring effective search of the solution space even for large-scale problems.

In summary, the MEA integrated with MAPPO significantly enhances the efficiency of multi objective optimization and the quality of the final solutions through its adaptive parameter adjustment mechanism. The resulting Pareto front exhibits marked improvements in both convergence and diversity.

6. Conclusions

This paper investigates the integrated optimization of handling equipment scheduling and stockyard spraying operations in coal ports, addressing the complex trade-off between operational efficiency and resource conservation by systematically modeling the end-to-end operational process, which encompasses train splitting plans, berth allocation, handling operations, stockyard allocation, and spraying operations. On this basis, we formulate a MILP model aimed at minimizing the total operating time for trains and ships, as well as reducing resource consumption costs, particularly those associated with water and electricity usage during spraying operations. This model captures the intricate interdependencies between equipment coordination, stockyard allocation, and environmental constraints, overcoming the limitations of fragmented optimization in traditional approaches.

To solve this NP-hard problem, we develop a novel hybrid algorithm that combines the global search strengths of the MEA framework with the adaptive decision-making capabilities of the MAPPO. The two-layer chromosome encoding structure maintains full coverage of the solution space: the lower layer encodes selection and sequencing of unloading and loading lines, while the upper layer captures the gene structures related to train splitting and berth allocation decisions. Through the establishment of three cooperative agents for crossover, mutation, and neighborhood search, a collaborative optimization strategy is formed to jointly guide the evolution of the population. The state representation combines the individual observational features of each agent with public population-level features, allowing agents to perceive both the local and global dynamics of the evolutionary process. During the training of the policy network, a global state is adopted to ensure coordinated decision-making across agents. The action space controls the selection of parents and operators, as well as the adjustment of probabilities, while the DRM provides statistical guidance to preserve or perturb gene structures in offspring. The composite reward mechanism couples global HV improvement with the local contributions of each agent, ensuring that exploration and exploitation are balanced during evolution. This MAPPO-driven collaborative adaptation enables the evolutionary operators to effectively steer the population toward globally optimal solutions while maintaining diversity.

Extensive computational experiments on a real-world coal port case validate the effectiveness of the proposed algorithm. A detailed comparison of two representative solutions on the Pareto front highlights the essential trade-off between efficiency and resource conservation. The case study shows that the OC solution achieves a 4.87% reduction in resource consumption with a 5.00% increase in total operating time compared to the ME solution, thereby demonstrating the algorithm's ability to balance dual objectives. Operational analyses further reveal that the ME solution prioritizes speed through concentrated loading-side operations, whereas the OC solution emphasizes long-term sustainability by means of balanced stockyard workloads and strategic stockyard allocation. This flexibility provides dispatchers with informed options that can be aligned with real-time priorities of either efficiency or sustainability. Beyond the case study, comparative analyses across ten instance scales consistently confirm the superiority of the proposed MEA-MAPPO algorithm, which outperforms NSGA-II, IDQN, MA-A2C, and MAPPO in both hypervolume and inverted generational distance, thereby validating its stronger convergence performance and enhanced solution diversity.

Future research could extend this study by incorporating uncertainties such as stochastic arrivals, fluctuating coal demand, and variations in operational capabilities caused by external factors into the scheduling model and by integrating more granular equipment coordination to enhance the practicality of the framework. Additionally, scaling the algorithm to multi-port scenarios could further advance its applicability in complex logistics networks. Overall, this study provides a theoretical and methodological foundation for smart and green port operations, contributing to the balance between efficiency and sustainability in bulk cargo handling.

Author Contributions: Conceptualization, Y.W. and S.H.; methodology, Y.W. and S.H.; software, Y.W. and Z.L.; validation, Y.W. and H.T.; formal analysis, H.T. and A.X.; investigation, S.H.; resources, Y.W.; data curation, Z.L. and A.X.; writing—original draft preparation, Y.W., and H.T.; writing—review and editing, Y.W. and S.H.; visualization, H.T.; supervision, S.H. and A.X.; project administration, Y.W. and S.H.; funding acquisition, S.H. All authors have read and agreed to the published version of the manuscript.

Funding: This research was funded by the Fundamental Research Funds for the Central Universities of Ministry of Education of China (2024JBZX038).

Data Availability Statement: Data are contained within the article.

Conflicts of Interest: The authors declare no conflicts of interest.

References

1. International Energy Agency. Coal 2024: Analysis and Forecast to 2026. Available online: https://www.iea.org/reports/coal-2024/executive-summary (accessed on 26 July 2025).
2. Liu, J.; Lyu, Y.; Wu, J.; Wang, J. Adoption Strategies of Carbon Abatement Technologies in the Maritime Supply Chain: Impact of Demand Information Sharing. *Int. J. Logist.-Res. Appl.* **2025**, *28*, 70–97. [CrossRef]
3. Lv, Y.; Zou, M.; Li, J.; Liu, J. Dynamic Berth Allocation under Uncertainties Based on Deep Reinforcement Learning towards Resilient Ports. *Ocean Coast. Manag.* **2024**, *252*, 107113. [CrossRef]
4. Xia, Z.; Guo, Z.; Wang, W.; Jiang, Y. Joint Optimization of Ship Scheduling and Speed Reduction: A New Strategy Considering High Transport Efficiency and Low Carbon of Ships in Port. *Ocean Eng.* **2021**, *233*, 109224. [CrossRef]
5. Li, H.; Zhao, J.; Jia, P.; Ou, H.; Zhao, W. Optimization of Bulk Cargo Terminal Unloading and Outbound Operations Based on a Deep Reinforcement Learning Framework. *J. Mar. Sci. Eng.* **2025**, *13*, 105. [CrossRef]
6. Golias, M.; Portal, I.; Konur, D.; Kaisar, E.; Kolomvos, G. Robust Berth Scheduling at Marine Container Terminals via Hierarchical Optimization. *Comput. Oper. Res.* **2014**, *41*, 412–422. [CrossRef]
7. Gambardella, L.M.; Mastrolilli, M.; Rizzoli, A.E.; Zaffalon, M. An Optimization Methodology for Intermodal Terminal Management. *J. Intell. Manuf.* **2001**, *12*, 521–534. [CrossRef]
8. Chen, L.; Bostel, N.; Dejax, P.; Cai, J.; Xi, L. A Tabu Search Algorithm for the Integrated Scheduling Problem of Container Handling Systems in a Maritime Terminal. *Eur. J. Oper. Res.* **2007**, *181*, 40–58. [CrossRef]
9. Hsu, H.-P.; Tai, H.-H.; Wang, C.-N.; Chou, C.-C. Scheduling of Collaborative Operations of Yard Cranes and Yard Trucks for Export Containers Using Hybrid Approaches. *Adv. Eng. Inform.* **2021**, *48*, 101292. [CrossRef]
10. Zeng, Q.; Yang, Z. Integrating Simulation and Optimization to Schedule Loading Operations in Container Terminals. *Comput. Oper. Res.* **2009**, *36*, 1935–1944. [CrossRef]
11. Cao, Y.; Yang, A.; Liu, Y.; Zeng, Q.; Chen, Q. AGV Dispatching and Bidirectional Conflict-Free Routing Problem in Automated Container Terminal. *Comput. Ind. Eng.* **2023**, *184*, 109611. [CrossRef]
12. Li, L.; Li, Y.; Liu, R.; Zhou, Y.; Pan, E. A Two-Stage Stochastic Programming for AGV Scheduling with Random Tasks and Battery Swapping in Automated Container Terminals. *Transp. Res. Part E-Logist. Transp. Rev.* **2023**, *174*, 103110. [CrossRef]
13. Homayouni, S.M.; Tang, S.H.; Ismail, N.; Ariffin, M.K.A.M.; Samin, R. A Hybrid Genetic-Heuristic Algorithm for Scheduling of Automated Guided Vehicles and Quay Cranes in Automated Container Terminals. In Proceedings of the CIE: 2009 International Conference on Computers and Industrial Engineering, Troyes, France, 6–9 July 2009; Volume 1–3, pp. 96–101.
14. Chen, X.; Liu, S.; Zhao, J.; Wu, H.; Xian, J.; Montewka, J. Autonomous Port Management Based AGV Path Planning and Optimization via an Ensemble Reinforcement Learning Framework. *OCEAN Coast. Manag.* **2024**, *251*, 107087. [CrossRef]
15. Liang, C.; Zhang, Y.; Dong, L. A Three Stage Optimal Scheduling Algorithm for AGV Route Planning Considering Collision Avoidance under Speed Control Strategy. *Mathematics* **2023**, *11*, 138. [CrossRef]
16. Yang, Y.; Sun, S.; He, S.; Jiang, Y.; Wang, X.; Yin, H.; Zhu, J. Research on the Multi-Equipment Cooperative Scheduling Method of Sea-Rail Automated Container Terminals under the Loading and Unloading Mode. *J. Mar. Sci. Eng.* **2023**, *11*, 1975. [CrossRef]
17. Chen, H.; Liu, W.; Oldache, M.; Pervez, A. Research on Train Loading and Unloading Mode and Scheduling Optimization in Automated Container Terminals. *J. Mar. Sci. Eng.* **2024**, *12*, 1415. [CrossRef]
18. Munuzuri, J.; Lorenzo-Espejo, A.; Pegado-Bardayo, A.; Escudero-Santana, A. Integrated Scheduling of Vessels, Cranes and Trains to Minimize Delays in a Seaport Container Terminal. *J. Mar. Sci. Eng.* **2022**, *10*, 1506. [CrossRef]
19. de Andrade, J.L.M.; Menezes, G.C. A Column Generation-Based Heuristic to Solve the Integrated Planning, Scheduling, Yard Allocation and Berth Allocation Problem in Bulk Ports. *J. Heuristics* **2023**, *29*, 39–76. [CrossRef]
20. Tang, X.; Jin, J.G.; Shi, X. Stockyard Storage Space Allocation in Large Iron Ore Terminals. *Comput. Ind. Eng.* **2022**, *164*, 107911. [CrossRef]
21. Ouhaman, A.A.; Benjelloun, K.; Kenne, J.P.; Najid, N. The Storage Space Allocation Problem in a Dry Bulk Terminal: A Heuristic Solution. *IFAC Pap.* **2020**, *53*, 10822–10827. [CrossRef]
22. van Vianen, T.; Ottjes, J.; Lodewijks, G. Simulation-Based Rescheduling of the Stacker-Reclaimer Operation. *J. Comput. Sci.* **2015**, *10*, 149–154. [CrossRef]
23. Xin, J.; Negenborn, R.R.; van Vianen, T. A Hybrid Dynamical Approach for Allocating Materials in a Dry Bulk Terminal. *IEEE Trans. Autom. Sci. Eng.* **2018**, *15*, 1326–1336. [CrossRef]
24. Bouzekri, H.; Alpan, G.; Giard, V. Integrated Laycan and Berth Allocation Problem with Ship Stability and Conveyor Routing Constraints in Bulk Ports. *Comput. Ind. Eng.* **2023**, *181*, 109341. [CrossRef]
25. Babu, S.A.K.I.; Pratap, S.; Lahoti, G.; Fernandes, K.J.; Tiwari, M.K.; Mount, M.; Xiong, Y. Minimizing Delay of Ships in Bulk Terminals by Simultaneous Ship Scheduling, Stockyard Planning and Train Scheduling. *Marit. Econ. Logist.* **2015**, *17*, 464–492. [CrossRef]

26. Fereidoonian, F.; Sadjadi, S.J.; Heydari, M.; Mirzapour Al-e-Hashem, S.M.J. A Timely Efficient and Emissions-Aware Multiobjective Truck-Sharing Integrated Scheduling Model in Container Terminals. *Proc. Inst. Mech. Eng. Part M-J. Eng. Marit. Environ.* **2024**, *238*, 982–1008. [CrossRef]
27. Zheng, Y.; Xu, M.; Wang, Z.; Xiao, Y. A Genetic Algorithm for Integrated Scheduling of Container Handing Systems at Container Terminals from a Low-Carbon Operations Perspective. *Sustainability* **2023**, *15*, 6035. [CrossRef]
28. Peng, Y.; Wang, W.; Liu, K.; Li, X.; Tian, Q. The Impact of the Allocation of Facilities on Reducing Carbon Emissions from a Green Container Terminal Perspective. *Sustainability* **2018**, *10*, 1813. [CrossRef]
29. Niu, Y.; Yu, F.; Yao, H.; Yang, Y. Multi-Equipment Coordinated Scheduling Strategy of U-Shaped Automated Container Terminal Considering Energy Consumption. *Comput. Ind. Eng.* **2022**, *174*, 108804. [CrossRef]
30. Peng, W.; Wang, D.; Qiu, H.; Chu, F.; Yin, Y. Integrated Optimization on Double-Side Cantilever Yard Crane Scheduling and Green Vehicle Path Planning at U-Shaped Yard. *IEEE Trans. Intell. Transp. Syst.* **2025**, *26*, 3684–3699. [CrossRef]
31. He, J.; Huang, Y.; Yan, W. Yard Crane Scheduling in a Container Terminal for the Trade-off between Efficiency and Energy Consumption. *Adv. Eng. Inform.* **2015**, *29*, 59–75. [CrossRef]
32. Lu, H.; Lu, X. Joint Optimization of Berths and Quay Cranes Considering Carbon Emissions: A Case Study of a Container Terminal in China. *J. Mar. Sci. Eng.* **2025**, *13*, 148. [CrossRef]
33. Jiang, X.; Zhong, M.; Shi, J.; Li, W. Optimization of Integrated Scheduling of Restricted Channels, Berths, and Yards in Bulk Cargo Ports Considering Carbon Emissions. *Expert Syst. Appl.* **2024**, *255*, 124604. [CrossRef]
34. Wang, W.; Guo, J.; Tian, Q.; Peng, Y.; Cao, Z.; Liu, K.; Peng, S. Stockyard Allocation in Dry Bulk Ports Considering Resource Consumption Reduction of Spraying Operations. *Transp. Res. Part E-Logist. Transp. Rev.* **2025**, *193*, 103816. [CrossRef]
35. Gong, L.; Huang, Z.; Xiang, X.; Liu, X. Real-Time AGV Scheduling Optimisation Method with Deep Reinforcement Learning for Energy-Efficiency in the Container Terminal Yard. *Int. J. Prod. Res.* **2024**, *62*, 7722–7742. [CrossRef]
36. Yue, L.; Fan, H. Dynamic Scheduling and Path Planning of Automated Guided Vehicles in Automatic Container Terminal. *IEEE-CAA J. Autom. Sin.* **2022**, *9*, 2005–2019. [CrossRef]
37. Zeng, Q.; Yang, Z.; Hu, X. A Method Integrating Simulation and Reinforcement Learning for Operation Scheduling in Container Terminals. *Transport* **2011**, *26*, 383–393. [CrossRef]
38. Zhong, R.; Wen, K.; Fang, C.; Liang, E. Real-Time Multi-Resource Jointed Scheduling of Container Terminals with Uncertainties Using a Reinforcement Learning Approach. In Proceedings of the 2022 13th Asian Control Conference, ASCC, Jeju, Republic of Korea, 4–7 May 2022; pp. 110–115.
39. Che, A.; Wang, Z.; Zhou, C. Multi-Agent Deep Reinforcement Learning for Recharging-Considered Vehicle Scheduling Problem in Container Terminals. *IEEE Trans. Intell. Transp. Syst.* **2024**, *25*, 16855–16868. [CrossRef]
40. Tang, G.; Guo, Y.; Qi, Y.; Fang, Z.; Zhao, Z.; Li, M.; Zhen, Z. Real-Time Twin Automated Double Cantilever Rail Crane Scheduling Problem for the U-Shaped Automated Container Terminal Using Deep Reinforcement Learning. *Adv. Eng. Inform.* **2025**, *65*, 103193. [CrossRef]
41. Zhu, J.; Zhang, W.; Yu, L.; Guo, X. A Novel Multi-Attention Reinforcement Learning for the Scheduling of Unmanned Shipment Vessels (USV) in Automated Container Terminals. *OMEGA-Int. J. Manag. Sci.* **2024**, *129*, 103152. [CrossRef]
42. Li, B.; Yang, C.; Yang, Z. Multiple Container Terminal Berth Allocation and Joint Operation Based on Dueling Double Deep Q-Network. *J. Mar. Sci. Eng.* **2023**, *11*, 2240. [CrossRef]
43. Li, C.; Wu, S.; Li, Z.; Zhang, Y.; Zhang, L.; Gomes, L. Intelligent Scheduling Method for Bulk Cargo Terminal Loading Process Based on Deep Reinforcement Learning. *Electronics* **2022**, *11*, 1390. [CrossRef]
44. Ai, T.; Huang, L.; Song, R.J.; Huang, H.F.; Jiao, F.; Ma, W.G. An Improved Deep Reinforcement Learning Approach: A Case Study for Optimisation of Berth and Yard Scheduling for Bulk Cargo Terminal. *Adv. Prod. Eng. Manag.* **2023**, *18*, 303–316. [CrossRef]
45. Cui, Y.; Geng, Z.; Zhu, Q.; Han, Y. Review: Multi-Objective Optimization Methods and Application in Energy Saving. *Energy* **2017**, *125*, 681–704. [CrossRef]

Journal of Marine Science and Engineering

MDPI

Article

Environmental and Economic Assessment of Alternative Marine Fuels for Bulk Carriers: A Comparative Analysis of Handymax, Panamax and Supramax Vessels

Georgios Charvalos [1,*], Athanasios Tzakis [1], Angelos Arvanitis [1], Sofia Peppa [2] and Christos Papadopoulos [1]

1 School of Naval Architecture and Marine Engineering, National Technical University of Athens, 15780 Athens, Greece; thanos.tzakis@hotmail.com (A.T.); arvagg1234@gmail.com (A.A.); chpap@mail.ntua.gr (C.P.)

2 Department of Naval Architecture, University of West Attica, 12243 Athens, Greece; speppa@uniwa.gr

* Correspondence: grcharvalos@mail.ntua.gr; Tel.: +30-694-848-0430

Abstract

In the present paper, a quantitative assessment of the effect of alternative fuel (LNG, LPG-B, LPG-P and MeOH) implementation in internal combustion engines in bulk carrier vessels on environmental compliance is presented. A fleet comprising 40 vessels across the Handymax, Panamax and Supramax classes is examined. By using LNG, the total fleet achieves environmental compliance up to 2030, with 52.5% of the fleet potentially achieving a minor superior energy ranking, while the EU ETS costs can be reduced by up to 24% compared to the case of burning conventional fuels. LPG-B and LPG-P demonstrated moderate improvements in the compliance period, with 50% to 87.5% and 52.5% to 97.5% surviving up to 2030, respectively. Reductions in the EU ETS costs were similar for these two fuels, with the reductions ranging from 3.3% to 12.1% for LPG-B and from 4.1% to 15.2% for LPG-P. Among all fuels, methanol showed the least improvement in extending the compliance period, with 52.5% to 67.5% of the fleet reaching 2030 with inferior to moderate CII ranks. The EU ETS cost reductions were low, ranging from 2.7% to 10%, with substantial fuel cost increases from 29.9% to 107%. The present study aims to assist ship owners/operators by providing a decision-support tool for bulk carrier alternative fuel pathways. Finally, it provides insights into the marine industry and shipping market regarding the future of the bulk carrier fleet in the context of decarbonization.

Keywords: decarbonization; bulk carriers; alternative fuels; CII; EU ETS; LNG; methanol; LPG

1. Introduction

As ongoing stringent environmental regulations push towards decarbonization, ship operators and owners are compelled to seek solutions to improve vessel energy efficiency and reduce fuel consumption and CO_2 emissions. Bulk carriers, vessels that are designed to transport unpackaged dry cargo, have been at the forefront of seaborne trade, representing the majority of the global fleet in terms of carrying capacity. Transporting large volumes of commodities, in 2024, they represented 42.7% of the global fleet, while as of February 2025, 14,180 bulk carriers were engaged in international trade [1].

Decarbonization is guided by the IMO's Revised GHG Strategy, which has set target checkpoints of at least a 20% (striving for 30%) reduction by 2030 and at least 70% (striving for 80%) by 2040, relative to 2008 levels. The strategy introduces important regulatory elements. The Energy Efficiency Design Index (EEDI) and the Energy Efficiency Existing Ship Index (EEXI) are technical standards for improving the energy performance of ships and reducing greenhouse gas (GHG) emissions. Moreover, the Carbon Intensity Indicator (CII) introduces an operational performance rating (A to E) that is based on the annual CO_2 emissions of the ship relative to its transport work, necessitating continuous improvement.

At the regional level, the European Union emissions trading system (EU ETS), as of 2024, covers CO_2 emissions from maritime transport. It requires shipping companies operating in the EU to monitor emissions and purchase allowances for 50% of the CO_2 emitted on intra-EU voyages and 100% during operations between EU ports. FuelEU, which entered into force in 2025, mandates a gradual reduction in the GHG intensity of onboard energy use, supporting the transition to alternative fuels produced from green and renewable energy.

In the wake of decarbonization, no solution is expected to dominate the market. Within the mixture of technological pathways, alternative fuels are an attractive option for reducing the carbon footprint. Zero-carbon fuels (hydrogen and ammonia) guarantee compliance and are foreseen to power the fleet in the long term, as they are currently hindered by technological, economic and infrastructure challenges. Thus, the use of alternative, low-carbon-content fuels is favored, and these fuels could serve as near-term transitional fuels. They include liquified natural gas, methanol (MeOH) and liquified petroleum gas (LPG).

According to data published by DNV [2], as of February 2025, 642 vessels are powered by LNG, with an additional 264 vessels having been ordered in 2024. Among these, 57 are bulk carriers, with 16 more on order [3]. The LNG-fueled fleet is anticipated to continue its steady expansion in the coming years, necessitating the development of robust and globally accessible bunkering infrastructure. In addition, 64 LNG bunkering vessels are presently in operation, with their capacities ranging from 1000 to 20,000 m^3, while 16 additional vessels are anticipated in 2028 [2]. As of 2024, 191 ports worldwide are equipped with operational LNG infrastructure, with 81 facilities being under construction.

The implementation of LNG has been examined from economic, safety and feasibility perspectives. In [4], Usiagu et al. conducted a review on the future role of LNG in decarbonization. Their analysis of LNG characteristics, environmental considerations, carbon reduction goals, policy factors, economic drivers, global production capacity and regional dynamics and geopolitical implications supports the view that LNG could serve as a transitional fuel towards future power generation from renewable energy sources. In [5], Schinas and Butler proposed a methodology for assessing the commercial viability of LNG as a marine fuel, including necessary incentives for its adoption. By analyzing contemporary regulations, fuel price statistics and industry data, they developed an economic model to compare fuel alternatives in terms of capital expenditure (CAPEX) and operational expenditure (OPEX). Their findings indicated that LNG-fueled vessels could reduce daily fuel costs by 20% to 30%, despite an increase in CAPEX ranging from 15% to 30%. In the absence of financial incentives for shipowners, retrofitting is primarily feasible for vessels operating in Emission Control Areas (ECAs) or on fixed routes. In [6], Wan et al. developed a model to evaluate LNG adoption across single or multiple regions. Combining quantitative and qualitative metrics, the model serves both as a policy development tool and as an adaptable framework for assessing emerging technologies in other economic sectors. Tam et al. [7] conducted a study on retrofitting a mini-Cape bulk carrier for dual-fuel LNG operation, focusing on cost efficiency and downtime reduction. The research identified significant challenges, with tank volume and placement being the primary constraints. It is further estimated that retrofit costs may exceed those of a conventional newbuilding by USD 8–10 million. Downtime was influenced by the yard's location, project duration, flag state requirements, classification society rules and the modularity of the LNG fuel system.

The impact of LNG within the European Union emissions trading system (EU ETS) was examined by He et al. [8]. Through a comparative analysis between estimated emission factors and data collected from an operational LNG engine, for an annual consumption of 2000 tons of fuel, a reduction in European Union Allowance (EUA) costs between USD 10,000 and 20,000 per vessel could be achieved. The authors emphasized the variability of emission factors under differing operational conditions and proposed the integration of methane emission coefficients into regulatory frameworks. Methane slip emissions were further studied by Psaraftis et al. [9], where the authors estimated a 174% rise in methane emissions between 2012 and 2018, which was attributed to a 30% increase in LNG use in marine engines.

LNG significantly enhances vessel energy efficiency. In [10], El-Manzalawy et al. demonstrated improvements to the Energy Efficiency Design Index (EEDI) of a containership using a dual-fuel LNG engine. Ejder [11] assessed LNG's potential in marine decarbonization by calculating the Carbon Intensity Indicator (CII) of a bulk carrier over eleven voyages spanning three years. The analysis covered navigation, port operations, power consumption and fuel usage, with the CII evaluated per voyage and annually. The results indicated that LNG could extend a vessel's lifespan by six years relative to conventional fuels. However, LNG alone remains insufficient to meet the IMO 2050 decarbonization targets, rendering it a viable short- to medium-term solution.

Environmental performance associated with LNG can be enhanced through the implementation of carbon capture systems. Carbon capture offers significant short-term improvements in a vessel's carbon footprint when compared with alternative fuels and fuel-saving technologies, contributing both technologically and competitively. Ros et al. [12] demonstrated that combining LNG with carbon capture can increase capture efficiency by utilizing LNG's cold energy to liquefy the CO_2 stream before combustion, constrained nevertheless, by the installed liquefaction capacity. In a parallel approach, Xia et al. [13] developed a thermodynamic model to reduce carbon emissions and enhance energy efficiency on a Newcastlemax bulk carrier. The proposed system integrates air separation, cold energy recovery, waste heat utilization and carbon capture. The results indicated a CO_2 capture rate surpassing the vessel's CO_2 production, due to oxygen-enriched combustion enabled by the air separation unit.

Methanol is considered a compelling alternative fuel option for future marine propulsion due to the ease of storage. As of March 2025, Clarkson data indicate that 54 vessels are equipped for methanol combustion, while 304 additional ships are under construction [14]. Furthermore, 121 existing vessels are suitable for retrofitting, and 586 methanol-ready newbuilds have been ordered. The global bunkering infrastructure remains nascent; according to the Methanol Institute, only ten operational bunkering sites exist worldwide. Nonetheless, several major ports are making strides in infrastructure development [15].

Despite methanol's promising outlook, comprehensive studies on vessel efficiency and performance under methanol combustion remain limited. This underscores the need for further ship design advancements to enable broader commercial adoption. In [16], Rachow utilized IMO draft reports and classification society guidelines to outline a safe design for methanol fuel systems in small tankers. In [17], Ammar conducted an economic–environmental analysis for a methanol-fueled containership, concluding that CO_2 emissions could be reduced by 18%, while reducing a vessel's speed could improve the economics of dual-fuel operation. In their corresponding works, Adami et al. [18,19] addressed the technical challenges of retrofitting Handymax and Panamax bulk carriers for dual-fuel methanol operation, highlighting the need for increased fuel storage. This necessitates trade-offs such as reduced endurance, frequent bunkering or a reduction in payload. Transporting cargo on medium-sized vessels results in a lower carbon footprint than smaller vessels, particularly when utilizing green or bio-methanol. Khaled et al. [20] evaluated MeOH–MDO dual-fuel use aboard cruise vessels, reporting emission reductions of CO_2 (25.7%), NO_x (38.46%), SO_x (45%), and PM (45%). Their analysis projected an annual savings of USD 17.5 million.

Despite LNG and MeOH offering reductions in carbon intensity, they alone cannot achieve the 2050 decarbonization targets. In research works, along with other energy technologies, they are considered as transitional fuels, supporting the gradual shift towards zero-carbon solutions. To highlight the importance of alternative fuels on decarbonization, Ampah et al. [21] retrieved and combined information from a plethora of documents on available databases. By analyzing the documents with R-studio, they identified LNG as the most researched alternative fuel, and confirmed increasing interest in methanol, ammonia and hydrogen. In [22], Ashrafi et al. highlighted the challenges associated with LNG, NH_3 and H_2 depending on the cost, infrastructure, availability and technological readiness.

Yan et al. [23] analyzed alternative fuel usage aboard a small bulk carrier in the Yangtze River. They concluded that green ammonia and green hydrogen hold the greatest decarbonization potential, exhibiting reductions of 78.8% and 91.3% respectively, while LNG and MeOH achieved 31.5% to 38% lower life cycle carbon emissions compared to diesel. In [24], Percic et al. conducted an analysis of the decarbonization pathways for short-sea Croatian passenger vessels, concluding that electrification is the optimum solution both in environmental and economical terms. The authors extended their work [25] by performing technical, environmental and economic analyses of alternative fuels on the Croatian inland fleet. By considering the ship operational profile and technical characteristics, the authors concluded that electrification is the optimum solution for passenger vessels, while methanol-powered systems, although hindered by the immature infrastructure, are the most cost-efficient solution for cargo ships. Zhao Yangfang. et al. [26] investigated the impact of alternative fuel transition on fuel costs, carbon emissions and retrofit costs. They claimed that bio-methanol and e-ammonia are expected to exhibit the lowest operational costs in the upcoming years, despite the fact that current investment in new ships remains low. The authors concluded that although no fuel stands out as a dominant future solution, only LNG and bio-diesel are favored due to their current infrastructure and small price fluctuations. Balcombe et al. [27] confirmed that LNG could serve as an intermediate solution towards decarbonization since it offers a moderate 10% reduction in GHG (affected by methane slip) and minor policy interventions.

Zhao et al. [28] proposed a power system and fuel optimization model for bulk carrier fleet decarbonization, identifying bio-LNG and MeOH as transitional fuels. Cumulative efficiency improvements can be achieved by reducing ballast water on new-built vessels, and by using solar panels and cold ironing on retrofitted vessels. In a similar approach [29], Xing et al., and Wang et al. [30], in their respective works, reaffirmed the transitional role of LNG, LPG and MeOH as having a significant environmental impact when the fuels are produced by green procedures, stressing the importance of fuel volume efficiency and storage. Loennechen et al. [31] developed a two-stage stochastic model for the Supramax fleet, identifying MeOH, LNG and e-ammonia as favorable options under renewable production scenarios. Taking into account vessel energy requirements, combining alternative fuels with wind power or carbon capture could further enhance compliance. Akman et al. [32], in a work proposed for bulk carrier preliminary design, confirmed the positive impact of LNG and MeOH on the EEDI of a vessel. On the conceptual design of a Handymax 35,000 t bulk carrier using alternative fuels, both fuels reduced the EEDI, with LNG exhibiting reduced values by 17.1% and 13.6% from diesel oil and MeOH, respectively. However, due to methane slip, GHG emissions were increased by 20%. Cost-effective options for a reduction in GHG emissions for a Supramax bulk carrier were investigated by Lagemann et al. [33].

The authors, by considering retrofit costs for each alternative fuel option and a tax penalty, concluded that LNG could offer significant reductions in GHG, with NH_3 produced from renewable electricity being the optimal solution for the vessel. Ahad Al-Enazi [34] et al., in an effort to address GHG holistically, developed a linear programming model to evaluate the optimal fuel option for LNG carriers, concluding that NH_3 and H_2/LNG blends are the most favored solutions. Al-Asmakh et al. [35] studied ship modifications for various fuels and concluded that LNG and H_2 offer strong environmental benefits. However, widespread adoption of H_2 is hindered by safety concerns, high costs and undeveloped infrastructure. In [36], Horvath et al. sought the most cost-effective combination of synthetic fuels for use in either ICEs or fuel cells. By comparing five that were synthetically produced and applied to both short-sea and deep-sea going vessels, they concluded that the most promising solution was the use of liquid H_2 in PEM fuel cells. Park et al. [37] assessed the decarbonization potential of alternative fuels and technologies for offshore supply vessels. LNG, with its well-developed infrastructure and a GHG reduction potential of 20%, along with hybrid propulsion systems, can serve as temporary solutions. Upon tackling significant challenges, H_2 and NH_3 can provide carbon elimination in the long term. Li et al. [38] reviewed fuel technologies and challenges, citing LNG slip, MeOH/H_2 storage difficulties and ammonia slip as key obstacles, highlighting the need for regulatory alignment and the importance of life cycle assessments for each technological solution. They advocate for optimization through the use of combustion control systems, thermal management optimization and multi-objective algorithms. LNG, combined with wind power, would be the most prominent short-term measure. Green ammonia, methanol and hydrogen are expected to become the dominant solutions in the long term upon the conduct of further research.

While abundant previous studies have examined alternative fuels from LCA and techno-economic views, a comprehensive comparative analysis of LNG, LPG variants and methanol across different bulk carrier classes under current regulatory frameworks remains limited. This study aims to determine which of these fuels can provide a suitable environmental compliance solution for the bulk carrier fleet.

The novelty of this work lies in evaluating the impact of alternative fuels (LNG, LPG-B, LPG-P and MeOH) on environmental compliance at the scale of an actual bulk carrier fleet rather than relying on a single custom ship as a case study. Moreover, the analysis is based on real consumption data from the fleet under consideration, providing a more accurate representation of the practical implications of adopting the selected fuels. This approach offers valuable insights into the suitability of different fuel pathways for bulk carriers, enabling shipowners and operators to use the findings as a decision-making tool when selecting retrofit options and compliance strategies for their fleets. Furthermore, the outcomes of this study can inform policymakers and stakeholders in shaping investment priorities and regulatory measures that effectively support the decarbonization of a bulk carrier fleet.

2. Methodology

In this work, for a fleet sample of 40 bulk carriers, the cases of LNG, LPG-B, LPG-P and MeOH are considered for use in internal combustion engines (ICEs). For each fuel, three application scenarios are examined as follows: alternative fuels only in auxiliary engines, alternative fuels only in the main engine, and alternative fuels in both main and auxiliary engines. These fuels were selected because such engines have been commercially available, rendering these fuels straightforward options. The combustion of the four selected fuels in the main engine and/or auxiliary engines is done because these engines are the ship's primary energy converters and fuel consumers. The main engine is a two-stroke marine diesel engine that is used for propulsion, while the auxiliary engines (A/Es) are four-stroke marine diesel engines that are used for electricity generation. The fleet sample used in this work, shown in Table 1, consists of four Handymax bulk carriers with a year of build (YOB) between 2009 and 2015, and 36 Panamax and Supramax vessels with a YOB ranging between 2006 and 2020. For each vessel, the deadweight (DWT), annual distance, heavy fuel oil (HFO), low sulfur fuel oil (LFO) and marine gas oil (MGO) annual consumption were available. The DWT ranged from 31,833 t to 61,644 t. All vessels are equipped with a marine two-stroke diesel engine for propulsion (main engine) and three four-stroke diesel engines used for electricity generation (auxiliary engines). The fleet consumed 48.13% of time at sea and spent 51.87% in port, at anchorage or undergoing repairs. The following assumptions were made:

1. The vessel operational profile is assumed constant for each of the following years.
2. Annual required CII reduction factors were obtained according to MEPC-83 [39].
3. The vessel compliance period is assumed to end on the third consecutive year that the ship is ranked as "D", or on the first year of "E" ranking; whichever occurs first.
4. The fuel storage and handling systems are not designed, and adequate space to facilitate the fuel is assumed to exist onboard the vessel. New-built or retrofitting costs are not accounted for.
5. Each fuel is considered for combustion in an ICE, and the engine thermal efficiency has been assumed (as it is a common practice) to be constant across the fuel options. The Fuel Emission Factors (EF) and Lower Heating Values (LHV) are presented in Table 2.
6. Alternative fuels are considered only for use in the main and auxiliary engines, as these are the ship's primary energy converters (propulsion and onboard electricity, respectively). Any other auxiliary machinery requiring combustion (e.g., auxiliary boiler) continues to burn conventional fuel.
7. Based on the online available data, the following fuel prices were assumed: HFO at 459.5 USD/t [40], LFO at 554 USD/t [40], MGO at 791 USD/t [40], LNG at 540 USD/t (average) [41–43], LPG-B at 490 USD/t [44], LPG-P at 520 USD/t [44] and MeOH at 530 USD/t (fossil fuel production) [45].
8. Each vessel is assumed to be equipped with a SO_x scrubber unit, combusting LFO or both. Since the dataset comprises annual fuel consumption records for heavy fuel oil (HFO), light fuel oil (LFO), and marine gas oil (MGO), the effects of the scrubber use on fuel consumption are directly included in the calculations through the fleet fuel consumption figures.
9. During the implementation of alternative fuels in the auxiliary engines, HFO and LFO are being combusted by the main engine. MGO is used as pilot fuel.
10. During the implementation of alternative fuels in the main engine, HFO and LFO are being combusted by the auxiliary engines. MGO is used as pilot fuel.
11. During the implementation of alternative fuels, both in the main and auxiliary engines, HFO and LFO are not used. MGO is used as pilot fuel.
12. According to the available information on the fleet sample, the vessels are trading within the EU at a percentage of 35% of their annual voyages, and 20% of these voyages occur between two EU ports (80% with only one EU port call). The EU ETS cost is assumed to be surrendered at 100%. The EUA price was considered 70 EUR/t [46].

Table 1. Fleet sample description.

Ship Type	Deadweight Range [t]	Year of Build Range	Average Speed Range [Knots]
Handymax	31,833–40,000	2009–2015	11–12
Panamax	40,000–82,000	2006–2020	9–14
Supramax	50,000–61,644	2006–2020	9–14

Table 2. Fuel Emission Factors (EF) and Lower Heating Value (LHV) (data taken from [47,48]).

Fuel	Emission Factor (EF) $[\frac{gr_{CO_2}}{gr_{fuel}}]$	LHV $[\frac{MJ}{kg}]$
HFO	3.114	40.5
LFO	3.151	41
MGO	3.206	42.7
LPG-P	3.0	46
LPG-B	3.03	46
LNG	2.750	49.1
MeOH	1.375	19.9

The following calculations were performed:

The attained CII of each vessel is directly calculated from the sample data as the ratio of CO_2 emissions [gr] produced to the transport work:

$$CII_{Att} = \frac{CO_2}{DWT \cdot Distance}\left[\frac{gr_{CO_2}}{t \cdot n.mile}\right], \tag{1}$$

The annual CO_2 emissions [gr] are obtained by:

$$CO_2 = \sum_{i=1}^{n} EF \cdot SFC_i, \tag{2}$$

where index i represents each fuel (n in total) that is used by each ship, SFC_i is the respective fuel consumption measured in metric tons, and EF is the fuel emission factor, provided in Table 1.

The required CII for the set is calculated according to the regulation as:

$$CII_{req} = \frac{100 - z}{100} 4745 \cdot DWT^{-0.622}, \tag{3}$$

where parameter z is a reduction factor of 5% for the year 2023 and 2% for each consecutive year. The A (superior) to E (inferior) ranking for bulk carriers is obtained by the rating boundaries set by the regulation [47]:

$$Ranking = \begin{cases} A,\ CII_{Att} < 0.86CII_{req} \\ B,\ 0.86CII_{req} \leq CII_{Att} < 0.94CII_{req} \\ C,\ 0.94CII_{req} \leq CII_{Att} < 1.06CII_{req} \\ D,\ 1.06CII_{req} \leq CII_{Att} < 1.18CII_{req} \\ E,\ CII_{Att} \geq 1.18CII_{req} \end{cases} \tag{4}$$

For each scenario, the alternative fuel accounts for 95% of the fuel energy, and the pilot fuel for 5%. The fuel energy, Ef [MJ], is assumed to remain constant and is obtained from the fuel quantities and their respective lower heating values (LHV, [MJ/kg]) as the following:

$$E_f = \sum_{i=1}^{n} SFC_i \cdot LHV_i \quad (5)$$

The total fuel cost (TFC [USD]) is calculated from the required quantities and fuel prices (p_i):

$$TFC = \sum_{i=1}^{n} SFC_i \cdot p_i \quad (6)$$

Finally, by multiplying the coefficients under assumption 12, the CO_2 amount produced and an EU ETS price at EUR 70, the EU ETS cost function is:

$$\text{ETS Cost } [€] = CO_2 \text{ [t]} \times 14.7 \text{ [€/t]}, \quad (7)$$

where 14.7 is a weighted average cost obtained from the EUA price, voyage coverage and compliance rate of 0.21. The calculation procedure is illustrated in Figure 1:

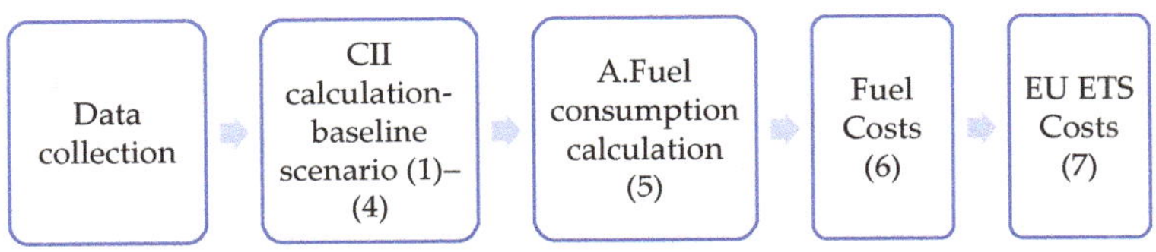

Figure 1. Calculation process.

3. Results and Discussion

In this section, the results of the work are presented. For illustrative purposes and page economy, the calculation procedure and results are presented for one vessel of the fleet sample. The demonstrated results include the CII ranking of the vessel, the summary tables of fuel consumption, the CO_2 emissions amount, the total fuel cost and the EU ETS cost. Then, a sensitivity analysis is presented in the figure, highlighting the dependence of the operational costs on the EU allowance price and the vessel operational profile. Afterwards, comparative figures of the CII ranking, total fuel cost and EU ETS cost are provided among the sample population.

3.1. Vessel Results Demonstration

To avoid repeating the results for all 40 ships, ship 26 is selected as a representative case to demonstrate all compliance and cost figures. It is a Panamax vessel with an EOC date of 2025. In Table 3, all the scenarios are numbered, and the amount of conventional and alternative fuels burnt from ship 26 is included.

In Figure 2, the CII ratings of ship 26 for each scenario are evident. Among all fuel cases, combustion of LNG in both the main and auxiliary engines provides the maximum life extension and guarantees environmental compliance beyond the year 2030. Combustion of LNG in the main engine can also render the ship compliant until 2032, while using LNG only in the auxiliary engine exhibits moderate improvements up to 2028. Apart from LNG, burning either LPG-P or LPG-B in all engines are the only two scenarios ensuring environmental compliance up to 2030; each other option exhibits conservative results with the ship being unreliable by the end of this decade. Among all fuel options, MeOH has the least significant effect on compliance, as the vessel performance is slightly improved, extending the compliance period only by 1 to 3 years, depending on the combustion applications.

Table 3. Scenarios and fuel consumption for ship 26.

Scenario		HFO [t]	LFO [t]	MGO [t]	Alternative Fuel [t]
Business as usual	1	4666	919	100	0
LNG in A/E	2	3400	669	146	1234
LNG in M/E	3	1266	249	223	3312
LNG in A/E and M/E	4	0	0	269	4546
LPG-B in A/E	5	3400	669	146	1296
LPG-B in M/E	6	1266	249	223	3479
LPG-B in A/E and M/E	7	0	0	269	4775
LPG-P in A/E	8	3400	669	146	1279
LPG-P in M/E	9	1266	249	223	3434
LPG-P in A/E and M/E	10	0	0	269	4713
MeOH in A/E	11	3400	669	146	2976
MeOH in M/E	12	1266	249	223	7990
MeOH in A/E and M/E	13	0	0	269	10,966

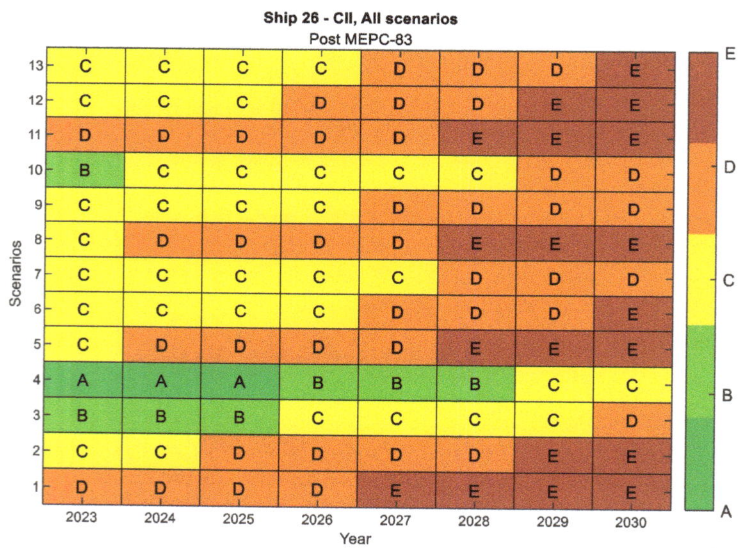

Figure 2. Ship 26: CII ratings for all scenarios.

Combining the results of Figures 2 and 3, MeOH exhibits the highest fuel costs compared to the baseline case of maintaining the use of conventional fuels. Although CO_2 emissions and, therefore, the EU ETS costs are reduced by 2.8% to 10.1% compared to the baseline case, MeOH exhibits the highest values among the four alternative fuel options.

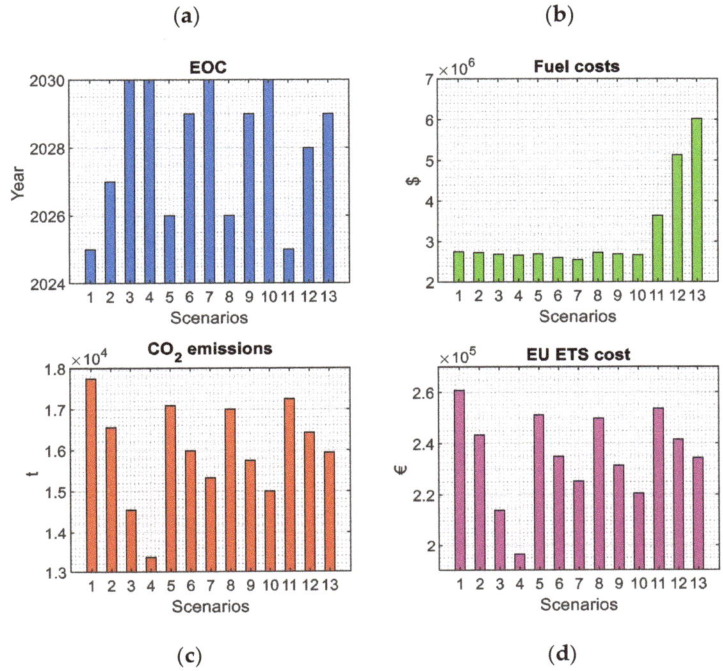

Figure 3. Ship 26: (**a**) EOC for all scenarios; (**b**) fuel costs for all scenarios; (**c**) CO_2 emissions for all scenarios; (**d**) EU ETS costs for all scenarios.

EU ETS Variation Analysis

Throughout Figures 4–7, for each fuel solution, different cases of the EU ETS cost values are presented. Apart from the selected case (assumption 12), four additional cases are considered for ship 26; trading within the EU by 35% with an allowance price of EUR 80 and EUR 100, and trading within the EU by 50% at the allowance prices of EUR 80 and EUR 100.

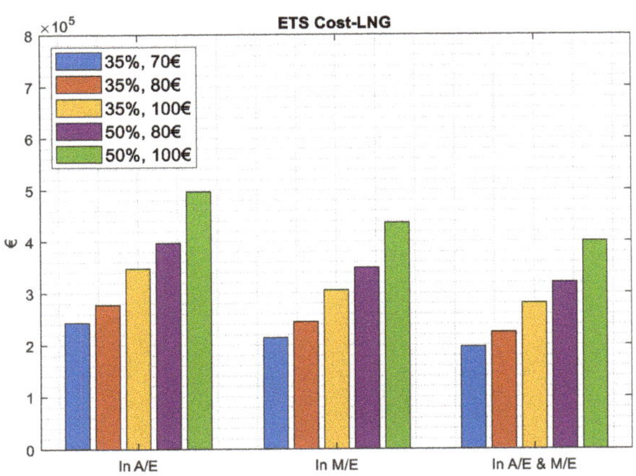

Figure 4. Ship 26: ETS costs for all LNG implementation scenarios.

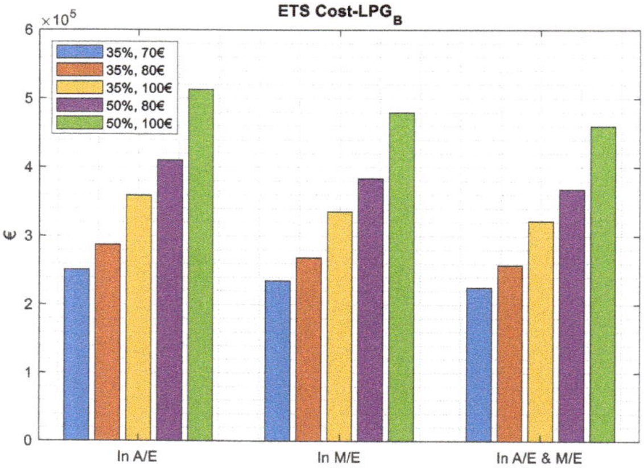

Figure 5. Ship 26: ETS costs for all LPG-B implementation scenarios.

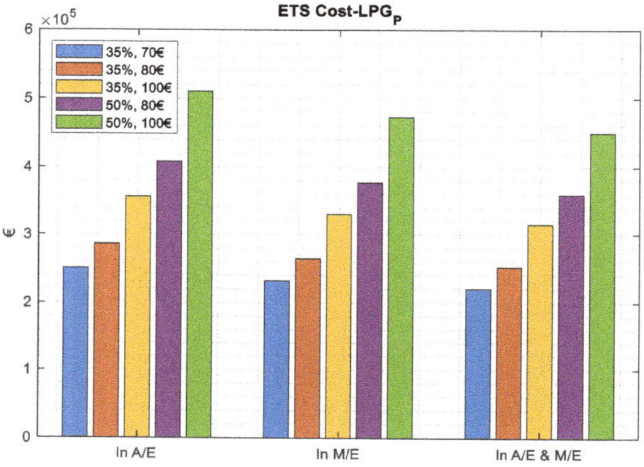

Figure 6. Ship 26: ETS costs for all LPG-P implementation scenarios.

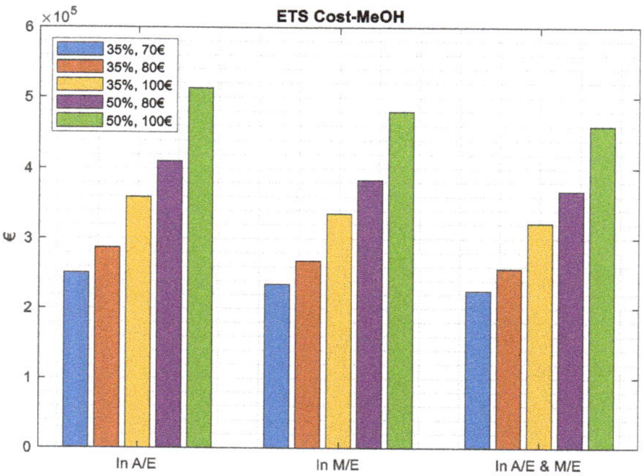

Figure 7. Ship 26: ETS costs for all MeOH implementation scenarios.

From this analysis, the EU ETS cost for LNG is estimated to range between EUR 196,447 and EUR 496,968. The corresponding values for LPG-B and LPG-P are EUR 225,350–EUR

512,704 and EUR 220,516–EUR 510,027, respectively, while for methanol, the cost ranges between EUR 234,311 and EUR 517,668. These results demonstrate that the EU ETS cost is highly sensitive to both the share of annual emissions generated within EU waters and the allowance price. This cost has a significant impact on vessel operations, as it represents a considerable share of annual revenue. Assuming a typical Panamax daily freight rate of 16,000 USD/day (with values ranging from 8500 to 28,500 USD/day [49]) and 175 sailing days annually (approximately 48% of the year), the annual revenue is between 1,487,500 and 4,900,000 USD. Under these conditions, the EU ETS cost accounts for approximately 4% to 35% of the annual revenue, underlining its critical influence on the vessel's economic performance.

3.2. Fleet Results Demonstration

In this section, the results of the calculation procedure are presented. First, the CII for each vessel is calculated, illustrating the expected compliance period of the vessel, shown in Figure 8. Then, in Figure 9, the end of compliance is presented as a function of the year of build and the deadweight. Afterwards, for each fuel, the improvements in CII in terms of the compliance period increase, CO_2 emission reduction, fuel consumption and cost, as well as the additional cost accounting for the EU ETS, are calculated. For illustrative purposes, the procedure is presented in full detail for the case of LNG; the CII ratings are presented through Figures 10–12, while for each implementation scenario, Tables 4 and 5 contain the fuel consumption and produced CO_2 amounts, fuel costs and EU ETS costs. For the remaining fuels, the results are included in the summary in Tables 6–9.

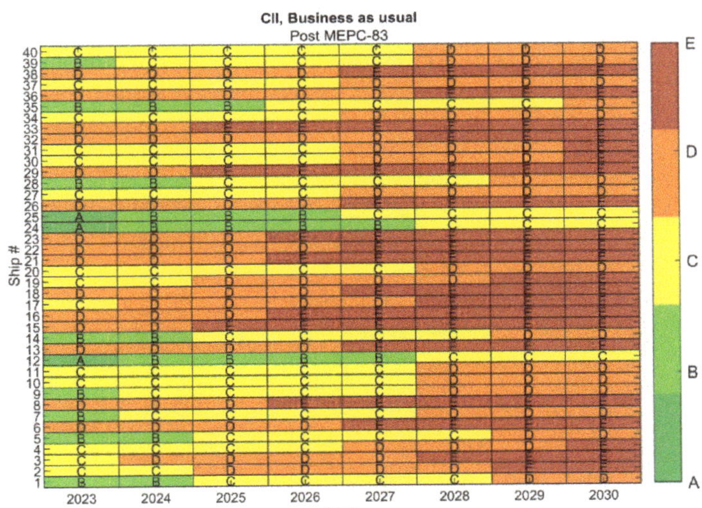

Figure 8. Fleet CII ratings for the baseline scenario.

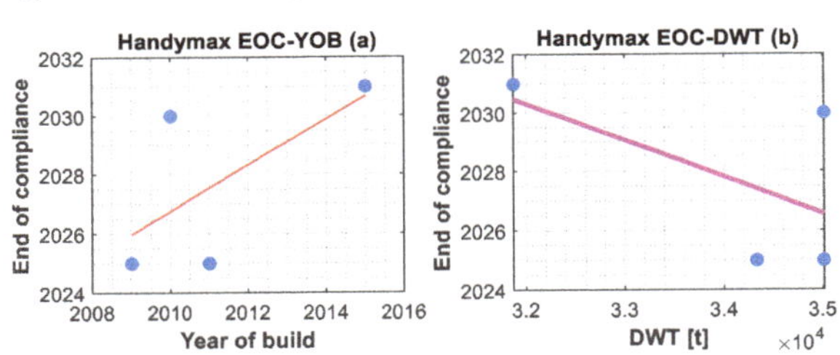

Figure 9. *Cont.*

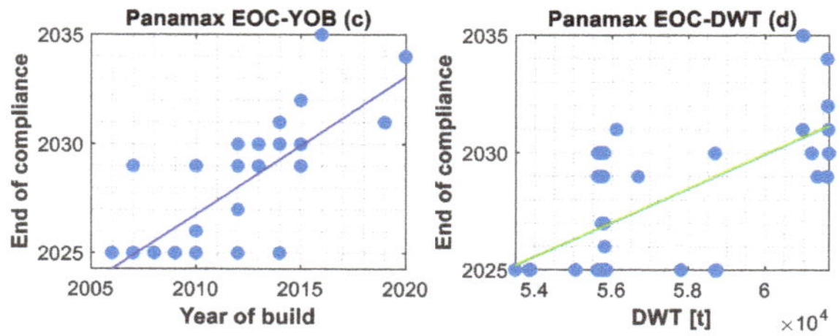

Figure 9. EOC vs. YOB, (**a**) Handymax and (**c**) Panamax, and EOC vs. DWT, (**b**) Handymax and (**d**) Panamax.

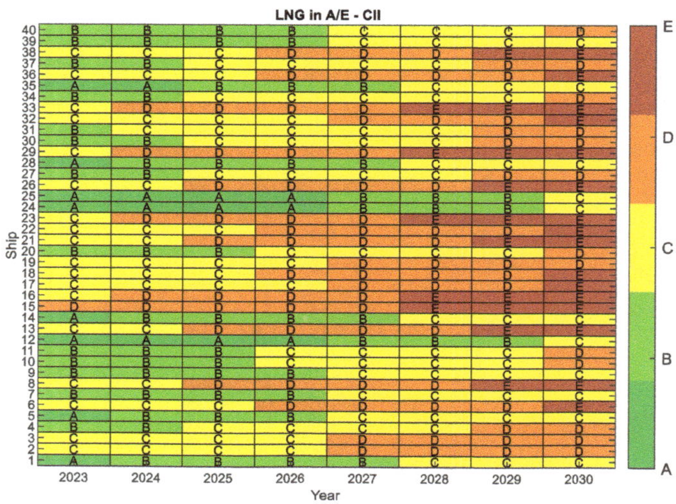

Figure 10. Fleet CII ratings for LNG in the auxiliary engines.

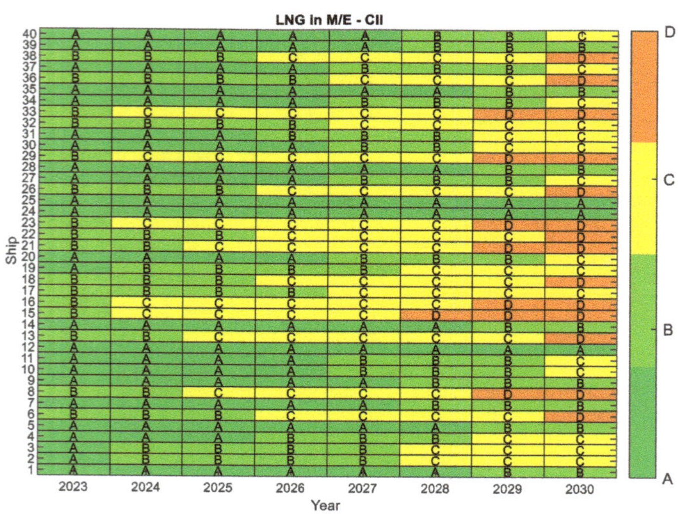

Figure 11. Fleet CII ratings for LNG in the main engine.

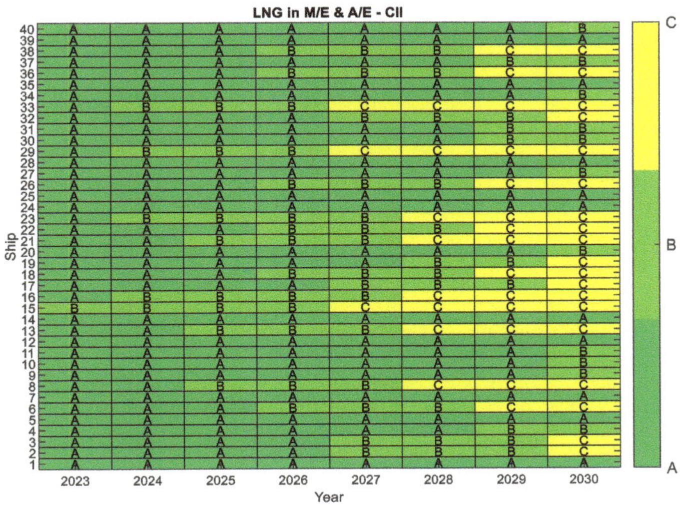

Figure 12. Fleet CII ranking for LNG in the main and auxiliary engines.

In Figure 8, the CII ranking for each vessel from 2023 to 2030 is presented. The reduction factors used, as stated in assumption 2, are based on the results of IMO MEPC-83. For the Handymax set, the average annual speed ranged between 11–12 knots, and the average end of compliance (EOC, end of the vessel compliance period) occurred after 16 years, measured from the year of build of the vessels. The Panamax-Supramax set also exhibited, on average, an EOC of 16 years, corresponding to annual achieved speeds from 9 to 14 knots.

In Figure 9, the end of compliance (EOC) of each vessel is shown as a function of the year of build (YOB) and the deadweight (DWT) for both the Handymax and Panamax-Supramax vessels in the sample. The results indicate that, on average, the vessels that were recently built exhibit delayed EOC dates. This strongly suggests the positive impact of EEDI regulation on ship design, demonstrating increased energy efficiency through an optimized hull design and reduced vessel power requirements. For the Panamax-Supramax set, a trend can be observed between the EOC and DWT; the compliance period is extended as the vessel's carrying capacity increases. However, the DWT is not the sole parameter determining vessel performance. The correlation between CII behavior and vessel size is influenced by power requirements, hull fouling conditions and operational parameters, including the number of voyages, average service speed during charter operations, time distribution at sea, idle periods and in port, as well as the percentage of time the ship operates in laden or ballast conditions. This trend agrees with the literature, which indicates that the carbon cost of transportation is reduced for larger vessels [18,19]. The same behavior is not observed for the Handymax vessels; however, no conclusion can be drawn since their number in the sample is limited.

In Figure 10, the CII values when LNG is combusted only in auxiliary engines are presented. Compared to the conventional fuel case, 40% of the sample achieves compliance up to 2030, whereas 27.5% of the fleet achieves a moderate energy efficiency ranking.

Table 4. Fuel consumption, costs and EU ETS for the baseline scenario and the use of LNG in auxiliary engines.

	Business as Usual						LNG in A/E						
Ship	**HFO [t]**	**LFO [t]**	**MGO [t]**	**Fuel Cost [USD]**	**CO_2 [t]**	**ETS [EUR]**	**HFO [t]**	**LFO [t]**	**MGO [t]**	**LNG [t]**	**Fuel Cost [USD]**	**CO_2 [t]**	**ETS [EUR]**
1	3595	1212	333	2,659,463	16,081	236,395	2620	883	309	1121	2,556,005	15,011	220,666
2	0	4919	204	2,886,109	16,155	237,473	0	3584	216	1135	2,825,037	15,108	222,084
3	2996	685	275	2,034,269	12,369	181,824	2183	499	251	861	1,951,231	11,544	169,702
4	771	4243	361	3,005,162	16,927	248,830	562	3091	333	1188	2,923,877	15,826	232,636
5	2709	1433	543	2,522,440	14,692	215,977	1974	1044	457	1027	2,417,007	13,722	201,714
6	3463	835	504	2,522,186	15,030	220,946	2523	609	429	1048	2,411,347	14,032	206,264
7	3809	375	272	2,250,367	13,914	204,542	2775	273	255	968	2,155,381	12,983	190,846
8	3726	1067	125	2,478,407	15,367	225,890	2715	777	155	1069	2,389,941	14,340	210,795
9	0	5039	136	2,898,548	16,313	239,795	0	3671	167	1146	2,841,671	15,255	224,245
10	2754	1054	215	2,075,147	12,585	184,997	2006	768	209	877	1,997,819	11,748	172,688
11	4113	0	422	2,306,933	14,161	208,163	2997	0	366	985	2,198,116	13,212	194,220
12	2426	2671	51	2,684,160	16,133	237,150	1767	1946	104	1126	2,610,762	15,066	221,468
13	1677	2316	208	2,252,022	13,187	193,850	1222	1688	206	922	2,183,575	12,319	181,094
14	2309	566	462	1,786,415	10,456	153,701	1682	413	380	730	1,702,634	9763	143,517
15	4326	0	481	2,455,753	15,014	220,703	3152	0	412	1044	2,338,422	14,009	205,929
16	3192	2263	247	2,981,018	17,864	262,608	2326	1649	254	1247	2,881,958	16,681	245,212
17	3476	1115	417	2,614,889	15,674	230,408	2533	813	368	1093	2,507,992	14,632	215,096
18	3421	415	96	1,947,406	12,267	180,331	2493	302	120	852	1,872,887	11,444	168,226
19	3253	440	327	2,063,335	12,566	184,727	2371	321	290	875	1,973,729	11,727	172,394
20	1617	1390	408	1,867,777	10,722	157,620	1178	1013	342	750	1,793,432	10,017	147,256

Table 4. *Cont.*

	Business as Usual						LNG in A/E						
Ship	HFO [t]	LFO [t]	MGO [t]	Fuel Cost [USD]	CO_2 [t]	ETS [EUR]	HFO [t]	LFO [t]	MGO [t]	LNG [t]	Fuel Cost [USD]	CO_2 [t]	ETS [EUR]
21	3419	624	23	2,004,792	12,686	186,481	2491	455	69	881	1,934,028	11,834	173,965
22	896	2619	483	2,261,940	12,592	185,099	653	1908	404	884	2,183,801	11,772	173,049
23	2335	3314	96	3,032,670	18,022	264,926	1702	2414	145	1260	2,951,508	16,834	247,458
24	4374	0	112	2,188,352	13,982	205,537	3187	0	139	970	2,098,703	13,040	191,684
25	3930	360	573	2,537,664	15,210	223,581	2864	262	480	1060	2,417,190	14,196	208,683
26	4666	919	100	2,827,706	17,746	260,865	3400	669	146	1234	2,725,150	16,557	243,382
27	3303	220	156	1,830,799	11,481	168,777	2407	160	161	798	1,755,541	10,711	157,446
28	3755	400	366	2,312,362	14,126	207,653	2736	292	325	983	2,210,854	13,182	193,777
29	3946	292	630	2,552,333	15,227	223,837	2876	213	521	1061	2,427,805	14,213	208,929
30	1726	3456	296	2,976,433	17,213	253,033	1258	2518	287	1205	2,889,744	16,084	236,442
31	4433	151	137	2,319,794	14,721	216,392	3230	110	160	1022	2,225,749	13,730	201,831
32	4231	1336	167	2,903,064	17,922	263,453	3083	974	196	1248	2,799,524	16,725	245,865
33	2460	2007	25	2,312,086	14,063	206,732	1793	1462	76	981	2,246,203	13,130	193,017
34	1398	3515	209	2,783,207	16,100	236,669	1019	2561	219	1128	2,708,805	15,045	221,166
35	4256	0	423	2,376,260	14,608	214,745	3101	0	368	1016	2,264,721	13,630	200,357
36	2349	2060	348	2,542,923	14,920	219,331	1711	1501	315	1043	2,453,549	13,936	204,863
37	2506	1825	224	2,391,142	14,276	209,852	1826	1330	222	996	2,310,588	13,331	195,960
38	2631	1370	481	2,400,633	14,049	206,527	1917	998	408	981	2,302,072	13,121	192,877
39	2878	2065	194	2,678,833	16,093	236,568	2097	1505	208	1123	2,591,519	15,027	220,892
40	293	3921	396	2,624,788	14,536	213,677	214	2857	349	1022	2,552,608	13,594	199,830

Table 5. Fuel consumption, costs and EU ETS for the use of LNG in the main engine and the use of LNG in all engines.

	LNG in M/E							LNG in M/E & A/E				
Ship	**HFO [t]**	**LFO [t]**	**MGO [t]**	**LNG [t]**	**Fuel Cost [USD]**	**CO_2 [t]**	**ETS [EUR]**	**MGO [t]**	**LNG [t]**	**Fuel Cost [USD]**	**CO_2 [t]**	**ETS [EUR]**
1	976	329	268	3009	2,472,808	13,209	194,179	244	4130	2,423,434	12,140	178,460
2	0	1335	236	3048	2,592,787	13,344	196,164	248	4184	2,454,957	12,298	180,782
3	813	186	211	2313	1,895,727	10,156	149,289	188	3174	1,862,788	9332	137,175
4	209	1151	287	3190	2,701,215	13,971	205,368	259	4378	2,569,076	12,870	189,185
5	735	389	310	2756	2,293,210	12,089	177,707	224	3783	2,219,741	11,120	163,461
6	940	227	303	2814	2,320,259	12,350	181,549	228	3862	2,266,203	11,353	166,882
7	1034	102	227	2598	2,115,955	11,414	167,782	211	3566	2,09,2558	10,483	154,095
8	1011	290	204	2871	2,341,190	12,610	185,365	233	3940	2,312,259	11,583	170,274
9	0	1367	219	3077	2,613,652	13,473	198,049	250	4223	2,478,333	12,415	182,503
10	747	286	198	2355	1,934,221	10,337	151,958	191	3232	1,896,477	9500	139,656
11	1116	0	271	2644	2,155,061	11,615	170,747	215	3629	2,129,509	10,668	156,816
12	658	725	193	3024	2,500,899	13,268	195,045	246	4151	2,435,700	12,202	179,364
13	455	629	203	2476	2,064,799	10,858	159,610	201	3399	1,994,310	9990	146,860
14	627	154	241	1960	1,624,476	8597	126,380	159	2689	1,578,092	7905	116,210
15	1174	0	296	2804	2,288,138	12,317	181,059	228	3849	2,258,297	11,313	166,300
16	866	614	265	3347	2,765,171	14,688	215,911	272	4594	2,695,863	13,505	198,522
17	943	303	287	2934	2,417,044	12,879	189,315	238	4027	2,363,070	11,838	174,015
18	928	113	161	2289	1,854,247	10,057	147,833	186	3141	1,843,184	9233	135,731
19	883	119	228	2349	1,922,253	10,315	151,628	191	3224	1,891,704	9476	139,304
20	439	377	230	2014	1,685,990	8831	129,812	164	2765	1,622,227	8127	119,460

Table 5. *Cont.*

			LNG in M/E					LNG in M/E & A/E				
Ship	HFO [t]	LFO [t]	MGO [t]	LNG [t]	Fuel Cost [USD]	CO_2 [t]	ETS [EUR]	MGO [t]	LNG [t]	Fuel Cost [USD]	CO_2 [t]	ETS [EUR]
21	928	169	146	2367	1,916,484	10,400	152,877	192	3248	1,906,072	9548	140,362
22	243	711	271	2373	2,012,659	10,392	152,768	193	3257	1,911,094	9574	140,732
23	634	899	226	3382	2,808,466	14,832	218,028	275	4641	2,723,577	13,644	200,563
24	1187	0	185	2605	2,098,269	11,452	168,346	212	3575	2,098,011	10,510	154,496
25	1067	98	324	2845	2,338,147	12,490	183,607	231	3905	2,291,238	11,478	168,726
26	1266	249	223	3312	2,689,090	14,553	213,927	269	4546	2,667,690	13,364	196,447
27	896	60	169	2142	1,736,433	9412	138,361	174	2940	1,725,093	8642	127,035
28	1019	109	255	2640	2,157,525	11,593	170,414	214	3623	2,125,877	10,650	156,548
29	1071	79	339	2849	2,344,020	12,506	183,840	231	3910	2,294,298	11,493	168,951
30	468	938	272	3236	2,711,791	14,184	208,500	263	4441	2,606,184	13,056	191,918
31	1203	41	200	2744	2,215,873	12,061	177,301	223	3766	2,210,012	11,071	162,744
32	1148	363	244	3350	2,735,586	14,710	216,236	272	4597	2,697,641	13,514	198,653
33	668	545	163	2634	2,167,650	11,558	169,907	214	3615	2,121,032	10,625	156,192
34	379	954	236	3028	2,539,055	13,269	195,053	246	4155	2,438,315	12,215	179,556
35	1155	0	276	2728	2,222,001	11,982	176,135	221	3743	2,196,648	11,004	161,760
36	637	559	260	2800	2,328,808	12,279	180,499	227	3843	2,254,779	11,295	166,041
37	680	495	219	2676	2,212,815	11,739	172,563	217	3672	2,154,792	10,794	158,678
38	714	372	286	2635	2,189,080	11,558	169,900	214	3616	2,122,024	10,630	156,265
39	781	560	231	3015	2,489,068	13,230	194,487	245	4138	2,428,267	12,164	178,816
40	80	1064	270	2743	2,337,053	12,008	176,515	223	3765	2,209,131	11,067	162,679

Table 6. Use of LNG as fuel: EOC, fuel and EU ETS cost comparison [%].

Compliance Period Extension					Fuel Cost			EU ETS Cost [70 EUR/t]		
ID	Base-Line	A/E	M/E	M/E and A/E	A/E	M/E	M/E and A/E	A/E	M/E	M/E and A/E
1	>2030	0			−1.9%	−5.1%	−7.0%	−6.7%	−17.9%	−24.5%
2	2027	2			−4.7%	−12.5%	−17.1%	−6.5%	−17.4%	−23.9%
3	2026	3			−1.7%	−4.5%	−6.1%	−6.7%	−17.9%	−24.6%
4	2029	1			−4.3%	−11.6%	−15.9%	−6.5%	−17.5%	−24.0%
5	>2030	0			−2.9%	−7.9%	−10.9%	−6.6%	−17.7%	−24.3%
6	2025	3			−2.2%	−5.9%	−8.1%	−6.6%	−17.8%	−24.5%
7	2031	0			−1.1%	−2.9%	−4.0%	−6.7%	−18.0%	−24.7%
8	2025	2			−1.2%	−3.2%	−4.4%	−6.7%	−17.9%	−24.6%
9	>2030	0			−4.5%	−12.2%	−16.8%	−6.5%	−17.4%	−23.9%
10	2030	0			−1.9%	−5.0%	−6.8%	−6.7%	−17.9%	−24.5%
11	2030	0			−1.1%	−3.1%	−4.2%	−6.7%	−18.0%	−24.7%
12	>2030	0			−2.4%	−6.5%	−9.0%	−6.6%	−17.8%	−24.4%
13	2025	2			−3.1%	−8.4%	−11.5%	−6.6%	−17.7%	−24.2%
14	>2030	0			−2.7%	−7.1%	−9.8%	−6.6%	−17.8%	−24.4%
15	2025	0			−1.3%	−3.4%	−4.6%	−6.7%	−18.0%	−24.6%
16	2025	1			−2.3%	−6.3%	−8.7%	−6.6%	−17.8%	−24.4%
17	2026	2			−2.1%	−5.7%	−7.8%	−6.6%	−17.8%	−24.5%
18	2025	3			−0.6%	−1.6%	−2.2%	−6.7%	−18.0%	−24.7%
19	2027	4			−1.5%	−4.1%	−5.6%	−6.7%	−17.9%	−24.6%
20	2030	0		>2030	−3.4%	−9.2%	−12.7%	−6.6%	−17.6%	−24.2%
21	2025	2			−0.5%	−1.4%	−2.0%	−6.7%	−18.0%	−24.7%
22	2025	3			−4.4%	−11.9%	−16.4%	−6.5%	−17.5%	−24.0%
23	2025	1			−2.8%	−7.5%	−10.3%	−6.6%	−17.7%	−24.3%
24	>2030	0			0.0%	0.0%	0.0%	−6.7%	−18.1%	−24.8%
25	>2030	0			−1.9%	−5.1%	−7.0%	−6.7%	−17.9%	−24.5%
26	2025	2			−0.8%	−2.1%	−2.9%	−6.7%	−18.0%	−24.7%
27	2029	1			−0.6%	−1.7%	−2.4%	−6.7%	−18.0%	−24.7%
28	>2030	0			−1.4%	−3.8%	−5.2%	−6.7%	−17.9%	−24.6%
29	2025	1			−2.0%	−5.4%	−7.4%	−6.7%	−17.9%	−24.5%
30	2029	1			−3.5%	−9.5%	−13.0%	−6.6%	−17.6%	−24.2%
31	2029	1			−0.3%	−0.7%	−1.0%	−6.7%	−18.1%	−24.8%
32	2025	4			−1.3%	−3.6%	−4.9%	−6.7%	−17.9%	−24.6%
33	2025	1			−2.0%	−5.5%	−7.5%	−6.6%	−17.8%	−24.4%
34	2030	1			−3.6%	−9.6%	−13.2%	−6.6%	−17.6%	−24.1%
35	>2030	0			−1.1%	−3.0%	−4.1%	−6.7%	−18.0%	−24.7%
36	2025	3			−2.9%	−7.9%	−10.8%	−6.6%	−17.7%	−24.3%
37	2029	1			−2.4%	−6.6%	−9.0%	−6.6%	−17.8%	−24.4%
38	2025	3			−2.8%	−7.6%	−10.4%	−6.6%	−17.7%	−24.3%
39	>2030	0			−2.3%	−6.2%	−8.4%	−6.6%	−17.8%	−24.4%
40	>2030	0			−4.8%	−12.8%	−17.6%	−6.5%	−17.4%	−23.9%

Table 7. Use of butane LPG as fuel: EOC, fuel and EU ETS cost comparison [%].

ID	Base-Line	Compliance Period Extension			Fuel Cost			EU ETS Cost [70 EUR/t]		
		A/E	M/E	M/E and A/E	A/E	M/E	M/E and A/E	A/E	M/E	M/E and A/E
1	>2030	2	0	0	−3.0%	−8.0%	−11.0%	−3.6%	−9.8%	−13.4%
2	2027	1	3	3	−5.6%	−15.1%	−20.7%	−3.4%	−9.2%	−12.7%
3	2026	2	4	4	−2.8%	−7.4%	−10.2%	−3.7%	−9.8%	−13.5%
4	2029	2	1	1	−5.3%	−14.3%	−19.6%	−3.5%	−9.3%	−12.8%
5	>2030	2	0	0	−4.0%	−10.7%	−14.7%	−3.6%	−9.6%	−13.2%
6	2025	2	4	5	−3.3%	−8.8%	−12.0%	−3.6%	−9.7%	−13.4%
7	2030	2	0	0	−2.2%	−5.9%	−8.1%	−3.7%	−9.9%	−13.6%
8	2025	0	3	5	−2.3%	−6.2%	−8.5%	−3.7%	−9.9%	−13.5%
9	2030	2	0	0	−5.5%	−14.8%	−20.3%	−3.4%	−9.3%	−12.7%
10	2030	2	0	0	−2.9%	−7.9%	−10.9%	−3.6%	−9.8%	−13.4%
11	2030	2	0	0	−2.3%	−6.1%	−8.4%	−3.7%	−9.9%	−13.6%
12	2034	1	0	0	−3.5%	−9.4%	−12.9%	−3.6%	−9.6%	−13.2%
13	2025	1	4	5	−4.2%	−11.2%	−15.3%	−3.6%	−9.5%	−13.1%
14	>2030	2	0	0	−3.7%	−10.0%	−13.7%	−3.6%	−9.7%	−13.3%
15	2025	0	2	4	−2.4%	−6.4%	−8.8%	−3.7%	−9.9%	−13.6%
16	2025	0	2	4	−3.4%	−9.2%	−12.6%	−3.6%	−9.7%	−13.3%
17	2027	2	3	3	−3.2%	−8.6%	−11.7%	−3.6%	−9.7%	−13.4%
18	2025	2	4	5	−1.7%	−4.7%	−6.4%	−3.7%	−10.0%	−13.7%
19	2025	2	5	5	−2.6%	−7.1%	−9.7%	−3.7%	−9.8%	−13.5%
20	2030	2	0	0	−4.5%	−12.0%	−16.4%	−3.5%	−9.5%	−13.1%

Table 7. *Cont.*

		Compliance Period Extension			Fuel Cost			EU ETS Cost [70 EUR/t]		
ID	Base-Line	A/E	M/E	M/E and A/E	A/E	M/E	M/E and A/E	A/E	M/E	M/E and A/E
21	2025	0	3	5	−1.7%	−4.5%	−6.2%	−3.7%	−10.0%	−13.7%
22	2025	2	4	5	−5.4%	−14.6%	−20.0%	−3.5%	−9.3%	−12.8%
23	2025	0	2	4	−3.8%	−10.3%	−14.2%	−3.6%	−9.6%	−13.2%
24	>2030	2	0	0	−1.2%	−3.2%	−4.4%	−3.7%	−10.0%	−13.8%
25	>2030	1	0	0	−3.0%	−8.0%	−11.0%	−3.7%	−9.8%	−13.4%
26	2025	1	4	5	−1.9%	−5.1%	−7.1%	−3.7%	−9.9%	−13.6%
27	2029	2	1	1	−1.8%	−4.8%	−6.6%	−3.7%	−10.0%	−13.7%
28	>2030	2	0	0	−2.5%	−6.8%	−9.3%	−3.7%	−9.9%	−13.5%
29	2025	0	2	4	−3.1%	−8.3%	−11.4%	−3.6%	−9.8%	−13.4%
30	2029	2	1	1	−4.5%	−12.2%	−16.7%	−3.5%	−9.5%	−13.0%
31	2029	2	1	1	−1.4%	−3.8%	−5.2%	−3.7%	−10.0%	−13.7%
32	2025	2	5	5	−2.5%	−6.6%	−9.0%	−3.7%	−9.8%	−13.5%
33	2025	0	2	4	−3.1%	−8.4%	−11.5%	−3.6%	−9.7%	−13.3%
34	2029	2	1	1	−4.6%	−12.4%	−17.0%	−3.5%	−9.5%	−13.0%
35	>2030	1	0	0	−2.2%	−6.0%	−8.2%	−3.7%	−9.9%	−13.6%
36	2025	2	5	5	−4.0%	−10.7%	−14.6%	−3.6%	−9.6%	−13.2%
37	2029	2	1	1	−3.5%	−9.4%	−13.0%	−3.6%	−9.7%	−13.3%
38	2025	1	4	5	−3.9%	−10.4%	−14.3%	−3.6%	−9.6%	−13.2%
39	2030	2	0	0	−3.4%	−9.0%	−12.4%	−3.6%	−9.7%	−13.3%
40	2030	1	0	0	−5.7%	−15.4%	−21.1%	−3.4%	−9.2%	−12.7%

Table 8. Use of propane LPG as fuel: EOC, fuel and EU ETS cost comparison [%].

	Compliance Period Extension				Fuel Cost			EU ETS Cost [70 EUR/t]		
ID	**Base-Line**	**A/E**	**M/E**	**M/E and A/E**	**A/E**	**M/E**	**M/E and A/E**	**A/E**	**M/E**	**M/E and A/E**
1	>2030	0	0	0	−1.9%	−22.4%	−7.1%	−2.7%	−7.3%	−10.0%
2	2027	1	3	3	−4.7%	−12.6%	−17.3%	−2.5%	−6.7%	−9.2%
3	2026	2	4	4	−1.7%	−4.6%	−6.3%	−2.7%	−7.3%	−10.0%
4	2029	1	1	1	−4.4%	−11.7%	−16.1%	−2.5%	−6.8%	−9.3%
5	>2030	0	0	0	−3.0%	−8.0%	−11.0%	−2.6%	−7.1%	−9.7%
6	2025	2	5	5	−2.2%	−6.0%	−8.2%	−2.7%	−7.2%	−9.9%
7	2030	0	0	0	−1.1%	−3.0%	−4.1%	−2.8%	−7.4%	−10.1%
8	2025	0	4	5	−1.2%	−3.3%	−4.6%	−2.7%	−7.4%	−10.1%
9	2030	0	0	0	−4.6%	−12.3%	−16.9%	−2.5%	−6.7%	−9.2%
10	2030	0	0	0	−1.9%	−5.1%	−7.0%	−2.7%	−7.3%	−10.0%
11	2030	0	0	0	−1.2%	−3.2%	−4.4%	−2.8%	−7.4%	−10.1%
12	2034	0	0	0	−2.5%	−6.6%	−9.1%	−2.7%	−7.1%	−9.8%
13	2025	1	4	5	−3.2%	−8.5%	−11.7%	−2.6%	−7.0%	−9.6%
14	>2030	0	0	0	−2.7%	−7.2%	−9.9%	−2.7%	−7.2%	−9.8%
15	2025	0	2	4	−1.3%	−3.5%	−4.8%	−2.8%	−7.4%	−10.1%
16	2025	0	3	5	−2.4%	−6.4%	−8.8%	−2.7%	−7.2%	−9.8%
17	2027	1	3	3	−2.1%	−5.8%	−7.9%	−2.7%	−7.2%	−9.9%
18	2025	2	5	5	−0.6%	−1.7%	−2.3%	−2.8%	−7.5%	−10.2%
19	2025	4	5	5	−1.6%	−4.2%	−5.8%	−2.7%	−7.3%	−10.1%
20	2030	0	0	0	−3.5%	−9.3%	−12.8%	−2.6%	−7.0%	−9.6%

Table 8. *Cont.*

		Compliance Period Extension			Fuel Cost			EU ETS Cost [70 EUR/t]		
ID	Base-Line	A/E	M/E	M/E and A/E	A/E	M/E	M/E and A/E	A/E	M/E	M/E and A/E
21	2025	0	4	5	−0.6%	−1.5%	−2.1%	−2.8%	−7.4%	−10.2%
22	2025	2	5	5	−4.5%	−12.0%	−16.5%	−2.5%	−6.8%	−9.3%
23	2025	0	3	5	−2.8%	−7.6%	−10.4%	−2.6%	−7.1%	−9.7%
24	>2030	0	0	0	−0.1%	−0.1%	−0.2%	−2.8%	−7.5%	−10.3%
25	>2030	0	0	0	−1.9%	−5.2%	−7.2%	−2.7%	−7.3%	−10.0%
26	2025	1	4	5	−0.8%	−2.2%	−3.0%	−2.8%	−7.4%	−10.2%
27	2029	1	1	1	−0.7%	−1.8%	−2.5%	−2.8%	−7.5%	−10.2%
28	>2030	0	0	0	−1.5%	−3.9%	−5.3%	−2.7%	−7.3%	−10.1%
29	2025	0	3	5	−2.0%	−5.5%	−7.5%	−2.7%	−7.3%	−10.0%
30	2029	1	1	1	−3.6%	−9.6%	−13.1%	−2.6%	−6.9%	−9.5%
31	2029	1	1	1	−0.3%	−0.8%	−1.1%	−2.8%	−7.5%	−10.3%
32	2025	2	5	5	−1.4%	−3.7%	−5.1%	−2.7%	−7.3%	−10.1%
33	2025	0	3	5	−2.1%	−5.6%	−7.6%	−2.7%	−7.2%	−9.9%
34	2029	1	1	1	−3.6%	−9.7%	−13.3%	−2.6%	−6.9%	−9.5%
35	>2030	0	0	0	−1.1%	−3.1%	−4.2%	−2.8%	−7.4%	−10.2%
36	2025	2	5	5	−3.0%	−8.0%	−10.9%	−2.6%	−7.1%	−9.7%
37	2029	1	1	1	−2.5%	−6.7%	−9.2%	−2.7%	−7.1%	−9.8%
38	2025	1	5	5	−2.9%	−7.7%	−10.6%	−2.7%	−7.1%	−9.8%
39	2030	0	0	0	−2.3%	−6.3%	−8.6%	−2.7%	−7.2%	−9.8%
40	2030	0	0	0	−4.8%	−12.9%	−17.7%	−2.5%	−6.7%	−9.2%

Table 9. Use of methanol as fuel: EOC, fuel and EU ETS cost comparison [%].

		Compliance Period Extension			Fuel Cost			EU ETS Cost [70 EUR/t]		
ID	Base-Line	A/E	M/E	M/E and A/E	A/E	M/E	M/E and A/E	A/E	M/E	M/E and A/E
1	>2030	0	0	0	29.9%	29.0%	80.2%	−2.70%	−7.3%	−10.0%
2	2027	1	2	3	23.6%	22.3%	63.5%	−2.50%	−6.7%	−9.2%
3	2026	2	3	4	30.4%	29.5%	81.6%	−2.72%	−7.3%	−10.0%
4	2029	1	1	1	24.4%	22.7%	65.5%	−2.53%	−6.8%	−9.3%
5	>2030	0	0	0	27.5%	25.5%	73.8%	−2.65%	−7.1%	−9.7%
6	2025	1	3	4	29.2%	27.6%	78.4%	−2.69%	−7.2%	−9.9%
7	2030	0	0	0	31.7%	31.1%	85.2%	−2.76%	−7.4%	−10.1%
8	2025	0	2	3	31.4%	31.5%	84.4%	−2.74%	−7.4%	−10.1%
9	2030	0	0	0	23.9%	22.7%	64.1%	−2.50%	−6.7%	−9.2%
10	2030	0	0	0	30.0%	29.3%	80.4%	−2.71%	−7.3%	−10.0%
11	2030	0	0	0	31.6%	30.4%	84.7%	−2.76%	−7.4%	−10.1%
12	2034	0	0	0	28.6%	28.6%	76.9%	−2.66%	−7.1%	−9.8%
13	2025	0	2	4	27.1%	26.2%	72.7%	−2.62%	−7.0%	−9.6%
14	>2030	0	0	0	28.2%	25.9%	75.6%	−2.67%	−7.2%	−9.8%
15	2025	0	0	2	31.3%	30.0%	84.0%	−2.75%	−7.4%	−10.1%
16	2025	0	1	2	28.8%	28.3%	77.4%	−2.67%	−7.2%	−9.8%
17	2027	0	2	3	29.4%	28.2%	78.9%	−2.70%	−7.2%	−9.9%
18	2025	1	3	4	32.8%	33.0%	88.1%	−2.78%	−7.5%	−10.2%
19	2025	3	5	5	30.7%	29.7%	82.4%	−2.73%	−7.3%	−10.1%
20	2030	0	0	0	26.4%	24.3%	70.9%	−2.61%	−7.0%	−9.6%

Table 9. *Cont.*

		Compliance Period Extension			Fuel Cost			EU ETS Cost [70 EUR/t]		
ID	Base-Line	A/E	M/E	M/E and A/E	A/E	M/E	M/E and A/E	A/E	M/E	M/E and A/E
21	2025	0	2	3	32.9%	33.4%	88.4%	−2.77%	−7.4%	−10.2%
22	2025	1	3	4	24.1%	21.6%	64.7%	−2.53%	−6.8%	−9.3%
23	2025	0	1	2	27.8%	27.6%	74.7%	−2.63%	−7.1%	−9.7%
24	>2030	0	0	0	34.1%	34.3%	91.6%	−2.81%	−7.5%	−10.3%
25	>2030	0	0	0	29.8%	28.1%	80.1%	−2.72%	−7.3%	−10.0%
26	2025	0	3	4	32.4%	32.6%	87.0%	−2.76%	−7.4%	−10.2%
27	2029	1	1	1	32.7%	32.5%	87.8%	−2.78%	−7.5%	−10.2%
28	>2030	0	0	0	31.0%	29.9%	83.1%	−2.74%	−7.3%	−10.1%
29	2025	0	1	2	29.6%	27.7%	79.5%	−2.71%	−7.3%	−10.0%
30	2029	1	1	1	26.2%	25.0%	70.3%	−2.59%	−6.9%	−9.5%
31	2029	1	1	1	33.6%	33.7%	90.1%	−2.80%	−7.5%	−10.3%
32	2025	2	4	5	31.1%	31.0%	83.6%	−2.73%	−7.3%	−10.1%
33	2025	0	1	2	29.6%	29.7%	79.4%	−2.68%	−7.2%	−9.9%
34	2029	1	1	1	26.0%	25.1%	69.9%	−2.58%	−6.9%	−9.5%
35	>2030	0	0	0	31.6%	30.5%	85.0%	−2.76%	−7.4%	−10.2%
36	2025	2	4	5	27.5%	26.3%	73.9%	−2.64%	−7.1%	−9.7%
37	2029	1	1	1	28.6%	27.9%	76.8%	−2.66%	−7.1%	−9.8%
38	2025	1	3	4	27.8%	26.0%	74.5%	−2.65%	−7.1%	−9.8%
39	2030	0	0	0	29.0%	28.5%	77.8%	−2.67%	−7.2%	−9.8%
40	2030	0	0	0	23.4%	21.2%	62.7%	−2.50%	−6.7%	−9.2%

In Figure 11, the CII ranking of the fleet is presented for the case of combusting LNG only in the main engine. It is evident that the total population of the sample achieves environmental compliance, with 70% of the measures reaching 2030 with at least "C" ranking values. Moreover, 25% of the ships, by 2030, will have achieved at least a minor superior ranking ("B").

A superior extension of the compliance period is achieved with the combustion of LNG in all engines, as illustrated in Figure 12. With this solution, for every vessel of the population of the sample, the compliance period is extended and reliability guaranteed; while 52.5% of the fleet exhibits at least a "B" ranking by 2030.

In Tables 4 and 5, the values of fuel consumption, fuel costs, CO_2 emissions and EU ETS costs are presented for the three LNG implementation scenarios. It is observed that fuel costs increase progressively with scenario complexity, with implementation in auxiliary engines considered the least complex option. This escalation in fuel costs is accompanied by a corresponding reduction in emission costs, as the lower carbon intensity of LNG compared to conventional fuels results in decreased CO_2 emissions and consequently lower EU ETS penalties. It is noteworthy that, due to fuel prices, lower fuel heating values and fuel consumption, the operational cost of each LNG scenario is lower than the operational cost of the conventional fuel case by 2.2%, 5.9% and 8.2% for each implementation case.

In Table 6, the results of all LNG scenarios are summarized and compared to the baseline case. By investing in LNG technology, depending on the implementation scenario, the vessel CII ranking can be sincerely improved and the compliance period can even exceed 2030. Correspondingly, the average annual fuel costs are reduced from 2.2% to 8.2%. The EU ETS costs are reduced for the three implementation scenarios by 6.6%, 17.8% and 24.4%. The combined effect of extended compliance periods and the reduced emission penalties based on the exhibited fuel cost demonstrates that LNG implementation provides both regulatory compliance benefits and economic advantages, with the optimal scenario selection depending on the vessel operator's risk tolerance and investment capacity.

In Table 7, the comparative results between LPG-B applications and the baseline scenario are included. When LPG-B is implemented in auxiliary engines, conservative improvements in the compliance period of each vessel are achieved, with only 50% of the fleet reaching 2030 and only 10% of the population entering the next decade with a "C" ranking. When LPG-P is implemented in the main engine, 67.5% of the fleet achieves compliance until 2030, while 40% of the fleet is ranked as "C" by 2030. The most significant effect on the extension of the compliance period for each vessel is achieved when LPG-B is burnt in all ICEs. Specifically, 87.5% of the vessels exhibit extended compliance periods up to 2030, with 35% of the ships reaching 2030 with at least moderate operational efficiency (ranking "C").

Annual fuel costs are decreased compared to the baseline case by an average value of 3.3%, 8.8% and 12.1% and are lower than the LNG case. These cost values correspond to a moderate reduction in EU ETS costs, ranging from 3.6% to 13.3%. While LPG-B does not achieve the dramatic compliance extensions observed with LNG, combusting it in the main engine or in all ICEs offers a balanced approach, combining reasonable environmental benefits with acceptable economic implications, rendering it an important candidate option for retrofitting.

Engine operation with LPG-P (Table 8) exhibits slightly improved attributes compared to the LPG-B case; however, these are not significant, as in the case of LNG. Compared to the baseline scenario, with LPG-P used in auxiliary engines, 52.5% of the vessels achieve compliance until 2030, while 17.5% of the group achieves a maximum CII ranking of "C". Improved compliance periods are evident for the case of burning in the main engine; 75% of the fleet achieves compliance up to 2030, with 47.5% of the vessels being ranked either as

"C" or "B". Significant improvements are exhibited when LPG-P is combusted in all ICES; 97.5% of the fleet survives until 2030, with 50% of the fleet achieving "C" and "B" rankings.

The fuel costs of each implementation case are, on average, lower than the baseline case by 2.3%, 6.1% and 8.3%. It is evident that the LPG-P cost reductions are similar to the LNG case and slightly lower than LPG-B. Consequently, the EU ETS costs are reduced compared to the conventional fuel case by a factor of 4.1% to 15.2%, depending on the implementation of the fuel. As in the case of LPG-B, propane LPG is a considerable solution for retrofitting bulk carriers, exhibiting moderate compliance period extensions and EU ETS cost reductions.

In Table 9, the summarized results for the case of using methanol are provided. Methanol demonstrates the least favorable performance among all the examined alternative fuels, showing limited improvements compared to either the LPG or LNG options. For the case of methanol in auxiliary engines, 52.5% of the fleet achieves CII compliance up to 2030, and only 10% of the vessels exhibit a CII ranking of "C". If methanol is combusted in the main engine, 57.5% of the fleet achieves environmental compliance up to 2030, and 27.5% of the vessels exhibit the maximum operational efficiency of the fleet and a "C" ranking. During the implementation of methanol in both the main and auxiliary engines, 67.5% of the fleet achieves a compliance period extension up to 2030, while only 30% of the fleet achieves a CII ranking of "C". The relatively modest compliance improvements reflect methanol's higher carbon intensity compared to gaseous fuels despite its renewable production potential.

The economic implications of adopting methanol are significant, exhibiting substantially increased fuel costs compared to the baseline scenario, from 29% to 107%, depending on the implementation scenario. This pronounced cost increase stems from methanol's lower energy density, requiring larger fuel volumes, combined with its current market pricing structure. The substantial fuel cost penalties correspond to only moderate decreases in the EU ETS costs, ranging from 2.7% to 10%, indicating an unfavorable cost–benefit ratio compared to other alternative fuel options.

Economic Evaluation

In this section, the economic Key Performance Indicator (KPI) of the payback period is evaluated for the implementation of each selected alternative fuel in both auxiliary and main engines in order to assess the economic impact of each retrofit option. The retrofit costs are assumed to be fixed across the 40-vessel fleet, and are approximated based on the literature values: USD 8,500,000 for LNG [7], USD 8,000,000 for both LPG options [50] and USD 6,000,000 for methanol [32]. To represent the annual vessel revenue, four daily freight levels are considered as follows: USD 14,000, 16,000, 18,000 and 20,000 for LNG, LPG-B and LPG-P, and USD 25,000, 28,000, 30,000 and 33,000 for methanol. The results are presented in Figure 13, showing the payback period values of 8, 10, 12 and 15 years. The payback period [years] is calculated by Equation (8):

$$PBP = \frac{TRC}{AR - TFC - ETS} \tag{8}$$

In Equation (8), TRC is the total retrofit cost in USD, while the annual revenue (AR) [USD] is obtained by the product of daily freight over 175 sailing days, as shown in Equation (9). A 1.16 EUR to USD currency is used.

$$AR = Dailyfreight \cdot Sailingdays. \tag{9}$$

The analysis indicates that, under current fuel prices, anticipated fuel consumption, and emissions trading system (ETS) penalties, retrofitting vessels to alternative fuels results in only marginal annual profits per vessel. Achieving a reasonable payback period requires an increase in the daily freight rate. Specifically, for an average daily freight of USD 16,000, only 7.5% of the fleet would achieve a 10-year payback period, whereas 75% of the fleet would require 15 years of operation to recover the investment. If the daily freight is increased to USD 18,000, 30% of the fleet would recover the investment within 10 years; at USD 20,000 per day, 80% of the fleet would exhibit a positive return within the same period.

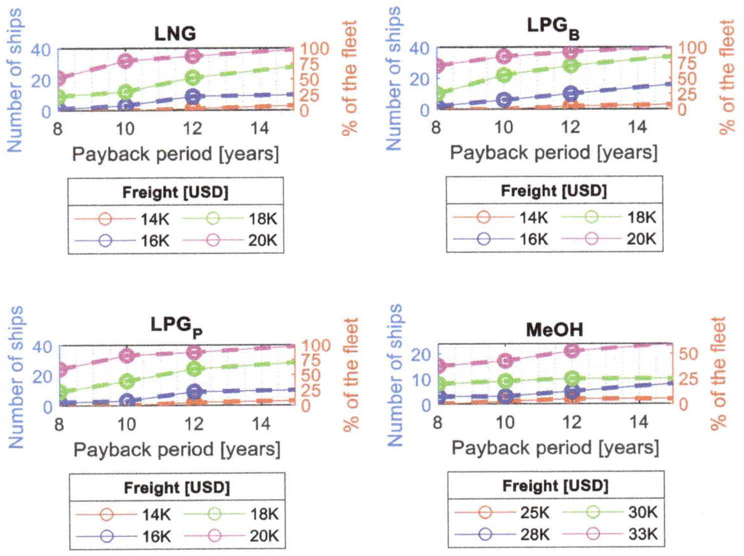

Figure 13. Payback period for each fuel, used in the main and auxiliary engines.

Slightly improved payback periods were observed for the LPG-B retrofits. At a daily freight of USD 16,000, 15% of the investments would be profitable within 10 years, while the corresponding percentages for daily freights of USD 18,000 and USD 20,000 rise to 55% and 85%, respectively. For the LPG-P retrofit, the 10-year payback percentages are 7.5%, 40% and 82.5% for the same freight levels.

In contrast, methanol retrofits require substantially higher daily freight rates to achieve profitability. To attain a 10-year payback period, the daily freight must reach at least USD 25,000 to allow only 2.5% of the fleet to break even. Even at a daily freight of USD 33,000, only 42.5% of the investments yield a positive return, underscoring the limited compatibility of bulk carrier operations with methanol fuel.

The financial viability of an investment is contingent upon the annual revenue of the vessel, which, in turn, is directly influenced by its operational profile. Optimizing the number of sailing days, minimizing idle periods, and reducing the duration of cargo operations and onboard fuel consumption may expedite the recovery of the initial investment.

Among the alternative fuel options evaluated, liquefied natural gas (LNG) emerges as the most advantageous solution in terms of both the compliance period extension under the Carbon Intensity Indicator (CII) and the cost reduction within the EU emissions trading system (EU ETS), particularly when fuel operational costs are taken into account. By comparison, butane LPG and propane LPG deliver moderate outcomes relative to LNG, yet they indicate potential for sustaining fleet reliability at least until 2030. Methanol, despite its status as the fuel with the lowest carbon content among those assessed, demonstrates the most conservative improvements in operational efficiency and EU ETS cost mitigation when applied to bulk carriers. These limited benefits are further constrained by methanol's considerable operational fuel costs, which currently restrict its competitiveness.

Compliance with the CII and the EU ETS does not encompass the full spectrum of forthcoming regulatory obligations. While EU ETS penalties may represent a substantial proportion of a vessel's annual income—reaching up to 34% of the freight value in certain scenarios, as presented in Section 3.1—this mechanism constitutes only one layer of an increasingly complex regulatory environment. Parallel frameworks, such as FuelEU Maritime, expand the scope of accountability by incorporating well-to-wake emissions, thus obligating shipowners to address the carbon footprint of fuel production. Under the forthcoming Global Fuel Standard (GFS), shipowners are likely to incur additional costs in the form of remedial units should the emission thresholds that are set be exceeded, thereby reinforcing the policy direction towards lower-carbon fuels and energy efficiency innovations. Consequently, the selection of an alternative fuel for retrofitting bulk carriers must be evaluated not only against the current compliance and operational criteria, but also with respect to evolving regulatory landscapes. While the fuels considered in this study provide meaningful emission reductions, they are ultimately positioned as transitional options within the broader trajectory towards a zero-carbon future. Given the accelerating regulatory stringency, their viability as standalone solutions beyond the 2040s is increasingly uncertain.

In practice, the adoption of alternative fuels can be complemented by energy efficiency measures, including wind-assisted propulsion, solar power installations, hybrid systems, hull and propeller optimization, energy harvesting devices and shipboard carbon capture technologies. These technical measures may increase the potential to significantly improve fleet sustainability when deployed in tandem with alternative fuels. Furthermore, operational strategies—including slow steaming, weather routing and enhanced voyage planning—can contribute to reducing carbon emissions and stabilizing compliance with future efficiency requirements. Ultimately, it is the integration of fuel, technological and operational solutions, rather than reliance on a single pathway, that will determine the long-term resilience of bulk carrier fleets.

In a multi-pathway future of net-zero emissions, coherent policymaking strategies will be essential to ensure a smooth and equitable transition towards carbon elimination. Guidance should be given on correlating the ship type to fuel type according to its operating regions. Financial incentives, such as tax exemptions, subsidies or preferential loan schemes, can support shipowners in retrofitting existing vessels or investing in new designs that are compatible with alternative fuels. Such measures would enhance the financial viability of investment projects and would assist in maintaining the commodities market price, which is directly related to the ship freight rates. Equally important is the development of alternative fuel infrastructure, which will enhance fuel availability and reduce the risks associated with early adoption. Additional measures should target energy providers, who must be incentivized to scale up green fuel production and secure reliable supply chains. Non-financial incentives can also play a significant role, including policies that prioritize low-emission vessels in port access, berth allocation or departure scheduling, thereby rewarding operational choices that align with decarbonization objectives.

4. Conclusions and Future Work

In this work, the effect on energy efficiency and environmental compliance of combusting LNG, LPG (propane and butane) and MeOH in ICE in medium-sized bulk carriers was investigated. A sample of the Handymax, Panamax and Supramax vessels with available operational data (their annual distances in nautical miles, fuel consumption in metric tons, and duration at sea, idle and at port) was used to determine which fuel pathway is suitable for maintaining fleet reliability. For each fuel, three implementation cases are considered as follows: combustion in auxiliary engines, combustion in the main engine and combustion in all ICEs. From the analysis, the following conclusions were drawn:

- End of compliance is delayed depending on the vessel's year of build. This trivial observation highlights the importance and positive effect of the EEDI regulation on ship efficiency. This implies that the increase in energy efficiency is a multi-level parameter, combining design optimization not only with fuel consumption and power generation, but also with hull and propeller design. Ultimately, maximizing ship energy efficiency is a prerequisite to the effective adoption of alternative fuels and decarbonization.
- A dependency of end-of-compliance on the deadweight of the vessel is observed for Panamax/Supramax vessels, implying that the transportation carbon cost is reduced with an increasing vessel size. However, the results for the Handymax vessels were inconclusive, which is likely due to the limited sample size.
- LNG demonstrates superior performance among all alternative fuels and is capable of extending the compliance period. By implementing LNG at the main engines, or at all ICEs, 100% of the fleet achieves environmental compliance up to 2030, with 25% and 52.5% of the ships having achieved at least a ranking of "B". The annual fuel costs are 2.2% to 8.2% lower than the baseline scenario of using conventional fuels, corresponding to a significant decrease in EU ETS costs between an increase of 6.6% to 24%. Based on the exhibited improvements, implementing LNG onboard is an attractive, compliance solution.
- LPG-B exhibited moderate improvements in the vessel compliance period and EU ETS costs when compared to LNG. The most significant effect of implementing LPG-B is observed when combusted in all ICEs, achieving compliance extensions up to 2030 for 87.5% of the fleet, with 35% of the ships being ranked at least "C". Fuel costs compared to the baseline scenario are reduced by 3.3% to 12.1%, depending on the implementation scenario, with a resulting cut-down in EU ETS costs from 3.6% to 13.3%. These fuels represent a balanced approach for operators seeking environmental benefits with a minimal economic impact.
- Similarly with LPG-B, LPG-P offers an extension of the compliance period up to 2030 for 52.5% to 97.5% of the fleet, while 17.5% to 50% of the fleet completes their compliance period with at least moderate energy rankings. Fuel cost reductions compared to the baseline scenario range between 2.3% to 8.3%, corresponding to a reduction in EU ETS of 4.1% to 15.2%.
- Vessels using methanol exhibit the least favorable performance among all alternatives, despite the low-carbon content of MeOH; 52.5% to 67.5% of the fleet achieves a compliance period extension up to 2030, with 10% to 30% of the vessels achieving a ranking of "C" at maximum. The fuel costs are substantially increased and, depending on the implementation case, they range between 29.9% to 107%, with a conservative reduction in the EU ETS of 2.7% to 10%. The unfavorable cost–benefit ratio makes methanol the least attractive option under current market conditions.
- Payback period calculations were conducted for the case of implementing the selected fuels in all engines. The analysis demonstrates an inverse relationship between the average daily freight rate and the retrofit payback period; as the freight rates increase, the time required to recover the initial investment is reduced. Indicatively, to achieve a 10-year payback period for 50% of the fleet, the required average daily freight rates would be USD 20,000 for LNG and LPG-P, USD 18,000 for LPG-B, while for MeOH, it would have to exceed USD 33,000. These increases would directly affect the commodities' market price. Reducing ship idle periods, duration of cargo operations or increasing the sailing days may improve the payback periods.

- The implementation scenario significantly affects performance outcomes, with the auxiliary engine applications providing minimal benefits, the main engine implementations yielding moderate improvements, and the combined main and auxiliary engine approaches delivering maximum compliance extensions across the selected fuels.
- When considering the retrofit of a bulk carrier to operate on alternative fuels, compliance with the CII regulation and the EU ETS should not be regarded as the vessel's sole regulatory obligation. An assessment of sustainability must also account for the FuelEU and the forthcoming Global Fuel Standard (GFS), which precisely evaluate the vessel's future GHG emissions profile on a well-to-wake basis. Therefore, the adoption of low-carbon fuels alone may prove insufficient; thus, combining them with energy efficiency technologies might enhance the vessel's environmental performance.
- Carbon emissions can also be mitigated through the implementation of operational measures. Strategies such as slow steaming and weather routing have the potential to enhance vessel performance while reducing the overall fuel consumption. When applied in combination with alternative fuels, these practices may further improve efficiency and strengthen compliance with the regulatory requirements.
- A transition to net-zero emissions depends on effective policymaking. Financial incentives for investment, such as tax exemptions, subsidies and preferential financing, can reduce barriers to retrofitting and new vessel designs. Expanding alternative fuel infrastructure is essential to ensure availability and reduce adoption risks. Non-financial measures, including port access privileges or berth prioritization for low-emission vessels, may further assist in the acceleration of the transition to a zero-carbon footprint.

This work could serve as a decision-making tool for ship owners and operators to understand the benefits and economic implications of alternative fuels onboard Handymax and Panamax vessels. The results demonstrate that fuel pathway selection should be based on operators' investment capacity, risk tolerance and long-term fleet strategy. Moreover, it provides insights into whether the implementation of alternative fuels on a global fleet scale would have beneficial impacts on shipping decarbonization efforts. This study reveals that while all alternative fuels offer environmental benefits, their economic viability varies significantly, with LNG providing the optimal balance between compliance improvements and cost implications.

This work focused on evaluating the environmental effects and economic implications of these fuels onboard Handymax and Panamax vessels when combusted in ICE. Fuel production carbon costs, energy efficiency technologies, zero-carbon fuels (such as ammonia and hydrogen) and detailed retrofit cost analyses were not included in this work and are intended as future expansions of this research beyond the bulk carrier classes.

Author Contributions: Conceptualization, G.C. and C.P.; methodology, G.C., A.T. and A.A.; software, G.C., A.T. and A.A.; validation, G.C., S.P. and C.P.; formal analysis, G.C.; investigation, G.C. and C.P.; resources, G.C., A.T., A.A., S.P. and C.P.; data curation, G.C. and C.P.; writing—original draft preparation, G.C.; writing—review and editing, G.C., S.P. and C.P.; visualization, G.C.; supervision, C.P.; project administration, G.C. and S.P. All authors have read and agreed to the published version of the manuscript.

Funding: This research received no external funding.

Data Availability Statement: The original contributions presented in this study are included in the article. Further inquiries can be directed to the corresponding author.

Conflicts of Interest: The authors declare no conflicts of interest.

Abbreviations

The following abbreviations are used in this manuscript:

AR	Ship annual revenue [USD]
CII_{Att}	Attained CII
CII_{Req}	Required CII
CO_2	Carbon dioxide
DWT	Deadweight [t]
EF	Fuel emission factor [gCO_2/g_{fuel}]
EOC	End of compliance
ETS	EU ETS total cost [EUR]
H_2	Hydrogen
HFO	Heavy fuel oil
LFO	Low sulfur fuel oil
LHV	Lower heating value [kJ/kg]
LNG	Liquified natural gas
LNG-B	Liquified petroleum gas, butane
LPG-P	Liquified petroleum gas, propane
MeOH	Methanol
MGO	Marine gas oil
NH_3	Ammonia
NO_x	Nitrogen oxide
p	fuel price
PBP	Payback period [years]
PM	Particulate matter
Ranking	CII ranking
SFC	Fuel mass consumed [t]
SO_x	Sulphur oxide
TFC	Total fuel cost [USD]
TRC	Total retrofit cost [USD]
YOB	Year of build

References

1. UN Trade and Development. Review of Maritime Transport 2024. Available online: https://unctad.org/publication/review-maritime-transport-2024 (accessed on 23 June 2025).
2. DNV. Rising LNG Demand: Overcoming Bunkering Challenges. Available online: https://www.dnv.com/expert-story/maritime-impact/rising-lng-demand-overcoming-bunkering-challenges/ (accessed on 23 June 2025).
3. Lloyd's List, Number of Ships Using LNG up 33% in 2024. Available online: https://www.lloydslist.com/LL1152328/Number-of-ships-using-LNG-up-33-in-2024#:~:text=There%20are%20638%20LNG%2Dfuelled%20ships%20in%20operation,orderbook%20*%2023%20Jan%202025.%20*%20News (accessed on 23 June 2025).
4. Usiagu, G.S.; Adekoya, O.O.; Okoli, C.E.; Daudu, C.D.; Ekemezie, I.O.; Ayorinde, O.B. LNG as a bridge fuel in the transition to renewable energy: A global perspective. *World J. Adv. Res. Rev.* **2024**, *21*, 742–749. [CrossRef]
5. Schinas, O.; Butler, M. Feasibility and commercial considerations of LNG-fueled ships. *Ocean Eng.* **2016**, *122*, 84–96. [CrossRef]
6. Wan, C.; Yan, X.; Zhang, D.; Yang, Z. A novel policy making aid model for the development of LNG fueled ships. *Transp. Res. Part A* **2016**, *119*, 29–44.
7. Tam, I.C.K.; Dev, A.K.; Ng, C.; Deltin, L. Concept design of a Bulk Carrier Retrofit with LNG fuel. In Proceedings of the ICMET OMAN, Muscat, Oman, 5–7 November 2019.
8. He, Y.; Yan, X.; Fan, A.; Yin, X.; Meng, Q.; Zhang, L.; Liu, P. Assessing the influence of actual LNG emission factors within the EU emissions trading system. *Transp. Policy* **2024**, *159*, 345–358. [CrossRef]
9. Psaraftis, H.N.; Kontovas, C.A. Decarbonization of maritime transport: Is there light at the end of the tunnel? *Sustainability* **2021**, *13*, 237. [CrossRef]
10. Elmanzalawy, M.A.; Elgohary, M.M.; Maged, M. Technical and environmental performance investigation of Marine Alternative fuels. In Proceedings of the 3rd International conference of Chemical, Energy and Environmental Engineering, Alexandria, Egypt, 19–21 July 2021.

11. Emir, E.; Samet, D.; Yasin, A. Decarbonization strategies in the maritime industry: An analysis of dual-fuel engine performance and the carbon intensity indicator. *Renew. Sustain. Energy Rev.* **2024**, *200*, 114587. [CrossRef]
12. Ros, J.A.; Skylogianni, E.; Doedée, V.; van den Akker, J.T.; Vredeveldt, A.W.; Linders, M.J.G.; Goetheer, E.L.V.; Monteiro, J.G.M.-S. Advancements in ship-based carbon capture technology on board of LNG-fueled ships. *Int. J. Greenh. Gas Control* **2022**, *114*, 103575. [CrossRef]
13. Xia, M.; Yao, S.; Li, C.; Ying, C.; Sun, J. Exergy, energy, economy analysis and multi-objective optimization of a comprehensive energy utilization system for LNG-powered ships based on zero-carbon emissions. *Case Stud. Therm. Eng.* **2024**, *53*, 103783. [CrossRef]
14. Riviera News. Available online: https://www.rivieramm.com/news-content-hub/news-content-hub/methanol-fuel-owners-should-prepare-ships-for-green-fuels-future-84495 (accessed on 24 June 2025).
15. The Methanol Institute. Available online: https://www.methanol.org/ (accessed on 23 June 2025).
16. Michael, R.; Steffen, L.; Bramastha, A.D. Analysis of the Requirement for the Ships Using Methanol as Fuel. *Int. J. Mar. Eng. Innov. Res.* **2018**, *3*, 58–68. [CrossRef]
17. Ammar Nader, R. An environmental and economic analysis of methanol fuel for a cellular container ship. *Transp. Res. Part D* **2019**, *69*, 66–76. [CrossRef]
18. Giorgia, A.; Massimo, F. Feasibility Analysis of a Methanol Fuelled Bulk Carrier. In Proceedings of the 4th International Conference on Modelling and Optimization of Ship Energy Systems, Delft, The Netherlands, 26–27 October 2023.
19. Giorgia, A.; Massimo, F. Methanol as a Marine Fuel for the Decarbonization of the Shipping Sector, Università degli Studi di Genova, Department of Marine, Electrical and Electronic Engineering. Available online: https://iris.unige.it/retrieve/75aeafa7-4acc-404e-af1a-40d93fdb330b/SDEWES24_FP_691.pdf (accessed on 24 June 2025).
20. Sallam, S.K.; Elgohary, M.M.; Seddiek, I.S. Methanol as an Eco-Environmental Alternative Fuel for Ships: A Case Study. *J. Mar. Sci. Eng.* **2023**, *31*, 2. [CrossRef]
21. Ampah, J.D.; Yusuf, A.A.; Afrane, S.; Jin, C.; Liu, H. Reviewing two decades of cleaner alternative marine fuels: Towards IMO's decarbonization of the maritime transport sector. *J. Clean. Prod.* **2021**, *320*, 128871. [CrossRef]
22. Ashrafi, M.; Lister, J.; Gillen, D. Toward a harmonization of sustainability criteria for alternative marine fuels. *Marit. Transp. Res.* **2022**, *3*, 100052. [CrossRef]
23. Yan, X.; He, Y.; Fan, A. Carbon footprint prediction considering the evolution of alternative fuels and cargo: A case study of Yangtze river ships. *Renew. Sustain. Energy Rev.* **2023**, *173*, 113068. [CrossRef]
24. Perčić, M.; Vladimir, N.; Fan, A. Life-cycle cost assessment of alternative marine fuels to reduce the carbon footprint in short-sea shipping: A case study of Croatia. *Appl. Energy* **2020**, *279*, 115848. [CrossRef]
25. Perčić, M.; Vladimir, N.; Fan, A. Techno-economic assessment of alternative marine fuels for inland shipping in Croatia. *Renew. Sustain. Energy Rev.* **2021**, *148*, 111363. [CrossRef]
26. Zhao, Y.; Liu, F.; Zhang, Y.; Wang, Z.; Song, Z.; Zan, G.; Wang, Z.; Guo, H.; Zhang, H.; Zhu, J.; et al. Economic Assessment of Maritime Fuel Transformation for GHG Reduction in the International Shipping Sector. *Sustainability* **2024**, *16*, 10605. [CrossRef]
27. Balcombe, P.; Brierley, J.; Lewis, C.; Skatvedt, L.; Speirs, J.; Hawkes, A.; Staffell, I. How to decarbonize international shipping: Options for fuels, technologies and policies. *Energy Convers. Manag.* **2019**, *182*, 72–88. [CrossRef]
28. Zhao, Y.; Ge, R.; Zhou, J. Decarbonization pathways for bulk vessels: Integrating power systems, fuels and control measures. *Ocean Eng.* **2024**, *300*, 117488. [CrossRef]
29. Xing, H.; Stuart, C.; Spence, S.; Chen, H. Alternative fuel options for low carbon maritime transportation: Pathways to 2050. *J. Clean. Prod.* **2021**, *297*, 126651. [CrossRef]
30. Wang, Z.; Dong, B.; Li, M.; Ji, Y.; Han, F. Configuration of Low-Carbon fuels green marine power systems in diverse ship types and Applications. *Energy Convers. Manag.* **2024**, *302*, 118139. [CrossRef]
31. Loennechen, O.; Fagerholt, K.; Lagemann, B.; Stålhane, M. Maritime fleet composition under future greenhouse gas emission restrictions and uncertain fuel prices. *Marit. Transp. Res.* **2024**, *6*, 100103. [CrossRef]
32. Akman, M.; Turan, B.İ.; Taşdemir, A.; Söğüt, M.Z. Alternative fuels for enhanced ship energy efficiency: A conceptual design study of a handymax bulk carrier. *Int. J. Hydrog. Energy* **2025**, *107*, 502–511. [CrossRef]
33. Lagemann, B.; Lindstad, E.; Fagerholt, K.; Rialland, A.; Erikstad, S.O. Optimal ship lifetime fuel and power system selection. *Transp. Res. Part D* **2022**, *102*, 103145. [CrossRef]
34. Al-Enazi, A.; Bicer, Y.; Okonkwo, E.C.; Al-Ansari, T. Evaluating the utilization of clean fuels in maritime applications: A techno-economic supply chain optimization. *Fuel* **2022**, *322*, 124195. [CrossRef]
35. Al-Asmakh, A.; Bicer, Y.; Al-Ansari, T. Alternative fuels and design modifications for environmentally sustainable marine vessels. *Ocean Eng.* **2025**, *330*, 121226. [CrossRef]
36. Horvath, S.; Fasihi, M.; Breyer, C. Techno-economic analysis of a decarbonized shipping sector: Technology suggestions for a fleet in 2030 and 2040. *Energy Convers. Manag.* **2018**, *164*, 230–241. [CrossRef]

37. Park, C.; Hwang, I.; Jang, H.; Jeong, B.; Ha, S.; Kim, J.; Jee, J. Comparative Analysis of Marine Alternative Fuels for Offshore Supply Vessels. *Appl. Sci.* **2024**, *14*, 11196. [CrossRef]
38. Li, Z.; Wang, K.; Liang, H.; Wang, Y.; Ma, R.; Cao, J.; Huang, L. Marine alternative fuels for shipping decarbonization: Technologies, applications and challenges. *Energy Convers. Manag.* **2025**, *329*, 119641. [CrossRef]
39. IMO Resolution MEPC400(83)—Amendments to the 2021 Guidelines on the Operational Carbon Intensity Reduction Factors Relative to Reference. Available online: https://www.imo.org/en/mediacentre/meetingsummaries/pages/mepc-83rd-session.aspx (accessed on 13 August 2025).
40. World Bunker Prices. Available online: https://shipandbunker.com/prices (accessed on 18 August 2025).
41. Ycharts Indicators, US Liquefied Natural Gas Exports Price. Available online: https://ycharts.com/indicators/us_natural_gas_liquefied_exports_price (accessed on 18 August 2025).
42. Federal Reserve Bank of St. Louis. Available online: https://fred.stlouisfed.org/series/PNGASJPUSDM (accessed on 18 August 2025).
43. Japan Agency for Energy and Metals and Mineral Resources [JOGMEC]. Available online: https://oilgas-info.jogmec.go.jp/nglng_en/index.html (accessed on 18 August 2025).
44. 3MCO. LPG Prices. Available online: https://3mgas.vn/news/saudi-aramco-lpg-prices-per-metric-tonne-mt-n153.html (accessed on 18 August 2025).
45. Methanex. Available online: https://www.methanex.com/our-products/about-methanol/pricing/ (accessed on 18 August 2025).
46. Trading Economics. Available online: https://tradingeconomics.com/commodity/carbon (accessed on 18 August 2025).
47. IMO Resolution MEPC.391(81). Available online: https://wwwcdn.imo.org/localresources/en/KnowledgeCentre/IndexofIMOResolutions/MEPCDocuments/MEPC.391(81).pdf (accessed on 13 August 2025).
48. FuelEU. Available online: https://eur-lex.europa.eu/legal-content/EN/TXT/PDF/?uri=CELEX:32023R1805 (accessed on 13 August 2025).
49. HandyBulk. Available online: https://www.handybulk.com/ship-charter-rates/ (accessed on 13 June 2025).
50. BW LPG. Available online: www.bwlpg.com/vlgc_fleet (accessed on 1 September 2025).

Journal of
Marine Science and Engineering

MDPI

Article

Case Study on Emissions Abatement Strategies for Aging Cruise Vessels: Environmental and Economic Comparison of Scrubbers and Low-Sulphur Fuels

Luis Alfonso Díaz-Secades [1,*], Luís Baptista [2] and Sandrina Pereira [2]

1 Department of Marine Science and Technology, University of Oviedo, 33003 Oviedo, Spain
2 Department of Marine Engineering, Escola Superior Náutica Infante D. Henrique (ENIDH), 2770-058 Paço de Arcos, Portugal; sandrinapereira@enautica.pt (S.P.)
* Correspondence: secadesalfonso@uniovi.es

Abstract

The maritime sector is undergoing rapid transformation, driven by increasingly stringent international regulations targeting air pollution. While newly built vessels integrate advanced technologies for compliance, the global fleet averages 21.8 years of age and must meet emission requirements through retrofitting or operational changes. This study evaluates, at environmental and economic levels, two key sulphur abatement strategies for a 1998-built cruise vessel nearing the end of its service life: (i) the installation of open-loop scrubbers with fuel enhancement devices, and (ii) a switch to marine diesel oil as main fuel. The analysis was based on real operational data from a cruise vessel. For the environmental assessment, a Tier III hybrid emissions model was used. The results show that scrubbers reduce SO_x emissions by approximately 97% but increase fuel consumption by 3.6%, raising both CO_2 and NO_x emissions, while particulate matter decreases by only 6.7%. In contrast, switching to MDO achieves over 99% SO_x reduction, an 89% drop in particulate matter, and a nearly 5% reduction in CO_2 emissions. At an economic level, it was found that, despite a CAPEX of nearly USD 1.9 million, scrubber installation provides an average annual net saving exceeding USD 8.2 million. From the deterministic and probabilistic analyses performed, including Monte Carlo simulations under various fuel price correlation scenarios, scrubber installation consistently shows high profitability, with NPVs surpassing USD 70 million and payback periods under four months.

Keywords: marine scrubbers; sulphur oxides; particulate matter; maritime air pollution; maritime transportation; decision support system

1. Introduction

Global energy demand continues to rise at an accelerating pace. Between 2013 and 2023, the average annual growth rate was approximately 1.3%; however, this figure surged to 2.2% in 2024 alone [1]. To meet this escalating demand, fossil fuels, namely coal, oil, and natural gas, remain the dominant energy sources, with projections from the International Energy Agency (IEA) indicating that their peak usage will not occur before 2030 [2]. This persistent reliance on fossil fuels plays a critical role in two pressing global challenges: the intensification of climate change and the deterioration of public health [3,4].

It is well established, both in the scientific community and among public and private stakeholders, that emissions resulting from the combustion of fossil fuels are directly linked to global warming. This is primarily due to the high carbon content of these fuels, which leads to significant carbon dioxide (CO_2) emissions during combustion [5]. This process adds an excess amount of CO_2 to the levels naturally present in the atmosphere. Similar patterns are observed with other well-known greenhouse gases, such as methane (CH_4) and nitrous oxide (N_2O) [6].

The impact of methane in climate change is a result of two distinct processes. In one hand, methane can be released directly into the atmosphere in its pure form from fossil sources, having a Global Warming Potential (GWP) of 29.8 over a 100-year time horizon [3]. On the other hand, the combustion of methane decreases its GWP, since the process converts it into carbon dioxide (2.75 g of CO_2 per gram of CH_4) and water. However, the combustion of methane can also result in the release of unburned methane to the atmosphere when an incomplete combustion of methane–air mixture happens, a phenomenon known as methane slip. In modern engines, methane slip can account for up to 6% of the total fuel consumed [7,8]. Such levels of unburned methane contribute significantly to global warming and may offset the environmental benefits gained by avoiding the use of more carbon-intensive fossil fuels.

In the case of N_2O, it contributes only a small fraction to total greenhouse gas emissions by volume, but its impact on global warming is disproportionately large due to its GWP, which is 273 times greater than that of CO_2 over a 100-year timescale [3]. This high GWP, combined with an atmospheric lifetime of approximately 116 ± 9 years, makes N_2O a significant contributor to climate change. Emissions of N_2O are projected to continue increasing in the coming years. While agriculture remains the primary source of these emissions, industrial sectors also play a notable role in their release [9].

Nitrogen oxides, primarily nitric oxide (NO) and nitrogen dioxide (NO_2), are not considered strong greenhouse gases themselves. However, they contribute indirectly to climate change and have a significant public health impact. Although NO_x lacks the radiative forcing strength of gases like CO_2 or CH_4, it plays an important role in the formation of tropospheric ozone, a potent greenhouse gas, particularly relevant in sectors such as aviation [10]. Additionally, NO_x emissions extend the atmospheric lifetime of methane, further amplifying its climate impact. From a health perspective, NO_2 is of particular concern due to its association with respiratory problems. It exacerbates conditions such as asthma and bronchitis, impairs lung function in children, and is linked to increased mortality rates [11].

In addition to nitrogen oxides, the combustion of fossil fuels poses further public health risks. These fuels often contain sulphur, which leads to the emission of sulphur oxides (SO_x) during combustion. These compounds are not only harmful on their own but also contribute to the formation of particulate matter (PM), posing serious risks to respiratory and cardiovascular health [12,13].

1.1. Pollutant Emissions from Maritime Sector

The maritime industry is not exempt from the challenges posed by emissions, both in terms of their contribution to global warming and their impact on public health. As a sector responsible for transporting approximately 80% of goods involved in global world trade, it is particularly sensitive to these issues [14]. This sensitivity arises from the sector's dependence on fossil fuels for propulsion combined with its high operational mobility, as vessels are capable of global voyages within days, yet often operate with outdated technologies, reflected by the average age of merchant vessel fleets being 21.8 years [15].

Traditionally, merchant vessels have relied on heavy fuel oil (HFO) for their main propulsion needs, using lighter marine diesel oil (MDO) in a marginal role, accounting for approximately 6% of total fuel consumption [16,17]. These vessels, which were predominantly designed to operate on high-sulphur fossil fuels, are associated with the emission of substantial quantities of pollutants, particularly SO_2, NO_x, and particulate matter. The International Maritime Organization (IMO), recognizing the environmental and health challenges posed by maritime emissions, began developing a regulatory framework to mitigate these impacts. Initial efforts focused on public health, with the adoption in 1997 of the Technical Code on Control of Emission of Nitrogen Oxides from Marine Diesel Engines. This Code was revised in 2008 and further amended during the 83rd session of the Marine Environment Protection Committee (MEPC 83). The latest amendments, which permit the use of multiple operational profiles for marine diesel engines, aim to improve emission control under varying engine loads. This will enable ships to optimize fuel consumption according to their specific operational profiles, thereby enhancing energy efficiency while ensuring compliance with NO_x emission limits [18].

Regarding SO_x emissions, IMO recognized that these emissions are directly dependent on the sulphur content of fuels. Consequently, regulating fuel sulphur content would effectively reduce SO_x emissions from the maritime sector. In line with this, the IMO implemented a global sulphur cap of 4.5% in 2008, which was subsequently lowered to 3.5% in 2012 [19]. Most recently, the IMO 2020 regulation further reduced the cap to the current limit of 0.5% for all seagoing vessels worldwide, with an even stricter limit of 0.1% enforced within designated Sulphur Emission Control Areas (SECAs) [20]. This reduction has significant public health benefits; however, some unintended consequences emerged in the early years of the sulphur cap. Sulphur oxides contribute to negative radiative forcing through aerosol formation, which has a cooling effect on the climate. Thus, reducing SO_x emissions may have diminished this cooling influence, with complex implications for global temperature trends [21,22].

Although the IMO initially focused on mitigating air pollutants with direct impacts on human health, such as NO_x, SO_x, and particulate matter, its current agenda has expanded to include the reduction in greenhouse gas (GHG) emissions from the maritime sector. To address this broader challenge, the IMO has established a comprehensive regulatory framework aimed at abating air pollution generated by international shipping. Regarding climate change mitigation, one of the IMO's key milestones is the 2023 IMO GHG Strategy, which sets a target of reaching net-zero GHG emissions from international shipping by or around 2050, with interim checkpoints in 2030 and 2040 [23].

Building upon the existing NO_x Technical Code and the IMO 2020 sulphur cap, the addition of the 2023 IMO GHG Strategy requires shipowners to comply with SO_x and NO_x limits while simultaneously reducing greenhouse gas emissions. Although new vessels will be designed with these regulations in mind, a critical question remains: How can aged vessels comply with these increasingly stringent requirements and remain profitable?

1.2. Exhaust Gas Cleaning Systems

For compliance with the NO_x Technical Code, shipowners typically employ Selective Catalytic Reduction (SCR). Additionally, other measures such as Fuel Water Emulsification (FEW), Direct Water Injection (DWI), and Exhaust Gas Recirculation (EGR) can also be implemented [24]. These alternatives help to mitigate the need for installing dedicated NO_x exhaust gas cleaning systems on the engine funnel, thereby reducing the backpressure generated by such systems [25].

Following the implementation of the IMO 2020 regulation, shipowners faced the decision of either switching to low-sulphur fuels or adopting alternative compliance strategies. The installation of exhaust gas cleaning systems, commonly referred to as scrubbers, has become one of the preferred solutions within the maritime industry. Scrubbers enable vessels already in service to meet SO_x emission limits without switching to low-sulphur fuels, primarily for economic reasons. Notably, within five years of installation, more than 95% of ships equipped with the most common scrubber systems achieve cost break-even [26]. While exhaust gas cleaning systems provide a convenient pathway to continue using low-cost residual fuels, concerns remain regarding the emission of other pollutants, such as CO_2, NO_x, and PM. This is in part motivated by the fact that the scrubber increases backpressure in the engine funnel, which raises the fuel consumption of the engines [25]. This raises critical questions about whether scrubbers effectively improve air quality and public health, while also contributing to the broader objective of maritime decarbonization [27,28]. Several studies have assessed the impact of scrubber installation on SO_x and PM emissions in merchant vessels. In 2013, Caiazzo et al. demonstrated that the use of an open-loop scrubber in a marine diesel engine operating on HFO led to a reduction in SO_2 emissions of approximately 93% [29]. This effectiveness was further validated by Fridell and Salo in 2016, who reported a reduction of about 99% in SO_2 emissions using a similar open-loop system [30]. Winnes et al. carried out a more comprehensive assessment in 2020 by studying a RoPax ferry equipped with a four-stroke engine and a scrubber, reporting a 99% reduction in SO_2 emissions but only a 40% reduction in PM emissions [31]. Yang et al. highlighted limitations in PM removal, showing that not all particulate matter is captured by the scrubber. The study, conducted on a container ship, confirmed a reduction of more than 95% in SO_x emissions, but only around 10% in PM emissions, underscoring the variable effectiveness of scrubbers in addressing different pollutant types [32]. However, research addressing their effects on NO_x and CO_2 emissions remains limited.

In addition to the potential increase in other air pollutant emissions, the use of scrubbers on ships to reduce SO_x emissions has raised environmental concerns due to the discharge of large volumes of polluted acidic washwater into the sea [26]. In 2023, the International Council on Clean Transportation reported the identification of up to 93 bans and restrictions related to scrubber washwater discharges [33]. These include outright prohibitions on any discharge of washwater or bleed-off from scrubbers, as well as conditional allowances based on specific criteria, with most restrictions targeting open-loop scrubbers. More recently, in June 2025, OSPAR, the cooperative mechanism involving 15 governments and the European Union to protect the marine environment of the North-East Atlantic, expressed serious concerns over hazardous discharges from ship scrubbers, particularly in coastal areas. During its Ministerial Meeting held in Vigo, OSPAR agreed to implement a phased ban on scrubber discharges in internal waters and port zones across the North-East Atlantic region [34]. Discharges from open-loop scrubbers will be prohibited in internal waters and port areas across the North-East Atlantic starting in July 2027, while discharges from closed-loop scrubbers will be banned by January 2029.

From an economic standpoint, open-loop scrubbers offer several advantages, including lower capital and operational expenditures (CAPEX and OPEX, respectively). By contrast, closed-loop systems retain the effluent onboard, and their operation is limited by the vessel's waste storage capacity. This constraint may compel shipowners to schedule additional port calls for effluent discharge, thereby incurring further operational costs. For this reason, closed-loop systems demand more onboard space and entail higher CAPEX and OPEX [35]. Hybrid scrubbers aim to mitigate the disadvantages of both configurations, offering greater operational flexibility, albeit with comparatively higher installation and operating costs [36].

Owners of older vessels must make strategic decisions to meet International Maritime Organization emission reduction targets without compromising economic viability. The initial focus of IMO regulations was the reduction of SO_x emissions, prompting many shipowners to retrofit their fleets with scrubbers. As the IMO now intensifies efforts to reduce GHG emissions, it is essential to assess whether scrubbers contribute meaningfully to GHG mitigation or if a transition to alternative fuels, such as low-sulphur marine diesel oil, is required. Although significant research has addressed SO_x abatement technologies, there remains a lack of comprehensive studies comparing the environmental and economic impacts of using MDO versus open-loop scrubbers, particularly for aging vessels. This study addresses this gap by presenting a novel comparative analysis of these two strategies on a case study vessel built in 1998, evaluating their effectiveness in reducing SO_x and other harmful emissions, as well as their economic implications. The rest of the paper is structured as follows: Section 2 presents the case study vessel and outlines the methodology used for the environmental and economic assessments. Section 3 discusses the results obtained from both the use of MDO and the scrubber installation scenarios. Finally, Section 4 summarizes the main conclusions of the study.

2. Materials and Methods

This paper presents a case study of a cruise vessel that initially operated using MDO and later retrofitted scrubbers in all its engines to comply with IMO SO_x emission regulations, enabling continued operation on the more economic HFO. The vessel's specifications are detailed in Table 1.

Table 1. Particulars of case study vessel.

Particular	Value
Type of Vessel	Cruise ship
Gross Tonnage	102,239 tons
Deadweight	9470 tons
Length Overall	272.8 m
Breadth Moulded	35.5 m
Diesel Generator	6 sets × Wärtsilä Sulzer ZA40S diesel generators
Generator Power	61.6 MW Tot. Nom. Power: 4 × 11.2 MW + 2 × 8.4 MW
Scrubber Type	6 × open-loop scrubber

The case study vessel was built in 1998 and typically operates between ports in the southern United States, primarily Miami, and various destinations throughout the Caribbean Sea. This operational profile is particularly relevant, as Miami lies within the North American Emission Control Area (ECA) designated by IMO. ECAs enforce stricter SO_X emission limits than the global standard, reducing the permissible sulphur content in marine fuels from 0.5% to 0.1%. The data for the environmental and economic assessments were collected from the ship's electronic logbook, during the month of April 2024. Although the vessel is aging, company officials intend to keep it in operation for an additional ten years. This projected service lifetime has therefore been adopted as the time horizon for the economic modelling.

2.1. Environmental Assessment

Currently, the assessment of pollutant emissions from marine vessels can be performed by using three different methods: bottom-up, top-down, or a combination of both [37,38]. These methods are outlined in the *Emission Inventory Guidebook 2023*, published by the European Environment Agency (EEA), more specifically in Chapter 1.A.3.d Navigation (shipping) [39]. The guidebook, first published in 1996, included a chapter on estimating emissions from navigation starting in 2009. Since then, it has been continuously updated and adopted by numerous scientists, also being referenced in the Fourth IMO Greenhouse Gas Study [16,39]. In the literature, these approaches are often referred to as Tier I, Tier II, and Tier III, differing in the emission assessment process and the geographic characterization of the ship [40].

The complete "bottom-up" method assesses emissions from an individual ship by considering its characteristics, such as type, construction date, cargo, engine power, and fuel consumption under specific load conditions. With this method, it is possible to estimate the near-instantaneous emission production of a vessel, helping to identify the main contributors to the emissions, offering a clear understanding of their impact. Since the method is data intensive, it is also expensive and time consuming [37,38].

The "top-down" method takes a broader perspective based on the highly aggregated information of a given shipping activity, relying on generalized factors such as fuel use statistics and engine types across ships to estimate emissions, as there is no information regarding specific circumstance or specific shipping activities. The advantage of the top-down approach, since it requires a small quantity of data, is that it is less expensive and less time consuming than the bottom-up approach [37,38].

The present study uses a hybrid method, which is a combination between the bottom-up and the top-down approaches, known also as the Tier III ship movement method. This hybrid method is generally recommended when detailed data on ship operations and technical characteristics (e.g., size, engine technology, installed power, fuel consumption, and operating hours in different activities) are available, as considered in the presented case study. This enables the estimation of emissions during open-sea navigation, port approach maneuvers, and docking [39]. Compared to Tier I or Tier II, this methodology ensures higher resolution and reliability in emission assessments and is aligned with international standards.

For commercial vessels, Tier III methodology calculates the emissions based on the engine's installed capacity and fuel consumption. Emissions for a trip are determined by summing the emissions from each segment of the journey [39].

$$E_{trip} = E_{dock} + E_{maneuvering} + E_{cruise} \tag{1}$$

Based on the data of fuel consumption in the three different navigation phases, cruise, dock, and maneuvering, the emissions can be calculated using specific fuel emission factors for each phase. In this sense, the pollutant emission calculation can be performed based on Equation (2) [39,40].

$$E_{trip} = \Sigma_p \left(FC_{j,m,p} \times EF_{f i,j,m,p} \right) \tag{2}$$

where E_{trip} represents the emissions during a complete journey (tons); FC is the fuel consumption (tons); EF is the emission factor (kg/ton); i is the pollutant type (NO_x, SO_x, CO_2, PM); m is the fuel type (HFO, MDO/MGO); j is the engine type (low, medium, and high speed); and p is the different phases of the journey (cruise, dock, and maneuvering).

For each level and emission calculation algorithm, it is necessary to use specific emission factors. While Tier I and II methodologies use tabulated emission factors, the more complex methodology, Tier III, adopted in the present study, requires the calculation of specific emission factors based on the present case load-dependent emission factors. It is important to note that variations in emission factor calculations can lead to discrepancies of up to 30% in total emissions [16].

The pollutant emissions (E_i) were determined using two primary methods. The energy-based approach estimates emissions based on engine power output (W_i), applying an energy-based emission factor (EF_e) in grams per kilowatt-hour (g/kWh) [16].

$$E_i = EF_e \times \dot{W}_i \tag{3}$$

In the present study $\dot{W}_i$ was obtained from the data collected on the ship.

This method is used for pollutants such as NO_x, CH_4, CO, N_2O, PM, and Non-Methane Volatile Organic Compounds (NMVOCs).

The fuel-based approach calculates emissions by multiplying hourly fuel consumption (FC_i) by a fuel-based emission factor (EF_f), expressed in grams of air pollutant per gram of fuel (g/g); it is used for CO_2 and SO_x emissions.

$$E_i = FC_i \times EF_f \tag{4}$$

where

$$FC_i = \dot{W}_i \times SFC_{base} \tag{5}$$

The baseline specific fuel consumption for main engines, auxiliary engines, and boilers represents the minimum specific fuel consumption along the load curve, indicating the point of maximum fuel efficiency for the engine [41]. In Table 2, the baseline SFC values used in this study are depicted.

Table 2. Specific fuel consumption [16].

Baseline SFC [g/kWh]	
Heavy Fuel Oil	205
Marine Diesel Oil	190

The methodologies and formulas used in this study were initially introduced in the IMO Third Greenhouse Gas Study [42] and later refined in the Fourth IMO Greenhouse Gas Study [16]. This revision integrated findings from literature reviews, engine manufacturers, research institutions, academic studies, and classification societies, providing a more accurate and up-to-date framework for emission calculations.

2.1.1. Carbon Dioxide

Based on the IMO proposals in document MEPC.1/Circ.684, Table 3 represents CO_2 emission factors for HFO and MDO.

Table 3. Emission factors for HFO and MDO [43].

Fuel	Emission Factor (gCO_2/gFuel)
HFO	3.114
MDO	3.206

The carbon dioxide emissions from the case study cruise vessel were calculated using Equations (4) and (5), which determine the vessel's total fuel consumption and the corresponding CO_2 emissions.

2.1.2. Nitrogen Oxides

Nitrogen oxide emissions result from the high combustion temperatures in the engines, which cause the oxidation of nitrogen present in the intake air as well as nitrogen particles in the fuel. The NO_x emission factor depends on the engine rate speed and the ship's Tier (i.e., the year when the engine was manufactured). In the present study, NO_x emission factors were based on the engine rotational speed, as referred to in Annex VI of IMO MARPOL Regulation 13. Table 4 presents the IMO regulatory limits for NO_x emissions, which vary according to the engine's rated speed.

Table 4. IMO Tier limits for NO_x emissions as a function of engine rated speed [44].

Tier	Ship Construction Date	Total Weighted Cycle Emissions Limit (gNO_x/kWh) n = rpm (Engine Rate Speed)		
		$n < 130$	$130 < n < 1999$	$n > 2000$
I	1 January 2000	17.0	$45n^{-0.2}$	9.8
II	1 January 2011	14.4	$36n^{-0.2}$	7.7
III	1 January 2016	3.4	$9n^{-0.2}$	1.96

n is the engine rate speed, which for the Wärtsilä Sulzer ZA40S diesel generators installed in the ship assessed, the alternators have a frequency of 60 Hz and 12 poles, which correspond to an engine rate speed of 514 rpm. Applying the Tier I coefficient for this engine category, the NO_x emission limit for the case study vessel engines is calculated to be 12.91 gNO_x/kWh.

2.1.3. Sulphur Oxides

Sulphur oxide emissions depend on the sulphur content of the fuel used. In this sense, the SO_x emission factor (g SO_x/g of fuel) is calculated as follows [16]:

$$EF_{f,SOx} = 2 \times 0.97753 \times S \tag{6}$$

where it is considered that 97.753% of the fuel sulphur content is converted to SO_x (with the remainder converted to sulphate/sulfite aerosol and classified as part of the particulate matter). The number "2" reflects the ratio of the molecular weight of SO_2 to sulphur, since the majority of SO_x is released as SO_2. The term S represents the sulphur content of the fuel assessed. In the present study the sulphur content of HFO is considered to be 3.5% and for MDO, it is 0.1% [45].

2.1.4. Particulate Matter

The emission factors for particulate matter are influenced by the sulphur content of the fuel. Table 5 presents the equations used to calculate the PM for engines running on heavy fuel oil or marine diesel oil [16].

Table 5. Particulate matter emission equations [16].

Fuel	Emission Factor (gPM/kWh)
HFO	$EF_e = 1.35 + SFC_i \times 7 \times 0.02247 \times (S - 0.0246)$
MDO	$EF_e = 0.23 + SFC_i \times 7 \times 0.02247 \times (S - 0.0024)$

The number 7 in the equations represents the molecular weight ratio between the sulphate of the particles and sulphur, while the value 0.02247 reflects the proportion of sulphur in the fuel that is converted to PM sulphate. Again, the term S represents the sulphur content of the fuel assessed.

In this study, the Tier III methodology was applied in the following stages:

1. Onboard data collection: Engine particulars, power output, load profiles, NO_x Tier classification, and fuel consumption for each phase (cruise, maneuvering, docking) were obtained from ship logs and manufacturer datasheets.
2. Determination of SFC of engines based on IMO baseline values corresponding to the engines type and rating.
3. Calculation of fuel and energy-based emission factors and pollutant emissions:

a. Fuel-based factors for CO_2 (based on IMO values; Table 3). Emission factors for HFO and MDO [44] (Table 3), along with SO_x (g/gFuel), as defined by Equation (6).

b. Energy-based factors for NO_x and PM (g/kWh): NO_x emissions were estimated using Tier I coefficients for engines operating within the $130 < n < 1999$ rpm range (see Table 4). PM emissions were calculated based on the equations presented in Table 5.

4. Total trip emissions were calculated by summing the three phase emissions: cruise, maneuvering, and docking.

2.2. Economic Assessment

While environmental considerations may justify the investment in the long term, shipowners, operating within the constraints of commercial viability, prioritize profitability and financial sustainability. Consequently, for a greener solution to be adopted within the sector, it must also demonstrate economic feasibility. In the case of scrubbers, their installation involves a substantial retrofit, which requires modifications such as cutting funnel bulkheads and integrating the scrubber units into the exhaust systems of each engine. This results in a significant initial capital investment. In addition to capital expenditure, there are also ongoing operational expenditures related to system operation and maintenance. Table 6 summarizes the estimated CAPEX and OPEX associated with scrubber installation in US dollars (USD) (data provided by the shipowner).

Table 6. Capital and operational expenditures of scrubbers.

Type	Expenditure	Unitary Cost (USD)	Total Cost (USD)
CAPEX	Scrubber for 11.4 MW engine	313,059.33	1,252,237.32
	Scrubber for 8.6 MW engine	308,252.33	616,504.66
		CAPEX TOTAL	1,868,741.98
OPEX	Operation and maintenance	108,427	108,427 USD/year
	Fuel cost—Case 1: HFO + scrubbers		18,394,500.68 per year
	Fuel cost—Case 2: MDO		26,612,116.38 per year

When deciding whether to switch from HFO to MDO to comply with IMO SOx regulations or to install scrubbers, shipowners must carefully evaluate the primary cost driver: fuel expenditure. Although the CAPEX for scrubber installation approaches USD 2 million, and the associated annual OPEX for scrubber operation and maintenance exceeds USD 108,000, these costs represent only a small fraction of the vessel's yearly fuel expenses. Based on the fuel consumption data, and assuming average fuel prices of USD 547 per ton for HFO and USD 885 per ton for MDO, the annual fuel costs are estimated at USD 18,394,500.68 for HFO and USD 26,612,116.38 for MDO [46]. Consequently, the annual differential in fuel costs between operating on MDO versus HFO is USD 8,217,615.69.

Both HFO and MDO are fossil fuels and thus correlated to some extent, but their prices have historically exhibited significant volatility and variable spreads, especially during periods of uncertainty or the introduction of stringent regulations. A notable example is the implementation of the IMO 2020 regulation, which increased the demand for MDO and led to a relative drop in the demand for HFO. Given the substantial impact of fuel prices on the total cost of each compliance strategy and considering that this is a long-term decision intended to span the remaining service life of the vessel, projected at ten years, the decision must be carefully weighed and supported by reliable data.

In this case, the economic assessment is based on the Net Present Value (NPV), to evaluate the long-term economic feasibility, and the Return on Investment (ROI), to measure the profitability of installing scrubbers, as compared to the alternative compliance strategy of switching from HFO to MDO. Additionally, the Internal Rate of Return (IRR) and the Discounted Payback Period (DPP) were calculated to provide a more comprehensive understanding of the investment's performance.

NPV is calculated using Equation (7):

$$NPV = \sum_{t=0}^{n} \frac{R_t}{(1+i)^t} \tag{7}$$

where R_t represents the net cash inflow or outflow at time, i is the discount rate, set at 10% in this analysis, and t is the time period, which in this case is set to 10 years, corresponding to the expected remaining operational life of the vessel.

The ROI is obtained using Equation (8):

$$ROI = \frac{FVI - IVI}{IVI} * 100 \tag{8}$$

where *FVI* represents the Final Value of the Investment, obtained by adding the initial CAPEX of the scrubber system to the accumulated net cash flows over the evaluation period, while *IVI* denotes the Initial Value of the Investment, corresponding to the upfront capital expenditure required for the installation of the scrubbers. The difference between FVI and IVI yields the net cash flows generated by the investment. The denominator of the ROI expression is the Initial Value of the Investment, which represents the total capital initially committed.

The IRR is obtained by setting NPV equal to zero, while DPP is calculated using the previously mentioned discount rate of 10% over a 10-year evaluation period.

While the calculation of NPV and the other aforementioned key economic indicators at a specific point in time, such as when data was collected onboard the case study vessel, may be sufficient to support preliminary decision-making, the significant influence of fuel prices, coupled with the high volatility of fossil fuel markets, requires a more comprehensive assessment. This is especially relevant when considering a long-term investment such as scrubber installation, expected to last for a ten-year period.

To address this, the present study conducted a deterministic grid sweep sensitivity analysis, which involved systematically evaluating combinations of HFO and MDO prices within the historical ranges shown in Table 7 [46]. This analysis aimed to assess the economic viability of scrubber installation under different fuel price scenarios by calculating the Net Present Value (NPV) for each price pair. For the scrubber installation to be considered economically viable, the NPV must remain positive. Otherwise, from both economic and environmental perspectives, switching to cleaner MDO fuel is preferred. This method allows for a robust understanding of how fluctuations in fuel prices impact investment decisions over the scrubber's expected lifetime.

Table 7. Price range of HFO and MGO in period of 2021–2025 (extracted from [46]).

Type of Fuel	Min Price (USD/ton)	Max Price (USD/ton)
HFO	443.25	732.25
MDO	694.0	1421.50

To assess the long-term economic performance of installing scrubbers versus switching to MDO, and given the potential variability in fuel prices over the vessel's remaining ten-year operational life and consequently in the price spread between HFO and MDO fuels, a risk-based economic assessment was conducted using a Monte Carlo simulation framework. This approach allows for the evaluation of investment feasibility across a wide range of plausible future fuel price scenarios. Fuel price uncertainty was modelled using 100,000 Monte Carlo iterations, based on historical price ranges of HFO and MDO presented in Table 7. To account for the interdependence between HFO and MDO prices under different market conditions, it was essential to incorporate statistical correlation between fuel types. For this purpose, a Gaussian copula, a statistical tool that models the joint behaviour of correlated variables by separately capturing their individual marginal distributions and their dependence structure (correlation), was employed to accurately represent the joint price dynamics of HFO and MDO. This approach enables the preservation of each fuel's distinct historical price distribution while realistically simulating their correlated movements under varying market conditions [47]. This method enables the generation of correlated random samples while maintaining the marginal distributions of HFO and MDO prices. This process involved the following steps:

1. Sampling from a bivariate standard normal distribution with a specified Pearson correlation coefficient (ρ).
2. Transforming each marginal using the standard normal cumulative distribution function (CDF) to obtain uniform variables.
3. Mapping the uniform variables to the historical price ranges of HFO and MDO using the inverse CDF of each fuel's empirical distribution.

The CDF plays a relevant role in the Gaussian copula framework by enabling the transformation of each variable's historical price data into uniform variables on the interval (0,1). For a given random variable X, the CDF is defined as follows:

$$F_X(x) = P(X \leq x) \tag{9}$$

This represents the probability that X takes a value less than or equal to x. This transformation facilitates the modelling of dependence separately from the marginal distributions. After introducing the desired correlation structure in the transformed space, the inverse CDF is applied to map the correlated uniform variables back to their original price distributions, preserving their historical characteristics while reflecting their joint behaviour.

According to the analysis of fuel prices for the 2021–2025 period, the Pearson correlation coefficient between HFO and MDO prices was found to be approximately ($\rho = -0.186$) with a p-value < 0.01. This result suggests a statistically significant, yet weak, negative correlation between the two fuels, indicating that when the price of one tends to increase, the other tends to decrease slightly. Despite its low magnitude, this inverse relationship may reflect underlying regulatory or market-driven dynamics affecting fuel demand. In addition to this observed correlation, two alternative scenarios were also modelled to capture more extreme but plausible market behaviours:

- A negative correlation ($\rho = -0.3$): This negative correlation reflects regulatory situations, such as IMO 2020 sulphur restrictions, where the demand shifts from HFO to MDO. When one price goes up, the other tends to go down, increasing their price spread.
- A moderately positive correlation ($\rho = 0.6$): This positive correlation represents normal market conditions where fuel prices generally move together because they share common cost drivers like crude oil prices.

For each iteration, the correlated pair of HFO and MDO prices was used to compute the annual fuel cost differential between the two compliance options. This differential, adjusted for the fixed annual OPEX and amortized CAPEX of the scrubber system, was treated as a constant net cash flow over the vessel's remaining 10-year operational life. The resulting cash flows were discounted at a rate of 10% and integrated into the Net Present Value (NPV) model, as defined in Equation (8). This probabilistic framework provides a robust basis for investment appraisal by incorporating both fuel price volatility and the statistical dependence between HFO and MDO prices under varying regulatory and market conditions [48,49].

To provide a clearer overview of the entire process, Figure 1 presents a flowchart outlining the methodology followed in this study.

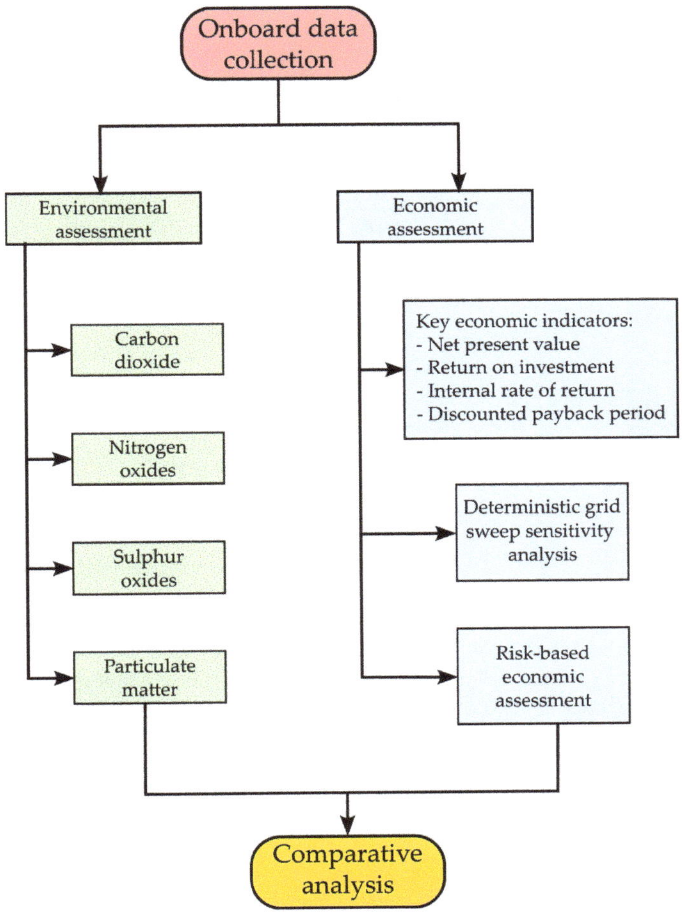

Figure 1. A flowchart of the methodological approach adopted in this study.

3. Results and Discussion

The results, in line with the methodology, are divided into two parts: an environmental and an economic assessment. Together, these analyses provide a comprehensive basis for decision-making, enabling shipowners to identify the most suitable emission reduction strategy from both environmental and economic standpoints, while ensuring compliance with IMO regulations on air pollution.

3.1. Environmental Assessment

To conduct this analysis, energy consumption data for a one-month period of the studied vessel's voyage was gathered. It was verified that the ship typically operates on routes between the United States and the eastern Caribbean countries. Using real-time voyage data collected during April 2024, it was verified that the vessel spent 71% of its time in transit, 24% docked or at anchor, and 5% performing port maneuvers. Fuel consumption calculation was based on the respective baseline specific fuel consumption (see Table 2).

Since no consumption data was available before the installation of the exhaust gas cleaning systems, there was a need to estimate this consumption. This fuel consumption was performed based on the specific characteristics of the vessel, which has diesel–electric propulsion, a given electrical energy need, and a specific fuel oil consumption.

Beyond the installation of scrubbers, the vessel is also equipped with fuel improvement devices (FIDs). These systems are designed to reduce the formation of nitrogen oxides and particulate matter by promoting fuel emulsification with small amounts of water or additives. The emulsification process leads to lower peak combustion temperatures, thereby mitigating thermal NO_x formation. Additionally, the built-in homogenizer enhances fuel atomization and uniformity, reducing sludge accumulation and promoting more efficient and complete combustion. As a result, soot and PM emissions are also reduced [50,51].

Table 8 presents the energy consumption associated with the three operational scenarios: (i) the use of HFO without scrubbers or FID, (ii) the use of HFO with scrubbers and FID, and (iii) the use of MDO.

Table 8. Vessel energy consumption by fuel type.

Fuel	Energy Consumption (Ton/Month)
HFO	2703.7
HFO + scrubbers + FID	2802.3
MDO	2505.9

The results indicate that the installation of scrubbers leads to an increase in energy consumption of approximately 3.52% from the baseline, primarily due to the additional power demand of the pumps and treatment systems. Conversely, switching from using HFO and scrubbers to MDO results in an estimated 7.31% reduction in energy consumption, owing to the superior combustion characteristics of distillate fuels. Taken together, these effects account for a total difference of 10.83% in energy efficiency between the HFO–scrubber configuration and the direct use of MDO. Although slightly lower, these figures are consistent with the findings of Karatuğ et al., who reported a fuel consumption increase ranging from 11% to 17% when scrubbers are employed [28].

Based on the methodology for emission calculation referred to in the previous section, and on the energy consumption estimations, pollutant emissions were calculated for the three considered scenarios. The results are depicted on Table 9.

Table 9. Emissions by fuel (ton/month).

Fuel	CO_2	NO_X	SO_X	PM
HFO	8419.3	170.30	185.01	22.23
HFO + scrubbers + FID	8726.5	176.52	5.75	20.73
MDO	8033.8	170.31	0.49	2.48

The results reveal that the installation of scrubbers in combination with FID has led to a substantial reduction in SO_X emissions, decreasing from 185.01 to 5.75 ton/month, equivalent to an approximate 96.89% reduction. In terms of PM emissions, a moderate decline was observed, from 22.23 to 20.73 ton/month, representing a reduction of about 6.71%. Similar emission abatement efficiencies were reported by Yang et al., who documented a 97% reduction in SO_x and a 10% decrease in $PM_{2.5}$ concentrations following scrubber implementation [32].

However, the calculations performed show that the installation of scrubbers has a negative impact on carbon dioxide and nitrogen oxide emissions. The emission of CO_2 increased from 8419.3 ton/month to 8726.5 ton/month, which represents an increase of 3.52%, similar to previous studies [32]. Regarding NO_x emissions, the increase was from 170.30 to 176.52 ton/month, representing a 3.65% increase. These adverse outcomes are consistent with the findings of Yang et al., who reported a 4.35% increase in NO_x emissions [32]. This rise is directly associated with the higher consumption of HFO, as vessels equipped with scrubbers are not obligated to switch to MDO when approaching coastal areas. Consequently, these ships tend to operate on HFO more frequently, contributing to elevated NO_x emissions.

It can be concluded that scrubbers are suitable to meet sulphur emission regulations but not to fulfil the IMO's decarbonization strategy due to the increase in fuel consumption. In this sense, and considering the substantial investments already made by shipowners with the installation of scrubbers, there is a need to integrate additional technologies to reduce greenhouse gas emissions.

The results from the scenario of using MDO instead of HFO show positive impacts in CO_2, SO_x, and PM emissions. When comparing both scenarios, the use of MDO as a substitute for HFO results in a reduction of SO_x emissions of more than 99% and of about 89% for PM. When comparing the MDO scenario with the HFO + scrubbers + FID scenario, it is interesting to verify that the use of MDO has lower emissions for both SO_x (more than 91% of reduction) and PM (about 88%).

Regarding the impact of the consumption of MDO on CO_2 emissions, it can be verified that it is also positive when compared with the other two scenarios assessed. For the HFO scenario, a reduction of 5% is reached and when compared with the scrubber scenario, the reduction is about 8%.

Regarding NO_x, the International Maritime Organization's MARPOL Annex VI establishes global NO_x emission limits (Tiers I and II), while the more stringent Tier III standards only apply within designated NO_x Emission Control Areas (NECAs). In parallel, SO_x regulations have also evolved, with stricter fuel sulphur limits of 0.1% in SO_x Emission Control Areas (SECAs), compared to the global cap of 0.5% [52]. While exhaust gas cleaning systems, such as the scrubbers analyzed in this study, are effective in reducing SO_x emissions, they do not mitigate NO_x. Therefore, additional NO_x reduction strategies are required to ensure compliance, particularly in Tier III Emission Control Areas. These include primary measures such as Exhaust Gas Recirculation (EGR), water-based techniques like Direct Water Injection (DWI) and fuel–water emulsions (FWE), as well as post-combustion technologies such as the Selective Catalytic Reduction (SCR). Among these, SCR is the most mature and widely implemented option, capable of achieving up to 90% NO_x reduction under appropriate conditions and enabling compliance with IMO Tier III standards [24].

However, care must be taken when integrating multiple exhaust gas treatment technologies, such as SCR units, scrubbers, and exhaust gas boilers, into the engine funnel. The combined presence of these systems can significantly increase backpressure, potentially forcing the engine to work harder to be able to expel the exhaust gases [25]. This may lead to higher fuel consumption, offsetting some of the environmental gains and becoming counterproductive from an energy efficiency perspective. In the case of NO_x, implementing DWI or FWE techniques may be a viable solution to improve its reduction while managing flow-related constraints [53]. The EGR option, on the other hand, is less favourable for marine applications due to maintenance challenges caused by fouling from exhaust gases [54]. With regard to SO_x control, one effective alternative to scrubber installation is the use of sulphur-free fuels, such as MDO, which inherently comply with SECA fuel sulphur requirements and avoid the space and operational complexity of additional equipment.

3.2. Economic Assessment

Following the environmental assessment, where it was demonstrated that using MDO leads to significantly lower particulate matter emissions compared to the continued use of HFO with scrubbers, given that scrubbers are largely ineffective in removing PM, the focus of the economic analysis shifts toward evaluating the viability of adopting MDO as the main fuel. The financial indicators presented in this section are therefore interpreted with the premise that MDO is the environmentally preferred option, and economic success is understood as achieving comparable or superior financial performance without installing scrubbers.

Initially, a set of standalone economic indicators, including Net Present Value, Return on Investment, Internal Rate of Return, and Discounted Payback Period, were calculated to assess the project's financial feasibility, using the prevailing prices of HFO and MDO at the time the data were collected. These economic indicators are depicted in Table 10.

Table 10. Economic indicators related to installation of scrubbers and use of HFO.

Economic Indicator	Value
Net Present Value	USD 47,958,712.14
Return on Investment (10 year)	4239%
Internal Rate on Return	434%
Discounted Payback Period	3.1 months

As a result of the IMO 2020 regulation on SO_x emissions, the global merchant fleet is now required to either switch to low-sulphur fuels or install exhaust gas cleaning systems to continue operating on cheaper residual fuels [55]. For the case study vessel, the installation of the scrubber system involved a CAPEX of just under USD 1.9 million. Considering the annual fuel cost differential between operating on MDO and HFO, calculated at USD 8,217,615.69 based on prevailing market prices at the time of data collection, both the initial CAPEX and the associated annual OPEX of the scrubber system are relatively minor compared to the potential cost savings. Table 10 summarizes the economic performance of scrubber installation over a projected ten-year operational period. The NPV is markedly positive, primarily due to the significant annual savings achieved by continuing to use HFO in combination with scrubbers rather than switching to MDO. The ROI over ten years reaches 4239%, while the IRR is calculated at 434%, both of which confirm the high profitability of the investment. Additionally, the DPP is just 3.1 months, underscoring the rapid recovery of the initial investment under the current assumptions.

It is important to highlight that the profitability of scrubber installation is highly sensitive to the price differential between HFO and MDO, which has historically exhibited significant volatility due to their independent market dynamics [56]. Consequently, relying on fixed price assumptions would be insufficient to capture the range of possible economic outcomes over the vessel's projected operational lifespan. To address this, a sensitivity analysis was carried out using a deterministic grid sweep approach. This analysis aims to evaluate the robustness of the investment decision and determine under which market conditions the use of MDO might become more economically viable than continuing to operate on HFO with scrubbers. The results of this analysis are presented in Figure 2.

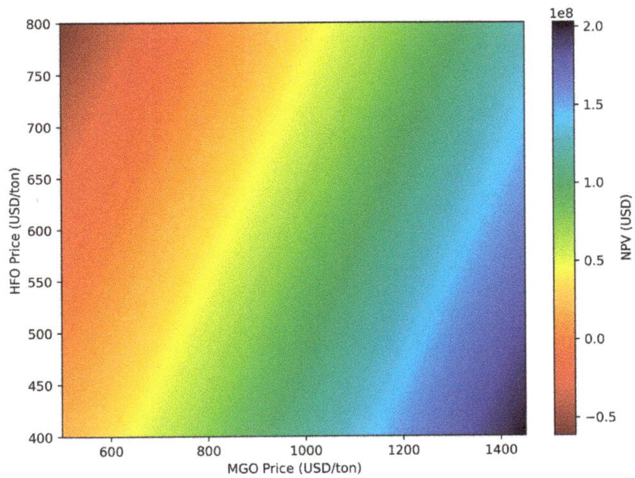

Figure 2. Grid sweep sensitivity analysis of HFO and MGO prices on NPV of scrubber installation.

The deterministic grid sweep analysis explored a range of HFO and MDO price combinations based on the ranges presented in Table 7. As shown in Figure 2, scrubber installation remains economically viable across the majority of scenarios. The only circumstances under which MDO emerges as the more favourable option occur when the price differential between MDO and HFO narrows substantially, or in the unlikely event that HFO becomes more expensive than MDO. Although MDO offers a marginal energy efficiency advantage, due to its slightly higher lower heating value, as indicated in Table 8, this is generally insufficient to counterbalance the substantial cost savings enabled by using HFO in combination with scrubbers. The capital and operating costs associated with the scrubber system, while not negligible, are relatively minor compared to the potential fuel cost savings over the vessel's remaining lifetime. Overall, the grid sweep analysis underscores the robustness of the investment decision in favour of scrubber installation across a broad spectrum of plausible market conditions, while also delineating the narrow set of fuel price scenarios under which switching to MDO may be justified from a financial standpoint. Zhu et al. similarly observed that fuel switching to MDO was economically feasible only when the HFO–MDO price differential surpassed USD 56 per ton [55].

Following the recommendation of Sheng et al. to exercise caution when interpreting the influence of fuel price differentials on scrubber adoption trends, a probabilistic analysis was conducted to complement the deterministic assessment of the HFO–MDO price spread [56]. This analysis employed the Monte Carlo method to account for price volatility and uncertainty. This approach aims to evaluate the likelihood of economic viability when choosing to operate with MDO instead of installing scrubbers and continuing to use HFO, accounting for market uncertainty. The correlation derived from 2021–2025 fuel price data was first analyzed, followed by the exploration of two additional correlation scenarios to capture alternative and more extreme market behaviours. Figure 3 presents the results based on the correlation obtained from actual fuel price data. Figure 4 illustrates the scenario, assuming a moderately positive correlation between HFO and MDO prices, while Figure 5 displays the outcome of the third scenario, which considers a negative correlation between the two fuel types.

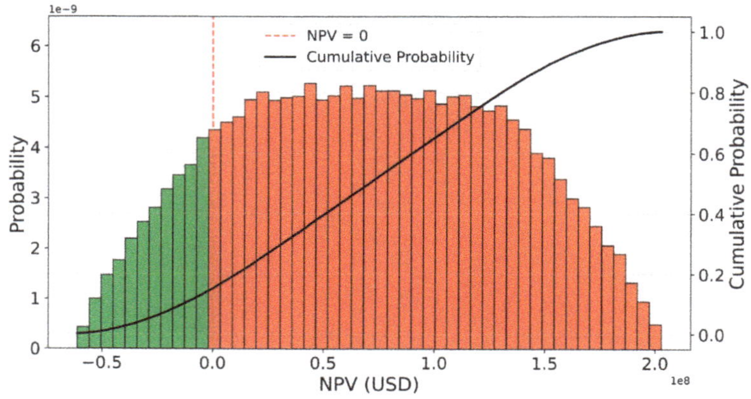

Figure 3. Results of Monte Carlo simulation for NPV of scrubber installation ($\rho = -0.186$).

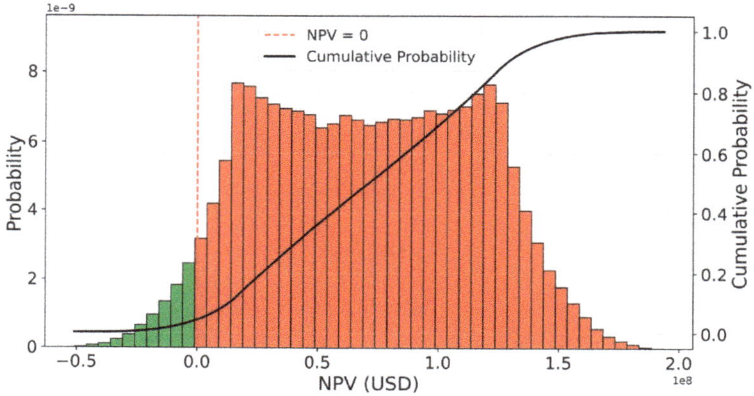

Figure 4. Results of Monte Carlo simulation for NPV of scrubber installation ($\rho = 0.6$).

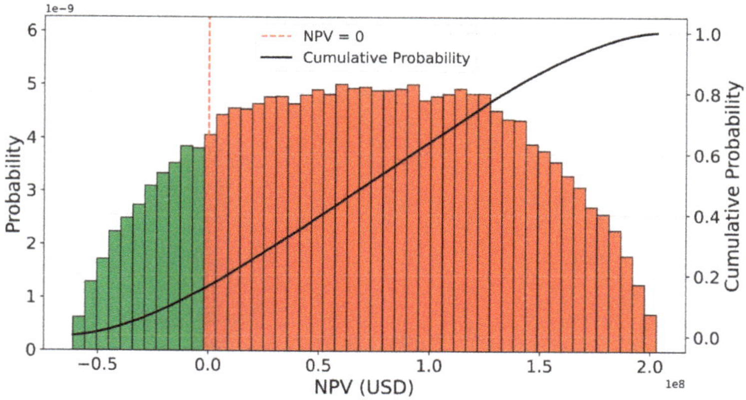

Figure 5. Results of Monte Carlo simulation for NPV of scrubber installation ($\rho = -0.3$).

The histograms display the probability density of NPV outcomes across the simulated range. The height of each bar represents the likelihood that the NPV falls within the corresponding interval. Taller bars indicate a higher probability density, meaning that a larger number of simulations yielded NPVs in that range. In the first scenario, Figure 3 ($\rho = -0.186$), bars corresponding to negative NPVs (displayed in green and associated with the strategy of switching to MDO) reach a maximum probability density of just over 4. In contrast, positive NPVs (shown in red and corresponding to the installation of scrubbers and continued use of HFO) peak near 5. Under these conditions, the probability of obtaining a positive NPV, indicating that scrubber installation is economically advantageous, is 84.87%. Conversely, the likelihood of achieving economic success by switching to MDO is only 4.3%. The average NPV is approximately USD 70,962,358, further supporting the financial viability of the scrubber-based compliance strategy under the assumed market conditions.

Figure 4 illustrates the distribution of NPV outcomes obtained from the Monte Carlo simulation under the assumption of a moderate positive correlation between HFO and MDO prices, aiming to reflect the fact that both fuels are derived from crude oil and share similar refining and distribution pathways. The red-shaded area denotes scenarios where the installation of scrubbers and continued use of HFO yields a positive NPV, while the green area indicates conditions under which switching to the less pollutant MDO becomes economically advantageous. In this second scenario, the green bars are relatively low, peaking at around 2.2, while the red bars exceed 7.5, indicating a much stronger concentration of positive NPV outcomes. Under this scenario, the probability of obtaining a positive NPV, indicating that scrubber installation is more economically favourable, is 95.7%. This leaves only a 4.3% probability of economic success when switching to MDO. The average NPV is USD 70,683,731, further substantiating the financial robustness of the scrubber-based compliance strategy within the defined market conditions.

Figure 5 presents the distribution of NPV results under a negative correlation between HFO and MDO prices, as modelled through Monte Carlo simulation. This scenario reflects a market condition in which the prices of the two fuels are not closely aligned and may diverge, for instance, if regulatory shifts or fuel switching trends create independent price dynamics. In the figure, the green-shaded area highlights scenarios in which switching to MDO yields a positive economic outcome, with a probability density peaking just below 4. In contrast, the red-shaded area represents cases where continuing to use HFO with a scrubber proves more financially advantageous, with the probability density exceeding 5. Under this correlation, the probability of achieving a positive NPV from scrubber installation is 83.65%, with an average NPV of USD 70,820,100. Conversely, the probability of an economically favourable outcome when switching to MDO is 16.35%. Although this represents a reduction in the certainty of scrubber profitability compared to the positively correlated case, the installation of scrubbers remains the preferred option in the vast majority of market scenarios.

The cumulative probability curves in Figures 4 and 5 provide additional insights into the risk profile associated with choosing MDO over the installation of scrubbers. These curves represent the probability that the NPV will fall below a given threshold, thereby offering a clear view of the economic risk distribution under each correlation scenario. In the positively correlated case, Figure 4, the cumulative distribution increases more gradually, with a relatively low probability of negative NPV outcomes. This reflects a high confidence level in achieving positive returns from scrubber installation. In contrast, the negatively correlated scenario, Figure 5, exhibits a noticeably steeper cumulative curve around the NPV = 0 threshold. This inflection reflects an increased likelihood of borderline or negative economic returns, implying that the relative advantage of scrubbers weakens under such market dynamics. The results align with previous studies, indicating that while scrubbers generally remain the more profitable option, the probability of favourable outcomes when switching to MDO rises in scenarios characterized by negatively correlated fuel prices [57].

The long-term abatement of air pollution in maritime transport will require a transition to alternative fuels. LNG remains a widely used option due to its technological maturity and reduced emissions, although concerns about methane slip persist. Renewable methanol is a promising alternative but is currently limited by availability and the need for engine modifications. Hydrogen and ammonia offer near-zero-emission potential, yet they face significant challenges related to storage, safety, and high production costs. Nuclear propulsion has also attracted renewed interest, but regulatory, safety, and societal concerns remain substantial. A combination of these fuels, supported by robust regulatory frameworks and technological advancements, is likely to define the future marine fuel mix and reduce the need for exhaust gas treatment systems.

4. Conclusions

The world is facing record-high temperatures and increasingly extreme weather events. Stakeholders, from all sectors, believe that these phenomena are related to the increase in greenhouse gas emissions. As international maritime transport accounts for 3% of global CO_2 emissions, the International Maritime Organization has set ambitious targets to reduce air pollution by at least 70%, striving for 80%, by 2040 and achieve net-zero GHG emissions by or around 2050. To meet these goals, the industry will need to make a transition to low-carbon and carbon-neutral fuels and to improve energy efficiency. New and ordered vessels are, already, being built based on these new IMO targets. The main problem is related with the fleet already in service, where shipowners need to make difficult decisions, namely, renew the fleet, retrofit the existing ships, switch to alternative fuels, or pay penalties such as EU ETS allowances or IMO Remedial Units.

Addressing the current gap in the literature concerning the environmental and economic comparison of low-sulphur MDO versus open-loop scrubbers, this study evaluates three operational scenarios applied to a 1998-built cruise ship: (i) HFO as baseline; (ii) HFO combined with an open-loop scrubber equipped with FID; and (iii) MDO use.

The environmental assessment results reveal that scrubber installation reduces SO_x emissions by approximately 97% (from 185.01 to 5.75 tons/month) and particulate matter by around 7%, from 22.23 to 20.72 tons/month. However, scrubbers increase CO_2 and NO_x emissions by approximately 3.52% and 3.65%, respectively, due to higher fuel consumption. In contrast, MDO usage lowers SO_x and PM emissions more effectively than scrubbers and reduces CO_2 emissions by about 7.3% compared to baseline, with no significant change in NO_x emissions. The findings from this case study corroborate previous works confirming the trade-offs associated with scrubber technology.

As previously noted, environmental assessments alone are insufficient for shipowners to make informed decisions regarding the adoption of emission reduction technologies. Economic considerations are also essential to ensure business profitability. In this context, the present study includes an economic assessment comparing the financial performance of installing exhaust gas cleaning systems with switching to marine diesel oil. Based on the prevailing prices of HFO and MDO at the time of data collection, key economic indicators were calculated: Net Present Value, Return on Investment, Internal Rate of Return, and Discounted Payback Period.

The results indicate that, under the considered price conditions, the installation of scrubbers represents a more economically favourable option than switching to MDO. However, given the volatility of fuel prices influenced by market dynamics and regulatory factors, a probabilistic Monte Carlo analysis and a deterministic grid sweep were conducted. These analyses considered the actual correlation derived from market data spanning 2021 to 2025, as well as hypothetical scenarios involving both positive and negative correlations between HFO and MDO prices. The results indicate that fuel costs dominate economic feasibility, and the expected HFO–MDO price spread disfavors switching to MDO under current market conditions.

This integrated decision-making framework demonstrates that while switching to MDO offers superior environmental performance, scrubber installation currently represents a more viable economic pathway for shipowners. Nonetheless, the increasing regulatory restrictions on open-loop scrubbers, particularly the forthcoming bans on discharge, limit their long-term applicability. Hence, future regulatory policies should support and incentivize a global transition toward low-emission or carbon-neutral fuels.

Future research should enhance decision-support tools by incorporating real-world operational data and extending comparative analyses to include a broader range of technical options, such as alternative fuels and energy efficiency technologies. This will better equip shipowners with comprehensive, context-specific guidance for sustainable fleet management.

Author Contributions: Conceptualization, L.B.; methodology, S.P. and L.A.D.-S.; software, L.A.D.-S.; formal analysis, S.P.; resources, L.B.; writing—original draft preparation, S.P. and L.A.D.-S.; writing—review and editing, L.B. All authors have read and agreed to the published version of the manuscript.

Funding: This research received no external funding.

Data Availability Statement: The data presented in this study are available on request from the author.

Acknowledgments: The authors gratefully acknowledge the financial support provided by the University of Oviedo and Banco Santander. The research stay of L.A. Díaz-Secades at the Escola Superior Náutica Infante D. Henrique (ENIDH) in Portugal was funded by Banco Santander through the "Ayudas Económicas de Movilidad de Excelencia para Docentes e Investigadores de la Universidad de Oviedo, 2025" program. The article was based on the MSc on Marine Engineering thesis of Juliana Almeida at ENIDH named "Steering toward a sustainable future: a comparative technical-economic review of maritime shipowner options" (in Portuguese).

Conflicts of Interest: The authors declare no conflicts of interest.

Abbreviations

The following abbreviations are used in this manuscript:

CAPEX	Capital Expenditure
CDF	Cumulative Distribution Function
DPP	Discounted Payback Period
DWI	Direct Water Injection
ECA	Emission Control Area
EGR	Exhaust Gas Recirculation
EU ETS	European Union Emissions Trading System
EEA	European Environment Agency
FID	Fuel Improvement Devices
FVI	Final Value of the Investment
FWE	Fuel–Water Emulsions
GWP	Greenhouse Warming Potential
HFO	Heavy Fuel Oil
IMO	International Maritime Organization
IRR	Internal Rate of Return

IVI	Initial Value of Investment
MDO	Marine Diesel Oil
NECA	NO_x Emission Control Areas
NMVOC	Non-Methane Volatile Organic Compounds
NPV	Net Present Value
OPEX	Operational Expenditure
PM	Particulate Matter
ROI	Return on Investment
SCR	Selective Catalytic Reduction
SECA	Sulphur Emission Control Areas
SFC	Specific Fuel Consumption
UNFCCC	United Nations Framework Convention on Climate Change

References

1. *Global Energy Review 2025*; International Energy Agency (IEA): Paris, France, 2025.
2. *World Energy Outlook 2024*; International Energy Agency (IEA): Paris, France, 2024.
3. *IPCC Climate Change 2021: The Physical Science Basis. Contribution of Working Group I to the Sixth Assessment Report of the Intergovernmental Panel on Climate Change*; Masson-Delmotte, V., Zhai, P., Pirani, A., Connors, S.L., Péan, C., Berger, S., Caud, N., Chen, Y., Goldfarb, L., Gomis, M.I., et al., Eds.; Cambridge University Press: Cambridge, UK; New York, NY, USA, 2021.
4. Khalaf, E.M.; Mohammadi, M.J.; Sulistiyani, S.; Ramírez-Coronel, A.A.; Kiani, F.; Jalil, A.T.; Almulla, A.F.; Asban, P.; Farhadi, M.; Derikondi, M. Effects of Sulfur Dioxide Inhalation on Human Health: A Review. *Rev. Environ. Health* **2024**, *39*, 331–337. [CrossRef]
5. Supran, G.; Rahmstorf, S.; Oreskes, N. Assessing ExxonMobil's Global Warming Projections. *Science* **2023**, *379*. [CrossRef] [PubMed]
6. Jones, M.W.; Peters, G.P.; Gasser, T.; Andrew, R.M.; Schwingshackl, C.; Gütschow, J.; Houghton, R.A.; Friedlingstein, P.; Pongratz, J.; Le Quéré, C. National Contributions to Climate Change Due to Historical Emissions of Carbon Dioxide, Methane, and Nitrous Oxide since 1850. *Sci. Data* **2023**, *10*, 155. [CrossRef]
7. Comer, B.; Beecken, J.; Vermeulen, R.; Sturrup, E.; Paschinger, P.; Osipova, L.; Gore, K.; Delahaye, A.; Verhagen, V.; Knudsen, B.; et al. *Fugitive and Unburned Methane Emissions from Ships (FUMES): Characterizing Methane Emissions from LNG-Fueled Ships Using Drones, Helicopters, and on-Board Measurements*; ICCT: Bellagio, Italy, 2024.
8. Álvarez, A.N.F.; Díaz-Secades, L.A. Enhancing Energy Efficiency in High-Speed Craft: A Trigeneration Energy Recovery System for LNG-Fuelled Vessels in the Mediterranean Sea. *Polish Marit. Res.* **2025**, *32*, 94–105. [CrossRef]
9. Tian, H.; Xu, R.; Canadell, J.G.; Thompson, R.L.; Winiwarter, W.; Suntharalingam, P.; Davidson, E.A.; Ciais, P.; Jackson, R.B.; Janssens-Maenhout, G.; et al. A Comprehensive Quantification of Global Nitrous Oxide Sources and Sinks. *Nature* **2020**, *586*, 248–256. [CrossRef]
10. Eastham, S.D.; Chossière, G.P.; Speth, R.L.; Jacob, D.J.; Barrett, S.R.H. Global Impacts of Aviation on Air Quality Evaluated at High Resolution. *Atmos. Chem. Phys.* **2024**, *24*, 2687–2703. [CrossRef]
11. Liu, Z.; Rieder, H.E.; Schmidt, C.; Mayer, M.; Guo, Y.; Winiwarter, W.; Zhang, L. Optimal Reactive Nitrogen Control Pathways Identified for Cost-Effective PM2.5 Mitigation in Europe. *Nat. Commun.* **2023**, *14*, 4246. [CrossRef]
12. Clappier, A.; Thunis, P.; Beekmann, M.; Putaud, J.P.; de Meij, A. Impact of SOx, NOx and NH3 Emission Reductions on PM2.5 Concentrations across Europe: Hints for Future Measure Development. *Environ. Int.* **2021**, *156*, 106699. [CrossRef]
13. Khaniabadi, Y.O.; Polosa, R.; Chuturkova, R.Z.; Daryanoosh, M.; Goudarzi, G.; Borgini, A.; Tittarelli, A.; Basiri, H.; Armin, H.; Nourmoradi, H.; et al. Human Health Risk Assessment Due to Ambient PM10 and SO_2 by an Air Quality Modeling Technique. *Process Saf. Environ. Prot.* **2017**, *111*, 346–354. [CrossRef]
14. *The Role of Shipbuilding in Maritime Decarbonisation*; OECD Publishing: Paris, France, 2025; ISBN 9789264319424.
15. *Review of Maritime Transport 2024*; UNCTAD: Geneva, Switzerland, 2024; ISBN 978-92-1-003206-3.
16. International Maritime Organization (IMO). Fourth IMO Greenhouse Gas Study. *Int. Marit. Organ.* **2021**, 951–952.
17. Díaz-Secades, L.A. Enhancement of Maritime Sector Decarbonization through the Integration of Fishing Vessels into IMO Energy Efficiency Measures. *J. Mar. Sci. Eng.* **2024**, *12*, 663. [CrossRef]
18. *Marine IMO Marine Environment Protection Committee (MEPC 83): Summary Report*; Lloyd's Register: London, UK, 2025.
19. *Resolution MEPC. 176(58) Revised MARPOL Annex VI*; International Maritime Organization (IMO): London, UK, 2010; Volume 9, pp. 76–99.
20. *Resolution MEPC. 328 (76) Amendments to the Annex of the Protocol of 1997–2021 Revised MARPOL Annex VI*; International Maritime Organization (IMO): London, UK, 2021; Volume 148, pp. 148–162.

21. Fuglestvedt, J.; Berntsen, T.; Eyring, V.; Isaksen, I.; Lee, D.S.; Sausen, R. Shipping Emissions: From Cooling to Warming of Climate—And Reducing Impacts on Health. *Environ. Sci. Technol.* **2009**, *43*, 9057–9062. [CrossRef]
22. Yuan, T.; Song, H.; Oreopoulos, L.; Wood, R.; Bian, H.; Breen, K.; Chin, M.; Yu, H.; Barahona, D.; Meyer, K.; et al. Abrupt Reduction in Shipping Emission as an Inadvertent Geoengineering Termination Shock Produces Substantial Radiative Warming. *Commun. Earth Environ.* **2024**, *5*, 281. [CrossRef]
23. Cret, L.; Baudry, M.; Lantz, F. How to Implement the 2023 IMO GHG Strategy? Insights on the Importance of Combining Policy Instruments and on the Role of Uncertainty. *Mar. Policy* **2024**, *169*, 106332. [CrossRef]
24. Zannis, T.C.; Katsanis, J.S.; Christopoulos, G.P.; Yfantis, E.A.; Papagiannakis, R.G.; Pariotis, E.G.; Rakopoulos, D.C.; Rakopoulos, C.D.; Vallis, A.G. Marine Exhaust Gas Treatment Systems for Compliance with the IMO 2020 Global Sulfur Cap and Tier III NOx Limits: A Review. *Energies* **2022**, *15*, 3638. [CrossRef]
25. Baldasso, E.; Mondejar, M.E.; Andreasen, J.G.; Rønnenfelt, K.A.T.; Nielsen, B.Ø.; Haglind, F. Design of Organic Rankine Cycle Power Systems for Maritime Applications Accounting for Engine Backpressure Effects. *Appl. Therm. Eng.* **2020**, *178*, 115527. [CrossRef]
26. Lunde Hermansson, A.; Hassellöv, I.-M.; Grönholm, T.; Jalkanen, J.-P.; Fridell, E.; Parsmo, R.; Hassellöv, J.; Ytreberg, E. Strong Economic Incentives of Ship Scrubbers Promoting Pollution. *Nat. Sustain.* **2024**, *7*, 812–822. [CrossRef]
27. Bendl, J.; Saraji-Bozorgzad, M.R.; Käfer, U.; Padoan, S.; Mudan, A.; Etzien, U.; Giocastro, B.; Schade, J.; Jeong, S.; Kuhn, E.; et al. How Do Different Marine Engine Fuels and Wet Scrubbing Affect Gaseous Air Pollutants and Ozone Formation Potential from Ship Emissions? *Environ. Res.* **2024**, *260*, 119609. [CrossRef]
28. Karatuğ, Ç.; Arslanoğlu, Y.; Guedes Soares, C. Feasibility Analysis of the Effects of Scrubber Installation on Ships. *J. Mar. Sci. Eng.* **2022**, *10*, 1838. [CrossRef]
29. Caiazzo, G.; Langella, G.; Miccio, F.; Scala, F. An Experimental Investigation on Seawater SO2 Scrubbing for Marine Application. *Environ. Prog. Sustain. Energy* **2013**, *32*, 1179–1186. [CrossRef]
30. Fridell, E.; Salo, K. Measurements of Abatement of Particles and Exhaust Gases in a Marine Gas Scrubber. *Proc. Inst. Mech. Eng. Part M J. Eng. Marit. Environ.* **2016**, *230*, 154–162. [CrossRef]
31. Winnes, H.; Fridell, E.; Moldanová, J. Effects of Marine Exhaust Gas Scrubbers on Gas and Particle Emissions. *J. Mar. Sci. Eng.* **2020**, *8*, 299. [CrossRef]
32. Yang, J.; Tang, T.; Jiang, Y.; Karavalakis, G.; Durbin, T.D.; Wayne Miller, J.; Cocker, D.R.; Johnson, K.C. Controlling Emissions from an Ocean-Going Container Vessel with a Wet Scrubber System. *Fuel* **2021**, *304*, 121323. [CrossRef]
33. Osipova, L.; Georgeff, E.; Comer, B. *Global Scrubber Washwater Discharges Under IMO's 2020 Fuel Sulfur Limit*; ICCT: Washington, DC, USA, 2021.
34. OSPAR Ministers Meet in Vigo to Take Concrete Measures to Enhance Protection of the North-East Atlantic Ocean. Available online: https://www.ospar.org/news/ministers-meet-in-vigo-to-take-concrete-measures-to-enhance-protection-of-the-north-east-atlantic-ocean (accessed on 22 July 2025).
35. den Boer, E.; Hoen, M. *Scrubbers—An Economic and Ecological Assessment*; IVL Swedish Environmental Research Institute Ltd.: Stockholm, Sweden, 2015.
36. Jee, J. Hybrid Multi-Criteria Decision-Making for Marine SOx Scrubber Systems. *J. Mar. Sci. Eng.* **2022**, *10*, 1599. [CrossRef]
37. Toscano, D.; Murena, F.; Quaranta, F.; Mocerino, L. Assessment of the Impact of Ship Emissions on Air Quality Based on a Complete Annual Emission Inventory Using AIS Data for the Port of Naples. *Ocean Eng.* **2021**, *232*, 109166. [CrossRef]
38. Topic, T.; Murphy, A.J.; Pazouki, K.; Norman, R. Assessment of Ship Emissions in Coastal Waters Using Spatial Projections of Ship Tracks, Ship Voyage and Engine Specification Data. *Clean. Eng. Technol.* **2021**, *2*, 100089. [CrossRef]
39. *Air Pollutant Emission Inventory Guidebook 2023*; European Environment Agency (EEA): Copenhagen, Denmaek, 2023.
40. Trozzi, C. Emission Estimate Methodology for Maritime Navigation. In Proceedings of the 9th International Emissions Inventory Conference, San Antonio, TX, USA, 27–30 September 2010.
41. Cooper, D.; Gustafsson, T. *Methodology for Calculating Emissions from Ships: 2. Emission Factors for 2004 Reporting*; Swedish Meteorological and Hydrological Institute: Norrköping, Sweden, 2004.
42. Smith, T.W.P.; Jalkanen, J.P.; Anderson, B.A.; Corbett, J.J.; Faber, J.; Hanayama, S.; O'Keeffe, E.; Parker, S.; Johansson, L.; Aldous, L.; et al. *Third IMO GHG Study 2014*; International Maritime Organization: London, UK, 2015.
43. *MEPC.1/Circ.684*; International Maritime Organization (IMO): London, UK, 2009.
44. International Maritime Organization (IMO). Nitrogen Oxides (NOx)—Regulation 13. Available online: https://www.imo.org/en/OurWork/Environment/Pages/Nitrogen-oxides-(NOx)----Regulation-13.aspx (accessed on 3 January 2022).
45. *MARPOL Annex VI, Regulation 14*; International Maritime Organization (IMO): London, UK, 2020.
46. Ship&Bunker Global 20 Ports Average Bunker Prices—Ship & Bunker. Available online: https://shipandbunker.com/prices/av/global/av-g20-global-20-ports-average (accessed on 28 June 2025).
47. Hawas, F.; Cifuentes, A. Valuation of Projects with Minimum Revenue Guarantees: A Gaussian Copula–Based Simulation Approach. *Eng. Econ.* **2017**, *62*, 90–102. [CrossRef]

48. Han, T.-C.; Wang, C.-M. Shipping Bunker Cost Risk Assessment and Management during the Coronavirus Oil Shock. *Sustainability* **2021**, *13*, 4998. [CrossRef]
49. Hofstadler, C.; Kummer, M. *The Fundamentals of Monte Carlo Simulations*; Springer: Berlin/Heidelberg, Germany, 2021; pp. 155–212.
50. Hasannuddin, A.K.; Yahya, W.J.; Sarah, S.; Ithnin, A.M.; Syahrullail, S.; Sugeng, D.A.; Razak, I.F.A.; Abd Fatah, A.Y.; Aqma, W.S.; Rahman, A.H.A.; et al. Performance, Emissions and Carbon Deposit Characteristics of Diesel Engine Operating on Emulsion Fuel. *Energy* **2018**, *142*, 496–506. [CrossRef]
51. Iakovenko, A.; Dunin, A.; Dushkin, P.; Savastenko, E.; Shatrov, M. The Influence of Mass Composition of Water-Fuel Emulsion on Ecological Characteristics of a Diesel Engine. *Energies* **2019**, *12*, 2689. [CrossRef]
52. Contini, D.; Merico, E. Recent Advances in Studying Air Quality and Health Effects of Shipping Emissions. *Atmosphere* **2021**, *12*, 92. [CrossRef]
53. Ni, P.; Wang, X.; Li, H. A Review on Regulations, Current Status, Effects and Reduction Strategies of Emissions for Marine Diesel Engines. *Fuel* **2020**, *279*, 118477. [CrossRef]
54. Díaz-Rodríguez, S.; Suárez, E.; Concheiro, M.; Paz-Penín, C. Experimental Study of the Ozone as an Anti-Fouling System in an Exhaust Gas Heat Exchanger. *Appl. Therm. Eng.* **2025**, *273*, 126538. [CrossRef]
55. Zhu, M.; Li, K.X.; Lin, K.-C.; Shi, W.; Yang, J. How Can Shipowners Comply with the 2020 Global Sulphur Limit Economically? *Transp. Res. Part D Transp. Environ.* **2020**, *79*, 102234. [CrossRef]
56. Sheng, D.; Jiang, J.; Wang, H.; Tan, Z.; Wang, Y. Optimal Compliance Choices for Ocean Carriers under the Sulphur Regulation. *Transp. Res. Part D Transp. Environ.* **2023**, *116*, 103639. [CrossRef]
57. Zis, T.P.V.; Cullinane, K.; Ricci, S. Economic and Environmental Impacts of Scrubbers Investments in Shipping: A Multi-Sectoral Analysis. *Marit. Policy Manag.* **2022**, *49*, 1097–1115. [CrossRef]

Journal of
*Marine Science
and Engineering*

MDPI

Article

Design and Innovative Application of Ship CCUS Technology Under the Requirements of Green Shipping

Niyu Zhang [1,2], Xinpeng Wang [1], Cong Wang [3], Zhongxiu Peng [3], Youxiao Chen [4], Lianhao Liu [4], Lijie Chen [4], Jiawei Hu [1], Wei Wei [1,*] and Xuesong Li [1,*]

1 CAS Key Laboratory of Low-Carbon Conversion Science and Engineering, Shanghai Advanced Research Institute, Chinese Academy of Sciences, Shanghai 201210, China; zhangniyu2024@sari.ac.cn (N.Z.); hujw@sari.ac.cn (J.H.)
2 University of Chinese Academy of Sciences, Beijing 100049, China
3 School of Maritime Economics and Management, Dalian Maritime University, Dalian 116026, China
4 CIMC Offshore Co., Ltd., Shenzhen 518000, China
* Correspondence: weiwei@sari.ac.cn (W.W.); lixuesong@sari.ac.cn (X.L.)

Abstract: In the context of green shipping, carbon capture, utilization, and storage (CCUS) technology is widely recognized as an effective approach for maritime emission reduction. However, existing onboard CO_2 capture systems (OCCS) face challenges including high retrofit complexity, operational limitations, significant deadweight tonnage loss, and compromising feasibility. This study addresses these issues through modular redesign and standardization of two core components: the capture device and CO_2 storage system. We developed clustered storage units to ensure secure CO_2 storage while maximizing economic efficiency, with optimal deployment determined through parameter optimization. The work further examines post-capture processing methods and implementation barriers for onboard CCUS, establishing practical pathways to advance decarbonization technologies and mature maritime CCUS industry chains.

Keywords: green shipping; ship emission reduction technology; carbon capture, utilization and storage (CCUS); onboard carbon dioxide capture system (OCCS); CO_2 storage; economic calculations

1. Introduction

Following the establishment of a global consensus on carbon emission reduction, various industries have implemented mitigation measures [1,2]. However, according to the IEA Global Energy Report 2025 [3,4], global CO_2 emissions from energy combustion and industrial processes have continued to rise due to increasing energy demand over the period 1900–2023 [5], reaching 37.6 gigatons (Gt) in 2024 (Figure 1a). In 2018, CO_2 emissions from shipping alone amounted to 1.056 billion tons (Figure 1b), representing approximately 3% of global carbon emissions [6], with a more recent update suggesting that this share may be approaching 6% [7]. According to the OECD 2023 report, container ships continue to contribute the highest share of CO_2 emissions among all ship types [8] (Figure 1c). Furthermore, the United Nations Conference on Trade and Development (UNCTAD), in its *Review of Maritime Transport 2024*, emphasizes the urgent need to accelerate decarbonization efforts to achieve the long-term goal of green shipping [9]. This concept focuses on transporting goods through methods that mitigate environmental impact and optimize resource consumption [10,11]. The International Maritime Organization (IMO) projects that, without effective intervention, CO_2 emissions from shipping could rise to 1.6 billion tons annually by the mid-21st century (Figure 1d).

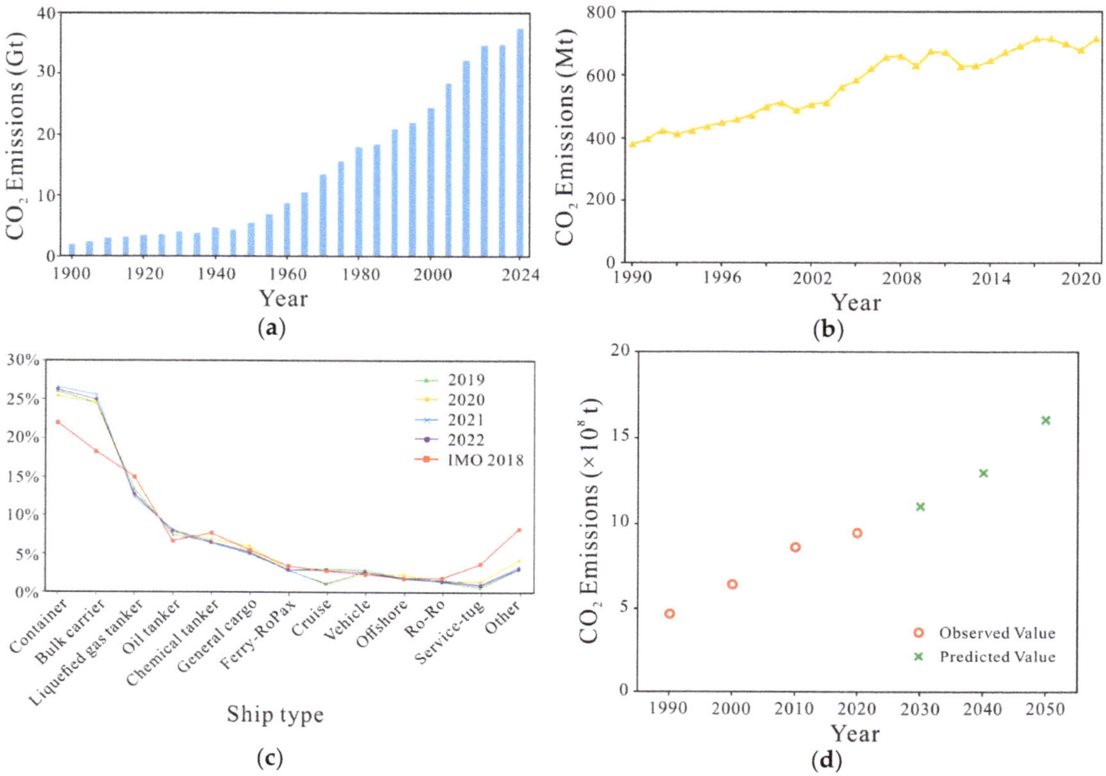

Figure 1. Status of global green shipping: (**a**) Global CO_2 emissions from energy combustion and industrial processes, 1900–2023 (Modified from reference [12]); (**b**) Total annual global CO_2 emissions from shipping, 1990–2021 (Modified from reference [1]); (**c**) Share (%) of global shipping CO_2 emissions by ship type, 2019–2022 (Modified from reference [8]); (**d**) Trends in CO_2 emissions from international shipping (Modified from reference [13]).

In recent years, major international organizations have been actively establishing regulatory frameworks to standardize green shipping practices [9,14–19]. The International Maritime Organization (IMO) introduced its *Initial Strategy on the Reduction of Greenhouse Gas Emissions from Ships* in 2018 [6,20], and subsequently released the *2023 IMO Strategy on Reduction of GHG Emissions from Ships*, which included the first revision of the guidelines on the full life-cycle greenhouse gas (GHG) intensity of marine fuels [21]. This revision also updated the global carbon reduction targets for the shipping sector, stating that "carbon intensity should be reduced by at least 40% by 2030 compared to 2008, and total annual GHG emissions from shipping should be reduced by at least 20% by 2030, with a view to achieving a 30% reduction; by at least 70% by 2040, with a view to achieving an 80% reduction; and ultimately to reach net-zero emissions around 2050." The IMO has also developed comprehensive guidelines for full life-cycle GHG calculations applicable to all types of marine fuels [22]. In addition, the European Union has incorporated the shipping sector into the EU Emissions Trading System (EU ETS) as of 2024 [23].

Under the requirements of green shipping, achieving the aforementioned carbon reduction targets necessitates the adoption of effective emission reduction technologies for ships. Currently, three primary categories of energy-saving and emission-reduction measures are employed in the maritime industry: improvements in energy efficiency, operational management strategies, and the adoption of alternative fuels [24]. Efforts have initially focused on upgrading the primary source of shipboard carbon emissions—the diesel engine—by exploring alternative propulsion technologies such as new energy systems, fuel cells, lithium batteries, and hybrid power systems. However, the relatively low energy density of batteries limits their ability to meet the high-power demands of ocean-going vessels, and their adoption also requires advanced charging infrastructure and substantial port-side support systems. These challenges make large-scale implementation difficult in the near term.

The maritime industry has shifted toward alternative clean fuels to replace fossil fuels and reduce substantial CO_2 emissions from combustion. Explored green fuels include liquefied natural gas (LNG), methanol, ammonia, hydrogen, and biofuel [7,17,25–36]. Notably, no single alternative energy source can simultaneously achieve high emission reduction rates and low costs. Compared to conventional fossil fuels, the overall emission reduction effect of these alternative fuels ranges from 20% to 95% [37]. For instance, green methanol and ammonia exhibit 60–95% emission reduction efficiencies, highlighting significant decarbonization potential, but their production costs remain high. The liquefied natural gas (LNG) industry is relatively mature, yet its emission reduction efficiency is only about 30%, limiting short-term benefits. As a result, the widespread adoption of alternative fuels is currently constrained. Future research and development may overcome existing production cost barriers.

According to the Intergovernmental Panel on Climate Change (IPCC), carbon capture, utilization, and storage (CCUS) is a foundational technology for achieving significant carbon reductions. This technology aims to capture CO_2 generated from energy production and industrial activities, either for direct utilization or long-term storage by injection into suitable geological formations. By doing so, it prevents CO_2 from entering the atmosphere and achieves substantial carbon emission reductions [38,39]. As a result, CCUS technology is currently one of the most viable short-term solutions for reducing carbon emissions from ships and meeting international climate targets [40–43].

The maturity of onshore carbon capture, utilization, and storage (CCUS) technologies positions ship-based CCUS as a promising approach for carbon reduction in the shipping industry. Multiple studies have verified its technical and economic feasibility through full-scale ship trials and simulation analyses [44–48]. Domestic and international researchers have conducted multidimensional studies on shipboard carbon capture technologies, covering applications of carbon solidification methods, optimized system capacity configuration, and energy consumption analysis [45,46,48,49]. These studies confirm that the technology can significantly reduce maritime carbon emissions in the short-term and meet International Maritime Organization (IMO) emission reduction targets [50–52]. Some research also indicates that integrating ship-based carbon capture systems with existing infrastructure is feasible, reducing additional investments while supporting zero-carbon emissions goals [53]. The selection of ship-based CCUS technology routes should prioritize efficiency and economy, with capture rate and energy consumption serving as key evaluation metrics. Some studies have optimized energy consumption and costs across system components [41,54,55]. Novel technical solutions—like solid-state carbon storage [56] and membrane capture-liquefaction systems—have been proposed to improve cost-effectiveness. Comparative analyses across vessel types (e.g., LNG carriers, oil tankers) indicate that CCUS technology is most economically viable for ships with high sailing speeds and freight values [57].

Onboard Carbon Capture Systems (OCCS), a key application of CCUS technology in the maritime sector, involve the separation and capture of CO_2 from ship exhaust gases, followed by storage and transport for either resource utilization or long-term sequestration underground or at the seabed, thereby achieving permanent CO_2 emission reductions [58]. According to an analysis by the China Classification Society (CCS), OCCS technology on ultra-large container ships offers greater economic advantages compared to clean energy alternatives when considering initial capital investment, incremental operating costs, and potential savings on carbon taxes [59]. Researchers have developed predictive models and proposed long-term implementation strategies for this technology [40,43,60–63].

Recent pilot projects have validated the real-world feasibility of ship-based carbon capture. For example, the ethylene carrier *Clipper Eris* was retrofitted with a 7 MW amine-based CO_2 capture system capable of removing about 70% of CO_2 from its main engine exhaust [64]. This project—a collaboration between Solvang, Wärtsilä, and others—includes onboard CO_2 liquefaction and storage for later offloading. Similarly, an advanced pilot onboard Crowley's MV *K-Storm* container ship [65] will field-test a modular carbon capture unit housed in two 40 ft deck containers (plus a 20 ft CO_2 tank), designed to minimize vessel modifications. This system is projected to capture approximately 1 tonne of CO_2 daily from the ship's exhaust. Industry consortia such as Project REMARCCABLE [66] (Stena Bulk, Global Centre for Maritime Decarbonisation, OGCI, etc.) are advancing plans to test CCUS on commercial medium-range tankers within the next few years. These pioneering demonstrations underscore that onboard carbon capture, utilization, and storage (CCUS) has transitioned from conceptual research to deployment on operational vessels. Insights from *Clipper Eris*, *K-Storm*, and similar initiatives directly inform the design and innovative application of ship-based CCUS technology under contemporary green shipping imperatives.

However, OCCS systems face a range of specialized requirements, primarily in two areas: capture equipment design, including technological integration, spatial arrangement, and stability under hull motion, and CO_2 storage device design [42,67,68]. Hence, how to optimize the CO_2 capture and storage system for ships through better design becomes one of the key issues.

The limitations of existing OCCS devices significantly hinder their widespread adoption: firstly, most OCCS systems require customization for specific carbon sources, increasing the complexity of equipment design, production, and operational costs. Conventional ship-based CO_2 capture systems typically rely on tall separation towers, resulting in prohibitively bulky installations that challenge spatial constraints aboard vessels. This lack of standardization also hampers system reusability across different scenarios and limits the development of an industrial supply chain. Secondly, current CO_2 capture devices fail to adequately address practical challenges such as large variations in flue gas flow, spatial constraints, and integration difficulties with existing ship systems. These systems also lack effective modular design and adaptable pipeline connection solutions, which are critical for practical implementation [41,69].

The limitations of current shipboard CO_2 storage systems are also significant, with cost being the primary concern: due to the high material and fabrication costs of high-pressure storage tanks, current CO_2 storage systems operate at relatively low pressures (around 1.5 MPa) [70]. Additionally, the tanks are large and bulky, occupying valuable space and reducing cargo capacity, thereby increasing transportation costs [41].

As illustrated in Figure 2, this paper presents a modular OCCS device designed for variable emission sources, achieved through the standardization and modularization of functional units. This design addresses the emission reduction needs and process flexibility requirements of mobile carbon sources such as ships, enhancing the adaptability of the OCCS to varying flue gas flow rates and conditions. It enables timely adjustments in equipment operation based on variations in gas volume and composition, facilitating the installation, maintenance, upgrades, and operation of the OCCS in diverse scenarios. By reducing production costs and technical barriers, this approach ultimately aids in the formation of a standardized equipment industry chain for OCCS, supporting mass production, installation, maintenance, and upgrades across various operational scenarios.

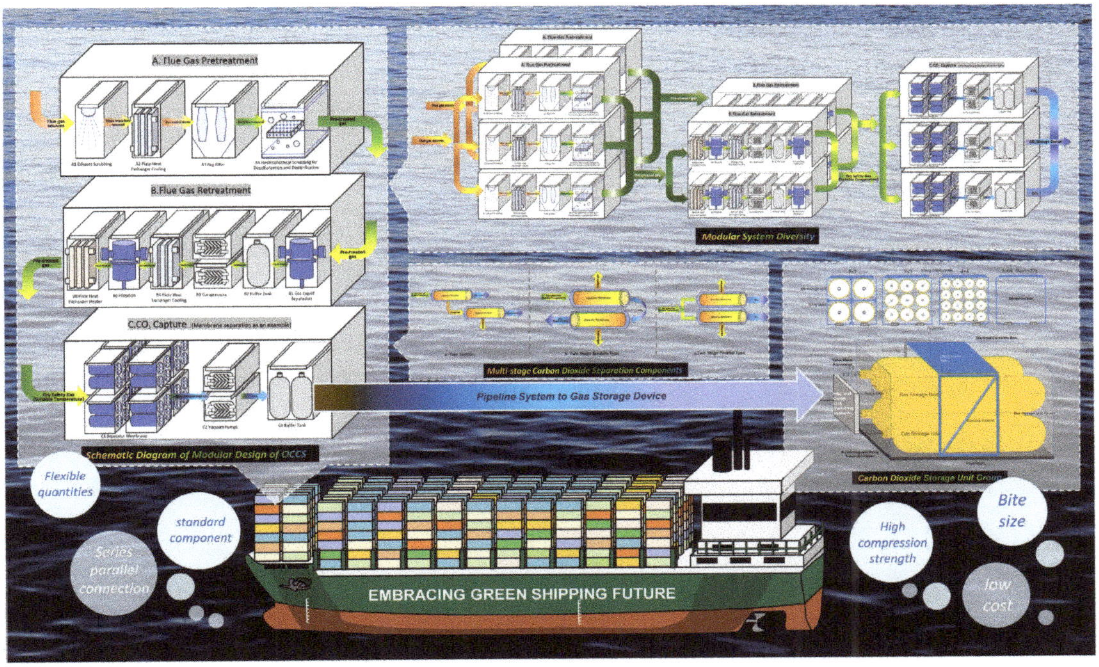

Figure 2. A schematic diagram of the general design of the ship's carbon dioxide capture and storage system.

To address the limitations of existing CO_2 storage devices, this paper proposes a "small-sized CO_2 storage unit group" that can be integrated into a standard container, while simultaneously improving the maximum limit of CO_2 storage pressure (exceeding 1.5 MPa) and reducing storage tank costs, overcoming the limitations of low-pressure CO_2 storage on ships. Additionally, the most appropriate CO_2 storage unit size and layout within the container are determined through economic analysis.

To address the limitations of conventional maritime CO_2 capture and storage systems this study proposes the following innovations: Section 2 introduces a modular ship-based CO_2 capture system designed for containerized integration, demonstrating adaptable assembly configurations and scalable gas-processing capabilities. Section 3 presents a compact, container-compatible CO_2 storage unit. By defining two critical performance metrics (cost per unit capacity and mass per unit capacity), this section derives an optimized design through parametric analysis and validates its economic superiority via case studies. Section 4 critically examines systemic challenges of maritime CCUS adoption, including technical implementation barriers, safety and regulatory protocols, logistical coordination gaps, and economic viability constraints. Section 5 synthesizes the study's contributions and outlines actionable pathways to advance the commercial maturity of ship-based CCUS technologies.

In summary, the design proposed in this paper demonstrates theoretical innovativeness and provides practical inspiration for advancing green ship carbon-reduction technologies.

2. Design of CO_2 Capture System for Marine Vessels

2.1. Analysis of Ship Flue Gas Components and Existing Capture Technology

Exhaust gases emitted from ships are primarily generated by the combustion of engine fuels. The chemical composition of these gases includes gaseous substances such as carbon dioxide (CO_2), carbon monoxide (CO), nitrogen oxides (NO_x), sulfur dioxide (SO_2), and volatile organic compounds (VOCs), along with particulate matter like black carbon (BC) and particulate organic matter (POM) [71]. Among these, CO_2 is the primary product of combustion. Several countries and regions have now established clear limits on ship emissions [7].

CO_2 emissions are typically calculated based on fuel consumption and serve as a benchmark for estimating other pollutants. Under normal combustion conditions, almost all of the carbon in fuel is converted to CO_2, with only a small fraction remaining in the form of CO and hydrocarbons (HC).

The main methods used to estimate emissions from ships are generally categorized into two approaches: the "top-down approach" [72,73] and the "bottom-up method" [74]. Table 1 [75–80] presents the annual fuel consumption and various types of emissions from ships.

Table 1. Ship emissions estimates and related fuel consumption1 (Modified according to references [75–80]).

Methods	Fuel Quantity/Mt	CO_2/Mt	NO_2/Mt	SO_2/Mt	Reference
Top-down methods	140~147	451	10.12	8.48	[75]
	200	634	\	8.50	[78]
Bottom-up methods	166	557	11.92	6.82	[79]
	280	813	21.38	12.03	[80]
	289	912	22.57	12.98	[76]

Under normal operating conditions, the exhaust manifold temperature of ocean-going vessels exceeds 500 °C, while the temperature after the waste heat recovery system remains above 150 °C. The exhaust pressure typically ranges between 0.1 and 0.2 MPa. The total exhaust gas volume is substantial, while the CO_2 partial pressure remains low. For most vessel types, CO_2 accounts for 3–7% of the exhaust volume, with a theoretical maximum of 15%. The exhaust gas also contains a high concentration of nitrogen (N_2 > 70%) and other gases [81].

According to the IPCC Carbon Sequestration Report [82], the four fundamental CO_2 capture technologies are pre-combustion capture, oxygen-enriched combustion capture, post-combustion capture, and industrial process capture systems. Common CO_2 capture methods for onshore power plants and industrial processes include chemical absorption, physical absorption, physical adsorption, membrane separation, and emerging low-temperature liquefaction and separation techniques. When applied to the limited space available on ships, chemical absorption, membrane separation, and cryogenic separation are considered the most promising options for CO_2 reduction [24,83]. Figure 3 below illustrates the general design of these three different CO_2 capturing processes. Table 2 compares these capture methods in terms of efficiency, energy demand, space/weight requirements, and operational considerations. Miller et al. [84] provided an economic comparison of various CO_2 capture methods, summarized in Table 3.

Chemical absorption suffers from bulky equipment, excessive energy consumption, and corrosion risks. The cryogenic separation method is extremely energy-intensive, carries safety risks of pipeline clogging due to dry ice, and remains insufficiently mature for current research applications.

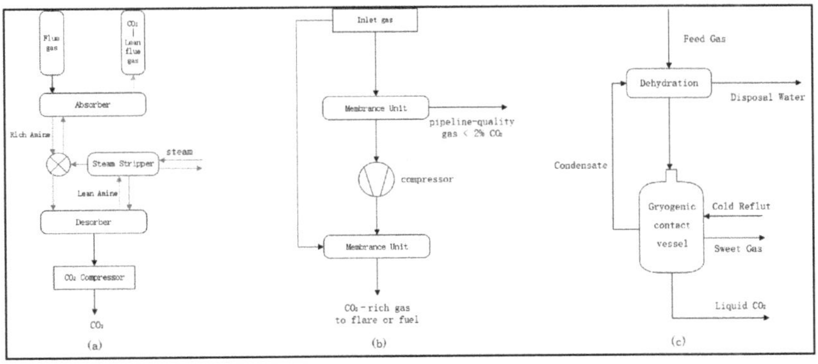

Figure 3. Schematic diagram illustrates the three main CO_2 capture processes: (**a**) chemical absorption with an amine absorber/desorber loop, (**b**) a membrane module array, (**c**) a cryogenic cooling unit.

Table 2. A summary of the key parameters of chemical absorption vs. membrane vs. cryogenic capture for ships.

Parameter	Chemical Absorption (Amine)	Membrane Separation	Cryogenic Separation
Typical CO_2 Capture Efficiency	High (80–90%+) [85,86]	Medium (40–60%) [87,88]	High 90%+ possible [88,89]
Energy Requirement/tCO_2	High (~3–4 GJ heat/t CO_2) [85]	Low (~150–250 kWh electricity/t CO_2) [87,88]	High (~400–600 kWh electricity/t CO_2 cooling) [89,90]
Footprint and Weight	Large towers, pumps, solvent storage tanks (bulky, heavy) [86]	Compact, modular units (small footprint) [87]	Compact cryocooler and storage tanks, moderate weight [89,90]
Operability Issues	Solvent corrosion, solvent degradation, large equipment requires maintenance, complex solvent management [85,86]	Membrane performance affected by impurities, moisture, periodic membrane replacement required [87,88]	Risk of CO_2 solidification (dry ice) causing blockages, managing extreme low temperatures, equipment reliability [89,90]
CO_2 Product Purity	High purity achievable (>99%) [86]	Moderate purity (80–90%), multiple stages needed for higher purity [88]	Very high purity achievable (>99%) [89]
Maturity Level	Mature, proven technology in industrial CCS; early marine demonstration projects (Clipper Eris, Stena Impero) [85,86]	Moderate maturity, mature on offshore platform. limited marine deployment experience [87,88]	Proven onshore in industrial settings; limited marine trials, emerging marine applications [89,90]

Table 3. Economic comparison of various CO_2 capture methods (Modified according to reference [84]).

Capture Methods	CO_2 Removal Rate/%	Decarbonization Cost/(USD·t^{-1})
Chemical absorption method	90	35
Membrane separation method	92	47
Pressure swing adsorption method	85	84

In comparison, despite the challenges of separation efficiencies of membrane separation methods, membrane systems are compact and chemical-free, which is a great advantage for confined spaces like ships. Therefore, the membrane separation method remains the primary candidates for ship CCUS in the near term.

Polyimide-based membranes [91,92] represent a high-performance material for CO_2 capture due to their exceptional CO_2 selectivity, permeability, thermal stability, and chemical-mechanical durability [93]. For marine applications, these membranes offer distinct advantages: outstanding resistance to salt spray corrosion, moisture-heat aging, and contamination; low water absorption; long-term reliability; and modular design feasibility. Compared to zeolite molecular sieves [94], polyimide membranes require less extensive flue gas pretreatment. Relative to metal-organic framework (MOF) membranes [95], they demonstrate superior thermal stability and enhanced tolerance to humidity, elevated temperatures, and harsh marine operating conditions. Thus, in the Section 2.2.1, polyimide membranes are selected as the exemplary material for system design.

2.2. Modularized Ship CO_2 Capture System Design

2.2.1. Functional Unit Standard Parts Design

Table 4 outlines the three primary modules of the carbon capture process proposed in this study, along with their respective internal functional units.

Table 4. List of three modules and each functional unit in OCCS design.

No.	Module	Functional Units
1	Flue gas pretreatment module	exhaust gas scrubbing unit, cooling unit, dust removal unit, desulfurization unit, denitrification unit
2	Flue gas reprocessing module	gas-liquid separation unit, buffer tank unit, compression unit, cooling unit, gas-liquid separation unit, filtration unit, buffer tank unit, electric heating unit;
3	CO_2 separation module	(taking the "membrane separation" method as an example) membrane module unit, vacuum pump unit, buffer tank unit

All functional units mentioned above are standardized in design, featuring uniform base dimensions, frame structures, and other components, enabling auxiliary units to be easily integrated into standard containers. The sequence and number of these functional units are not rigidly fixed; they can be flexibly adjusted based on the characteristics of the incoming gas. Units can be added, removed, or rearranged in series or parallel configurations, as the modular container system is pre-designed with reserved space for various processing components, allowing for adaptable reconfiguration. The foundational schematic of the overall design is illustrated in Figure 4.

The flue gas, transported via pipeline at a very high temperature (typically 300 °C or higher), enters Module A for pretreatment. The process involves sprinkler scrubbing and coil cooling to achieve preliminary temperature reduction and overall heat energy decrease. The flue gas then passes through a dedusting unit and a desulphurization-denitrification unit to remove impurities before entering Module B. In Module B, a gas-liquid separator removes most water from the gas, a compressor increases gas pressure, and a coil cooler cools the gas to prevent overheating. A reserved coil heater can be activated if the gas temperature is too low, ensuring suitable temperature and pressure conditions before the gas enters the CO_2 separation unit.

- Module A. Flue gas pretreatment (Figure 5)

A1. Tail Gas Washing Unit: This unit removes major impurities from the incoming flue gas, adjusts its composition, and prevents particulate matter from clogging pipelines and downstream components. For instance, when the flue gas contains high concentrations of acidic gases, a tailored scrubbing solution can be applied in the washing section to remove acidic and soluble impurities via spray treatment. Combined with a graded filtration system (coarse, medium, and fine), the unit effectively eliminates large dust particles, fine particulates, and micro-impurities, thereby protecting the integrity of the subsequent membrane module.

A2. Cooling Unit: This unit lowers the temperature of the incoming flue gas to minimize water vaporization and prevent high temperatures from compromising equipment, piping, and valve integrity. For example, a high-efficiency air-cooled or water-cooled heat exchanger can be employed to bring the gas to an optimal temperature for subsequent filtration and treatment.

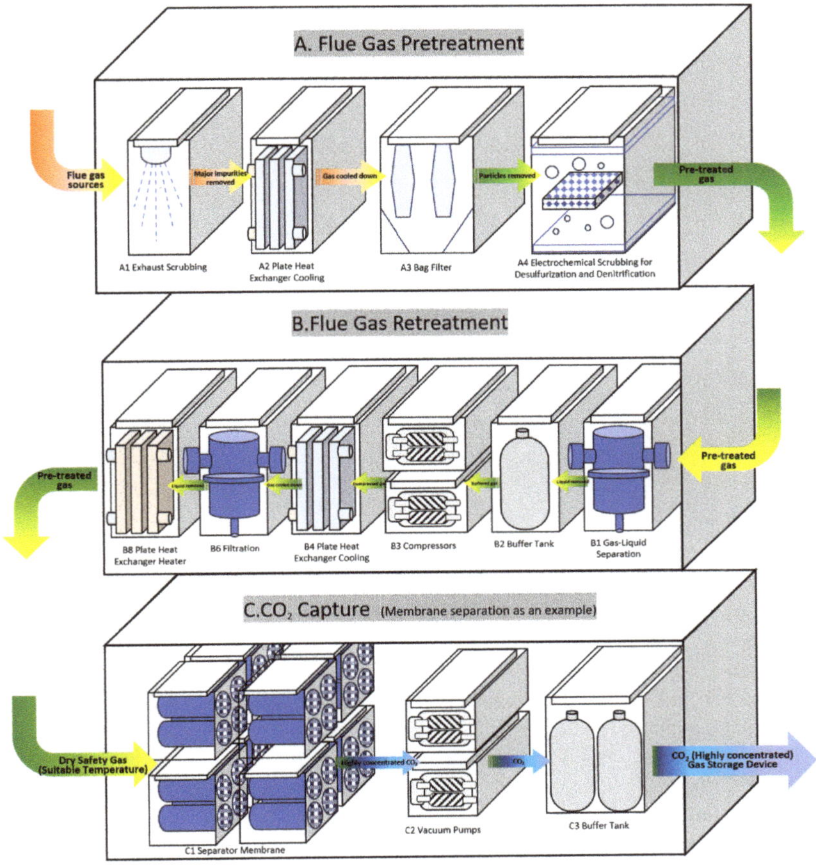

Figure 4. Schematic diagram of modularized design of OCCS.

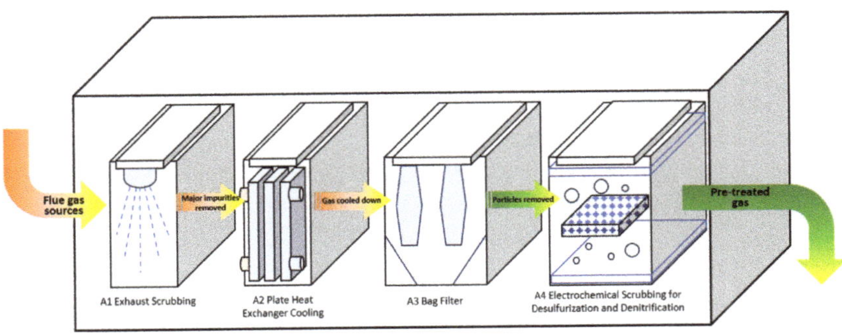

Figure 5. Schematic diagram of module A: flue gas pretreatment.

A3. Dedusting Unit: This unit ensures the effective removal of fine particulate matter from the flue gas. A multi-stage electrostatic precipitator combined with bag-type dust collectors may be used to achieve high-efficiency particulate capture.

A4. Desulfurization and Denitrification Units: These units target the removal of acidic gases and nitrogen oxides from the flue gas to prevent rapid corrosion of downstream systems. In scenarios where sulfur and nitrogen content is elevated, advanced treatment technologies such as membrane absorption coupling and electrochemical scrubbing can be employed. These methods utilize membrane separation and electrochemical oxidation-reduction reactions to reduce sulfur and nitrogen levels, followed by micro-channel selective catalytic reduction (SCR) to convert residual compounds into harmless nitrogen and water.

- Module B. Flue gas reprocessing (Figure 6)

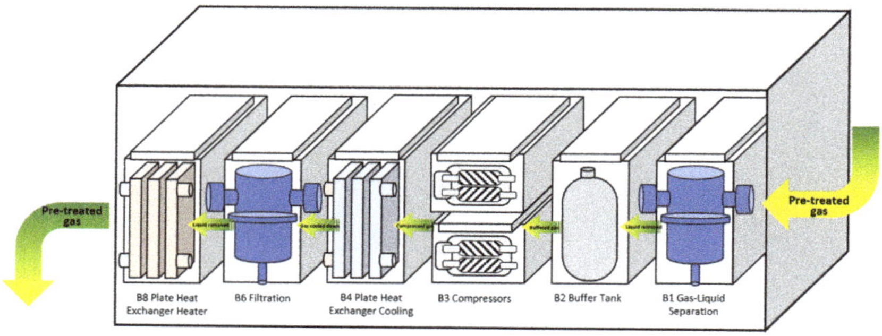

Figure 6. Schematic diagram of Module B: flue gas reprocessing.

B1. Gas-Liquid Separation Unit: Used to initially separate gas and entrained liquid impurities. The purified gas continues to downstream processing, while the separated liquid is periodically discharged into a designated recovery tank via external piping.

B2. Buffer Tank Unit: This unit stabilizes the pressure and flow rate of the gas stream to ensure smooth operation of subsequent components.

B3. Compression Unit: Used to increase the gas pressure to meet CO_2 separation requirements. Typically, an integrated scroll or screw compressor is utilized; under specific conditions, a multi-stage compressor may also be considered.

B4. Cooling Unit: A heat exchanger is employed to further reduce the gas temperature, thereby enhancing the efficiency of subsequent CO_2 separation processes.

B5. Secondary Gas-Liquid Separation Unit: This stage removes residual water droplets from the gas stream, ensuring minimal moisture content.

B6. Filtration Unit: High-precision filters are used to eliminate residual fine particulate matter and ensure the cleanliness of the gas.

B7. Buffer Tank Unit: Provides additional pressure stabilization to maintain operational consistency before CO_2 separation.

B8. Heating Unit: The gas may be appropriately heated based on process requirements to meet the temperature requirements for the next step of membrane separation.

- Module C. CO_2 Membrane Separation (Figure 7)

This stage applies to various CO_2 separation technologies, including chemical adsorption, physical adsorption, membrane separation, and low-temperature distillation. Considering the spatial and operational constraints of shipboard environments, this section adopts the membrane separation method for the demonstration of the modular standard parts process design.

C1. Membrane Module: This unit employs composite membrane materials with high CO_2 selectivity and permeability. A common choice is a polyimide-based membrane doped with specific metal ions or nanoparticles to enhance separation performance.

The membrane module design is flexibly configurable based on the required CO_2 capture purity, flue gas composition, and flow rate. Key design features such as internal gas distributors and deflectors ensure uniform gas distribution across the membrane surface, mitigate concentration polarization, and enhance CO_2 separation efficiency and membrane stability.

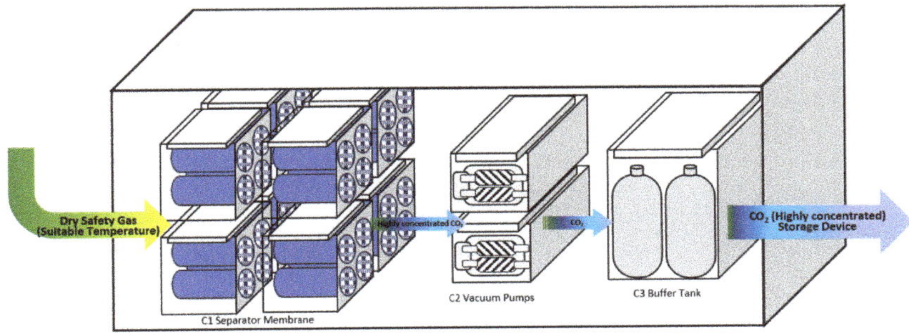

Figure 7. Schematic diagram of Module C: CO_2 Membrane Separation.

C2. Vacuum Pump: Activated under conditions of low inlet pressure, volume, or CO_2 concentration. It assists in drawing the gas stream through the membrane, improving separation performance under suboptimal conditions.

C3. Buffer Tank Unit: Temporarily stores the captured CO_2 and serves as a pressure buffer to stabilize flow before compression or storage stages.

To improve overall energy efficiency and reduce operating costs, waste heat recovery or other integrated heat exchange systems [96] may be incorporated into the design where feasible.

After the CO_2 capture system, the CO_2 in the buffer tank C3 has reached the purity required for capture, and it can be connected to the transport pipework onboard the ship via a double-valve connector after being compressed into a stable liquid phase and then transported to the CO_2 storage unit.

To support the proposed modular OCCS design, physical fabrication of the system's key components has been carried out in collaboration with China International Marine Containers (Group) Co., Ltd. (Shenzhen, China) As of the time of writing, major units—including the membrane separation housings, absorption modules, CO_2 liquefaction chambers, and pressure vessels—have been individually manufactured and dry-tested. Also, module-to-module mechanical integration has been iteratively tested to evaluate physical interface compatibility, structural tolerances, and operational layout. Membrane selection is a key part in this system, tested membranes are purchased from different suppliers. See Figure 8 below for the schematic design of this system.

The system employs a two-stage membrane separation configuration to progressively concentrate CO_2 from an initial ~7% concentration in the exhaust gas. Stage 1 involves compression, heat exchange, and a primary membrane unit, achieving ~20–30% CO_2 capture. The permeate stream from this stage (containing ~30–40% CO_2) undergoes further compression before entering Stage 2 for secondary separation, yielding product gas with over 50% CO_2 concentration. Vacuum pumps are integrated to enhance the membrane driving force, while partial gas recycling improves overall efficiency. Buffer tanks stabilize system pressure and facilitate modular integration.

In parallel, ongoing research focuses on membrane performance screening and long-term operational validation under a separate track. Upon completion, these findings will be reported in a dedicated publication addressing system efficiency, material selection, and performance stability, with explicit reference to the design framework outlined herein.

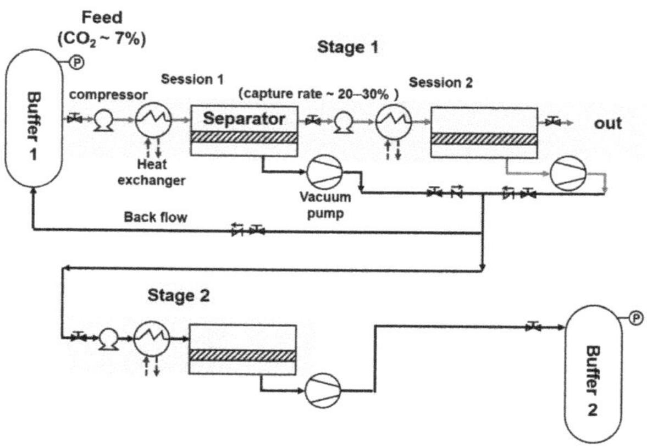

Figure 8. Process verification diagram for a membrane-based modular CO_2 capture system.

2.2.2. Diversity in System Construction

- Module C. Lap Diversity for CO_2 Separation (Membrane Separation as an Example)

Depending on the specific application scenario, this module can be used with multi-stage, multi-stage parallel or series-connected membrane module units. As an example, the concept of "multi-stage CO_2 separation modules" is introduced with two-stage, two-stage series and two-stage parallel CO_2 separation modules:

1. Two-stage: As illustrated in Figure 9, the two-stage structure is typically used to increase the total amount of carbon dioxide separated and the proportion of carbon dioxide separated from the mixture. First, after a series of treatments, the gas enters the first-stage CO_2 separation unit. Driven by a certain pressure or concentration difference, CO_2 begins to permeate through the first-stage membrane module unit to the opposite side of the membrane, while the remaining gas is retained on the feed side. The separated CO_2 gas is drawn by a vacuum pump into the buffer tank unit and then injected directly into the storage module. The retained gas from the first stage is directed via pipeline to the second-stage CO_2 separation unit for secondary separation of any remaining CO_2.
2. Two-stage tandem connection: As illustrated in Figure 10, the tandem structure is primarily used to improve the purity of separated CO_2. After the gas undergoes a series of treatments, it enters the first stage of the CO_2 separation unit. Driven by a pressure or concentration difference, the CO_2 begins to permeate through the first stage membrane module, while the remaining gas is left outside the membrane. Since this retained gas is already clean, it can be uniformly discharged through the external exhaust pipeline. Depending on the pressure on both sides of the membrane, the vacuum pump unit may be activated to accelerate CO_2 filtration. The gas is then buffered in the tank and sent to the second stage of the CO_2 separation unit. The second stage operates similarly to the first stage, further filtering and purifying the CO_2. It is important to note that the materials used for each stage of the CO_2 separation process can vary and be customized based on the components and filtration sequence.
3. Two-stage parallel connection: As illustrated in Figure 11, the parallel connection structure is primarily used to reduce CO_2 separation processing time and increase the amount of CO_2 captured simultaneously. When the incoming gas flow rate increases significantly, this method allows for an increased total gas handling capacity of the system. It can be understood as a direct parallel connection of two primary membrane separation units.

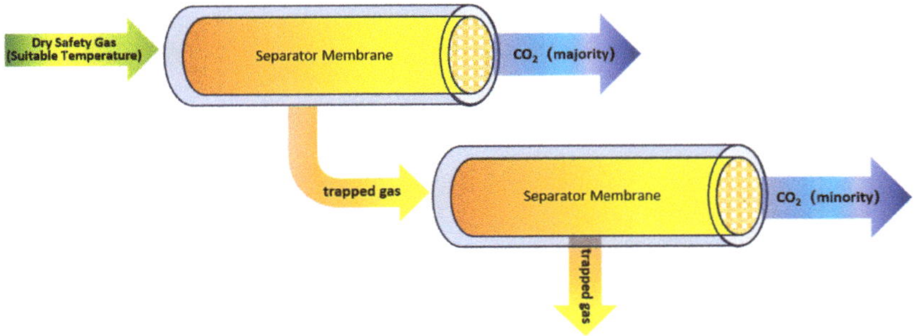

Figure 9. Schematic diagram of two-section membrane separation method.

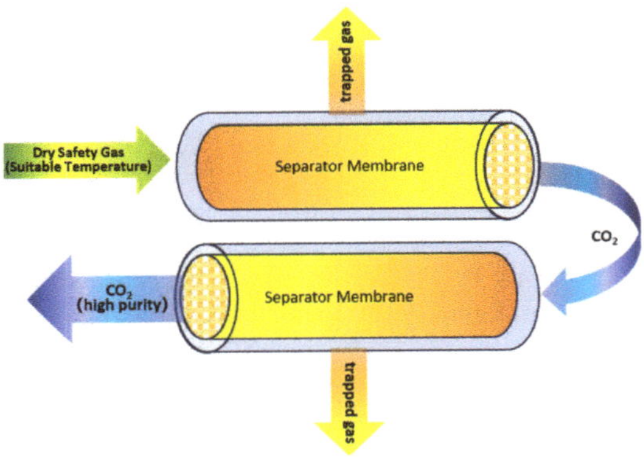

Figure 10. Schematic diagram of two-stage tandem membrane separation method.

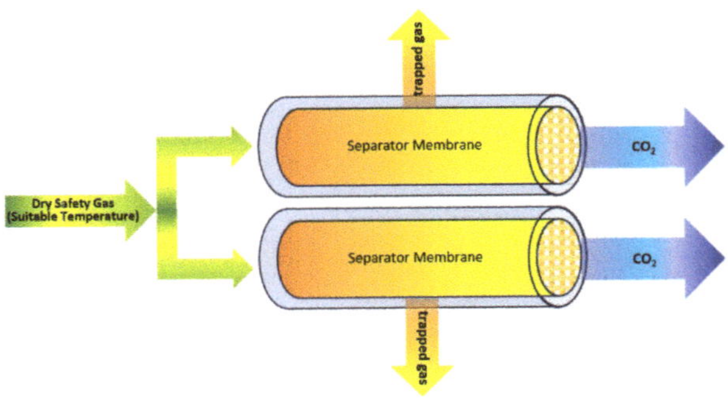

Figure 11. Schematic diagram of two-stage parallel membrane separation method.

In special application scenarios, a combination of multi-stage and series-parallel configurations can be used to further enhance both the purity and capture rate of CO_2. Each stage of the membrane material can be selected individually based on the actual gas composition, flow rate, and target capture purity. Additionally, the inclusion of three-way valves and piping between membrane modules allows for flexible strategy adjustments and synergistic operation.

- Diversity of the CO_2 capture system

As illustrated in Figure 12, due to variations in application scenarios, gas composition, and flow rate, the standard modules designed in this paper allow for flexible series and parallel configurations of different modules in the CO_2 capture system. This enables adjustment of the output of individual functional units, reordering of unit sequences, and modification of connection methods and numbers across modules. As a result, the system achieves both large-scale flue gas processing and high-precision CO_2 capture without being constrained by fixed processing capacities, and without the need to compromise between throughput and accuracy under special conditions.

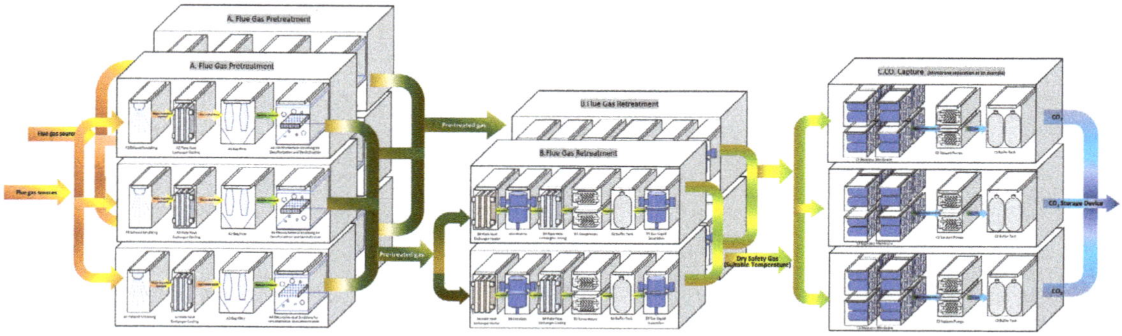

Figure 12. Schematic diagram of series-parallel combination lap of OCCS system.

2.2.3. Advantage: Modular Design and Standardized Components

1. Standardization advantages: Standardized dimensions can facilitate serial and parallel integration, easy installation on fixed-size base supports, and the formation of a "containerized" complete process system. This allows for flexible disassembly, assembly, maintenance, and adjustment across different ship scenarios, ultimately reducing industrial production costs and promoting the development of a dedicated equipment processing industry chain.
2. Economic and efficient: The modular design significantly reduces equipment manufacturing, installation, and maintenance costs, improving long-term operational economy.
3. High flexibility, small volume, and compactness: Modular design enables each unit to be flexibly assembled or disassembled based on actual flue gas emissions, space layout, and operational needs in different applications.
4. High-quantity, high purity, high-speed: Through optimized flue gas pretreatment and multi-stage membrane module separation, the system can efficiently capture CO_2 from the incoming flue gas, enhance both the capture rate and purity.

3. Design of CO_2 Storage and Transportation System for Marine Vessels

A study [70] indicates that as liquefaction pressure increases, the cost of liquefaction and pumping systems decreases, but the cost of storage tanks and CO_2 carriers rises. Therefore, high-pressure storage tank costs directly affect marine CO_2 storage.

Current Pilot OCCS systems often use a single-tank storage. Although a single tank may seem more material-efficient in terms of wall-to-volume ratio, it presents significant limitations in terms of handling, maintenance, and safety.

Therefore, to resolve these challenges, this study proposes a modular CO_2 storage unit compatible with standardized containers, which can ensure secure storage while maximizing economic efficiency. Modular vessels can be manufactured using seamless steel pipes, improving safety and reducing production complexity and cost. The smaller-unit design also improves operational flexibility and fault isolation, since a failure in one vessel does not compromise the entire system. By replacing conventional large-scale storage tanks with a standardized container framework, this design enables scalable deployment, rapid installation, and adaptability to diverse storage volumes and pressure-phase conditions.

This study rigorously analyzes material costs, pressure resistance, weight, and structural integrity of the CO_2 storage unit, introducing two critical performance metrics—'cost per unit capacity' and 'weight per unit capacity'—to compare designs across material choices and dimensional configurations, ultimately identifying the optimal unit dimensions that balance safety and cost-effectiveness. Furthermore, the system integrates structural reinforcements, valve-meter assemblies, and auxiliary components to ensure operational stability, thereby improving the efficiency and adaptability of maritime CO_2 storage systems while complying with high-pressure safety standards and minimizing costs.

3.1. Design of Small-Sized CO_2 Storage Units and Their Fittings

The CO_2 storage units are interconnected via a modular high-strength framework fabricated from lightweight, corrosion-resistant, and pressure-tolerant advanced materials, such as aluminum alloys or high-strength steel alloys. Additionally, the incorporation of composite materials—such as carbon fiber-reinforced polymer matrix composites—leverages their high strength-to-weight ratio to minimize unit weight and structural load while maintaining pressure integrity. Each unit is securely anchored within a standardized container, forming an integrated assembly that ensures structural stability during transit. This configuration prevents displacement, collisions, or hazardous scenarios, even under mechanical stresses such as vibrations or impacts.

3.1.1. CO_2 Storage Unit Group Tank Structure

The CO_2 storage unit design must accommodate CO_2 phase behavior. As CO_2 is typically liquefied for large-scale transport, storage temperature and pressure must be maintained between the triple point (−56.6 °C, 0.518 MPa) and critical point (30.98 °C, 7.377 MPa) to prevent phase transitions [70,97,98]. Ambient temperature fluctuations induce internal pressure variations, necessitating precise control systems. To ensure secure storage while maximizing economic efficiency, this study proposes a high-pressure CO_2 storage device compatible with standardized containers (Figure 13). The vessel features a maximum pressure tolerance of 7.35 MPa (operational pressure < 7 MPa), achieved through reduced tank diameter. This design simultaneously enhances structural integrity, lowers fabrication costs, and enables flexible spatial configuration through modular assembly within the containerized framework.

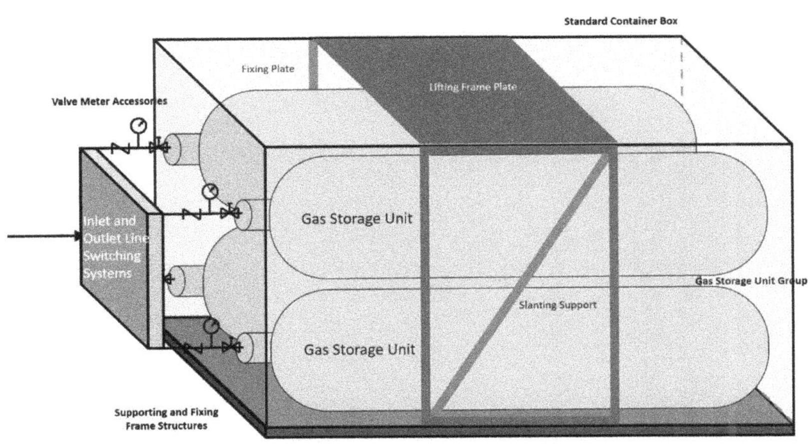

Figure 13. Schematic design of containerized CO_2 storage unit set.

The CO_2 storage units are configured in two primary orientations: horizontal grid-based arrays and vertical stacks. Deployment quantities are adaptable based on the stored gas's target pressure and phase conditions. For horizontal configurations, modular scaling permits array dimensions ranging from 1×1 to 20×20 units. Vertical single-layer arrangements accommodate 2–120 units, with capacity determined by the storage unit diameter to maintain structural integrity.

The quantity of CO_2 storage units within a modular group can be dynamically optimized based on the stored gas's pressure, phase state, and operational requirements. For instance, systems storing high-pressure liquefied gases may scale unit quantities proportionally to accommodate large-capacity storage demands, whereas configurations handling gases with significant pressure fluctuations often minimize unit counts to amplify the pressure-carrying capacity of individual units.

3.1.2. Auxiliary Safety Devices

Safety is paramount in high-pressure vessel design. As illustrated in Figure 14, this system incorporates auxiliary safety mechanisms—including valve-gauge assemblies and structural reinforcements—to mitigate risks for modular storage units during dynamic maritime operations.

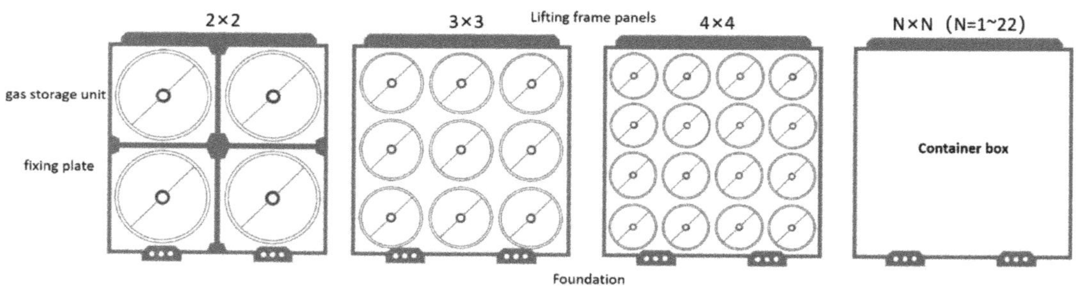

Figure 14. Schematic sketch of container frame and horizontal arrangement.

The modular gas storage unit, designed as a cylindrical pressure vessel, integrates pressure sensors, flow sensors, liquid-level detectors, safety-relief valves, and an emergency venting system. These components transmit real-time telemetry (temperature, pressure, liquid level) to centralized monitoring systems while autonomously initiating emergency protocols (e.g., pressure relief, alarms) during abrupt fluctuations.

The compact gas storage unit is secured within the container via a reinforced framework (supports, bases, fixed plates, inclined braces, and lifting plates) to ensure structural stability. Vibration-damping interfaces (e.g., rubber gaskets) at support-unit contact surfaces mitigate mechanical stresses induced by maritime turbulence. Additionally, continuous sensor-based monitoring ensures real-time detection of operational anomalies, enabling prompt safety interventions.

3.2. Comparison of Economic Safety of CO_2 Storage Unit Size

3.2.1. Comparison Parameters

In this paper, the following two economic parameters need to be introduced to the design of CO_2 storage unit size:

1. Cost per Unit Capacity: the cost of tank construction and operation per unit of storage capacity.
2. Mass per Unit Capacity: the mass of the tank per unit capacity.

The calculation of these two parameters is governed by Equations (1) and (2):

$$\text{Cost per Unit Capacity} = P \times (M_1 + M_2 \times a)/(V_1 + V_2) \quad (1)$$

$$\text{Mass per Unit Capacity} = \rho \times (M_1 + M_2)/(V_1 + V_2) \quad (2)$$

where: P is the Unit price of steel materials (yuan: CNY currency); M_1 is the quality of the cylinder; M_2 is the quality of storage tank heads; M_1 is the volume of the cylinder; M_2 is the volume of storage tank heads; a is the price cost factor of tank heads; ρ is the density of steel.

The cost per unit capacity directly affects the CO_2 storage cost and thus the overall operating cost of the ship, the weight per unit capacity takes up the effective cargo capacity of the ship and affects the operating income of the ship, and a larger weight per unit capacity will increase the resistance of the ship's voyage and increase the cost of fuel consumption, so it is hoped that, under the premise of guaranteeing the safe storage of CO_2, and by taking into account the material of the storage tanks, the strength of the materials, the price, and other factors, it is possible to measure the following Minimum unit capacity cost and unit capacity weight.

3.2.2. Comparison Results and Explanation of the Layout of CO_2 Storage Units

- Comparison of materials for CO_2 storage unit

The vessel-wall thickness calculation is governed by Equation (3):

$$t = (P \times R)/(\sigma \times \varphi - 0.6 \times P) + C \quad (3)$$

where: t is the wall thickness (mm); P is the design pressure (MPa); R is the inner radius of the container (mm); σ is the permissible stress of the material (MPa) determined by the material properties and operating temperatures; φ is the coefficient of welded joints (no welding is 1, welded joints are usually 0.85~1.0); 0.6P is the pressure correction; C is the corrosion margin (mm).

Taking into account the fluctuations and differences in international steel market prices, we have selected the Chinese steel market in Asia as the focus of our recent market research. A comprehensive market survey was conducted on six candidate tank materials—Q370, Q420, Q690, SA517, Q490RW, and Q580R [99–101]—encompassing parameters such as price, tensile strength, density, and yield strength. The findings are tabulated in Table 5. Preliminary analysis indicates that the SA517 model exhibits the optimal cost-performance ratio, although detailed engineering calculations remain imperative before finalizing the material selection.

Table 5. Parameters of six candidate tank materials.

Material	Price (yuan/t)	Strength (MPa)	Yield Strength (MPa)	Welding Coefficient
Q370	4200	220	≥370	0.955
Q420	4500	237	≥420	
Q690	6000	287	≥690	
SA517	4800	331	550~860	
Q490RW	5000	254	≥490	
Q580R	5500	287	≥580	

The same design constraints (pressure, geometry, temperature) were applied to the six aforementioned steel materials and the Cost per Unit Capacity and Mass per Unit Capacity were calculated for all materials using Equations (1)–(3). As illustrated in Figure 15, SA517-

grade steel exhibits a Cost per Unit Capacity of 7989.55 yuan/m^3 and a Mass per Unit Capacity of 1664.49 kg/m^3—both representing the lowest values among the materials. This demonstrates its exceptional economic performance relative to other options. Consequently, the subsequent design of the CO_2 storage tank in this paper will select SA517 for further optimization, including layout design.

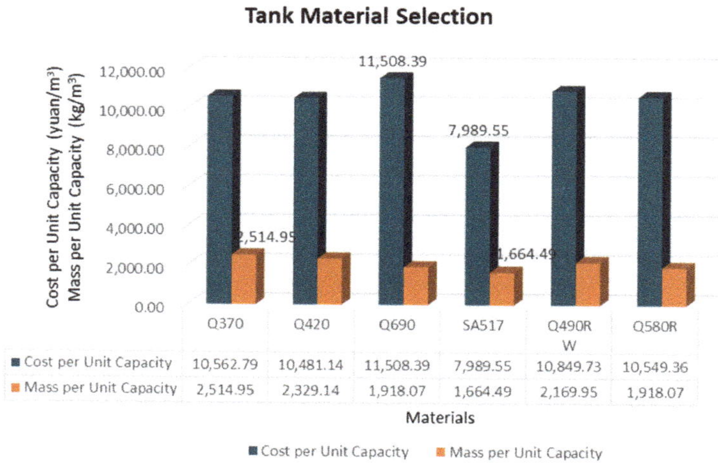

Figure 15. Results of material selection for CO_2 storage unit.

- CO_2 storage unit size (diameter, height) design comparison

This study proposes the integration of compact CO_2 storage units within standardized containers. The design process prioritizes spatial configuration (horizontal and vertical orientation) and modular scalability to optimize container utilization. Using standardized containers as modular frameworks, cost-effectiveness metrics were evaluated across varying layouts and unit dimensions. This analysis identified the optimal CO_2 storage unit size, balancing structural efficiency and economic viability.

Horizontal Layout: Assuming a working pressure of 7 MPa and a maximum design pressure of 7.35 MPa, the CO_2 storage units are arranged in a horizontal grid configuration. Each unit is equipped with hemispherical headers of uniform dimensions at both ends. To identify the optimal design scheme, we evaluate and optimize the number of storage units, the inner and outer diameters of the tanks, and the length of the cylindrical body. This optimization is conducted for two standard container sizes (20-inch and 40-inch), with a focus on minimizing both the cost per unit capacity and the mass per unit capacity.

The optimization calculation results are visualized in Figure 16. Under the specified conditions and horizontal network configuration, the most economically efficient layout for the 20-inch container is a 3 × 3 arrangement of CO_2 storage units, each with a diameter of 690 mm and a cylindrical height of 4700 mm. In contrast, for the 40-inch container, a 2 × 2 arrangement with a unit of 1030 mm in diameter and 4300 mm in cylindrical height is both safer and more cost-effective. Overall, the 2 × 2 (ϕ: 1030 mm, H: 4300 mm), 3 × 3 (ϕ: 690 mm × H: 4700 mm), and 4 × 4 (ϕ: 510 mm × H: 4800 mm) configurations are considered the most reasonable layout options based on a balance of safety and economic performance.

Vertical Placement: Assuming a working pressure of 7 MPa and a maximum design pressure of 7.35 MPa, the CO_2 storage units are arranged in a single-layer vertical configuration. Each unit is designed with a hemispherical head at the top and a flat bottom. To determine the optimal design scheme, we optimize the number of storage units, the inner and outer diameters of the tanks, and the length of the cylindrical body. This optimization is carried out for two standard container sizes (20-inch and 40-inch), with consideration given to two key performance indicators: tank mass per unit capacity and tank cost per unit capacity.

(a) (b)

Figure 16. Comparison of CO_2 storage unit sizes (horizontal layout in container): (**a**) for 20-inch container; (**b**) for 40-inch container.

The optimization results are presented in Figure 17. Under the specified conditions, the most favorable outcome is achieved with CO_2 storage units having a diameter of 690 mm and a cylindrical height of 1455 mm, arranged in a single vertical layer of 16 units. Alternative configurations—such as units with a diameter of 1030 mm and a cylindrical height of 1285 mm (10 units), or units with a diameter of 510 mm and a cylindrical height of 1545 mm (21 units)—yield slightly less optimal, yet still viable, economic performance.

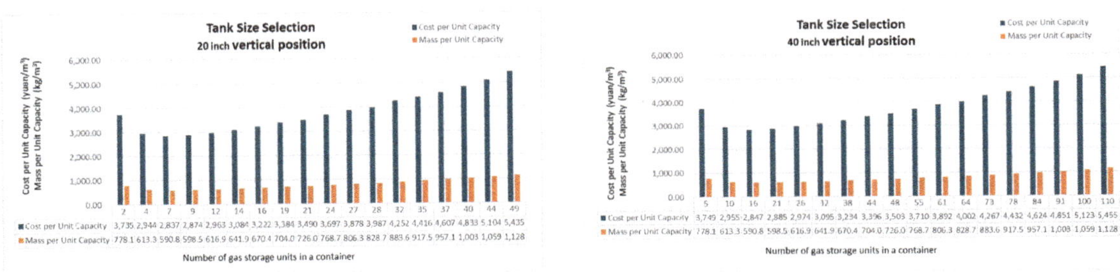

(a) (b)

Figure 17. Comparison of CO_2 storage unit sizes (vertical single layer layout in container): (**a**) for 20-inch container; (**b**) for 40-inch container.

3.2.3. Case Analysis

Through comprehensive market research and database queries [102–105], this study has compiled many existing ship data and operational insights—encompassing ship tonnage, energy consumption, average daily CO_2 emissions, standard container capacity, and voyage duration—to establish the parameter specifics of the ocean-going vessel X, as tabulated in Table 6. These data facilitate the calculation and comparison of different CO_2 storage schemes for the vessel.

Taking into account the target CO_2 avoidance rate (set at 30%) and a voyage duration of 20 days, it is necessary to store a minimum of 8347 tons of CO_2. For each scheme, the CO_2 storage temperature and pressure are specified, followed by calculating the storage density using the CO_2 phase diagram (These calculations are supported by REFPROP [106], a mature thermodynamic and transport property calculation software developed by the National

Institute of Standards and Technology (NIST) of the United States. Widely adopted in both industry and academia, this software is used to calculate and predict the thermodynamic properties of various working fluids, encompassing physical properties, thermodynamic properties, and transport properties). Finally, the required CO_2 storage space for each scheme is derived from fundamental mass-volume formulas. Three existing design options are considered, and the calculation basis and results for each are presented below in Equation (4) and Table 7.

Table 6. Relevant parameters of the case vessel.

Gross Tonnage/(t)	TEU [1] Standard	TEU Actual	Average Daily CO_2 Emission/(t per Day)
148,667	14,000	8900	292.13

[1] TEU: Twenty-Foot Equivalent Unit.

Table 7. Comparative analysis of CO_2 storage options for case ships.

Option	Large C-Tank	20-Inch Containerized Storage Tank (One)	20-Inch Containerized CO_2 Storage Unit (2 × 2)
Operating temperature and pressure	238 K 19 bar	238 K 19 bar	243 K 40 bar
CO_2 storage density	1143.8 Kg/m^3	1143.8 Kg/m^3	1085 Kg/m^3
storage space required	7298 m^3	7298 m^3	7694 m^3
Parameters of equipment	C-Tank 3750 m^3	Single Tank Storage Volume 18 m^3	Single Tank Storage Volume 6 m^3
Number of equipment	Two C-Tank	406 TEU	320 TEU
Additional occupancy rate of container space	7.61%	4.56%	3.6%
Annual cost of lost deadweight ton (USD)	12,191,220	7,305,120	5,767,200

Annual DWT (Deadweight Tonnage) loss cost:

$$C_{lost} = TEU \times \rho \times c_{TEU}^{lost} \times \varphi \quad (4)$$

where: TEU represents the total container capacity of the vessel; ρ denotes the loss rate of deadweight tonnage; C_{TEU}^{lost} is the cost of lost cargo space (according to Korberg [36], the value of 900 EUR/TEU/trip was obtained in this case); and φ indicates the frequency of the liner's voyages per year (φ = 20 in this case).

As shown in Table 7, with the same CO_2 storage capacity (8347 tonnes), this case compares three different CO_2 storage units: a large C-Tank, an individual containerized tank, and a storage unit system within containers.

1. C-Tank: Due to its large dimensions, this tank must be pre-installed at a designated hull position, leading to significant deadweight tonnage loss. In the early stage of CO_2 storage, substantial free space exists inside the tank, resulting in an additional container space occupancy rate of 7.61%. The annual cost of deadweight tonnage loss amounts to USD 12,191,220.
2. Individual Containerized Tank: Calculations show that 406 TEU tanks meet the total CO_2 storage requirement, with an additional space occupancy rate of 4.56%. The annual cost of lost deadweight ton is USD 7,305,120.
3. Small Storage Units in Containers: Due to the more efficient use of container space, 320 TEU with small storage units satisfy the storage requirement, achieving an additional space occupancy rate of only 3.6%. The annual deadweight tonnage loss cost is USD 5,767,200.

Comparisons have shown that the small CO_2 storage units system proposed in this study, integrated into standard-size containers, demonstrates a significantly lower additional occupancy rate and deadweight loss cost compared to the C-Tank. Specifically, its annual cost of lost deadweight ton is 52.2% lower than the C-Tank, highlighting its economic superiority.

Furthermore, these small-sized storage units provide other operational advantages, such as easier lifting and unloading during docking, enabling more flexible and timelier reallocation of cargo space.

3.3. Advantage: Small Size CO_2 Storage Unit Group Design

1. High use flexibility: By reducing the tank diameter, a higher-pressure design is achieved, enabling the device to safely store CO_2. Compared to conventional large-diameter CO_2 storage tanks, the system offers more stable and reliable pressure-bearing performance, effectively mitigating safety risks associated with stress concentration within the tank structure. Additionally, the number of CO_2 storage units can be flexibly adjusted, allowing the system to be optimized based on actual gas pressure and phase transitions.
2. Efficient space utilization and convenient transportation: Multiple CO_2 storage units are integrated into standard containers, optimizing the use of container space, enabling a compact layout, and enhancing space utilization. The standard container design also facilitates lifting and transportation, making full use of existing logistics and transportation networks and infrastructure. This design enables efficient transfer, thereby improving equipment mobility and adaptability.
3. Almost no downtime maintenance period: In the event of a failure of an individual container CO_2 storage unit, it can be bypassed, allowing continued operation with minimal disruption. The equipment maintenance time is extremely short, making it more convenient compared to traditional large CO_2 storage tanks. Additionally, this design has almost no impact on the vessel's normal navigation, CO_2 capture, or storage operations.

4. Discussion

This subsection critically examines unresolved technical and operational challenges in maritime CCUS systems, including practical implementation constraints and design compatibility, safety and environmental risks of onboard CO_2 storage, CO_2 offloading logistics, infrastructure, and risks, empirical economic data and feasibility analysis, advantages and disadvantages of modality, regulatory framework and incentives, and integration with broader decarbonization strategies.

4.1. Practical Implementation Constraints and Design Compatibility

Deploying CCUS on ships poses significant practical challenges [107] that must be addressed in design and operation. Key constraints include:

1. Space and Weight Limitations: Onboard space and deck area is limited, installing CCUS unit can encroach on cargo holds or deck area, especially in retrofits. Careful naval architecture assessments are needed to integrate the CCUS system without exceeding stability and strength limits. Newbuild designs can plan for this from the beginning, whereas retrofitting existing ships is far more challenging, requiring creative use of available spaces and perhaps relocating or upgrading other equipment to make room.
2. Power and Utility Integration: CO_2 capture systems demand considerable energy and utility support (steam, cooling water, electricity, etc.). The ship's existing engines and generators must accommodate the parasitic load of carbon capture. These integration

issues mean that comprehensive engineering and possibly engine re-tuning are needed when retrofitting CCUS.

3. Operational Complexity and Crew Training: Crew members will need to learn the normal operation of the capture plant, routine maintenance, such as replacing filters or membranes, and respond to emergencies such as leaks or high CO_2 levels. Updating safety procedures and documentation is essential for safe operation
4. Regulatory and Classification Compliance: Since regulations and classification standards are still catching up, shipowners may be reluctant to invest without assured compliance pathways. The ongoing development of IMO's framework is expected to provide more clarity in the near future.

By recognizing and addressing these practical constraints—spatial, integrational, human, and regulatory—designers can better ensure that shipboard CCUS systems are compatible with vessel operations. Early pilots have highlighted these issues. For instance, the K-Storm project [108] aims to minimize onboard modifications by using self-contained modules. Despite these challenges, a small number of ships are already running CCUS systems safely, thanks to comprehensive training and risk management practices.

4.2. *Safety and Environmental Risks of Onboard CO_2 Storage*

Storing CO_2 onboard at high-pressure and low temperature introduces specific safety and environmental risks. There is an explosion risk associated with pressurized CO_2 tanks. Therefore, tanks must have over-pressure protection such as pressure relief valves or rupture discs. CO_2 is also odorless, colorless, and an asphyxiant gas. A leak in an enclosed compartment could quickly create a life-threatening atmosphere. Thus, crew training on CO_2 leak response is critical. As discussed, pure CO_2 has the peculiarity of a triple point at 5.17 bar, −56.6 °C [109]. When the storage tank or associated piping falls below these conditions, dry ice will form. Solid CO_2 can cause blockages in valves or lines and can also lead to pressure spikes as it warms and re-sublimates in confined spaces. Even though CO_2 is not toxic, a sudden large release in port or an enclosed waterway could lower air quality and, if it dissolves, locally acidify the water.

Overall, the storage of CO_2 on ships demands engineering and operational controls comparable to those for LPG/LNG carriers, with knowledge on the unique properties of CO_2. Although formal regulations are still in development, industry consensus holds that onboard CO_2 storage risks can be managed to an acceptable level, analogous to other hazardous cargoes.

4.3. *CO_2 Offloading Logistics, Infrastructure, and Risks*

Once CO_2 is captured onboard, a critical challenge is how to offload and handle the CO_2 at port. The current infrastructure at ports worldwide is largely inadequate to receive and manage captured CO_2 in large quantities. A recent joint study by the Global Centre for Maritime Decarbonisation (GCMD) [110], Lloyd's Register and others identified low port readiness as a major bottleneck to the adoption of onboard carbon capture. While the technologies for transferring liquid CO_2 exist, safe and efficient operational procedures have not yet been demonstrated at scale. As shown in Figure 18, there are a few pathways for offloading captured CO_2 from a vessel:

1. Ship-to-Shore: The ideal scenario is a ship arriving at a port equipped with a CO_2 reception facility. The vessel would connect to a shore-based tank via insulated cryogenic transfer hoses or loading arms. However, only a few ports involved in pilot projects are installing the refrigerated storage tanks, vaporizers, and pumps needed. Most ports do not have plans to install shore-based tanks until upstream supply and

downstream subsurface or usage is clear. Shipowners are depending on a developed discharging network for their onboard CCUS installation.

2. Ship-to-Ship: Another approach is transferring the CO_2 to a smaller shuttle vessel or barge outfitted to carry LCO_2, which then delivers it to a storage hub. This avoids complicated terminal retrofits at every port and instead relies on a few CO_2 aggregation vessels shuttling to sequestration sites. This study ranked ship-to-ship and ship-to-shore via an intermediary vessel as the top options for scaling up offloading, since they reduce port-side bottlenecks.
3. Containerized (ISO Tank) Offloading: In this approach, captured CO_2 is liquefied onboard and pumped into portable tank containers (usually 20- or 40-foot ISO tanks) which are then landed by crane at port and transported by truck or rail. This method is highly flexible and leverages existing container logistics, requiring minimal new infrastructure at the port aside from perhaps a refrigeration hookup to keep the CO_2 cold. It is already in use in one system: the Value Maritime OCCS [111], which fills a swap-out CO_2 absorption cartridge that is simply offloaded at port.

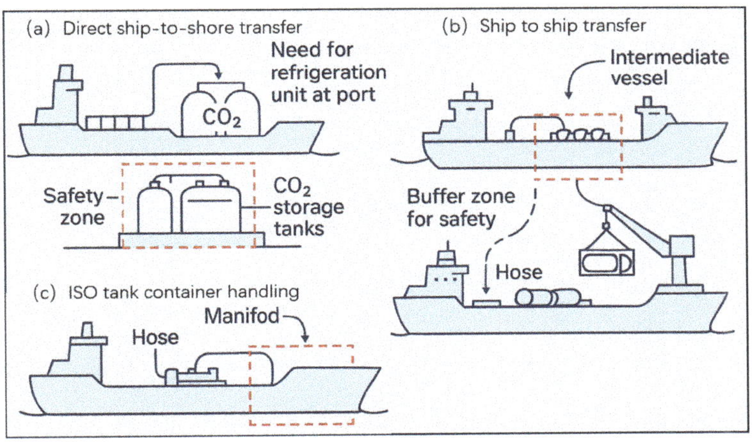

Figure 18. A schematic illustration of the possible offloading setups and the associated flow of CO_2 to its destination.

In any offloading scenario, infrastructure for onward transport or storage of the CO_2 is vital. The GCMD survey of planned CO_2 hubs worldwide shows that port infrastructure for handling CO_2 must be scaled up along with these projects to achieve economies of scale. Until large storage or utilization sites are operational, ships might be limited to discharging at a few specialized ports (e.g., terminals associated with carbon storage projects in the North Sea or Gulf of Mexico [112,113]).

Despite these challenges, some early operations have been successful. Notably, in 2024, Evergreen Marine reported [114] that its 14,000 TEU Ever Top containership offloaded captured CO_2 to a shore facility for reuse, and was able to deduct this CO_2 from its emissions for IMO Carbon Intensity rating purposes. CO_2 offloading is currently a weak link in CCUS chain, but initiatives are underway to close this gap by building infrastructure and establishing safe operating guidelines.

4.4. Empirical Economic Data and Feasibility Analysis

A crucial aspect of onboard CCUS is its economic viability. At the current stage, empirical data is limited to pilot projects and techno-economic studies, but these provide valuable insights. Generally, onboard carbon capture is found to increase fuel consumption, energy penalty and entail high capital and operating costs, leading to a significant cost per tonne of CO_2 emission reduced. Here, we compile data from recent projects and studies to quantify these trade-offs.

In Table 8, Stena Impero Pilot (REMARCCABLE Project) [115] is a First-of-a-kind retrofit pilot; relatively high abatement cost due to pioneering nature and small-scale demonstration. Costs are expected to decrease significantly with larger scale and standardization. VLCC Newbuild [116] Study completed theoretical analysis projecting cost efficiencies for large-scale implementation; cost per ton significantly reduced due to economies of scale and optimized ship integration. Wärtsilä OCCS System [117] (Clipper Eris) reported significantly lower estimated abatement costs through industrial scaling and optimized integration; limited detailed financial data publicly disclosed. Value Maritime [111] Modular System compact modular solution targeting small ships, offering low complexity and minimal integration challenges, though detailed economic data remain limited.

Table 8. Comparison of Onboard Carbon Capture and Storage (OCCS) performance and economic data from recent studies and projects.

Project or Study	Ship Type/Size	CO_2 Capture Rate	Fuel Penalty	Cap Ex Cost (USD)	Estimated Abatement Cost (USD/ton CO_2)
Stena Impero Pilot (REMARCCABLE Project)	MR (Medium Range) tanker	~30% [118]	~10% increase in fuel consumption [118]	13.6 million	~769 [119]
VLCC Newbuild Feasibility Study (The Mærsk Mc-Kinney Møller Center)	Very Large Crude Carrier (VLCC)	~75%	~8% (assumed)	Not stated	220–290
Wärtsilä OCCS System on Ethane Carrier (Clipper Eris)	Ethane Carrier (Clipper Eris)	~70%	Not publicly detailed	No data	~60 [120]
Value Maritime System (small-scale modular)	Small Tanker (~10,000–20,000 DWT)	~20% [121]	Negligible (uses auxiliary boilers)	Lower Cap Ex, but no details	Not available

The REMARCCABLE engineering study (2024) [122] retrofitted a carbon capture system on the medium-range tanker Stena Impero. The calculated abatement cost for this prototype was approximately USD 769 per tonne of CO_2. This figure is high—by comparison, many shore-based industrial CCS projects target costs in the USD 50–150/ton range—reflecting the early stage and small-scale of maritime CCUS.

At a different scale, a scenario analysis by the Mærsk Mc-Kinney Møller Center (2022) [123] found that for a VLCC newbuild, which can more easily accommodate a large capture system, the CO_2 abatement cost could range USD 220–290 per ton for ~75% net emissions reduction. This result brings CO_2 capture into a range where a combination of carbon pricing and efficiency gains could make it viable.

Beyond the capital expenditure of installing the equipment, operating costs can be significant. Major components of OPEX for an onboard capture system should include fuel penalty, solvent and consumables, maintenance and crew labor, CO_2 offloading and logistics.

Overall, current empirical data indicate that shipboard CCUS is technically feasible but economically challenging today. The cost per ton of CO_2 is presently high, but there is a credible trajectory for cost reduction. Key to improving economics will be scaling up system size, optimizing current system, introducing new approaches with efficiency advantages, such as Modular systems.

4.5. Trade-Offs and Design Considerations for Modular CCUS Systems

Modular ship CCUS systems package the capture equipment into compact units that can be more easily installed, removed, or scaled. This approach offers clear benefits for retrofitting, flexibility and cost reduction, but it also comes with trade-offs in performance and ship integration. Here, we discuss the implications of modular design versus fully integrated design.

1. Ease of Retrofit and Installation: Modular CCUS units are typically self-contained skids that can be fabricated and tested onshore, then lifted onto the vessel.
2. Cost Considerations: Modular units, especially if containerized, open up the possibility of manufacturing in series, turning CO_2 capture systems into a commodity that can be produced in factories at lower unit cost. This could drive down prices significantly if demand scales up.
3. Deadweight tonnage consumption: Traditional systems are bulky with irregular shapes, often featuring tall towers or tanks. In contrast, modular container-integrated capture systems offer greater practicality and are often preferred by cost-conscious shipowners.
4. Performance and Efficiency: A potential drawback of self-contained modular systems is that they might not be as optimized or efficient as a fully integrated system. For example, a containerized capture module might carry its own small CO_2 compressor and chiller powered by an independent diesel genset or the ship's electrical grid, rather than tapping directly into the ship's main engine waste heat or excess steam.
5. Operational management flexibility: A modular system's failure affects only part of the capture capacity, whereas a single integrated system is a single point of failure for capture functionality. This redundancy aspect may make modular attractive for reliability.

The industry is currently experimenting with both approaches. Clipper Eris (Solvang/Wärtsilä) [120] represents an integrated retrofit. In contrast, K-Storm (Crowley/Carbon Ridge) is a clear modular retrofit, aiming for minimal ship impact and easy installation [124]. Another example, Eastern Pacific Shipping's tanker Pacific Cobalt, will use a Value Maritime module, essentially bolting on a prefabricated capture and filter unit to achieve about 40% CO_2 capture [125]. This is low-hanging fruit—it may not maximize capture, but its relatively low complexity.

4.6. Regulatory Framework and Incentives (IMO, EU ETS, Etc.)

The regulatory landscape for shipping CCUS is rapidly evolving, as international and regional bodies recognize the need to accommodate and encourage carbon capture in maritime decarbonization.

4.6.1. IMO's Developing Framework

At present, there are no explicit IMO rules that either mandate or comprehensively regulate onboard carbon capture systems. However, the IMO has moved quickly in the past two years to acknowledge OCCS. In 2021–2023, various IMO committees (MPC, MSC) began discussions on how to handle novel GHG reduction methods, including CCUS. Most significantly, in April 2025 during MEPC 83, the IMO approved a work plan to develop a full regulatory framework for onboard carbon capture and storage [126]. Also, the IMO's Working Group on GHG has been tasked to figure out how to account for captured CO_2 in the IMO DCS (Data Collection System) [127] and CII (Carbon Intensity Indicator) schemes [128]. By around 2028, the IMO is expected to roll out: safety guidelines and accounting mechanisms so that captured CO_2 can be deducted from a ship's reported emissions. Already, IMO's revised GHG Strategy (2023) [129] includes "mid-term measures should encourage innovation like CCUS", viewing it as a complementary measure alongside alternative fuels. This top-level recognition means any carbon pricing or fuel standard the IMO implements could include credits for OCCS [130].

4.6.2. EU Regulations—Emissions Trading System (ETS)

The EU has been more immediately prescriptive by extending its Emissions Trading System to shipping. From 2024, large ships must account for their CO_2 emissions on voyages to, from, and between EU ports, and surrender carbon allowances accordingly [130]. ETS creates a direct monetary incentive: every tonne of CO_2 one can legally declare as captured and stored saves the cost of an allowance [130]. It is important that the infrastructure to realize this must exist, and the EU recognizes this and is funding CO_2 transport and storage projects that could interface with shipping.

Moreover, Norway has its own aggressive climate targets and is currently funding pilots like Clipper Eris. The UK is considering how to treat captured CO_2 in its MRV system. In absence of full regulations, classification societies and some flag administrations are issuing interim standards. Lloyd's Register's first OCCS class notation shows that class can formalize requirements for construction and trials [131]. ABS has released guide-lines on safety and is likely to offer similar notations. Amendments in 2009 to the London Protocol now also allow cross-border transport of CO_2 for the purpose of geological storage [132]. All these effort supports the improvement of CO_2 credit mechanisms and economic incentive systems.

4.7. Integration with Broader Decarbonization Strategies

The IMO's 2050 goals cannot be met by one technology alone, which needs to combine CCUS with other measures, such as alternative fuels, energy efficiency, operational changes, and possibly carbon offsetting. In this section, we discuss how onboard CCUS can integrate with and complement other decarbonization approaches for ships.

4.7.1. Bridge Strategy: Complementary to Alternative Fuels

The industry is exploring new fuels like LNG, methanol, ammonia, hydrogen, biofuels, etc., each with pros and cons. While fuels like green ammonia and hydrogen could eventually become the best zero-carbon options, building the necessary production and fueling infrastructure will take time and heavy investment. Meanwhile, ships can still run on traditional fuels or a mix of fuels, using CCUS technology to trap the CO_2 emissions. This 'bridge' approach helps reduce emissions early on without waiting for the perfect fuel. Industry experts recognize that adding CCUS to existing ships is a smart way to cut emissions now, whereas many alternative fuels are more suited for new ships or vessels that can be converted. For example, installing CCUS on current ships now can help meet 2030 emission goals, with plans to retire or upgrade these ships to zero-emission vessels by 2035.

4.7.2. Integrate with Efficiency Measures

Energy efficiency measures will reduce the amount of fuel burned and hence CO_2 produced through better hull designs, propulsion improvements, wind-assisted devices, speed optimization, etc. Reducing the amount of exhaust CO_2 helps lighten the load on a CCUS system. When there is less CO_2 to capture, a smaller or less powerful system might do the job, or the existing one could run more efficiently, saving energy. In practice, shipowners will likely do both: take steps to improve fuel efficiency and use CCUS to catch most of the remaining CO_2. Combining these approaches can lead to bigger reductions than doing either alone. Also, slowing down the ship means less exhaust flow and potentially cooler exhaust. This might make capturing CO_2 easier because the gas stays in the absorber longer, and waste heat could be more readily available relative to engine output.

4.7.3. Integrate with Operational Strategies

If carbon costs are introduced, an operator might do cost-optimization: only capture CO_2 until the marginal cost of capture equals the levy cost of emitting. Basically, they would only capture CO_2 until the expense of capturing equals what they would pay if they emitted it anyway. Think of it like how power plants sometimes only run carbon capture systems when carbon prices are high. In a broader strategy, using CCUS becomes just one of several tools to control emissions—alongside things like scheduling when and where they operate, choosing routes that avoid high-emission areas, and other operational decisions.

4.7.4. Integrate with Infrastructure and End-Use Integration

If ports develop CO_2 utilization hubs, then a ship capturing CO_2 essentially becomes part of a larger carbon circular economy. This synergy amplifies the impact of CCUS beyond just the ship itself and could attract cross business investment.

In summary, using onboard carbon capture alongside other strategies like switching to alternative fuels, improving energy efficiency, making operational tweaks, and possibly offsetting carbon is the way forward. The key here is combining these methods because ships are complex systems, and decarbonization is not just about one fix. As industry rules and regulations become clearer and more established, industry players can find the best mix of solutions that work for their circumstances.

5. Conclusions

Global carbon emission reduction efforts are increasing, shipping as one of the important sources of carbon emissions, there is an urgent need for technological upgrading in response to the IMO and other international organizations to reduce emissions targets, OCCS technology came into being, this paper focuses on the existing two major problems of OCCS technology, capture equipment and CO_2 storage device to design technology upgrades to provide design assistance for the eventual formation of OCCS batch standardized equipment industry chain. The key contributions of this paper are summarized as follows:

(1) This paper designs a "modular OCCS device", which is more flexible to meet the emission reduction demand and process flexibility adjustment of such mobile carbon sources as ships. Modular design of standardized parts improves the adaptability of OCCS to different flue gas flow rates and working conditions. The series-parallel assembly method is convenient to meet the requirements of incoming gas volume and working conditions, and facilitates the installation, maintenance, upgrading and use in various scenarios, thus reducing the cost of industrial production in the long run.

(2) This paper breaks the cost limitation of low-pressure storage of CO_2 in ships by designing a "small-size CO_2 storage units" that can be loaded into a standard container, and improves high-pressure storage safety with guaranteed economy. Adopting standard container box as the outer frame instead of traditional large-size storage tanks, it can realize universal transportation, rapid deployment and flexible deployment of quantity, and meet the requirements of different CO_2 storage volume and gas pressure phase state. Considering the material cost, pressure strength, weight and other factors of the CO_2 storage device, we introduce the two core parameters of "cost per unit capacity" and "mass per unit capacity" to compare the design of different materials and sizes of the CO_2 storage unit, and finally calculate the safest and most economical size of the CO_2 storage unit.

(3) When the CO_2 storage units are arranged in a horizontal grid configuration, the 2×2 (ϕ: 1030 mm, H: 4300 mm), 3×3 (ϕ: 690 mm $\times$ H: 4700 mm), and 4×4 (ϕ: 510 mm $\times$ H: 4800 mm) configurations are considered the most reasonable layout options based on a balance of safety and economic performance. When the CO_2

storage units are arranged in a single-layer vertical configuration, the most favorable outcome is achieved with CO_2 storage units having a diameter of 690 mm and a cylindrical height of 1455 mm.

Future work will report the experimental validation results, including system operational stability, capture rate, and membrane optimization strategies, in a dedicated follow-up paper referencing this foundational design study.

In addition, this paper thoroughly discusses unresolved technical and operational challenges in maritime CCUS systems. In summary, the design proposed in this paper demonstrates theoretical innovativeness and offers practical inspiration for advancing green ship carbon reduction technologies. As for how the CO_2 captured by ships can be properly disposed of, how to ensure that it is no longer released into the atmosphere, and how to achieve the most convenient and efficient storage and resource utilization should be given enough attention. It is believed that under the background of green ship, CCUS technology will be used in synergy with other ship emission reduction technologies, and make a bottom-up guarantee for carbon emission reduction of ships.

Author Contributions: Conceptualization, N.Z. and X.L.; methodology, N.Z. and X.L.; validation, C.W., Z.P., Y.C., L.L., L.C., J.H. and W.W.; writing—original draft preparation, N.Z., X.L. and X.W.; writing—review and editing, N.Z., X.L. and X.W.; visualization, N.Z., C.W. and Z.P.; supervision, X.L., C.W. and W.W. All authors have read and agreed to the published version of the manuscript.

Funding: This research was supported by the Key R&D Program of Shandong Province, China (2022CXGC020303).

Data Availability Statement: The original contributions presented in this study are included in the article. Further inquiries can be directed to the corresponding author.

Conflicts of Interest: Authors Youxiao Chen, Lianhao Liu and Lijie Chen are employed by the company CIMC Offshore Co., Ltd. The remaining authors declare that the research was conducted in the absence of any commercial or financial relationships that could be construed as a potential con-flict of interest.

Abbreviations

The following abbreviations are used in this manuscript:

CCUS	Carbon dioxide Capture, Utilization and Storage
OCCS	Onboard Carbon Capture Systems
LNG	Liquefied Natural Gas
TEU	Twenty-Foot Equivalent Unit
DWT	Deadweight Tonnage

References

1. Crippa, M.; Guizzardi, D.; Banja, M.; Solazzo, E.; Muntean, M.; Schaaf, E.; Pagani, F.; Monforti-Ferrario, F.; Olivier, J.; Quadrelli, R.; et al. CO_2 Emissions of All World Countries. JRC Publications Repository. Available online: https://publications.jrc.ec.europa.eu/repository/handle/JRC130363 (accessed on 28 April 2025).
2. Chen, S.; Miao, C.; Zhang, Q. Understanding the Evolution of China's Green Shipping Policies: Evidence by Social Network Analysis. *J. Clean. Prod.* **2024**, *482*, 144204. [CrossRef]
3. IEA. Global CO_2 Emissions from Energy Combustion and Industrial Processes and Their Annual Change, 1900–2023—Charts—Data & Statistics. Available online: https://www.iea.org/data-and-statistics/charts/global-CO2-emissions-from-energy-combustion-and-industrial-processes-and-their-annual-change-1900-2023 (accessed on 1 May 2025).
4. IEA. CO_2 Emissions in 2023—Analysis. Available online: https://www.iea.org/reports/CO2-emissions-in-2023 (accessed on 28 April 2025).
5. Luo, X.; Wang, M. Study of Solvent-Based Carbon Capture for Cargo Ships through Process Modelling and Simulation. *Appl. Energy* **2017**, *195*, 402–413. [CrossRef]

6. Fourth Greenhouse Gas Study 2020. Available online: https://www.imo.org/en/OurWork/Environment/Pages/Fourth-IMO-Greenhouse-Gas-Study-2020.aspx (accessed on 28 April 2025).
7. Wang, Y.; Cao, Q.; Liu, L.; Wu, Y.; Liu, H.; Gu, Z.; Zhu, C. A Review of Low and Zero Carbon Fuel Technologies: Achieving Ship Carbon Reduction Targets. *Sustain. Energy Technol. Assess.* **2022**, *54*, 102762. [CrossRef]
8. OECD. CO_2 Emissions from Global Shipping. Available online: https://doi.org/10.1787/bc2f7599-en (accessed on 1 May 2025).
9. Review of Maritime Transport 2024 I UN Trade and Development (UNCTAD). Available online: https://unctad.org/publication/review-maritime-transport-2024 (accessed on 1 May 2025).
10. Green Shipping: Innovations, Strategies, And Trends. Available online: https://www.gocomet.com/blog/what-is-green-shipping/ (accessed on 24 May 2025).
11. Green Ship: Green Technologies in Shipping Industry. Available online: https://marine-digital.com/article_green_ship (accessed on 24 May 2025).
12. IEA. CO_2 Emissions from Energy Combustion and Industrial Processes, 1900–2021—Charts—Data & Statistics. Available online: https://www.iea.org/data-and-statistics/charts/CO2-emissions-from-energy-combustion-and-industrial-processes-1900-2021 (accessed on 1 May 2025).
13. Joung, T.-H.; Kang, S.-G.; Lee, J.-K.; Ahn, J. The IMO Initial Strategy for Reducing Greenhouse Gas(GHG) Emissions, and Its Follow-up Actions towards 2050. *J. Int. Marit. Saf. Environ. Aff. Shipp.* **2020**, *4*, 1–7. [CrossRef]
14. DNV. Nordic Roadmap Conference Brings Together Partners to Accelerate Decarbonization Through Green Shipping Corridors. Available online: https://www.dnv.com/news/nordic-roadmap-conference-brings-together-partners-to-accelerate-decarbonization-through-green-shipping-corridors-235672/ (accessed on 24 May 2025).
15. Tiuttu, T.-A. Green Shipping: What It Is & Why Is It Important. SDG Monitor. Available online: https://www.sdgmonitor.co/post/green-shipping-what-it-is-why-is-it-important (accessed on 24 May 2025).
16. World Economic Forum. Green Infrastructure Can Protect Ports from Growing Environmental Risks. Here's How. Available online: https://www.weforum.org/stories/2022/08/ports-green-infrastructure-environmental-risks-climate/ (accessed on 24 May 2025).
17. Future Fuels Nordic—A Nordic Roadmap for Introduction of Zero Emission Fuels in Shipping. Available online: https://futurefuelsnordic.com/ (accessed on 24 May 2025).
18. Green Shipping Corridors. C40 Cities. Available online: https://www.c40.org/what-we-do/scaling-up-climate-action/ports-and-shipping/green-shipping-corridors/ (accessed on 24 May 2025).
19. Announcements at COP29—Green Shipping Challenge. Available online: https://greenshippingchallenge.org/announcements-at-cop29/ (accessed on 24 May 2025).
20. Hua, R.; Yin, J.; Wang, S.; Han, Y.; Wang, X. Speed Optimization for Maximizing the Ship's Economic Benefits Considering the Carbon Intensity Indicator (CII). *Ocean Eng.* **2024**, *293*, 116712. [CrossRef]
21. IMO Framework on Life Cycle GHG Intensity of Marine Fuels (LCA). Available online: https://www.imo.org/en/OurWork/Environment/Pages/Lifecycle-GHG---carbon-intensity-guidelines.aspx (accessed on 2 May 2025).
22. Bilgili, L.; Ölçer, A.I. IMO 2023 Strategy-Where Are We and What's next? *Mar. Policy* **2024**, *160*, 105953. [CrossRef]
23. Reducing Emissions from the Shipping Sector-European Commission. Available online: https://climate.ec.europa.eu/eu-action/transport/reducing-emissions-shipping-sector_en (accessed on 2 May 2025).
24. Xing, H.; Spence, S.; Chen, H. A Comprehensive Review on Countermeasures for CO_2 Emissions from Ships. *Renew. Sustain. Energy Rev.* **2020**, *134*, 110222. [CrossRef]
25. Melideo, D.; Desideri, U. The Use of Hydrogen as Alternative Fuel for Ship Propulsion: A Case Study of Full and Partial Retrofitting of Roll-on/Roll-off Vessels for Short Distance Routes. *Int. J. Hydrog. Energy* **2024**, *50*, 1045–1055. [CrossRef]
26. Ustolin, F.; Campari, A.; Taccani, R. An Extensive Review of Liquid Hydrogen in Transportation with Focus on the Maritime Sector. *J. Mar. Sci. Eng.* **2022**, *10*, 1222. [CrossRef]
27. Mao, X.; Ying, R.; Yuan, Y.; Li, F.; Shen, B. Simulation and Analysis of Hydrogen Leakage and Explosion Behaviors in Various Compartments on a Hydrogen Fuel Cell Ship. *Int. J. Hydrog. Energy* **2021**, *46*, 6857–6872. [CrossRef]
28. Schwarzkopf, D.A.; Petrik, R.; Hahn, J.; Ntziachristos, L.; Matthias, V.; Quante, M. Future Ship Emission Scenarios with a Focus on Ammonia Fuel. *Atmosphere* **2023**, *14*, 879. [CrossRef]
29. Fu, H.B.; Shang, G.F.; Lin, J.; Hu, Y.J.; Hu, Q.Q.; Guo, L.; Zhang, Y.C.; Chen, J.M. Fractional Iron Solubility of Aerosol Particles Enhanced by Biomass Burning and Ship Emission in Shanghai, East China. *Sci. Total Environ.* **2014**, *481*, 377–391. [CrossRef]
30. Feng, Y.; Zhu, H.; Dong, Z. Simultaneous and Global Optimizations of LNG-Fueled Hybrid Electric Ship for Substantial Fuel Cost, CO_2, and Methane Emission Reduction. *IEEE Trans. Transp. Electrif.* **2023**, *9*, 2282–2295. [CrossRef]
31. Hwang, S.; Jeong, B.; Jung, K.; Kim, M.; Zhou, P. Life Cycle Assessment of LNG Fueled Vessel in Domestic Services. *J. Mar. Sci. Eng.* **2019**, *7*, 359. [CrossRef]
32. Svanberg, M.; Ellis, J.; Lundgren, J.; Landälv, I. Renewable Methanol as a Fuel for the Shipping Industry. *Renew. Sustain. Energy Rev.* **2018**, *94*, 1217–1228. [CrossRef]

33. Ushakov, S.; Stenersen, D.; Einang, P.M. Methane Slip from Gas Fuelled Ships: A Comprehensive Summary Based on Measurement Data. *J. Mar. Sci. Technol.* **2019**, *24*, 1308–1325. [CrossRef]
34. Bicer, Y.; Dincer, I. Clean Fuel Options with Hydrogen for Sea Transportation: A Life Cycle Approach. *Int. J. Hydrog. Energy* **2018**, *43*, 1179–1193. [CrossRef]
35. Rehbein, M.C.; Meier, C.; Eilts, P.; Scholl, S. Mixtures of Ammonia and Organic Solvents as Alternative Fuel for Internal Combustion Engines. *Energy Fuels* **2019**, *33*, 10331–10342. [CrossRef]
36. Korberg, A.D.; Brynolf, S.; Grahn, M.; Skov, I.R. Techno-Economic Assessment of Advanced Fuels and Propulsion Systems in Future Fossil-Free Ships. *Renew. Sustain. Energy Rev.* **2021**, *142*, 110861. [CrossRef]
37. Prussi, M.; Yugo, M.; De, P.L.; Padella, M.; Edwards, R.; Lonza, L. *JEC Well-to-Tank Report v5*; JRC Publications Repository; Publications Office of the European Union: Luxembourg, 2020.
38. IPCC. *Global Warming of 1.5°C: IPCC Special Report on Impacts of Global Warming of 1.5°C above Pre-Industrial Levels in Context of Strengthening Response to Climate Change, Sustainable Development, and Efforts to Eradicate Poverty*, 1st ed.; Cambridge University Press: Cambridge, UK, 2022.
39. Calvin, K.; Dasgupta, D.; Krinner, G.; Mukherji, A.; Thorne, P.W.; Trisos, C.; Romero, J.; Aldunce, P.; Barrett, K.; Blanco, G.; et al. *IPCC, 2023: Climate Change 2023: Synthesis Report*; Intergovernmental Panel on Climate Change (IPCC): Geneva, Switzerland, 2023.
40. Sheng, X.; Lin, S.; Liang, W.; Xie, H.; Liu, M. Optimal Long-Term Planning of CCUS and Carbon Trading Mechanism in Offshore-Onshore Integrated Energy System. *Appl. Energy* **2025**, *379*, 124983. [CrossRef]
41. Wu, H.; Zhang, X.; Wu, Q.; Zhou, X.; Yue, S. Readiness and Challenges of Carbon Capture Technologies Based on the Shipping Industry. *Mar. Pollut. Bull.* **2024**, *207*, 116878. [CrossRef]
42. Wang, Z.; Lu, M.; Dong, S.; Tang, M.; Yan, X.; Li, K.; Yang, H. Study of Ship-Based Carbon Capture Optimization Considering Multiple Evaluation Factors and Main Engine Loads. *J. Clean. Prod.* **2024**, *478*, 143996. [CrossRef]
43. Tsimplis, M.; Noussia, K. The Use of Ships within a CCUS System: Regulation and Liability. *Resour. Conserv. Recycl.* **2022**, *181*, 106218. [CrossRef]
44. Zhao, T.; Li, R.; Zhang, Z.; Song, C. Current Status of Onboard Carbon Capture and Storage (OCCS) System: A Survey of Technical Assessment. *Carbon Capture Sci. Technol.* **2025**, *15*, 100402. [CrossRef]
45. Stec, M.; Tatarczuk, A.; Iluk, T.; Szul, M. Reducing the Energy Efficiency Design Index for Ships through a Post-Combustion Carbon Capture Process. *Int. J. Greenh. Gas Control* **2021**, *108*, 103333. [CrossRef]
46. Wang, H.; Zhou, P.; Wang, Z. Reviews on Current Carbon Emission Reduction Technologies and Projects and Their Feasibilities on Ships. *J. Marine. Sci. Appl.* **2017**, *16*, 129–136. [CrossRef]
47. Wang, H.; Zhou, P.; Wang, Z. Experimental and Numerical Analysis on Impacts of Significant Factors on Carbon Dioxide Absorption Efficiency in the Carbon Solidification Process. *Ocean Eng.* **2016**, *113*, 133–143. [CrossRef]
48. Fang, S.; Xu, Y.; Li, Z.; Ding, Z.; Liu, L.; Wang, H. Optimal Sizing of Shipboard Carbon Capture System for Maritime Greenhouse Emission Control. *IEEE Trans. Ind. Appl.* **2019**, *55*, 5543–5553. [CrossRef]
49. Zhang, Y.; Xu, H.; Xu, J.; Zhang, X. Advancements and Challenges in Chemical Absorption Technologies for Shipborne Carbon Capture Applications: Absorbent Development, Improvement of Absorption Towers, and System Integration. *J. Energy Chem.* **2025**, *106*, 880–910. [CrossRef]
50. Kotzampasakis, M. Maritime Emissions Trading in the EU: Systematic Literature Review and Policy Assessment. *Transp. Policy* **2025**, *165*, 28–41. [CrossRef]
51. Oh, J.; Anantharaman, R.; Zahid, U.; Lee, P.; Lim, Y. Process Design of Onboard Membrane Carbon Capture and Liquefaction Systems for LNG-Fueled Ships. *Sep. Purif. Technol.* **2022**, *282*, 120052. [CrossRef]
52. Buirma, M.; Vleugel, J.; Pruyn, J.; Doedée, V.; Schott, D. Ship-Based Carbon Capture and Storage: A Supply Chain Feasibility Study. *Energies* **2022**, *15*, 813. [CrossRef]
53. García-Mariaca, A.; Llera-Sastresa, E. Review on Carbon Capture in ICE Driven Transport. *Energies* **2021**, *14*, 6865. [CrossRef]
54. Awoyomi, A.; Patchigolla, K.; Anthony, E.J. Process and Economic Evaluation of an Onboard Capture System for LNG-Fueled CO_2 Carriers. *Ind. Eng. Chem. Res.* **2020**, *59*, 6951–6960. [CrossRef]
55. Feenstra, M.; Monteiro, J.; van den Akker, J.T.; Abu-Zahra, M.R.M.; Gilling, E.; Goetheer, E. Ship-Based Carbon Capture Onboard of Diesel or LNG-Fuelled Ships. *Int. J. Greenh. Gas Control* **2019**, *85*, 1–10. [CrossRef]
56. Zhou, P.; Wang, H. Carbon Capture and Storage—Solidification and Storage of Carbon Dioxide Captured on Ships. *Ocean Eng.* **2014**, *91*, 172–180. [CrossRef]
57. Güler, E.; Ergin, S. An Investigation on the Solvent Based Carbon Capture and Storage System by Process Modeling and Comparisons with Another Carbon Control Methods for Different Ships. *Int. J. Greenh. Gas Control* **2021**, *110*, 103438. [CrossRef]
58. Wang, H. Analyzing the Current State and Future Development of Carbon Capture, Utilization and Storage (CCUS) Strategies and Technologies in China. *J. Phys. Conf. Ser.* **2023**, *2608*, 012061. [CrossRef]
59. China Classification Society Releases 'Shipping Low-Carbon Development Outlook 2023'. Available online: https://www.ccs.org.cn/ccswz/articleDetail?id=202312081257422748 (accessed on 1 May 2025).

60. El-Kady, A.H.; Amin, M.T.; Khan, F.; El-Halwagi, M.M. Analysis of CO_2 Pipeline Regulations from a Safety Perspective for Offshore Carbon Capture, Utilization, and Storage (CCUS). *J. Clean. Prod.* **2024**, *439*, 140734. [CrossRef]
61. Vakili, S.; Manias, P.; Armstrong, L.-M.; Turnock, S.; Teagle, D.A.H. Technical, Economic, and Environmental Assessment of CO_2 Ship Transport in Carbon Capture and Storage. *J. Environ. Manag.* **2025**, *373*, 123919. [CrossRef]
62. Li, J. Accelerating the Offshore CCUS to Carbon-Neutral China. *Fundam. Res.* **2025**, *5*, 715–726. [CrossRef] [PubMed]
63. Al Baroudi, H.; Awoyomi, A.; Patchigolla, K.; Jonnalagadda, K.; Anthony, E.J. A Review of Large-Scale CO_2 Shipping and Marine Emissions Management for Carbon Capture, Utilisation and Storage. *Appl. Energy* **2021**, *287*, 116510. [CrossRef]
64. Vessel Characteristics: Ship CLIPPER ERIS (LPG Tanker) Registered in Norway. Available online: https://www.marinetraffic.com/en/ais/details/ships/shipid:5868406/mmsi:257083750/imo:9834727/vessel:CLIPPER_ERIS (accessed on 26 May 2025).
65. Vessel Characteristics: Ship STORM (Container Ship) Registered in Antigua Barbuda. Available online: https://www.marinetraffic.com/en/ais/details/ships/shipid:364287/mmsi:305350000/imo:9389435/vessel:STORM (accessed on 26 May 2025).
66. GCMD. PROJECT REMARCCABLE (Realising Maritime Carbon Capture to Demonstrate the Ability to Lower Emissions). Available online: https://gcformd.cldy.asia/projects/project-remarccable/ (accessed on 26 May 2025).
67. Lim, Y.K.; Lee, S.G.; Ko, M.; Park, K.; Lee, J.M. Operational Strategy of Pre-Cooling Process of CO_2 Storage Tank in CCS Ship Transportation Using Model-Based Optimization. *Chem. Eng. Res. Des.* **2016**, *109*, 770–779. [CrossRef]
68. Zhang, T.; Gao, H. Experimental investigation on performance of absorption refrigeration unit under ship swing condition. *J. Eng. Thermophys.* **2019**, *40*, 246–251.
69. Visonà, M.; Bezzo, F.; d'Amore, F. Techno-Economic Analysis of Onboard CO_2 Capture for Ultra-Large Container Ships. *Chem. Eng. J.* **2024**, *485*, 149982. [CrossRef]
70. Seo, Y.; Huh, C.; Lee, S.; Chang, D. Comparison of CO_2 Liquefaction Pressures for Ship-Based Carbon Capture and Storage (CCS) Chain. *Int. J. Greenh. Gas Control* **2016**, *52*, 1–12. [CrossRef]
71. Sarbanha, A.-A.; Larachi, F.; Taghavi, S.-M.; Thiboutot-Rioux, M.; Boudreau, A.; Dugas, G. Mitigation of Ship Emissions: Overview of Recent Trends. *Ind. Eng. Chem. Res.* **2023**, *62*, 1707–1724. [CrossRef]
72. Gan, L.; Che, W.; Zhou, M.; Zhou, C.; Zheng, Y.; Zhang, L.; Rangel-Buitrago, N.; Song, L. Ship Exhaust Emission Estimation and Analysis Using Automatic Identification System Data: The West Area of Shenzhen Port, China, as a Case Study. *Ocean Coast. Manag.* **2022**, *226*, 106245. [CrossRef]
73. Chen, Y.-F.; Guo, N.; Wu, D.-S.; Hu, R. Numerical Investigation on Immiscible Displacement in 3D Rough Fracture: Comparison with Experiments and the Role of Viscous and Capillary Forces. *Adv. Water Resour.* **2018**, *118*, 39–48. [CrossRef]
74. Vutukuru, S.; Dabdub, D. Modeling the Effects of Ship Emissions on Coastal Air Quality: A Case Study of Southern California. *Atmos. Environ.* **2008**, *42*, 3751–3764. [CrossRef]
75. Corbett, J.J.; Fischbeck, P.S.; Pandis, S.N. Global Nitrogen and Sulfur Inventories for Oceangoing Ships. *J. Geophys. Res. Atmos.* **1999**, *104*, 3457–3470. [CrossRef]
76. Eyring, V.; Köhler, H.W.; van Aardenne, J.; Lauer, A. Emissions from International Shipping: 1. The Last 50 Years. *J. Geophys. Res. Atmos.* **2005**, *110*, D17305. [CrossRef]
77. Lauer, A.; Eyring, V.; Hendricks, J.; Jöckel, P.; Lohmann, U. Global Model Simulations of the Impact of Ocean-Going Ships on Aerosols, Clouds, and the Radiation Budget. *Atmos. Chem. Phys.* **2007**, *7*, 5061–5079. [CrossRef]
78. Endresen, Ø.; Sørgård, E.; Behrens, H.L.; Brett, P.O.; Isaksen, I.S.A. A Historical Reconstruction of Ships' Fuel Consumption and Emissions. *J. Geophys. Res. Atmos.* **2007**, *112*, D12301. [CrossRef]
79. Endresen, Ø.; Sørgård, E.; Sundet, J.K.; Dalsøren, S.B.; Isaksen, I.S.A.; Berglen, T.F.; Gravir, G. Emission from International Sea Transportation and Environmental Impact. *J. Geophys. Res. Atmos.* **2003**, *108*, 4560. [CrossRef]
80. Corbett, J.J.; Koehler, H.W. Updated Emissions from Ocean Shipping. *J. Geophys. Res. Atmos.* **2003**, *108*, 4650. [CrossRef]
81. Qu, Z.; Kong, Z.; Yin, H.; Zhang, D.; Du, M. Study on carbon capture technology and development trend of exhaust gas from ocean-going ships. *J. Environ. Eng. Technol.* **2024**, *14*, 17–24.
82. Carbon Dioxide Capture and Storage—IPCC. Available online: https://www.ipcc.ch/report/carbon-dioxide-capture-and-storage/ (accessed on 3 May 2025).
83. Balcombe, P.; Brierley, J.; Lewis, C.; Skatvedt, L.; Speirs, J.; Hawkes, A.; Staffell, I. How to Decarbonise International Shipping: Options for Fuels, Technologies and Policies. *Energy Convers. Manag.* **2019**, *182*, 72–88. [CrossRef]
84. Miller, D.C.; Litynski, J.T.; Brickett, L.A.; Morreale, B.D. Toward Transformational Carbon Capture Systems. *AIChE J.* **2016**, *62*, 2–10. [CrossRef]
85. Rochelle, G.T. Amine Scrubbing for CO_2 Capture. *Science* **2009**, *325*, 1652–1654. [CrossRef]
86. Habibic, A. "World's First" Ship with Full-Scale Onboard CCS Ready for Pilot Testing. Offshore Energy. Available online: https://www.offshore-energy.biz/worlds-first-ship-with-full-scale-onboard-ccs-ready-for-pilot-testing/ (accessed on 26 May 2025).
87. Favre, E. Membrane Processes and Postcombustion Carbon Dioxide Capture: Challenges and Prospects. *Chem. Eng. J.* **2011**, *171*, 782–793. [CrossRef]

88. Scholes, C.A.; Stevens, G.W.; Kentish, S.E. Membrane Gas Separation Applications in Natural Gas Processing. *Fuel* **2012**, *96*, 15–28. [CrossRef]
89. Song, C.; Liu, Q.; Deng, S.; Li, H.; Kitamura, Y. Cryogenic-Based CO_2 Capture Technologies: State-of-the-Art Developments and Current Challenges. *Renew. Sustain. Energy Rev.* **2019**, *101*, 265–278. [CrossRef]
90. Tuinier, M.J.; van Sint Annaland, M.; Kuipers, J.A.M. A Novel Process for Cryogenic CO_2 Capture Using Dynamically Operated Packed Beds—An Experimental and Numerical Study. *Int. J. Greenh. Gas Control* **2011**, *5*, 694–701. [CrossRef]
91. Jankowski, A.; Grabiec, E.; Nocoń-Szmajda, K.; Marcinkowski, A.; Janeczek, H.; Wolińska-Grabczyk, A. Polyimide-Based Membrane Materials for CO_2 Separation: A Comparison of Segmented and Aromatic (Co)Polyimides. *Membranes* **2021**, *11*, 274. [CrossRef] [PubMed]
92. Mehrabi, M.; Vatanpour, V. Polyimide-Based Separation Membranes for Liquid Separation: A Review on Fabrication Techniques, Applications, and Future Perspectives. *Mater. Today Chem.* **2024**, *35*, 101895. [CrossRef]
93. CO_2 Removal from Natural Gas & Hydrocarbon Recovery—MTRINC. MTR Industrial Separations. Available online: https://www.mtrinc.com/natural-gas/CO2-removal-from-natural-gas/ (accessed on 1 June 2025).
94. Donald Carruthers, J.; Petruska, M.A.; Sturm, E.A.; Wilson, S.M. Molecular Sieve Carbons for CO_2 Capture. *Microporous Mesoporous Mater.* **2012**, *154*, 62–67. [CrossRef]
95. Demir, H.; Aksu, G.O.; Gulbalkan, H.C.; Keskin, S. MOF Membranes for CO_2 Capture: Past, Present and Future. *Carbon Capture Sci. Technol.* **2022**, *2*, 100026. [CrossRef]
96. Ros, J.A.; Skylogianni, E.; Doedée, V.; van den Akker, J.T.; Vredeveldt, A.W.; Linders, M.J.; Goetheer, E.L.; Monteiro, J.G.M. Advancements in Ship-Based Carbon Capture Technology on Board of LNG-Fuelled Ships. *Int. J. Greenh. Gas Control* **2022**, *114*, 103575. [CrossRef]
97. Dong, C.-Y.; Cho, K.-H. An Optimally Evolved Connective Ratio of Neural Networks That Maximizes the Occurrence of Synchronized Bursting Behavior. *BMC Syst Biol* **2012**, *6*, 23. [CrossRef]
98. Ghorbani, B.; Miansari, M.; Zendehboudi, S.; Hamedi, M.-H. Exergetic and Economic Evaluation of Carbon Dioxide Liquefaction Process in a Hybridized System of Water Desalination, Power Generation, and Liquefied Natural Gas Regasification. *Energy Convers. Manag.* **2020**, *205*, 112374. [CrossRef]
99. Wang, J.; Li, L.; Qu, G.; Cheng, M.; Ren, Y.; Lu, P.; Zhang, T.; Wang, Y.; Wu, F. Advanced Materials for CO_2 Capture and Storage: A Comprehensive Review of Current Progress and Future Prospects. *J. Environ. Chem. Eng.* **2025**, *13*, 115881. [CrossRef]
100. Materials of CO_2 Storage Tanks—Jianshen Tank. Available online: https://www.jianshentank.com/storagetank/materials-of-co2-storage-tanks.html (accessed on 6 June 2025).
101. A Complete Guide to Tank Materials (Updated for 2020). GSC Tanks. Available online: https://www.gsctanks.com/tank-materials/ (accessed on 6 June 2025).
102. IMO | Global Integrated Shipping Information System (GISIS). Available online: https://gisis.imo.org/Public/Default.aspx (accessed on 6 June 2025).
103. Shipping Intelligence Network. Available online: https://sin.clarksons.net/ (accessed on 6 June 2025).
104. Tian, Z.; Zhou, Y.; Zhang, Y.; Gao, W. Design Principle, 4E Analyses and Optimization for Onboard CCS System under EEDI Framework: A Case Study of an LNG-Fueled Bulk Carrier. *Energy* **2024**, *295*, 130985. [CrossRef]
105. Sinotech. Ecochlor. Available online: https://ecochlor.com/sinotech-ccs/ (accessed on 6 June 2025).
106. Welcome to REFPROP's Documentation—REFPROP 10.0 Documentation. Available online: https://refprop-docs.readthedocs.io/en/latest/ (accessed on 6 June 2025).
107. Storrs, K.; Lyhne, I.; Drustrup, R. A Comprehensive Framework for Feasibility of CCUS Deployment: A Meta-Review of Literature on Factors Impacting CCUS Deployment. *Int. J. Greenh. Gas Control* **2023**, *125*, 103878. [CrossRef]
108. K STORM, Container Ship, IMO 9389435. Available online: https://www.balticshipping.com/vessel/imo/9389435 (accessed on 25 May 2025).
109. Trædal, S.; Stang, J.H.G.; Snustad, I.; Johansson, M.V.; Berstad, D. CO_2 Liquefaction Close to the Triple Point Pressure. *Energies* **2021**, *14*, 8220. [CrossRef]
110. GCMD. Our Publications. Available online: https://gcformd.org/our-publications/ (accessed on 1 June 2025).
111. Value Maritime's OCCS Achieves First LR Class Notation on EPS Tanker. Available online: https://bunkermarket.com/value-maritimes-occs-achieves-first-lr-class-notation-on-eps-tanker/ (accessed on 25 May 2025).
112. North Sea Port Reports Modest Growth in 2024 World Cargo News. Available online: https://www.worldcargonews.com/ports-terminals/2025/01/north-sea-port-reports-modest-growth-in-2024/ (accessed on 26 May 2025).
113. Ros, M.; Read, A.; Uilenreef, J.; Limbeek, J. Start of a CO_2 Hub in Rotterdam: Connecting CCS and CCU. *Energy Procedia* **2014**, *63*, 2691–2701. [CrossRef]
114. WorldCargo News. Evergreen Vessel Captures CO_2 at Sea. Available online: https://www.worldcargonews.com/news/2024/10/evergreens-container-ship-captures-CO2-at-sea/ (accessed on 26 May 2025).

115. OGCI | Project REMARCCABLE | Report. Available online: https://www.ogci.com/project-remarccable (accessed on 30 May 2025).
116. Report: VLCC Best Business Case for Onboard Carbon Capture | News | Motorship. Available online: https://www.motorship.com/carbon-capture/report-vlcc-best-business-case-for-onboard-carbon-capture/1476926.article (accessed on 28 May 2025).
117. Onboard Carbon Capture and Storage: Pioneering Solution for Ships. Available online: https://www.wartsila.com/marine/products/exhaust-treatment/carbon-capture-and-storage (accessed on 28 May 2025).
118. Case Study: Onboard CCS Would Result in 10% Fuel Penalty. Available online: https://www.marinelink.com/news/case-study-onboard-ccs-result-fuel-517971 (accessed on 30 May 2025).
119. Tanker Operator Final Report Released of Onboard Carbon Capture Pilot on mr Tanker Stena Impero. Available online: https://www.tankeroperator.com/news/final-report-released-of-onboard-carbon-capture-pilot-on-mr-tanker-stena-impero/14484.aspx (accessed on 30 May 2025).
120. Wärtsilä Completes Commercial Launch of Onboard Carbon Capture Solution. Available online: https://article.wn.com/view/2025/05/08/W_rtsil_Completes_Commercial_Launch_of_Onboard_Carbon_Captur/ (accessed on 30 May 2025).
121. What Are the Key Considerations for Onboard Carbon Capture? | Hellenic Shipping News Worldwide. Available online: https://www.hellenicshippingnews.com/what-are-the-key-considerations-for-onboard-carbon-capture/ (accessed on 30 May 2025).
122. Kok, C. Engineering Study Charts Potential of Carbon Capture Technology to Help Decarbonize Shipping. GCMD. Available online: https://gcformd.org/engineering-study-charts-potential-of-carbon-capture-technology-to-help-decarbonize-shipping/ (accessed on 28 May 2025).
123. Maritime Decarbonization Strategy 2022. Maersk Zero Carbon Shipping. Available online: https://www.zerocarbonshipping.com/publications/maritime-decarbonization-strategy (accessed on 28 May 2025).
124. Mandra, J.O. Crowley to Trial Carbon Ridge's Onboard Carbon Capture Solution. Offshore Energy. Available online: https://www.offshore-energy.biz/crowley-to-trial-carbon-ridges-onboard-carbon-capture-solution/ (accessed on 30 May 2025).
125. Desai, C. Eastern Pacific Shipping. Available online: https://www.epshipping.com.sg/eps-and-value-maritime-install-carbon-capture-solution-on-mr-tanker/ (accessed on 30 May 2025).
126. Marine Environment Protection Committee (MEPC 83), 7 to 11 April 2025. Available online: https://www.imo.org/en/MediaCentre/MeetingSummaries/Pages/MEPC-83rd-session.aspx (accessed on 30 May 2025).
127. IMO Data Collection System (DCS). Available online: https://www.imo.org/en/OurWork/Environment/Pages/Data-Collection-System.aspx (accessed on 1 June 2025).
128. DNV. Carbon Intensity Indicator (CII): INSIGHTS & Support. Available online: https://www.dnv.com/maritime/insights/topics/CII-carbon-intensity-indicator/ (accessed on 1 June 2025).
129. 2023 IMO Strategy on Reduction of GHG Emissions from Ships. Available online: https://www.imo.org/en/OurWork/Environment/Pages/2023-IMO-Strategy-on-Reduction-of-GHG-Emissions-from-Ships.aspx (accessed on 28 April 2025).
130. EU Emissions Trading System (EU ETS)—European Commission. Available online: https://climate.ec.europa.eu/eu-action/eu-emissions-trading-system-eu-ets_en (accessed on 30 May 2025).
131. First LR Class Notation for Onboard Carbon Capture System Assigned to Eastern Pacific Shipping Tanker | LR. Available online: https://www.lr.org/en/knowledge/press-room/press-listing/press-release/2024/first-lr-class-notation-for-onboard-carbon-capture-system-assigned-to-eastern-pacific-shipping-tanker/ (accessed on 30 May 2025).
132. London Convention (LDC,LC) and London Protocol (LP). Available online: https://www.imo.org/en/KnowledgeCentre/IndexofIMOResolutions/Pages/LDC-LC-LP.aspx (accessed on 30 May 2025).

Journal of
Marine Science and Engineering

MDPI

Article

Optimal Alternative Fuel Selection for Dual-Fuel Ships Under FuelEU Maritime Regulations: Environmental and Economic Assessment

Cong Wang [1], Zhongxiu Peng [1,*], Jianming Yang [1], Niyu Zhang [2,3], Ke Li [4] and Xuesong Li [2,3]

[1] School of Maritime Economics and Management, Dalian Maritime University, Dalian 116026, China; wangcong2017@dlmu.edu.cn (C.W.); yjm0491@dlmu.edu.cn (J.Y.)
[2] CAS Key Laboratory of Low-Carbon Conversion Science and Engineering, Shanghai Advanced Research Institute, Chinese Academy of Sciences, Shanghai 201210, China; zhangniyu2024@sari.ac.cn (N.Z.); lixuesong@sari.ac.cn (X.L.)
[3] University of Chinese Academy of Sciences, Beijing 100049, China
[4] Shanghai Marine Diesel Engine Research Institute, Shanghai 201108, China; like@smderi.com.cn
* Correspondence: pzxkadima@163.com

Abstract: To address greenhouse gas (GHG) emissions from the maritime sector, the European Union (EU) has introduced the FuelEU Maritime regulation to incentivize ships to adopt diversified compliance pathways and energy solutions. This study aims to determine the optimal alternative fuel configurations for dual-fuel ships of different types under environmental, economic, and regulatory constraints. An integrated environmental and cost assessment model from a well-to-wake (WtW) perspective to systematically evaluate the environmental benefits and economic feasibility of fossil-based, bio-based, and renewable electricity-based alternative fuels applied in dual-fuel ships. By incorporating the PROMETHEE II method within a multi-criteria decision analysis (MCDA) framework, together with the CRITIC objective weighting method, the study enables a robust ranking of alternative fuel configurations across three key dimensions: environmental performance, cost feasibility, and regulatory compliance. The results indicate that, regardless of ship type, the very low sulfur fuel oil (VLSFO) + marine gas oil (MGO) and VLSFO + methanol (MEOH) combinations fail to meet the GHG intensity targets for 2025–2050. Only the VLSFO + electrolytic liquid hydrogen (E-LH_2) and VLSFO + electrolytic ammonia (E-NH_3) configurations are compliant. Although e-fuels incur the highest annual costs, the EU compliance penalty associated with fossil fuels increases exponentially. In contrast, e-fuels retain long-term cost advantages, ultimately driving a sector-wide transition toward e-fuel-dominated energy structures by 2050. Their superior environmental performance and regulatory compatibility emerge as the core drivers of the maritime energy transition.

Keywords: FuelEU Maritime; dual-fuel ships; alternative fuel selection; environmental assessment; economic assessment

1. Introduction

Greenhouse gas (GHG) emissions from the maritime sector account for nearly 3% of global total emissions, making it a critical issue in the context of international climate governance [1,2]. In response, the International Maritime Organization (IMO) released the "2023 IMO Strategy on Reduction of GHG Emissions from Ships", aiming to regulate GHG emissions from shipping through an increasingly complex and dynamically evolving regulatory framework [3,4]. To align with the IMO's strategic objectives, regional bodies such as the European Union (EU) have introduced complementary regulatory initiatives. The EU has legally committed to achieving climate neutrality by 2050 and, in July 2021, launched the "Fit for 55" legislative package to support this target [5,6]. As part of this package, the maritime sector has been incorporated into the EU Emissions Trading System (ETS) starting in 2024. Furthermore, the FuelEU Maritime regulation is scheduled to take effect on 1 January 2025, with the aim of accelerating the decarbonization of maritime transport [4].

The FuelEU Maritime regulation mandates that all commercial ships with a gross tonnage above 5000—regardless of flag—engaged in the transport of passengers or cargo must progressively reduce their GHG intensity when calling at EU ports. The regulation sets a clearly defined reduction trajectory: a 2% decrease starting in 2025, reaching an 80% reduction target by 2050. FuelEU Maritime adopts a technology-neutral and performance-based compliance framework [7]. It utilizes a well-to-wake (WtW) life-cycle approach to account for emissions of carbon dioxide (CO_2), methane (CH_4), and nitrous oxide (N_2O), thereby enabling ships to comply through diverse energy system configurations and operational profiles. The scope of application varies by voyage type: for intra-EU voyages between ports of EU Member States, the regulation applies to the total energy used throughout the voyage; for voyages involving outermost EU regions or between EU and non-EU ports, it applies to 50% of the energy used during the journey [8].

With the implementation of the FuelEU Maritime regulation, ships are under dual pressure to improve energy efficiency and reduce emissions, positioning the adoption of alternative fuels as a key strategy to meet compliance challenges. Complementing alternative fuels is the deployment of innovative marine propulsion systems, among which dual-fuel engines have gained particular attention for their enhanced flexibility and adaptability to a variety of fuel types [9]. In the context of marine alternative fuels, liquefied natural gas (LNG) exhibits lower carbon intensity compared to conventional fossil fuels and is regarded as a promising short-term option. However, fossil-based LNG is unlikely to meet the stringent requirements of deep decarbonization in maritime transport [10,11]. In contrast, bio-LNG derived from biomass offers renewable and environmentally friendly advantages, making it a greener alternative [10]. Methanol is another widely recognized low-carbon fuel, notable for its ease of storage and transportation, and can be produced from both natural gas and biomass sources [11]. Non-fossil-based methanol is considered a strong medium- to long-term candidate for maritime decarbonization [12]. Hydrotreated vegetable oil (HVO), currently the most compatible direct substitute for conventional fuels, remains costly and is still under development [11]. Hydrogen-based e-fuels, such as electrolytic liquid hydrogen (E-LH_2) and ammonia (E-NH_3) from renewable electricity, are regarded as long-term solutions due to their zero-carbon characteristics [13]. Nevertheless, their large-scale application in the shipping industry faces complex technological, logistical, and infrastructural challenges. The production of e-fuels is heavily reliant on high-cost water electrolysis technologies. For liquid hydrogen, large-scale storage and transport are hindered by high energy consumption for liquefaction and boil-off losses. In the case of liquid ammonia, the development of efficient catalysts and the minimization of energy losses during cracking remain critical obstacles [14,15]. High infrastructure costs and the need for large-scale deployment are shared challenges for both E-LH_2 and E-NH_3 [14,16]. Moreover, one of the most significant barriers to widespread adoption is the high production cost of e-fuels, which is substantially greater than that of conventional fuels [11].

A substantial body of research has emerged evaluating the environmental and economic implications of marine alternative fuels. Wang et al. conducted a comprehensive analysis of the properties of LNG, hydrogen, methanol, ammonia, and biofuels, and developed a life-cycle assessment (LCA) framework for marine applications. They reviewed the existing LCA literature with a focus on environmental impacts and economic viability [10]. Biu et al. proposed a life-cycle cost (LCC) framework to evaluate the full LCA performance of innovative marine dual-fuel engines. Their findings indicate that, under baseline scenarios, dual-fuel engines outperform conventional diesel engines in terms of cost-effectiveness [17]. Lee et al. further conducted an LCA of nine alternative fuel production pathways and integrated carbon pricing analysis to assess their economic performance. The results identified biomass-based Fischer–Tropsch diesel (FT-diesel), e-methanol, and E-NH_3 as the most environmentally favorable options [11]. In a comparative study of 22 alternative fuel pathways, Law et al. highlighted that biofuels offer a balanced trade-off in terms of cost, availability, and technological readiness level (TRL), although challenges remain in mitigating non-GHG pollutants. In contrast, hydrogen and ammonia demonstrated the highest energy consumption and economic cost burdens [18]. These studies have laid a solid foundation for the environmental and economic assessment of alternative marine fuels, establishing relatively systematic LCA and LCC frameworks. However, most of the existing literature treats fuels as the primary object of analysis, lacking an in-depth exploration of ship–fuel compatibility and long-term policy adaptability.

Several scholars have assessed the competitiveness of marine fuels within the context of the European Union's regulatory framework. Solakivi et al. evaluated the cost implications and long-term feasibility of low-carbon and carbon-neutral marine fuels under the EU's "Fit for 55" legislative package. Their findings suggest that the most cost-effective regulatory compliance pathway involves a gradual transition from fossil fuels to biofuels, ultimately shifting toward e-fuels [19]. Flodén et al. focused on the emission reduction potential of different ship types within the EU ETS, analyzing associated mitigation measures, modal shifts, and cost variations [20]. Vierth et al. estimated the impact of various EU ETS expansion scenarios and revisions to the Energy Taxation Directive (ETD) on fuel costs for cargo ships calling at Swedish ports, simulating how increased fuel costs would affect transport and logistics costs, port choice, transport modes, and ship selection [2]. To explore the potential decarbonization pathways for various ship types in Sweden's maritime sector, Trosvik and Brynolf developed a scenario-based modeling tool to assess the effects of EU ETS, FuelEU Maritime, and related policy instruments [21]. Within the context of EU climate policy, numerous studies have conducted scenario-based simulations focusing on carbon pricing mechanisms and regulations such as FuelEU Maritime, providing quantitative support for evaluating policy impacts. Nevertheless, these analyses tend to emphasize fuel pathways or macro-level trends in the maritime sector, while overlooking the heterogeneity in responses across ship types and technical configurations. Moreover, they often fall short in identifying actionable, ship-specific optimal configurations.

Beyond the single-dimensional assessment of environmental and economic feasibility, scholars have increasingly advocated for the incorporation of multi-criteria decision analysis (MCDA) into fuel pathway evaluation, emphasizing the need for integrated consideration of environmental performance, economic viability, and technological maturity [22]. Ren and Liang developed a comprehensive sustainability evaluation framework for marine alternative fuels, proposing a fuzzy group MCDA model to assess methanol, LNG, and hydrogen as candidate fuels [23]. Soltani Motlagh et al. applied a hybrid AHP–TOPSIS approach to analyze shipowners' and stakeholders' preferences regarding various emission reduction options [24]. Moshiul et al. introduced TOPSIS as a decision-support tool to holistically evaluate the performance of alternative fuels across environmental, technological, and economic dimensions [25]. Strantzali et al. proposed an MCDA framework based on the PROMETHEE II ranking method, incorporating economic, technological, environmental, and social criteria to facilitate comprehensive evaluation and decision-making for marine alternative fuel adoption [26]. The above-mentioned studies have incorporated MCDA methods into the evaluation of marine fuels, achieving notable progress in balancing environmental, economic, and technological dimensions. However, most existing methods rely heavily on subjective expert weighting and lack a systematic integration and quantitative analysis of the regulatory compliance dimension, which limits their applicability in policy-oriented evaluations.

Despite the growing body of research on alternative marine fuels, existing studies have yet to establish a systematic evaluation framework aligned with the tightening regulatory landscape of the EU. In particular, the current literature lacks comprehensive LCA that account for both the environmental impacts and economic performance of alternative fuels across diverse ship types. Moreover, few studies provide a robust basis for identifying optimal fuel-ship configurations to support practical decision-making. Although works such as [2,19] have examined cost implications under EU carbon policy scenarios, their analyses are limited to single-criterion assessments and omit MCDA approaches. Similarly, while [20,21] consider multiple ship categories, they do not adopt a full WtW life-cycle assessment, thereby limiting their ability to evaluate the compatibility between fuel options and ship operational profiles.

Addressing these gaps, this study makes two key contributions: (1) It develops an integrated WtW life-cycle assessment framework that incorporates both environmental and economic dimensions, specifically tailored to evaluate dual-fuel propulsion systems under the constraints of the FuelEU Maritime regulation. Unlike previous studies that often assess fuel properties in isolation from ship-specific characteristics, this research systematically analyzes the life-cycle GHG and cost feasibility of fossil-based, bio-based, and renewable electricity-based fuels across various ship types. This approach enhances the practical relevance and contextual applicability of the evaluation results; (2) Building upon detailed environmental and economic modeling, this study introduces an integrated MCDA approach combining the CRITIC objective weighting method and the PROMETHEE II ranking method. This robust evaluation framework accounts for environmental performance, cost feasibility, and regulatory compliance, thereby improving the model's adaptability to tightening carbon intensity targets under EU regulations and strengthening its support for evidence-based decision-making in maritime energy transitions.

2. Method and Data

2.1. Method

This study focuses on dual-fuel ships equipped with alternative fuels. First, an environmental impact assessment model and an annual total cost estimation model are developed from a WtW perspective. Second, based on the annual GHG intensity limits stipulated in the FuelEU Maritime regulation, the compliance surplus and potential penalty costs of each fuel option are calculated. Third, three core quantitative indicators—WtW GHG intensity, annual total cost, and compliance surplus—are selected. The CRITIC objective weighting method is employed to determine the relative importance of each criterion, and the PROMETHEE II MCDA method is used to assess the adaptability of various fuel alternatives across different ship types, ultimately identifying the optimal fuel choice. Figure 1 illustrates the analytical framework of this study, which aims to provide a comprehensive WtW assessment of alternative fuels—encompassing environmental impact, annual economic performance, and regulatory penalty costs—while supporting fuel selection decisions tailored to specific ship types.

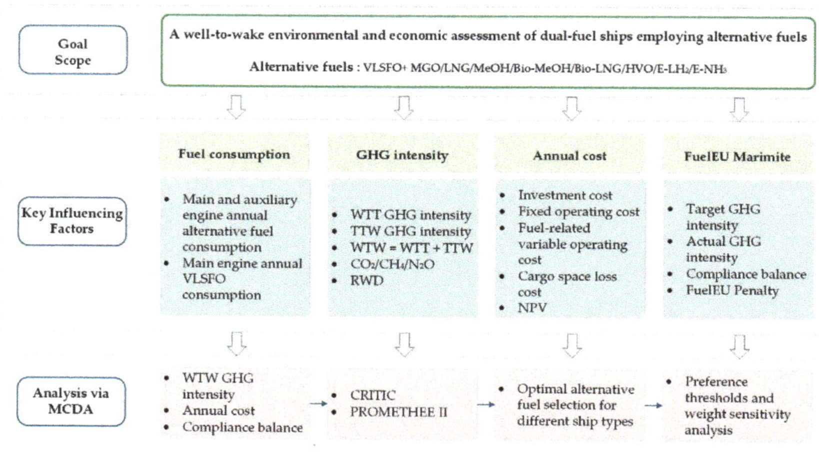

Figure 1. Analytical framework.

2.1.1. Calculation of Fuel Consumption for Dual-Fuel Ships

This study comprehensively considers the operational status of ships under two distinct conditions: at-sea navigation and port berthing. Fuel consumption is estimated based on specific fuel consumption rates. During navigation, the main engine load factor is estimated using the propeller law, as adopted in [27], which expresses the load as the cube of the ratio between average operational speed and design speed, thereby reflecting the load characteristics of different ship types under real-world conditions. Given the weak correlation between auxiliary engine load and ship speed, a fixed load factor of 50% is uniformly applied, in accordance with [27,28]. The fuel usage mode for both main and auxiliary engines also follow the approach in [27]. Specifically, dual-fuel ships are assumed to operate with 95% alternative fuel and 5% very low sulfur fuel oil (VLSFO) as ignition fuel for the main engine, while the auxiliary engines are assumed to use alternative fuels exclusively throughout the voyage, with marine gas oil (MGO) included in the alternative fuel category. Notably, ref. [27,28] also provide a substantial number of model input parameters, which serve as key data sources supporting the computational setup of this study.

Auxiliary engine fuel consumption is initially calculated based on the use of MGO as the reference fuel. When other alternative fuels are employed, the equivalent energy consumption is determined through conversion based on their respective lower calorific values (LCV), ensuring energy-equivalent substitution. According to reference [27], the hourly fuel consumption of the main engine (F_{ME}) and the auxiliary engine (F_{AE}) is calculated as follows:

$$F_{ME} = \left(\frac{v_{ave}}{v_{design}}\right)^3 \times P_{ME}^{design} \times SFOC_{ME} \times 10^{-6} \quad (1)$$

$$F_{AE} = LF_{AE} \times P_{AE}^{design} \times SFOC_{AE} \times 10^{-6} \quad (2)$$

where v_{ave} is the average sailing speed of the ship (kn); v_{design} is the ship's design speed (kn); P_{ME}^{design} is the design power of the main engine (kW); $SFOC_{ME}$ is the specific fuel consumption of the main engine, set at 206 g/kWh [28]; LF_{AE} is the auxiliary engine load factor, set to be 0.5 [27,28]; P_{AE}^{design} is the design power of the auxiliary engine (kW); and $SFOC_{AE}$ is the specific fuel consumption of the auxiliary engine, set at 221 g/kWh [28]. The annual consumption of alternative fuel for the main engine (F_{ME}^{ALT}), the annual consumption of VLSFO for pilot fuel (F_{ME}^{VLSFO}), and the annual consumption of alternative fuel for the auxiliary engine (F_{AE}^{ALT}) are calculated as follows:

$$F_{ME}^{ALT} = \frac{95\% \times F_{ME} \times LCV^{VLSFO}}{LCV^{ALT}} \times T^{sail} \tag{3}$$

$$F_{ME}^{VLSFO} = 5\% \times F_{ME} \times T^{sail} \tag{4}$$

$$F_{AE}^{ALT} = \frac{F_{AE} \times LCV^{MGO}}{LCV^{ALT}} \times \left(T^{berth} + T^{sail}\right) \tag{5}$$

where LCV^{VLSFO} is the lower calorific value of VLSFO (MJ/kg), LCV^{ALT} is the lower calorific value of the alternative fuel (MJ/kg), and LCV^{MGO} is the lower calorific value of MGO (MJ/kg). T^{sail} and T^{berth} represent the ship's annual sailing time and berthing time, respectively, measured in hours. The annual total consumption of alternative fuel (F_{total}^{ALT}) and the annual total consumption of VLSFO (F_{total}^{VLSFO}) are given by

$$F_{total}^{ALT} = F_{ME}^{ALT} + F_{AE}^{ALT} \tag{6}$$

$$F_{total}^{VLSFO} = F_{ME}^{VLSFO} \tag{7}$$

2.1.2. Well-to-Wake GHG Intensity Calculation Model

This study calculates the annual average GHG emission intensity of dual-fuel ships based on the full life-cycle fuel pathway, encompassing both the well-to-tank (WtT) and tank-to-wake (TtW) stages. The assessment includes three major GHGs: CO_2, CH_4, and N_2O. Excluding the use of shore power during port stays, this study adaptively modifies the relevant formulas provided in ANNEX I of the FuelEU Maritime regulation, as referenced in [29], to better reflect the operational conditions of dual-fuel ships. Specifically, the calculation of TtW emission intensity ($gCO_2eq/gFuel$) consists of the following two components. Combustion emissions: These refer to the GHG emissions resulting from the combustion of alternative fuels in both main and auxiliary engines, as well as the VLSFO ignition fuel used in the main engine. These emissions are converted into carbon dioxide equivalents (CO_2eq). Unburned emissions: These include methane and nitrous oxide emissions arising from the use of alternative fuels in main and auxiliary engines due to incomplete combustion or leakage. These emissions are also converted into CO_2eq. The corresponding formulas, along with definitions of each variable and computational details, follow the methodology described in [29].

$$CO_{2eqTtW_{ALT,ME}} = \left(C_{fCO_2,ALT,ME} \times GWP_{CO_2}\right) + \left(C_{fCH_4,ALT,ME} \times GWP_{CH_4}\right) + \left(C_{fN_2O,ALT,ME} \times GWP_{N_2O}\right) \tag{8}$$

$$CO_{2eqTtW_{ALT,AE}} = \left(C_{fCO_2,ALT,AE} \times GWP_{CO_2}\right) + \left(C_{fCH_4,ALT,AE} \times GWP_{CH_4}\right) + \left(C_{fN_2O,ALT,AE} \times GWP_{N_2O}\right) \tag{9}$$

$$CO_{2eqTtW_{VLSFO,ME}} = \left(C_{fCO_2,VLSFO,ME} \times GWP_{CO_2}\right) + \left(C_{fCH_4,VLSFO,ME} \times GWP_{CH_4}\right) + \left(C_{fN_2O,VLSFO,ME} \times GWP_{N_2O}\right) \tag{10}$$

$$CO_{2eqTtW_{slip,ALT,ME}} = \left(C_{sfCO_2,ALT,ME} \times GWP_{CO_2}\right) + \left(C_{sfCH_4,ALT,ME} \times GWP_{CH_4}\right) + \left(C_{sfN_2O,ALT,ME} \times GWP_{N_2O}\right) \tag{11}$$

$$CO_{2eqTtW_{slip,ALT,AE}} = \left(C_{sfCO_2,ALT,AE} \times GWP_{CO_2}\right) + \left(C_{sfCH_4,ALT,AE} \times GWP_{CH_4}\right) + \left(C_{sfN_2O,ALT,AE} \times GWP_{N_2O}\right) \quad (12)$$

where $C_{fCO_2,ALT,ME}$, $C_{fCH_4,ALT,ME}$, $C_{fN_2O,ALT,ME}$, $C_{fCO_2,ALT,AE}$, $C_{fCH_4,ALT,AE}$, $C_{fN_2O,ALT,AE}$, $C_{fCO_2,VLSFO,ME}$, $C_{fCH_4,VLSFO,ME}$, and $C_{fN_2O,VLSFO,ME}$ denote the emission factors (gGHG/gFuel) of CO_2, CH_4, and N_2O released during the combustion of alternative fuels in the main engine and auxiliary engine and of VLSFO in the main engine, respectively. Similarly $C_{sfCO_2,ALT,ME}$, $C_{sfCH_4,ALT,ME}$, $C_{sfN_2O,ALT,ME}$, $C_{sfCO_2,ALT,AE}$, $C_{sfCH_4,ALT,AE}$, and $C_{sfN_2O,ALT,AE}$ related terms represent the emission factors (gGHG/gFuel) of unburned methane and nitrous oxide leakage associated with alternative fuels in the main and auxiliary engines, with $C_{sfCO_2,ALT,ME} = C_{sfCO_2,ALT,AE} = C_{sfN_2O,,ALT,ME} = C_{sfN_2O,ALT,AE} = 0$, $C_{sfCH_4,ALT,ME} = C_{sfCH_4,ALT,AE} = 1$. No unburned emissions occur during the combustion of VLSFO in the main engine.

The 100-year global warming potential (GWP) values for CO_2, CH_4, and N_2O are denoted by GWP_{CO_2}, GWP_{CH_4}, and GWP_{N_2O}, and are set at 1, 28, and 265, respectively [30].

Equations (13)–(15) are used to calculate the annual average GHG emission intensity (gCO_2eq/MJ) for dual-fuel ships. This calculation encompasses both the upstream GHG emission intensity—arising from the fuel's extraction, production, transportation, and bunkering processes—and the direct GHG emissions resulting from fuel combustion in ship engines and potential leakage. The final value of $GHGIE_{actual}$ is obtained by summing these two components.

$$WTT = \frac{F_{total}^{ALT} \times CO_{2eqWtT_{ALT}} \times LCV^{ALT} + F_{total}^{VLSFO} \times CO_{2eqWtT_{VLSFO}} \times LCV^{VLSFO}}{F_{total}^{ALT} \times LCV^{ALT} + F_{total}^{VLSFO} \times LCV^{VLSFO}} \quad (13)$$

$$TTW = \frac{\begin{array}{l} F_{ME}^{ALT} \times \left[\left(1 - \frac{1}{100}c_{slipME}\right) \times CO_{2eqTtW_{ALT,ME}} + \frac{1}{100}c_{slipME} \times CO_{2eqTtW_{slip,ALT,ME}}\right] \times 10^3 \\ +F_{ME}^{VLSFO} \times CO_{2eqTtW_{VLSFO,ME}} \times 10^3 \\ +F_{AE}^{ALT} \times \left[\left(1 - \frac{1}{100}c_{slipAE}\right) \times CO_{2eqTtW_{ALT,AE}} + \frac{1}{100}c_{slipAE} \times CO_{2eqTtW_{slip,ALT,AE}}\right] \times 10^3 \end{array}}{F_{total}^{ALT} \times LCV^{ALT} \times RWD_{ALT} + F_{total}^{VLSFO} \times LCV^{VLSFO} \times RWD_{VLSFO}} \quad (14)$$

$$GHGIE_{actual} = WTT + TTW \quad (15)$$

where c_{slipME}, c_{slipAE} represent the fuel leakage rates for the main engine and auxiliary engine, respectively. The parameter RWD is a key adjustment factor defined in the FuelEU Maritime regulation, introduced to incentivize the use of renewable energy fuels. For fuels derived from non-biological origins, a multiplier of 2 may be applied to the GHG Intensity calculation during the period from 1 January 2025 to 31 December 2033. Otherwise, RWD is set to 1 [29].

2.1.3. Annual Cost Calculation Model

The annual cost C_{total} primarily consists of four components: annualized capital investment cost $C_{investment}$, fixed operating cost of the ship $C_{OPEX-fixed}$, fuel-related variable operating cost $C_{OPEX-varible}$, and the opportunity cost associated with cargo space loss C_{lost}. The calculation is expressed in Equation (16) as follows:

$$C_{total} = C_{investment} + C_{OPEX-fixed} + C_{OPEX-varible} + C_{lost} \quad (16)$$

$$C_{investment} = \frac{r(1+r)^{life}}{(1+r)^{life}-1} \times CAPEX \tag{17}$$

Equation (17) represents the capital recovery factor formula [31], which is used to annualize the one-time CAPEX over the ship's operational lifetime. Here, $life$ denotes the ship lifetime, which is set to be 30 years in this study [32]; r is the discount rate, set at 5% according to reference [27]; and $CAPEX$ represents the capital investment cost of the ship [31].

$$C_{OPEX-fixed} = (1+\delta) \times \frac{CAPEX}{life} \tag{18}$$

$$C_{OPEX-varible} = F_{total}^{ALT} \times f_{ALT} + F_{total}^{VLSFO} \times f_{VLSFO} \tag{19}$$

The fixed operating cost of a ship varies significantly with ship size. Equation (18) presents the calculation method for fixed operating costs [27], where δ denotes the ratio of fixed operating cost to average annual shipbuilding cost. This ratio is ship-size dependent and exhibits a strong correlation with ship scale [27,33]. Equation (19) defines the annual variable operating cost—specifically, the annual fuel cost. In this equation, f_{ALT} represents the price of the alternative fuel, and f_{VLSFO} denotes the price of VLSFO.

$$C_{lost} = TEU \times \rho \times c_{TEU}^{lost} \times \varphi \tag{20}$$

$$C_{lost} = DWT \times \rho \times c_{dwt}^{lost} \times \left(T^{berth} + T^{sail}\right) \tag{21}$$

When alternative fuels are adopted, the lower energy density relative to conventional fuels may require fuel storage tank volumes that are 1.8 to 4.5 times larger to maintain equivalent voyage ranges, thereby encroaching on available cargo space [27]. The cost of lost cargo capacity is calculated using differentiated approaches based on ship type. Equation (20) applies to containerships, while Equation (21) is used for other ship types. In these equations, ρ represents the cargo space loss rate; TEU is the nominal container capacity of the containership (in twenty-foot equivalent units); c_{TEU}^{lost} is the unit cost per lost TEU; φ is the annual voyage frequency for containerships; DWT is the deadweight tonnage of non-container ships; and c_{dwt}^{lost} is the unit cost per lost deadweight tonnage.

2.1.4. FuelEU Penalty Calculation Model

ANNEX IV of the FuelEU Maritime regulation provides the methodology for calculating the EU compliance balance and associated penalty costs, as expressed in Equations (22) and (23) [29]:

$$\begin{aligned} Compliance\ balance \\ &= \left(GHGIE_{target} - GHGIE_{actual}\right) \\ &\times \left(F_{total}^{ALT} \times 10^3 \times LHV^{ALT} + F_{total}^{VLSFO} \times 10^3 \times LHV^{VLSFO}\right) \end{aligned} \tag{22}$$

$$\begin{aligned} FuelEU\ Penalty \\ &= \begin{cases} \frac{|Compliance\ Balance|}{GHGIE_{actual} \times 41000} \times 2400, GHGIE_{target} < GHGIE_{actual} \\ 0, GHGIE_{actual} \leq GHGIE_{target} \end{cases} \end{aligned} \tag{23}$$

where $GHGIE_{target}$ denotes the regulatory GHG intensity limit (gCO_2eq/MJ) for energy used onboard the ship; $GHGIE_{actual}$ is the calculated annual average GHG intensity of energy used onboard during the relevant reporting period; and *Compliance balance* is the compliance balance, calculated as the absolute value of the difference between $GHGIE_{target}$

and $GHGIE_{actual}$, expressed in grams of CO_2 equivalent (gCO_2eq). *FuelEU Penalty* refers to the monetary penalty imposed under the FuelEU Maritime regulation, expressed in euros (EUR). The penalty is calculated based on a reference cost of EUR 2400 per ton of VLSFO-equivalent emissions, assuming 1 ton of VLSFO corresponds to 41,000 MJ of energy.

2.1.5. PROMETHEE II Method

MCDA is commonly used to support the selection of optimal compromise solutions based on multiple, often conflicting, criteria. PROMETHEE is a preference ranking method designed to rank and select a finite set of alternatives under such conflicting criteria [34]. In this study, the PROMETHEE II method is applied to a set of fuel configuration alternatives composed of VLSFO, MGO, and seven emerging alternative fuels (including LNG, methanol, ammonia, etc.), with the aim of identifying the optimal substitute fuel option. To objectively determine the weights of the evaluation criteria, the CRITIC method is employed. This method posits that the decision-making information entropy of a criterion increases with its degree of dispersion (standard deviation) and decreases with its correlation (covariance) with other criteria [35]. Moreover, the PROMETHEE II method requires the specification of preference thresholds for each criterion, reflecting the decision-maker's sensitivity to differences among alternatives. To avoid the uncertainty introduced by subjective threshold assignment, this study adopts an automated approach: the 75th percentile of the normalized difference distribution for each criterion is extracted as its preference threshold. This enhances the reasonableness and adaptability of the preference function in distinguishing between alternatives. To evaluate the impact of changes in preference thresholds and criterion weights on the ranking results, a sensitivity analysis is further conducted in Section 3.4. This includes both single-criterion perturbation tests and Monte Carlo simulations of weight variations, thereby verifying the robustness of the decision outcomes under different parameter settings.

Step 1: The decision matrix is normalized using Equations (24) and (25), corresponding to beneficial (profit-type) and non-beneficial (cost-type) criteria, respectively. Here, m denotes the number of criteria, n represents the number of alternatives, and x_{ij} is the performance value of the i alternative with respect to the j criterion.

$$x'_{ij} = \frac{x_{ij} - minx_{ij}}{maxx_{ij} - minx_{ij}} \tag{24}$$

$$x'_{ij} = \frac{maxx_{ij} - x_{ij}}{maxx_{ij} - minx_{ij}} \tag{25}$$

Step 2: The CRITIC method is employed to objectively determine the weights of the criteria. This method evaluates the objective importance of each criterion by considering both the contrast intensity (i.e., variability) of the criterion and its conflict (i.e., correlation) with other criteria [35]. The detailed procedure is defined by Equations (26)–(29), which involve computing the standard deviation of each criterion, followed by the assessment of conflict intensity, information content, and final objective weights.

$$\sigma_j = \sqrt{\frac{\sum_{i=1}^{n}\left(x'_{ij} - \overline{x}_j\right)^2}{n-1}} \tag{26}$$

$$f_j = \sum_{i=1}^{m}\left(1 - |r_{ij}|\right) \tag{27}$$

$$C_j = \sigma_j \times f_j \tag{28}$$

$$\omega_j = \frac{C_j}{\sum_{j=1}^{m} C_j} \tag{29}$$

where x'_{ij} is the normalized value of the i alternative with respect to the j criterion; σ_j is the sample standard deviation of criterion j, representing the contrast intensity; $\overline{x}_j$ is the mean value of criterion j; f_j denotes the conflict measure of criterion j with all other criteria; r_{ij} is the correlation coefficient between criterion i and j criterion, measured using the Pearson correlation coefficient; C_j represents the amount of information conveyed by criterion j; and ω_j is the final normalized weight of criterion j.

Step 3: A pairwise comparison is performed among all alternatives based on their criterion values, to determine the evaluative differences between each alternative and all others [34].

$$d_j(i,k) = g_j(i) - g_j(k) \tag{30}$$

where $d_j(i,k)$ represents the evaluation difference between alternative i and alternative k with respect to criterion j, and $g_j(\cdot)$ denotes the preference function for criterion j, applied to the normalized evaluation difference.

Step 4: The preference function is calculated. Brans and Vincke [36] proposed six types of preference functions. In this study, two types are adopted: the Usual Criterion and the V-shape with Indifference Criterion, as defined in Equations (31) and (32).

$$F_j(i,k) = \begin{cases} 0, d_j(i,k) \leq 0 \\ 1, d_j(i,k) > 0 \end{cases} \tag{31}$$

$$F_j(i,k) = \begin{cases} 0, d_j(i,k) \leq q \\ \frac{d_j(i,k)-q}{p}, q < d_j(i,k) \leq q+p \\ 1, d_j(i,k) > q+p \end{cases} \tag{32}$$

where $q \geq 0$ is the indifference threshold—below which no preference is expressed—set to 0 in this study, and $p > 0$ is the linear preference threshold, defining the interval $(q, q+p)$ within which the preference strength increases linearly. In this study, p is defined as the third quartile of the normalized evaluative differences.

Step 5: Compute the aggregated global preference index [34].

$$\pi(i,k) = \sum_{j=1}^{m} \omega_j F_j(i,k) \tag{33}$$

where $\pi(i,k)$ denotes the aggregated preference index of alternative i over alternative k.

Step 6: Calculate the positive outranking flow and the negative outranking flow [34]. The positive outranking flow $\phi^+(i)$ reflects the degree to which alternative i outranks other alternatives, while the negative outranking flow $\phi^-(i)$ reflects the degree to which alternative i is outranked by the others.

$$\phi^-(i) = \frac{1}{n-1} \sum \pi(x,i) \tag{34}$$

$$\phi^+(i) = \frac{1}{n-1} \sum \pi(i,x) \tag{35}$$

Step 7: Calculate the net outranking flow $\phi(i)$ [34], which is used to rank the alternatives. A higher net outranking flow value indicates a stronger overall performance of the alternative.

$$\phi(i) = \phi^+(i) - \phi^-(i), \phi(i) \in [-1,1] \tag{36}$$

2.2. Data

2.2.1. Case Study Ships

In accordance with the FuelEU Maritime regulation, this study selects five representative ship categories—tankers, bulk carriers, containerships, general cargo ships, and ferries—covering a total of 17 sub-types of ships with gross tonnage exceeding 5000. Based on actual operational data, the sailing and berthing durations within EU waters were extracted in line with the regulatory scope, serving as inputs for evaluating the environmental and economic performance of ships under the FuelEU Maritime framework. The main parameters are summarized in Table 1.

Table 1. Main parameters of the selected ships.

Ship Category	Ship Subtype	DWT/ TEU	Main Engine Power (kW)	Auxiliary Engine Power (kW)	Design Speed (kn)	Average Speed (kn)	Sailing Time Within EU Waters (h)	Berthing Time at EU Ports (h)	Newbuilding Price (Million EUR)
Tanker	UL/VLCC	296,174	22,200	4620	14.8	11.4	4913	529	115.92
	Suezmax	147,630	18,200	4400	14.0	11.8	3963	2487	80.96
	Aframax	108,553	14,070	2310	15.3	9.9	1630	858	68.54
	Panamax	71,573	22,400	4200	16.0	9.8	5453	3410	54.74
	Handysize	37,993	9960	1776	14.2	11.2	4110	3662	45.08
	Small tanker	6870	2400	945	10.0	8	4272	4585	16.56
Bulker	Capesize	181,709	18,660	2400	11.5	10.8	579	486	68.08
	Panamax	81,290	9930	2400	14.3	11.1	2461	1237	34.04
	Handymax	82,094	9710	2400	14.0	10.9	984	1247	34.04
	Handysize	21,353	9720	2400	17.5	10	3092	3942	19.32
Containership	Post-Panamax	18,270	59,360	15,000	19.0	15.3	2649	773	196.42
	Neo-Panamax	14,074	72,240	13,790	24.2	15.1	3921	698	140.76
	Intermediate	6750	57,099	9000	25.5	15.8	2769.5	1242	81.88
General Cargo	General Cargo I	9500	4000	1030	13.0	10.7	3312	1491	17.00
	General Cargo II	10,952	3360	1170	12.0	10.5	2031	1605	17.94
Ferry	Ro-Pax	12,662	33,600	5850	23.0	17.5	4759.12	3473.9	60.68
	Pass/Car	6133	31,200	7840	21.0	19	6567.71	1803	45.25

Note: (a) General Cargo I refers to ships with a deadweight tonnage (DWT) between 5000 and 9999; General Cargo II refers to those with a DWT greater than 10,000. (b) Data sources: https://www.clarksons.net.cn (accessed on 5 March 2025) and https://www.myvessel.cn (accessed on 5 March 2025).

2.2.2. Fuel Data

This study adopts a WtW perspective to analyze three categories of energy carriers: conventional fossil fuels, biomass-based fuels, and renewable electricity-based fuels. A dual-fuel internal combustion engine is selected as the propulsion system, with the combination of VLSFO and MGO serving as the reference configuration. When MGO is replaced by alternative fuels, both the newbuilding price adjustment factor (reflecting changes in construction cost) and the associated deadweight loss during operation must be considered. Detailed fuel parameters are provided in Table 2. The loss of cargo deadweight results in cargo space opportunity costs, which vary depending on ship type and voyage length; the unit loss costs are shown in Table 3.

Table 2. Fuel-related parameters.

Fuel Type	LCV (MJ/kg)	WtT Emission Factor (gCO_2eq/MJ)	TtW Emission Factor			Newbuilding Price Adjustment Factor	Cargo Deadweight Loss Rate for This Fuel (%)
			(CO_2 gCO_2eq/g Fuel)	CH_4 (gCH_4eq/g Fuel)	N_2O (gN_2Oeq/g Fuel)		
VLSFO	41 [37]	13.2 [38]	3.114 [39]	0.00005 [39]	0.00018 [39]	1	–
MGO	42.7 [38]	14.4 [38]	3.206 [39]	0.00005 [39]	0.00018 [39]	1	–
LNG	49.1 [38]	18.5 [40]	2.750 [39]	0 [39]	0.00011 [39]	1.23 [41]	1 [27]
MEOH	19.9 [38]	31.3 [40]	1.375 [38]	0 [38]	0.00018 [38]	1.1 [42]	2 [27]
Bio-LNG	50 [19]	−25 [38]	2.750 [38]	0 [38]	0.00011 [38]	1.23 [41]	1 [27]
Bio-MEOH	20 [38]	−55.4 [38]	1.375 [38]	0 [38]	0.00018 [38]	1.1 [42]	2 [27]
HVO	44 [19]	−33.4 [38]	3.115 [38]	0.00005 [38]	0.00018 [38]	1 [43]	–
E-LH_2	120 [44]	0 [40]	0 [39]	0 [38]	0.00018 [38]	1.58 [45]	5 [27]
E-NH_3	18.6 [38]	0 [40]	0 [40]	0 [38]	0.00018 [38]	1.25 [45]	3 [27]

Note: (a) The WtT emission factor for VLSFO is based on the corresponding value for LFO. (b) "–" indicates no cargo deadweight loss.

Based on reference [7], this study establishes annual GHG intensity target values for the period 2025–2050. Fuel price trajectories are determined with reference to the trends outlined in [19,21], incorporating differentiated annual adjustment strategies. Fossil fuels and HVO follow a progressive price increase path, while biofuels such as Bio-LNG and Bio-MEOH adopt a gradual price reduction scheme. Renewable electricity-based fuels are modeled using a stepwise cost-decline trajectory toward 2050. Specific values are presented in Table 4. The 2025 baseline prices for VLSFO and MGO are estimated based on historical average port prices from 2020 to 2024 obtained from the Clarkson database. The USD-to-EUR exchange rate is converted using the fixed value as of 24 March 2025.

Table 3. Unit cargo loss cost by ship type.

Voyage Length	Short	Medium	Long
Tanker (€/dwt/day)	0.59 [46]	0.33 [46]	0.12 [46]
Bulker (€/dwt/day)	0.30 [46]	0.21 [46]	0.08 [46]
Ferry (€/m^3 of space loss)	6 [47]	8 [47]	10 [47]
General cargo (€/dwt/day)	0.1 [47]	–	–
Containership (€/TEU/trip)	600 [47]	900 [47]	1100 [47]

Note: "–" indicates that the unit cargo loss cost for this ship type is not affected by voyage length.

Table 4. Annual GHG intensity targets and fuel prices.

Year	GHG Intensity Target	Fuel Price (EUR/t)								
		VLSFO	MGO	LNG	MEOH	Bio-LNG	Bio-MEOH	HVO	E-LH_2	E-NH_3
2025	89.34	540.14	731.5	543.60 [4]	460 [48]	1383.90 [4]	1193 [49]	1000 [50]	5059.54 [51]	1072.52 [51]
2030	85.69	514.46	847	665.03	575	1231	1152	1031.12	5299.2	872.71
2035	77.94	692.58	885	715.22	582	1138.5	1147.5	1177.86	4857.6	804.26
2040	62.90	870.7	923	752.87	589	1046	1143	1324.6	4416	735.82
2045	34.64	1015.30	935.5	790.51	585.5	980	1125	1467.97	3974.4	663
2050	18.23	1159.89	948	828.15	582	914	1107	1611.33	3532.8	590.36

3. Results

3.1. Environmental Assessment of Alternative Fuels

According to the FuelEU Maritime regulation [29], renewable fuels of non-biological origin are granted time-dependent reward factors. As a result, the evaluation of WtW GHG intensity is divided into two periods: 2025–2030 and 2035–2050. Figures 2–6 present the WtW GHG intensity results for alternative fuels across five ship types for the period 2025–2030, with horizontal lines indicating the annual GHG intensity targets from 2025 to 2050. It is evident that, among the alternative fuel options, the VLSFO + MEOH combination exhibits the highest WtW GHG intensity, even exceeding that of the baseline VLSFO + MGO. This is primarily due to the higher WtT intensity of methanol compared to MGO, which offsets the TtW advantages. Both VLSFO + MGO and VLSFO + MEOH fail to meet the GHG intensity targets for 2025 and beyond. While VLSFO + LNG offers moderate improvement, it is projected to become non-compliant by 2035. Although the WtT GHG intensity of biofuels such as VLSFO + HVO and VLSFO + BIO-LNG offers notable advantages, their relatively high TtW emissions are expected to pose compliance challenges beyond 2040. By comparison, the VLSFO + Bio-MEOH configuration demonstrates a well-balanced performance across both WtT and TtW stages—second only to e-fuels—resulting in a significantly lower overall WtW GHG intensity than other biofuel options. This finding aligns with the conclusions of related studies [11,12], confirming the viability of Bio-MEOH as a medium- to long-term decarbonization pathway in maritime transport. E-fuel options such as VLSFO + E-LH_2 and VLSFO + E-NH_3, characterized by their ultra-low emissions profiles, can meet the long-term GHG intensity reduction targets set for the 2025–2050 period. Among them, E-LH_2 is considered the most environmentally favorable option and is prioritized by policymakers for promotion—an assessment that is consistent with the perspective presented in [52].

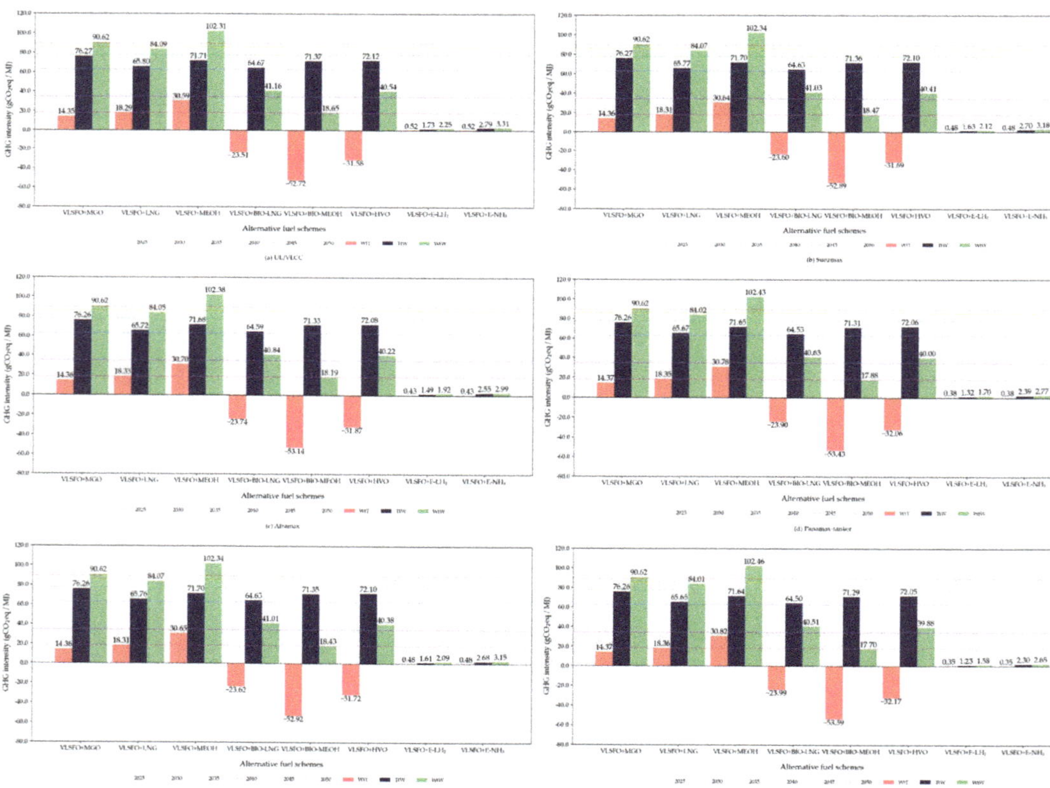

Figure 2. WtW GHG intensity of tanker ships during 2025–2030.

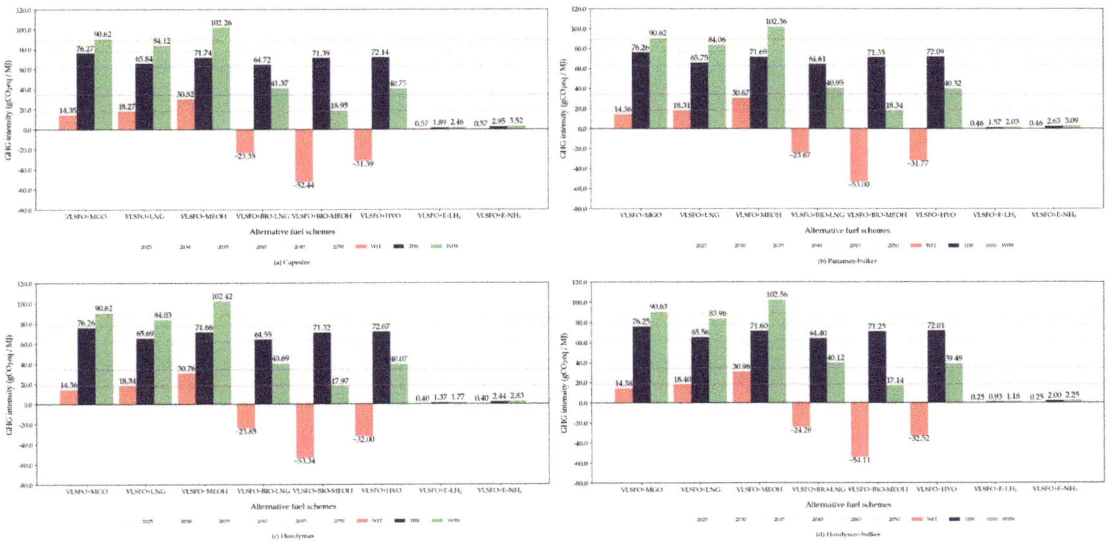

Figure 3. WtW GHG intensity of bulker ships during 2025–2030.

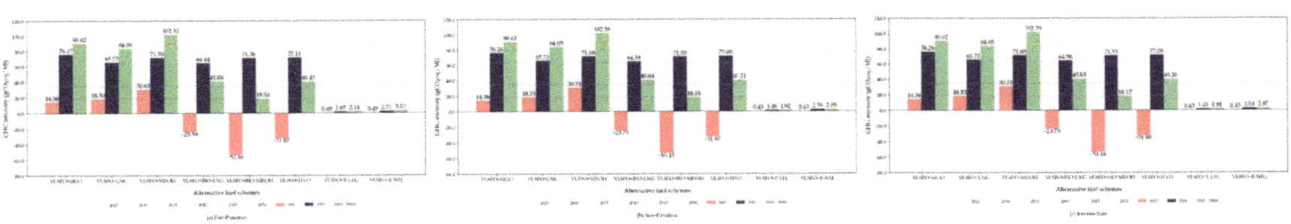

Figure 4. WtW GHG intensity of containerships during 2025–2030.

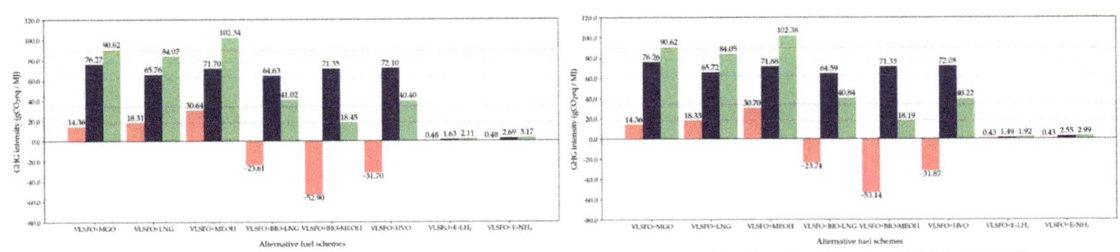

Figure 5. WtW GHG intensity of general cargo ships during 2025–2030.

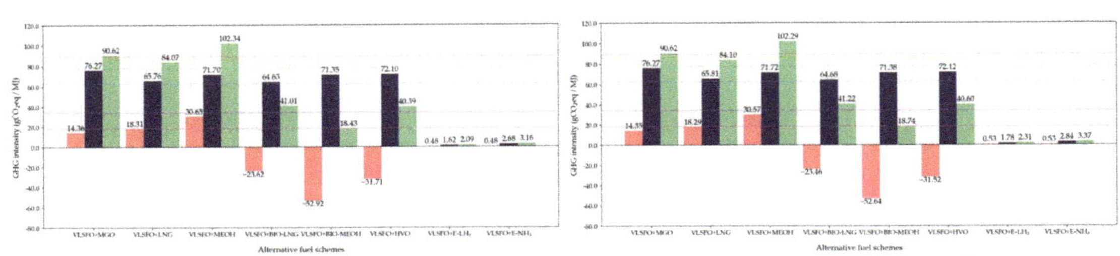

Figure 6. WtW GHG intensity of ferries during 2025–2030.

Figures A1–A5 illustrate the WtW GHG intensity results for alternative fuels across five ship types during the 2035–2050 period. As the reward factor only applies to E-LH_2 and E-NH_3, a noticeable increase—approximately twofold—in their WtW GHG intensity can be observed. Nevertheless, both fuel combinations still meet the 2050 GHG intensity targets.

The comparison across different ship types also reveals pronounced structural differences in WtW GHG intensity, which stem from the combined effects of ship design parameters, operational conditions, and emission profiles. For instance, the highest value appears in the VLSFO + MEOH scenario for the Handysize bulker, reaching 102.56 gCO_2eq/MJ, while the lowest is also found in the same ship type under the VLSFO + E-LH_2 configuration, at just 1.18 gCO_2eq/MJ. Moreover, the VLSFO + Bio-MEOH combination enables compliance with 2025–2050 GHG targets for several ship types, including Aframax, Panamax, and Small tankers, as well as Handymax and Handysize bulkers. However, this compliance is not observed across all ship types. This variation highlights the need for policy considerations that tailor alternative fuel configurations to the specific characteristics of different ship types. Regardless of ship category, VLSFO + MEOH consistently exhibits the highest WtW GHG intensity, while VLSFO + E-LH_2 remains the lowest. The overall compliance trends of different fuels remain largely consistent across ship types, which is fundamentally determined by the intrinsic properties of each fuel.

3.2. Economic and Compliance Assessment of Alternative Fuels

Figures 7–11 present the annual cost and EU compliance penalty cost for five ship types. From these figures, it is evident that, between 2025 and 2050, annual cost trends vary significantly across different ship–fuel combinations. According to reference [53], fuel costs typically account for approximately 50% of a ship's operational costs (excluding capital expenditure and cargo handling), making them the primary driver of annual cost variation. For all ship types, the cost of fossil fuels increases gradually over time, leading to a corresponding rise in annual costs, albeit at a relatively moderate rate. Among the evaluated configurations, VLSFO + LNG consistently demonstrates strong performance across most ship types. From the shipowner's economic perspective, it frequently ranks among the lowest annual cost options. LNG also receives the highest overall ranking, with evaluation results broadly consistent with the analysis presented in [52]. While fossil fuels show moderate cost growth, alternative fuels exhibit more dynamic trends in both pricing and cost competitiveness over time. In particular, the price of e-fuels declines over time, resulting in decreasing annual costs. However, despite this trend, fossil fuels still maintain significantly lower annual costs compared to e-fuels. Notably, the VLSFO + E-LH_2 combination remains the most expensive option in ship types such as Aframax, Capesize, and Handymax throughout the 2025–2050 period. Among biofuels, HVO demonstrates relatively strong economic feasibility before 2035, with annual costs slightly higher than those of MGO and LNG in most ship types, followed by a steady increase thereafter. Conversely, the annual cost of Bio-LNG decreases steadily after 2035. Bio-MEOH, however, shows poor economic performance, with annual costs surpassing those of e-fuels across ships such as RoPax and Pass/Car throughout the entire analysis period.

In terms of compliance assessment, consistent with the environmental evaluation, the VLSFO + MEOH combination exhibits the highest WtW GHG intensity, leading to high EU compliance penalties as early as 2025. For fossil fuel blends such as VLSFO + MGO and VLSFO + LNG, significant increases in compliance penalties begin around 2030 and 2035, respectively, reflecting a tightening regulatory stance on high-emission fuels. Among biofuels, VLSFO + Bio-LNG begins to incur notable penalties in certain ship types by 2050. In contrast, e-fuels—owing to their low WtW GHG intensity—demonstrate a clear advantage in compliance cost avoidance across the entire 2025–2050 horizon. The varying responses of different ship types to fuel-related economic and compliance factors highlight the need for differentiated, ship-specific, and phased policy strategies in future regulatory design.

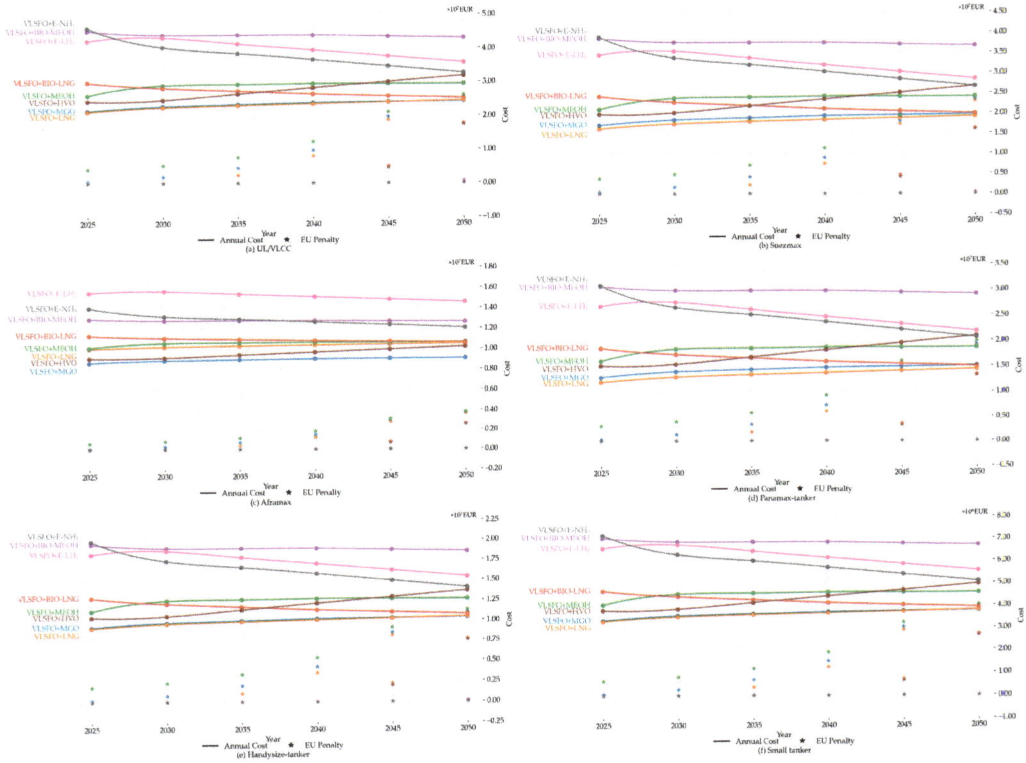

Figure 7. Annual cost and EU compliance penalty cost for tanker ships.

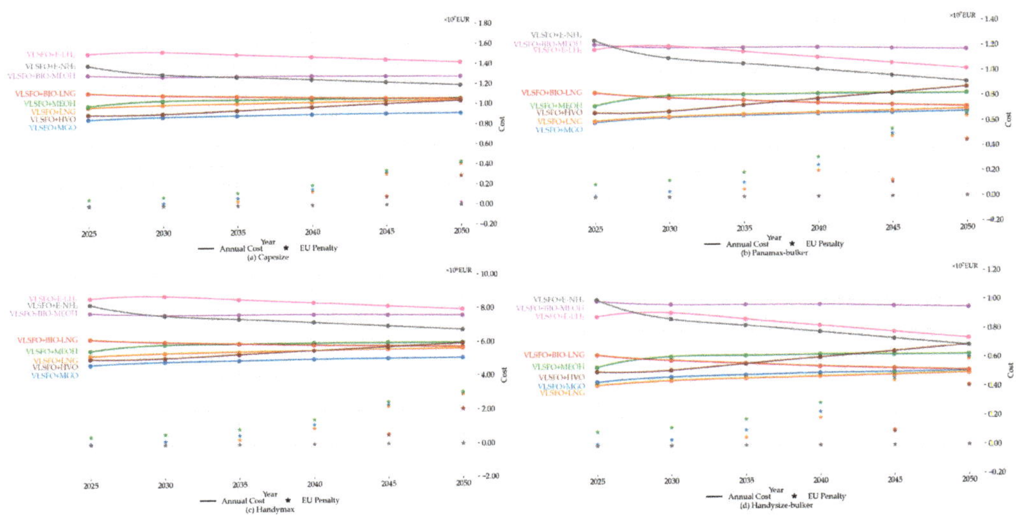

Figure 8. Annual cost and EU compliance penalty cost for bulker ships.

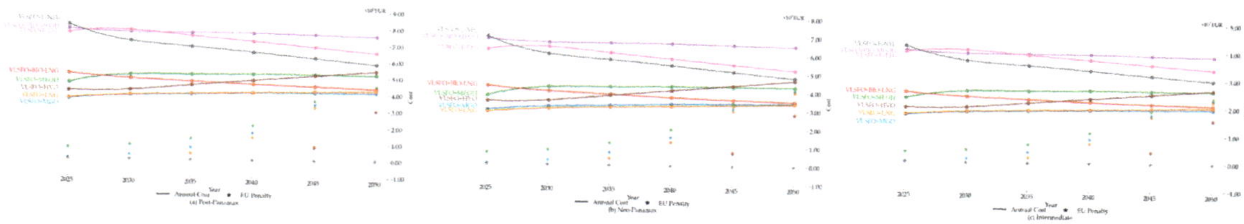

Figure 9. Annual cost and EU compliance penalty cost for containerships.

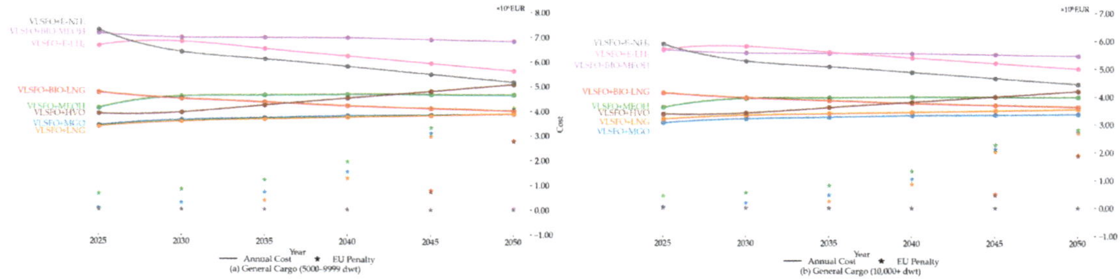

Figure 10. Annual cost and EU compliance penalty cost for general cargo ships.

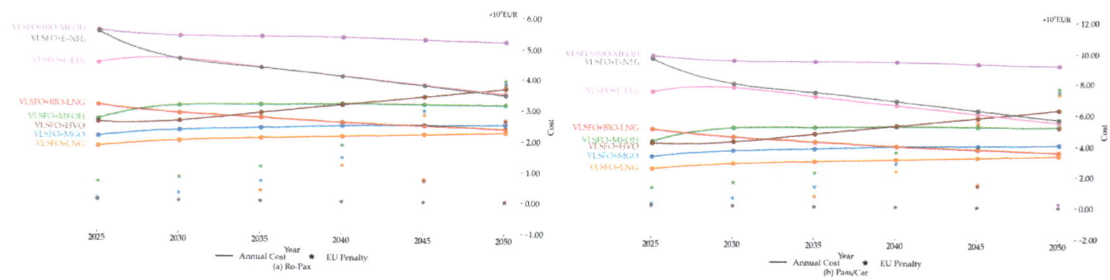

Figure 11. Annual cost and EU compliance penalty cost for ferries.

3.3. Optimal Selection of Alternative Fuel Options

This study applies the PROMETHEE II multi-criteria decision-making method to rank alternative fuel options for 17 ship subtypes. Three quantitative indicators are selected to represent the dimensions of environmental performance, cost, and regulatory compliance: WtW GHG intensity, annual cost, and compliance surplus. The CRITIC method is used to determine the objective weights of each criterion, based on both contrast intensity and inter-criteria conflict. Table 5 presents the optimal alternative fuel options identified for different ship types over the period 2025–2050. Results show that the rankings remain relatively stable across all ship types between 2025 and 2035, while greater diversity emerges from 2040 to 2050. In 2025 and 2035, all ship types—except for Pass/Car—select VLSFO + HVO as the optimal fuel option, primarily due to its relatively low WtW GHG intensity and moderate annual cost, making it the dominant choice. From 2040 onward, ships such as Suezmax and Panamax begin to shift toward VLSFO + Bio-LNG as the preferred alternative, while Ro-Pax and Pass/Car ships transition to e-fuels. By 2050, most ship types ultimately select e-fuels as their optimal transition fuel. This overall trend is consistent with the alternative fuel transition pathway outlined in [11]. However, due to the incorporation of ship-type heterogeneity in this study, the specific results exhibit a certain degree of divergence, reflecting the nuanced differences in fuel performance and feasibility across ship types. Notably, VLSFO + MGO and VLSFO + MEOH are not selected as the optimal option for any ship type, underscoring the overall advantages of biofuels and e-fuels in terms of environmental, economic, and compliance performance.

Table 5. Optimal alternative fuel choices for ships from 2025 to 2050.

Ship/Year	2025	2030	2035	2040	2045	2050
UL/VLCC	VLSFO + HVO	VLSFO + HVO	VLSFO + HVO	VLSFO + HVO	VLSFO + BIO-LNG	VLSFO + E-NH_3
Suezmax	VLSFO + HVO	VLSFO + HVO	VLSFO + HVO	VLSFO + BIO-LNG	VLSFO + BIO-LNG	VLSFO + E-NH_3
Aframax	VLSFO + HVO	VLSFO + HVO	VLSFO + HVO	VLSFO + HVO	VLSFO + HVO	VLSFO + HVO
Panamax	VLSFO + HVO	VLSFO + HVO	VLSFO + HVO	VLSFO + BIO-LNG	VLSFO + BIO-LNG	VLSFO + E-LH_2
Handysize	VLSFO + HVO	VLSFO + HVO	VLSFO + HVO	VLSFO + HVO	VLSFO + BIO-LNG	VLSFO + E-NH_3
Small tanker	VLSFO + HVO	VLSFO + HVO	VLSFO + HVO	VLSFO + HVO	VLSFO + BIO-LNG	VLSFO + E-NH_3
Capesize	VLSFO + HVO	VLSFO + HVO	VLSFO + HVO	VLSFO + HVO	VLSFO + HVO	VLSFO + E-NH_3
Panamax	VLSFO + HVO	VLSFO + HVO	VLSFO + HVO	VLSFO + HVO	VLSFO + BIO-LNG	VLSFO + E-NH_3
Handymax	VLSFO + HVO	VLSFO + HVO	VLSFO + HVO	VLSFO + HVO	VLSFO + HVO	VLSFO + E-NH_3
Handysize	VLSFO + HVO	VLSFO + HVO	VLSFO + HVO	VLSFO + BIO-LNG	VLSFO + BIO-LNG	VLSFO + E-NH_3
Post-Panamax	VLSFO + HVO	VLSFO + HVO	VLSFO + HVO	VLSFO + HVO	VLSFO + BIO-LNG	VLSFO + E-NH_3
Neo-Panamax	VLSFO + HVO	VLSFO + HVO	VLSFO + HVO	VLSFO + BIO-LNG	VLSFO + BIO-LNG	VLSFO + E-NH_3
Intermediate	VLSFO + HVO	VLSFO + HVO	VLSFO + HVO	VLSFO + HVO	VLSFO + BIO-LNG	VLSFO + BIO-LNG
General Cargo I	VLSFO + HVO	VLSFO + HVO	VLSFO + HVO	VLSFO + HVO	VLSFO + BIO-LNG	VLSFO + E-NH_3
General Cargo II	VLSFO + HVO	VLSFO + HVO	VLSFO + HVO	VLSFO + HVO	VLSFO + E-NH_3	VLSFO + E-NH_3
Ro-Pax	VLSFO + HVO	VLSFO + HVO	VLSFO + HVO	VLSFO + E-LH_2	VLSFO + E-LH_2	VLSFO + E-LH_2
Pass/Car	VLSFO + E-LH_2	VLSFO + HVO	VLSFO + E-LH_2	VLSFO + E-LH_2	VLSFO + E-LH_2	VLSFO + E-LH_2

3.4. Sensitivity Analysis

In the PROMETHEE II method, this study adopts the V-shape with indifference criterion preference function for the two indicators—annual cost and compliance surplus. The preference threshold p is one of the key parameters influencing the ranking outcomes. To assess the impact of variations in p on the stability of rankings, 20 evenly spaced values were set within the range [0.1, 1] for each indicator. Sensitivity curves were then constructed using Kendall's Tau correlation coefficient to capture the ranking consistency under these perturbations.

The results indicate that the ranking structure is highly sensitive to the preference threshold: in certain intervals, even small variations in p can significantly alter the ranking outcomes. However, within specific sub-ranges, the Kendall's Tau values stabilize and approach 1, suggesting that parameter values within these intervals yield greater consistency and robustness in ranking, thus providing quantitative support for informed parameter selection.

Furthermore, recognizing the potential uncertainty associated with the CRITIC-derived criterion weights in practical applications, this study incorporates weight settings as a second core dimension of the sensitivity analysis. A Monte Carlo simulation with 10% perturbation amplitude and 1000 iterations was conducted to evaluate the ranking robustness of each alternative under varying weight configurations. The standard deviation of net flow values is employed as the primary metric for evaluating sensitivity to weight fluctuations, with lower values indicating reduced susceptibility to weight changes and enhanced robustness of the ranking outcomes.

Using the UL/VLCC ship type as an illustrative case, Figures 12–17 present the sensitivity analysis results for different years between 2025 and 2050. For the annual cost indicator, the recommended p values are primarily distributed in the 0.38–0.81 range, within which Kendall's Tau values remain consistently close to 1, indicating high ranking stability. Similarly, for the compliance surplus indicator, the recommended p values fall mainly within the 0.29–0.76 interval, also demonstrating strong robustness.

In the weight perturbation analysis, the VLSFO + MEOH configuration exhibits persistently high standard deviations in net outranking flow across the entire time period, reflecting strong sensitivity to weight variations and an inability to achieve stable rankings—consistent with the findings in Table 5, where this option never emerges as the top-ranked solution. In contrast, VLSFO + HVO shows the lowest standard deviation values during the 2025–2040 period, indicating robust and stable ranking outcomes. Although VLSFO + Bio-LNG and VLSFO + E-NH_3 do not always exhibit the minimum standard deviation, their relatively low variability and consistently favorable performance across multiple ship types suggest strong adaptability and potential advantages across key evaluation dimensions.

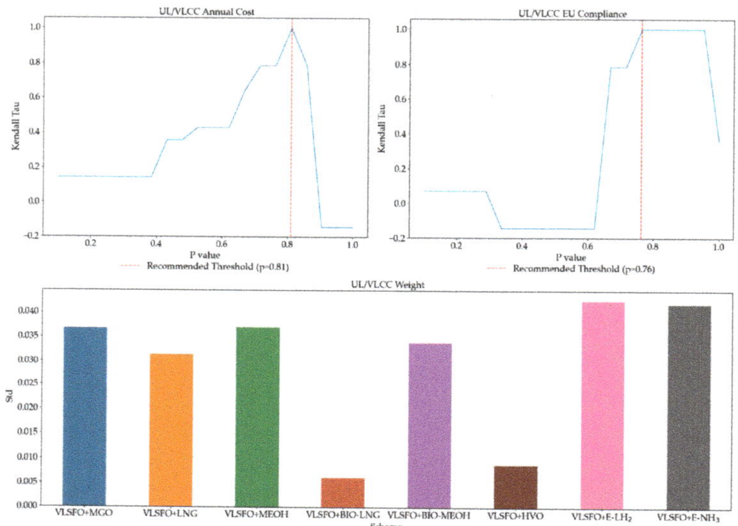

Figure 12. Sensitivity analysis of UL/VLCC ships in 2025.

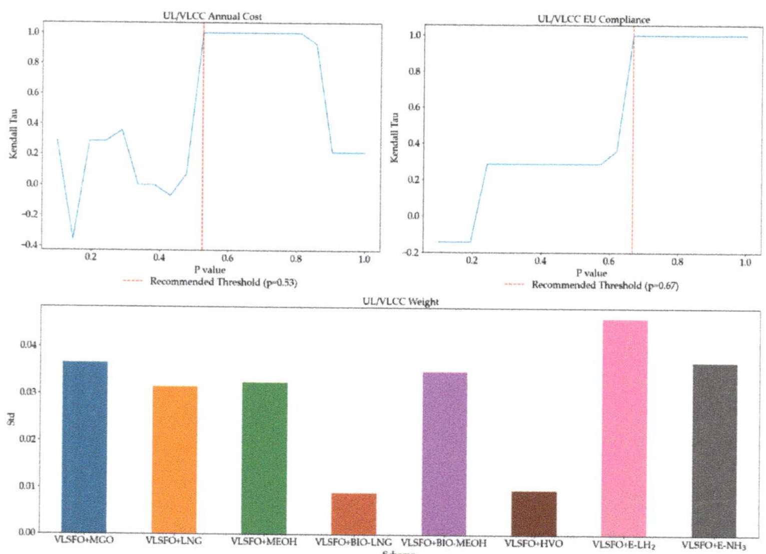

Figure 13. Sensitivity analysis of UL/VLCC ships in 2030.

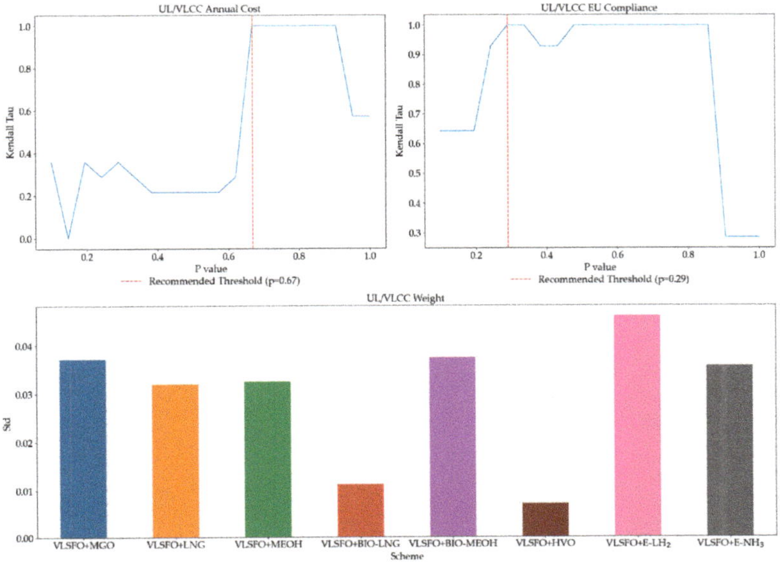

Figure 14. Sensitivity analysis of UL/VLCC ships in 2035.

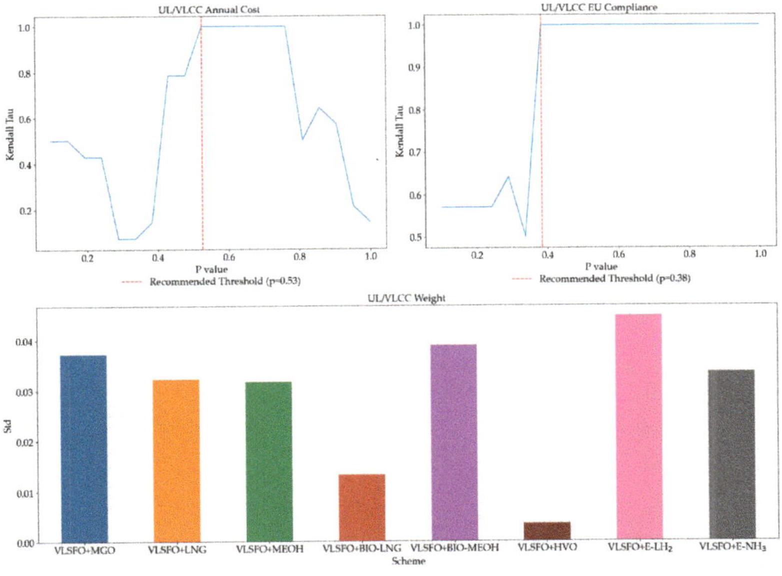

Figure 15. Sensitivity analysis of UL/VLCC ships in 2040.

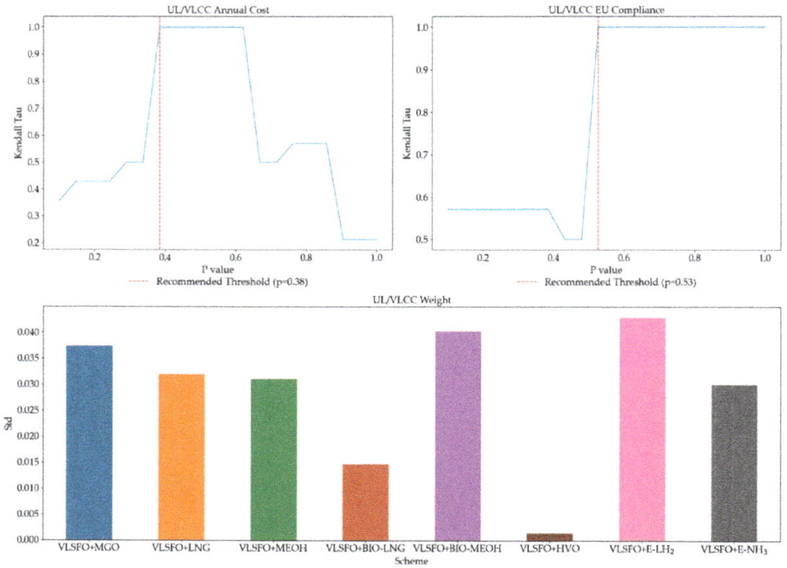

Figure 16. Sensitivity analysis of UL/VLCC ships in 2045.

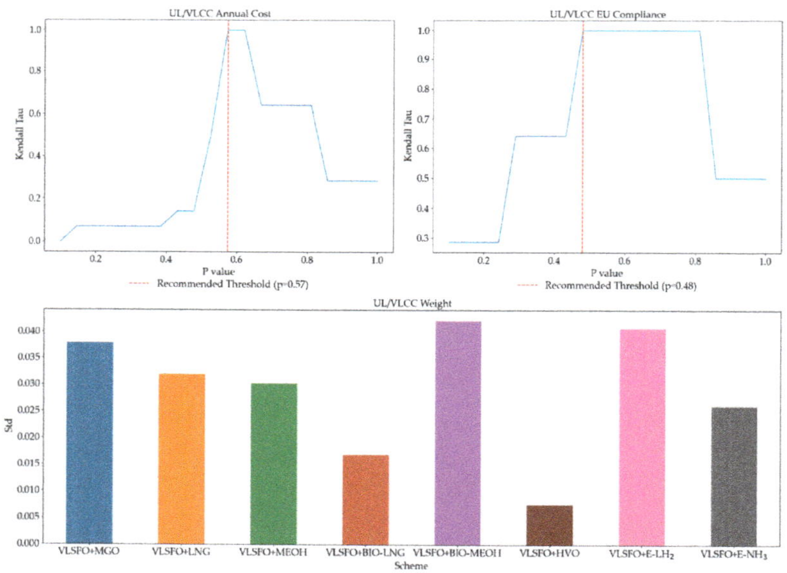

Figure 17. Sensitivity analysis of UL/VLCC ships in 2050.

4. Conclusions and Discussion

This study addresses the critical question of how to determine the optimal alternative fuel configurations for different ship types under the constraints of the FuelEU Maritime regulation. It proposes a multi-criteria evaluation framework that incorporates life-cycle thinking and aligns with regulatory requirements. Through a systematic assessment, the study identifies the preferred fuel pathways for 17 representative subtypes across five major categories of ships from 2025 to 2050. The findings provide scientific support for shipping companies, shipbuilders, and policymakers in planning newbuilds, retrofitting existing ships, and developing fuel infrastructure. In particular, the results offer quantitative guidance for practical decision-making in the maritime energy transition, especially in the context of tightening carbon intensity targets and growing fuel diversification. The key findings are summarized as follows:

(1) Fuel type is the decisive factor that influences WtW GHG intensity and regulatory compliance, while ship-type differences have a relatively limited impact. E-fuels (VLSFO + E-LH_2/E-NH_3), due to their ultra-low life-cycle emissions, are the only solutions that enable full compliance with the 2025–2050 GHG reduction targets across all ship types. In contrast, conventional fuels (VLSFO + MGO/MEOH) and biofuels (HVO/BIO-LNG) fail to achieve long-term compliance due to excessive emissions in either the WtT or TtW stages;

(2) Although fossil fuels exhibit relatively low short-term annual costs, their compliance penalty costs increase sharply starting around 2030, ultimately eroding their long-term cost advantage. E-fuels (E-LH_2/E-NH_3), despite high initial costs, gain competitiveness by 2050 due to declining fuel prices and persistently low compliance penalties. Biofuels, by contrast, display transitional characteristics: they are more economically viable than e-fuels in the near to medium term but offer a limited window of environmental compliance. This suggests that biofuels can serve as an interim solution to balance economic and emission reduction objectives before a full-scale transition to e-fuel systems;

(3) In the short term, biofuels—particularly HVO—emerge as the most favorable options; however, by 2050, e-fuels (E-LH_2/E-NH_3) become the optimal choices for most ship types due to their superior environmental compliance. Life-cycle carbon performance and regulatory compatibility are the decisive factors shaping the competitiveness of alternative fuels. Conventional fuel options such as VLSFO + MGO and VLSFO + MEOH are systematically phased out, as they fail to strike a balance between economic feasibility and emissions reduction.

While this study develops a foundational assessment framework across environmental, economic, and regulatory dimensions, future research may expand in the following directions to enhance decision-support capabilities:

(1) Integrate technological maturity and social acceptability into the evaluation framework to establish a more comprehensive decision-making model that captures technical, economic, and societal dimensions;

(2) Develop dynamic, time-sequenced models informed by market supply–demand dynamics and policy evolution, employing Monte Carlo simulations or machine learning to forecast alternative fuel price fluctuations and to enhance the credibility of cost-effectiveness assessments.

(3) Move beyond the assumption of fixed ship operating profiles by quantifying the regulatory and environmental impacts of flexible strategies, such as slow steaming and multi-fuel blending, on GHG intensity and compliance costs.

Author Contributions: Conceptualization, C.W. and Z.P.; methodology, C.W., Z.P. and J.Y.; software, Z.P., J.Y. and N.Z.; validation, X.L., K.L. and N.Z.; writing—original draft preparation, C.W., Z.P. and J.Y.; writing—review and editing, C.W., Z.P. and X.L.; visualization, J.Y. and N.Z.; supervision, C.W., X.L. and K.L. All authors have read and agreed to the published version of the manuscript.

Funding: This research was supported by Key R&D Program of Shandong Province, China (2022CXGC020303), the National Social Science Fund of China: Research on the institutional dilemma and policy design of state-owned enterprises carrying out disruptive innovation (24BJL091), and the Fundamental Research Funds for the Central Universities (3132024302).

Data Availability Statement: The original contributions presented in this study are included in the article. Further inquiries can be directed to the corresponding author.

Conflicts of Interest: The authors declare no conflicts of interest.

Nomenclature

List of Abbreviations

GHG	greenhouse gas
EU	European Union
WtW	well-to-wake
MCDA	multi-criteria decision analysis
VLSFO	very low sulfur fuel oil
MGO	marine gas oil
MEOH	methanol
E-LH_2	electrolytic liquid hydrogen
IMO	International Maritime Organization
ETS	Emissions Trading System
CO_2	carbon dioxide
CH_4	methane
N_2O	nitrous oxide
LNG	liquefied natural gas
HVO	hydrotreated vegetable oil
LCA	life-cycle assessment
LCC	lLife-cycle cost
FT-diesel	Fischer–Tropsch diesel
TRL	technological readiness level
ETD	Energy Taxation Directive
LCV	lower calorific values
TtW	tank-to-wake
GWP	global warming potential
CAPEX	capital expenditure
EUR	euros

Symbols

F_{ME}	hourly fuel consumption of the main engine, t/h
F_{AE}	hourly fuel consumption of the auxiliary engine, t/h
v_{ave}	average sailing speed of the ship, kn
v_{design}	design speed of the ship, kn
P_{ME}^{design}	design power of the main engine, kW
$SFOC_{ME}$	specific fuel consumption of the main engine, g/kWh
LF_{AE}	auxiliary engine load factor
P_{AE}^{design}	design power of the auxiliary engine, kW
$SFOC_{AE}$	specific fuel consumption of the auxiliary engine, g/kWh
$1F_{ME}^{ALT}$	annual consumption of alternative fuel for the main engine, t
$1F_{ME}^{VLSFO}$	annual consumption of VLSFO for pilot fuel, t
F_{AE}^{ALT}	annual consumption of alternative fuel for the auxiliary engine, t
LCV^{VLSFO}	lower calorific value of VLSFO, MJ/kg
LCV^{ALT}	lower calorific value of the alternative fuel, MJ/kg
LCV^{MGO}	lower calorific value of MGO, MJ/kg
T^{sail}	annual sailing time of the ship, h
T^{berth}	annual berthing time of the ship, h
F_{total}^{ALT}	annual total consumption of alternative fuel, t
$C_{fCO_2,ALT,ME}$, $C_{fCH_4,ALT,ME}$, $C_{fN_2O,ALT,ME}$, $C_{fCO_2,ALT,AE}$, $C_{fCH_4,ALT,AE}$, $C_{fN_2O,ALT,AE}$, $C_{fCO_2,VLSFO,ME}$, $C_{fCH_4,VLSFO,ME}$, and $C_{fN_2O,VLSFO,ME}$	emission factors of CO_2, CH_4, and N_2O released during combustion of alternative fuels in the main engine and auxiliary engine, gGHG/gFuel

$C_{sfCO_2,ALT,ME}$, $C_{sfCH_4,ALT,ME}$, $C_{sfN_2O,ALT,ME}$, $C_{sfCO_2,ALT,AE}$, $C_{sfCH_4,ALT,AE}$, and $C_{sfN_2O,ALT,AE}$	emission factors of unburned methane and nitrous oxide leakage associated with alternative fuels in the main and auxiliary engines, gGHG/gFuel
GWP_{CO_2}, GWP_{CH_4}, GWP_{N_2O}	the 100-year global warming potential values for CO_2, CH_4, and N_2O
WTT	annual well-to-tank GHG emission intensity, gCO_2eq/MJ
TTW	annual tank-to-wake GHG emission intensity, gCO_2eq/MJ
$GHGIE_{actual}$	annual well-to-wake GHG emission intensity of the ship, gCO_2eq/MJ
c_{slipME}, c_{slipAE}	fuel leakage rates for the main engine and auxiliary engine
RWD	a key adjustment factor defined in the FuelEU Maritime regulation
C_{total}	annual cost, €
$C_{investment}$	annualized capital investment cost, €
$C_{OPEX-fixed}$	fixed operating cost of the ship, €
$C_{OPEX-varible}$	fuel-related variable operating cost, €
C_{lost}	opportunity cost associated with cargo space loss, €
$life$	ship lifetime
r	discount rate
$CAPEX$	capital investment cost of the ship, €
δ	ratio of fixed operating cost to average annual shipbuilding cost
f_{ALT}	price of the alternative fuel, €
f_{VLSFO}	price of VLSFO, €
ρ	cargo deadweight loss rate
TEU	nominal container capacity of the containership
c_{TEU}^{lost}	unit cost per lost TEU, €/TEU/trip
φ	annual voyage frequency for containerships, trip
DWT	deadweight tonnage of non-container ships
c_{dwt}^{lost}	unit cost per lost deadweight ton, €/tonne of space loss/h
$GHGIE_{target}$	regulatory GHG intensity limit for energy used onboard the ship, gCO_2eq/MJ
$GHGIE_{actual}$	calculated annual average GHG intensity of energy used onboard during the relevant reporting period, gCO_2eq/MJ
$Compliance\ balance$	compliance balance, calculated as the absolute value of the difference between an
x_{ij}	the performance value of the i alternative with respect to the j criterion
x'_{ij}	normalized value of the i alternative with respect to the j criterion
σ_j	sample standard deviation of criterion j
$\overline{x}_j$	the mean value of criterion j
f_j	conflict measure of criterion j with all other criteria
r_{ij}	correlation coefficient between criterion i and j criterion
C_j	amount of information conveyed by criterion j
ω_j	the final normalized weight of criterion j
$d_j(i,k)$	the evaluation difference between alternative i and alternative k with respect to criterion j
$g_j(\cdot)$	the preference function for criterion j
$\pi(i,k)$	the aggregated preference index of alternative i over alternative k
$\phi^-(i)$	the negative outranking flow
$\phi^+(i)$	the positive outranking flow
$\phi(i)$	the net outranking flow

Appendix A

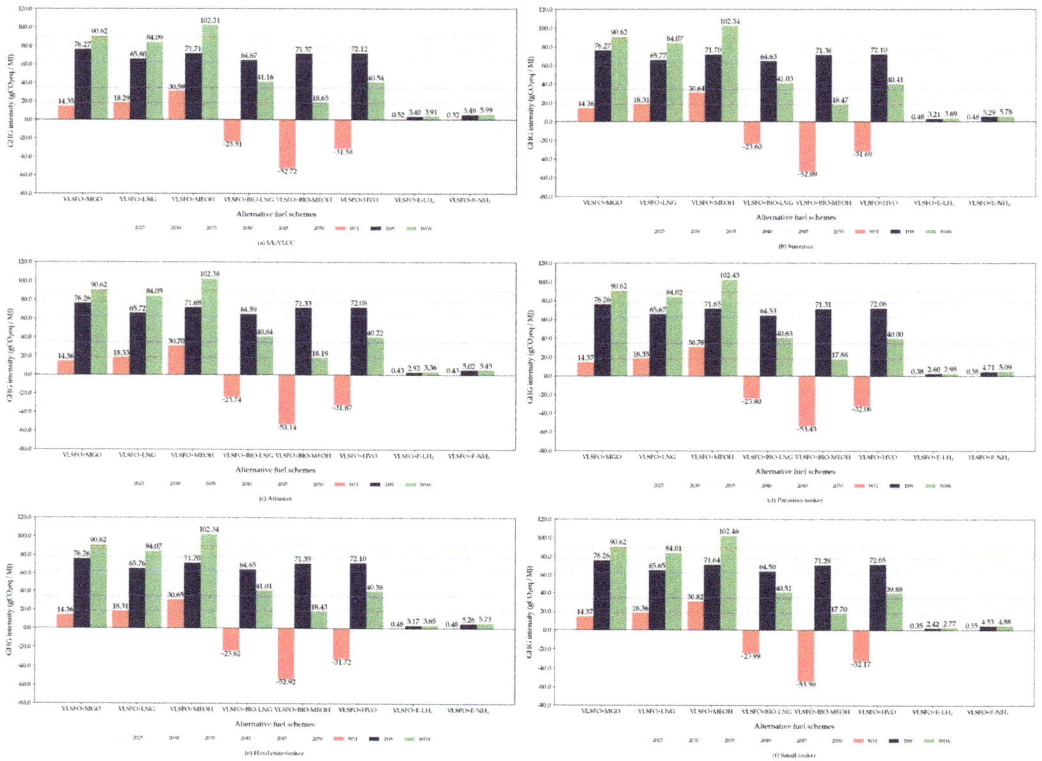

Figure A1. WtW GHG intensity of tanker ships from 2035 to 2050.

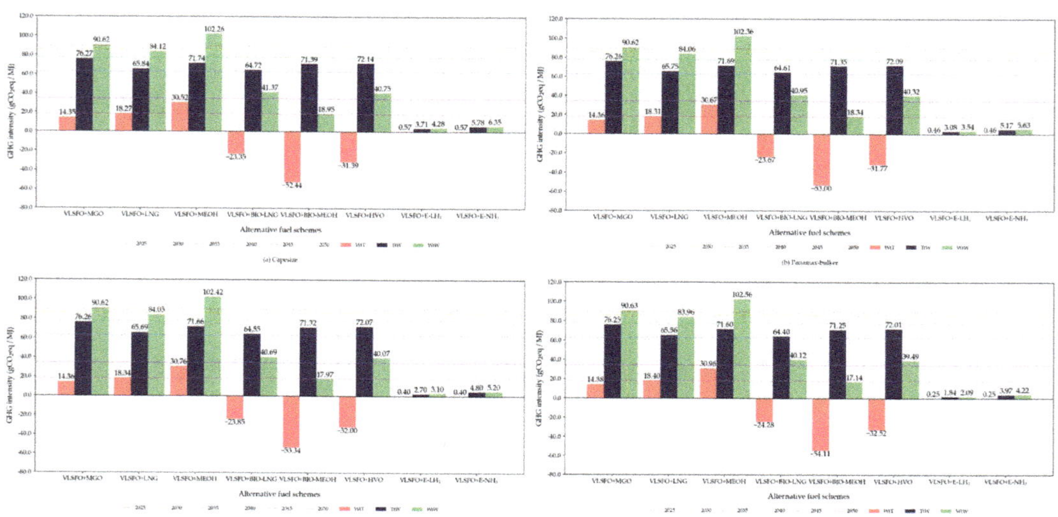

Figure A2. WtW GHG intensity of bulker ships from 2035 to 2050.

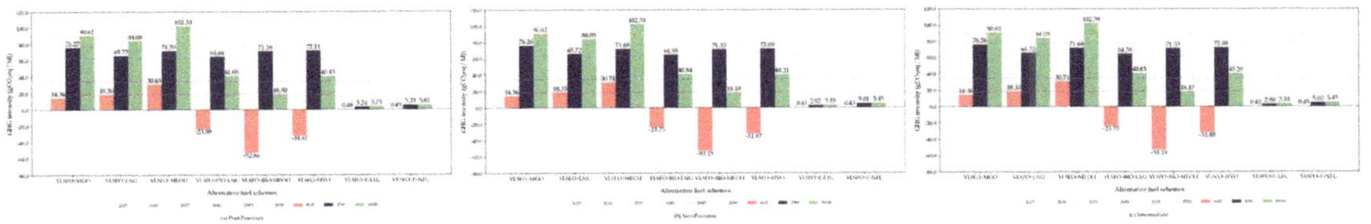

Figure A3. WtW GHG intensity of containerships from 2035 to 2050.

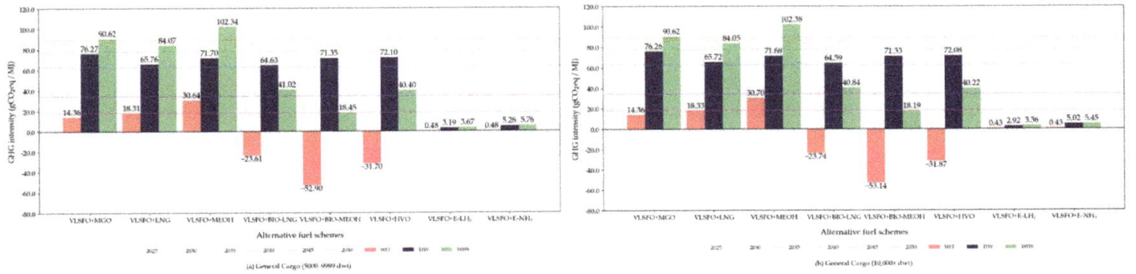

Figure A4. WtW GHG intensity of general cargo ships from 2035 to 2050.

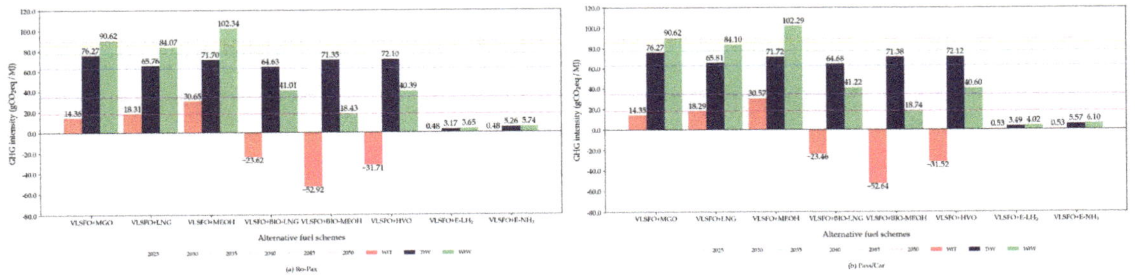

Figure A5. WtW GHG intensity of ferries from 2035 to 2050.

References

1. International Maritime Organization. Fourth Greenhouse Gas Study 2020. Available online: https://www.imo.org/en/OurWork/Environment/Pages/Fourth-IMO-Greenhouse-Gas-Study-2020.aspx (accessed on 20 April 2025).
2. Vierth, I.; Ek, K.; From, E.; Lind, J. The cost impacts of Fit for 55 on shipping and their implications for Swedish freight transport. *Transp. Res. Part A Policy Pract.* **2024**, *179*, 103894. [CrossRef]
3. International Maritime Organization. Revised GHG Reduction Strategy for Global Shipping Adopted. Available online: https://www.imo.org/en/MediaCentre/PressBriefings/pages/Revised-GHG-reduction-strategy-for-global-shipping-adopted-.aspx (accessed on 20 April 2025).
4. DNV. FuelEU Maritime—Requirements, Compliance Strategies, and Commercial Impacts. Available online: https://www.dnv.com/maritime/publications/fueleu-maritime-white-paper-download/ (accessed on 27 February 2025).
5. European Union. Official Journal of the European Union. Available online: https://eur-lex.europa.eu/legal-content/EN/TXT/?uri=OJ:L:2021:243:TOC (accessed on 20 April 2025).
6. European Union. Fit for 55 Package. Available online: https://www.europarl.europa.eu/RegData/etudes/BRIE/2022/733513/EPRS_BRI(2022)733513_EN.pdf (accessed on 20 April 2025).
7. European Commission. Decarbonising Maritime Transport—FuelEU Maritime. Available online: https://transport.ec.europa.eu/transport-modes/maritime/decarbonising-maritime-transport-fueleu-maritime_en (accessed on 27 February 2025).
8. European Commission. Questions and Answers on Regulation (EU) 2023/1805 on the Use of Renewable and Low-Carbon Fuels in Maritime Transport, and AMENDING DIRECTIVE 2009/16/EC. Available online: https://transport.ec.europa.eu/transport-modes/maritime/decarbonising-maritime-transport-fueleu-maritime/questions-and-answers-regulation-eu-20231805-use-renewable-and-low-carbon-fuels-maritime-transport_en (accessed on 27 February 2025).

9. MAN. MAN B&W ME-LGIP Dual-Fuel Engines. Available online: https://www.man-es.com/docs/default-source/document-sync-archive/b-w-me-lgip-dual-fuel-engines-dual-fuel-technology-reshapes-the-future-two-stroke-engine-operation-eng.pdf?sfvrsn=6e76dfc0_4 (accessed on 27 February 2025).
10. Wang, Y.; Xiao, X.; Ji, Y. A Review of LCA Studies on Marine Alternative Fuels: Fuels, Methodology, Case Studies, and Recommendations. *J. Mar. Sci. Eng.* **2025**, *13*, 196. [CrossRef]
11. Lee, H.; Lee, J.; Roh, G.; Lee, S.; Choung, C.; Kang, H. Comparative Life Cycle Assessments and Economic Analyses of Alternative Marine Fuels: Insights for Practical Strategies. *Sustainability* **2024**, *16*, 2114. [CrossRef]
12. Xing, H.; Stuart, C.; Spence, S.; Chen, H. Alternative fuel options for low carbon maritime transportation: Pathways to 2050. *J. Clean. Prod.* **2021**, *297*, 126651. [CrossRef]
13. Zincir, B. Environmental and economic evaluation of ammonia as a fuel for short-sea shipping: A case study. *Int. J. Hydrogen Energy* **2022**, *47*, 18148–18168. [CrossRef]
14. Ratnakar, R.R.; Gupta, N.; Zhang, K.; van Doorne, C.; Fesmire, J.; Dindoruk, B.; Balakotaiah, V. Hydrogen supply chain and challenges in large-scale LH2 storage and transportation. *Int. J. Hydrogen Energy* **2021**, *46*, 24149–24168. [CrossRef]
15. Ishaq, H.; Crawford, C. Review of ammonia production and utilization: Enabling clean energy transition and net-zero climate targets. *Energy Convers. Manag.* **2024**, *300*, 117869. [CrossRef]
16. Brouzas, S.; Zadeh, M.; Lagemann, B. Essentials of hydrogen storage and power systems for green shipping. *Int. J. Hydrogen Energy* **2025**, *100*, 1543–1560. [CrossRef]
17. Bui, K.Q.; Perera, L.P.; Emblemsvåg, J. Life-cycle cost analysis of an innovative marine dual-fuel engine under uncertainties. *J. Clean. Prod.* **2022**, *380*, 134847. [CrossRef]
18. Law, L.C.; Foscoli, B.; Mastorakos, E.; Evans, S. A Comparison of Alternative Fuels for Shipping in Terms of Lifecycle Energy and Cost. *Energies* **2021**, *14*, 8502. [CrossRef]
19. Solakivi, T.; Paimander, A.; Ojala, L. Cost competitiveness of alternative maritime fuels in the new regulatory framework. *Transp. Res. Part D Transp. Environ.* **2022**, *113*, 103500. [CrossRef]
20. Flodén, J.; Lars, Z.; Anastasia, C.; Rasmus, P.; Erik, F.; Julia, H.; Johan, R.; Woxenius, J. Shipping in the EU emissions trading system: Implications for mitigation, costs and modal split. *Clim. Policy* **2024**, *24*, 969–987. [CrossRef]
21. Trosvik, L.; Brynolf, S. Decarbonising Swedish maritime transport: Scenario analyses of climate policy instruments. *Transp. Res. Part D Transp. Environ.* **2024**, *136*, 104457. [CrossRef]
22. Deniz, C.; Zincir, B. Environmental and economical assessment of alternative marine fuels. *J. Clean. Prod.* **2016**, *113*, 438–449. [CrossRef]
23. Ren, J.; Liang, H. Measuring the sustainability of marine fuels: A fuzzy group multi-criteria decision making approach. *Transp. Res. Part D Transp. Environ.* **2017**, *54*, 12–29. [CrossRef]
24. Soltani Motlagh, H.R.; Issa Zadeh, S.B.; Garay-Rondero, C.L. Towards International Maritime Organization Carbon Targets: A Multi-Criteria Decision-Making Analysis for Sustainable Container Shipping. *Sustainability* **2023**, *15*, 16834. [CrossRef]
25. Moshiul, A.M.; Mohammad, R.; Hira, F.A. Alternative Fuel Selection Framework toward Decarbonizing Maritime Deep-Sea Shipping. *Sustainability* **2023**, *15*, 5571. [CrossRef]
26. Strantzali, E.; Livanos, G.A.; Aravossis, K. A Comprehensive Multicriteria Evaluation Approach for Alternative Marine Fuels. *Energies* **2023**, *16*, 7498. [CrossRef]
27. Zou, J.; Yang, B. Evaluation of alternative marine fuels from dual perspectives considering multiple vessel sizes. *Transp. Res. Part D Transp. Environ.* **2023**, *115*, 103583. [CrossRef]
28. Fan, L.; Gu, B.; Luo, M. A cost-benefit analysis of fuel-switching vs. hybrid scrubber installation: A container route through the Chinese SECA case. *Transp. Policy* **2020**, *99*, 336–344. [CrossRef]
29. European Commission. Regulation (EU) 2023/1805 of the European Parliament and of the Council of 13 September 2023 on the Use of Renewable and Low-Carbon Fuels in Maritime Transport, and Amending Directive 2009/16/EC (Text with EEA relevance). Available online: https://eur-lex.europa.eu/legal-content/EN/TXT/?uri=CELEX:32023R1805#ntr4-L_2023234EN.01004801-E0004 (accessed on 27 February 2025).
30. European Commission. Directive (EU) 2018/2001 of the European Parliament and of the Council of 11 December 2018 on the Promotion of the Use of Energy from Renewable Sources (Recast) (Text with EEA Relevance.). Available online: https://eur-lex.europa.eu/eli/dir/2018/2001/oj/eng (accessed on 27 February 2025).
31. Zhang, W.; He, Y.; Wu, N.; Zhang, F.; Lu, D.; Liu, Z.; Jing, R.; Zhao, Y. Assessment of cruise ship decarbonization potential with alternative fuels based on MILP model and cabin space limitation. *J. Clean. Prod.* **2023**, *425*, 138667. [CrossRef]
32. Wang, Y.; Wright, L.A. A Comparative Review of Alternative Fuels for the Maritime Sector: Economic, Technology, and Policy Challenges for Clean Energy Implementation. *World* **2021**, *2*, 456–481. [CrossRef]
33. Tran, N.K.; Haasis, H.-D. An empirical study of fleet expansion and growth of ship size in container liner shipping. *Int. J. Prod. Econ.* **2015**, *159*, 241–253. [CrossRef]

34. Behzadian, M.; Kazemzadeh, R.B.; Albadvi, A.; Aghdasi, M. PROMETHEE: A comprehensive literature review on methodologies and applications. *Eur. J. Oper. Res.* **2010**, *200*, 198–215. [CrossRef]
35. Diakoulaki, D.; Mavrotas, G.; Papayannakis, L. Determining objective weights in multiple criteria problems: The critic method. *Comput. Oper. Res.* **1995**, *22*, 763–770. [CrossRef]
36. Brans, J.P.; Vincke, P. Note—A Preference Ranking Organisation Method. *Manag. Sci.* **1985**, *31*, 647–656. [CrossRef]
37. Ship & Bunker. Integr8: VLSFO Calorific Value, Pour Point and Competitiveness with LSMGO. Available online: https://shipandbunker.com/news/world/662089-integr8-vlsfo-calorific-value-pour-point-and-competitiveness-with-lsmgo (accessed on 23 March 2025).
38. Sustainable Ships. FuelEU Maritime Guide. Available online: https://www.sustainable-ships.org/rules-regulations/fueleu (accessed on 23 March 2025).
39. International Maritime Organization. Guidelines on Life Cycle GHG Intensity of Marine Fuels. Available online: https://wwwcdn.imo.org/localresources/en/OurWork/Environment/Documents/annex/MEPC%2080/Annex%2014.pdf (accessed on 23 March 2025).
40. Lindstad, E.; Lagemann, B.; Rialland, A.; Gamlem, G.M.; Valland, A. Reduction of maritime GHG emissions and the potential role of E-fuels. *Transp. Res. Part D Transp. Environ.* **2021**, *101*, 103075. [CrossRef]
41. Ashrafi, M.; Lister, J.; Gillen, D. Toward a harmonization of sustainability criteria for alternative marine fuels. *Marit. Transp. Res.* **2022**, *3*, 100052. [CrossRef]
42. MAN. The Benefits of Methanol. Available online: https://www.man-es.com/discover/the-benefits-of-methanol (accessed on 23 March 2025).
43. DNV. Biofuels in Shipping—Current Market and Guidance on Use and Reporting. Available online: https://www.dnv.com/maritime/publications/Biofuels-in-shipping-white-paper-2025-download/ (accessed on 23 March 2025).
44. Wang, Y.; Cao, Q.; Liu, L.; Wu, Y.; Liu, H.; Gu, Z.; Zhu, C. A review of low and zero carbon fuel technologies: Achieving ship carbon reduction targets. Sustain. *Energy Technol. Assess.* **2022**, *54*, 102762. [CrossRef]
45. Lagemann, B.; Lagouvardou, S.; Lindstad, E.; Fagerholt, K.; Psaraftis, H.N.; Erikstad, S.O. Optimal ship lifetime fuel and power system selection under uncertainty. *Transp. Res. Part D Transp. Environ.* **2023**, *119*, 103748. [CrossRef]
46. Clarksons Research. Shipping Intelligence Weekly 20/12/2024. Available online: https://www.clarksons.net.cn/api/download/SIN/download/718365d9-1918-4b7b-91fb-b92ba474fd91?downloadInline=true (accessed on 27 March 2025).
47. Korberg, A.D.; Brynolf, S.; Grahn, M.; Skov, I.R. Techno-economic assessment of advanced fuels and propulsion systems in future fossil-free ships. *Renew. Sustain. Energy Rev.* **2021**, *142*, 110861. [CrossRef]
48. ABS. View of The Value Chain. Available online: https://absinfo.eagle.org/acton/attachment/16130/f-c04f4eaf-a121-4560-9e15-5a56ee443f32/1/-/-/-/-/sustainability-outlook-iii-21055-web.pdf (accessed on 27 March 2025).
49. Methanol Institute. Economic Value of Methanol for Shipping Under FuelEU Maritime and EU ETS. Available online: https://www.methanol.org/wp-content/uploads/2024/09/ECONOMIC-VALUE-OF-METHANOL-FOR-SHIPPING-PAPER_final.pdf (accessed on 27 March 2025).
50. Lorenzi, G.; Baptista, P.; Venezia, B.; Silva, C.; Santarelli, M. Use of waste vegetable oil for hydrotreated vegetable oil production with high-temperature electrolysis as hydrogen source. *Fuel* **2020**, *278*, 117991. [CrossRef]
51. ABS. Beyond the Horizon: Carbon Neutral Fuel Pathways and Transformational Technologies. Available online: https://ww2.eagle.org/en/interactive/interactive-publications/current/outlook-2024.html (accessed on 27 March 2025).
52. Hansson, J.; Månsson, S.; Brynolf, S.; Grahn, M. Alternative marine fuels: Prospects based on multi-criteria decision analysis involving Swedish stakeholders. *Biomass Bioenergy* **2019**, *126*, 159–173. [CrossRef]
53. Notteboom, T.E.; Vernimmen, B. The effect of high fuel costs on liner service configuration in container shipping. *J. Transp. Geogr.* **2009**, *17*, 325–337. [CrossRef]

Journal of
Marine Science
and Engineering

MDPI

Article

Toward Smart and Sustainable Port Operations: A Blue Ocean Strategy Approach for the Spanish Port System

Nicoletta González-Cancelas [1], Juan José Guil López [2], Javier Vaca-Cabrero [1,*] and Alberto Camarero-Orive [1]

[1] Department of Transport, Territorial and Urban Planning Engineering, ETSI Caminos, Canales y Puertos, Universidad Politécnica de Madrid, Calle Profesor Aranguren, 3, 28040 Madrid, Spain; ncioleta.gcancelas@upm.es (N.G.-C.)
[2] MSc Civil Engineering, Universidad Politécnica de Madrid, Calle Profesor Aranguren, 3, 28040 Madrid, Spain
* Correspondence: j.vaca@upm.es

Abstract: The digital transformation of the maritime sector, driven by Industry 4.0, is reshaping port operations toward smarter and more sustainable models. This paper analyzed the implementation of Port 4.0 technologies in the Spanish port system through the lens of the Blue Ocean Strategy. By redefining competitive boundaries and applying tools such as the Four Actions Framework and value innovation curves, the study proposes a new strategic vision where ports collaborate rather than compete. Key enabling technologies (such as Big Data, IoT, AI, and Blockchain) were assessed for their capacity to optimize energy use, reduce emissions, and enhance operational efficiency. The findings highlight the potential for a unified, data-driven port ecosystem that creates a new uncontested market space for Spanish ports while promoting environmental and economic sustainability.

Keywords: Blue Ocean Strategy; Port 4.0; digitalization; maritime sustainability; energy efficiency; green port operations

1. Introduction

The maritime and port sector stands as one of the fundamental pillars of the global economy, enabling the circulation of goods, raw materials, and energy on a planetary scale. However, while port infrastructure has historically served as a backbone for industrial expansion, its strategic positioning has become increasingly challenged by the confluence of complex dynamics [1]. These include rising environmental pressures, technological disruption, increasing urban–port tensions, and a growing demand for multimodal, sustainable, and intelligent logistics chains.

In Europe, and particularly in Spain, port systems operate in a delicate balance between public mandate and competitive market logic as well as between innovation initiatives and institutional inertia [2]. In this context, achieving systemic transformation requires more than the addition of digital tools or energy projects; it demands a redefinition of value, governance, and purpose. While initiatives such as the European Green Deal and national Port 4.0 frameworks offer a normative reference, the strategic orientation of port authorities often remains reactive and fragmented.

This paper addresses the need for a cohesive, anticipatory model capable of overcoming these constraints by leveraging the port system not as a collection of isolated infrastructures, but as a coordinated, strategic ecosystem.

The Spanish port system is composed of 28 autonomous port authorities, each with distinct investment policies, digital strategies, and degrees of technological maturity. This

https://doi.org/10.3390/jmse13050872

atomization generates significant challenges in terms of redundancy, operational inefficiencies, and institutional misalignment [3]. While some ports (such as Barcelona, Valencia, or Algeciras) have developed advanced digital platforms and sustainability programs, others remain dependent on outdated administrative systems and lack interoperability [2].

More critically, competition between ports under the same national umbrella has led to suboptimal infrastructure allocation and missed opportunities for synergies [4]. The existing regulatory framework allows each port authority a high level of independence, but in doing so, it also restricts the capacity to act collectively in response to national or European imperatives.

Moreover, digitalization efforts have often been undertaken in parallel rather than collaboratively, exacerbating fragmentation. This structural dispersion has a direct impact on the capacity of the system to attract non-traditional actors, innovate across sectors, or lead sustainable urban–port integration. It also weakens the ability of ports to function as coordinated nodes in a global smart logistics network [5]. Thus, while each port may pursue its own strategy, the absence of a shared strategic architecture significantly undermines the performance and transformative potential of the system as a whole.

In response to these structural challenges, this paper proposes the adoption of a Blue Ocean Strategy (BOS) approach as a means to reconfigure the Spanish port system from within [6]. Rather than engaging in incremental improvements or isolated modernization projects, the BOS invites a strategic shift from competing within existing boundaries to creating new value spaces [7].

Applied to ports, this means breaking away from traditional rivalry and instead constructing an ecosystem based on collaboration, digital interoperability, and shared innovation [8]. The BOS framework offers specific tools for this transformation including the Value Curve, ERRC Matrix, PEC Map, and Non-Customer Analysis [9].

These tools help redefine not only the services that ports offer, but also the actors they engage with and the outcomes they prioritize. Through this approach, ports are no longer seen as logistical endpoints, but as platform orchestrators capable of catalyzing green, digital, and territorial innovation [10]. Such a reframing is crucial not only for boosting competitiveness, but also for ensuring an alignment with the global sustainability goals and national strategic agendas. By focusing on systemic redesign rather than isolated performance indicators, the BOS approach provides a roadmap for reconciling efficiency, innovation, and resilience in the maritime and port domain [11].

Despite the relevance of the port sector to national economies and global trade, the strategic transformation of port systems has received limited attention in both the academic literature and institutional reform [12]. Most studies have tended to focus on individual ports, benchmarking digital maturity or measuring environmental performance in isolation [13]. While valuable, these approaches often miss the systemic dimension of port governance, particularly in contexts where multiple authorities operate under a shared institutional umbrella [14].

There has been a notable lack of strategic design methods applied to the port system as a whole. The Blue Ocean Strategy has been widely adopted in private-sector innovation but is rarely applied to public infrastructure systems or national-scale networks [15]. This paper addresses that gap by adapting BOS tools to a port ecosystem with public ownership, regulated competition, and strong ties to urban and environmental dynamics.

It also contributes to the policy discussion by offering a structured pathway for aligning innovation, investment, and governance in line with the Spanish Port 4.0 initiative and the European Union's digital and climate agendas [3]. As such, the paper situates itself at the intersection of strategic design, public-sector transformation, and sustainable logistics innovation [2].

The objective of this paper was to design and propose a strategic transformation model for the Spanish port system, based on the principles and tools of the Blue Ocean Strategy. The approach aims to shift the system away from fragmented competition and toward a collaborative, interoperable, and sustainability-driven ecosystem.

In order to achieve this, the paper follows a four-phase methodological structure. First, it conducts a diagnosis of the current "red ocean" conditions characterizing the Spanish port sector. Second, it applies core BOS tools to identify the value gaps, latent opportunities, and innovation leverage points. Third, it formulates a strategic vision based on BOS principles, ensuring alignment with user needs, technological feasibility, and stakeholder adoption. Finally, it consolidates these elements into a coherent model of the Blue Ocean Strategy formulation for the port system. Each section of the paper builds sequentially on the previous one, combining conceptual frameworks with practical diagnostics and strategic design outputs. The conclusions reflect on the implications of this model for policy, port governance, and future research, positioning this work as a foundational reference for port innovation strategies in complex institutional contexts.

2. Literature Review

2.1. Strategic Approaches in Port Governance

Port governance has evolved significantly in recent decades, with increasing attention paid to models that balance efficiency, public interest, and global competitiveness. Traditionally, the literature has distinguished between landlord, tool, and service port models, with the landlord model prevailing across Europe including Spain [16]. However, beyond this typology, scholars have highlighted the importance of governance as a dynamic, strategic function, and not merely as an administrative control [17]. In this view, ports are no longer neutral facilitators of trade, but strategic actors embedded in complex territorial, logistic, and digital ecosystems [18].

Recent contributions have expanded this debate by emphasizing multilevel governance, particularly in the context of port–city integration and the emergence of metropolitan logistics platforms [19]. Studies have shown that governance fragmentation among port authorities can hinder long-term investment planning, policy coherence, and the development of shared digital infrastructures [20]. Nonetheless, most academic analyses remain descriptive or diagnostic, offering limited prescriptive frameworks to reconfigure governance logics or strategic alignments across port networks [21].

In [22], the authors discussed governance mechanisms and cross-border digital investment risks in emerging port ecosystems, highlighting the importance of coordinated management structures—an approach that resonates with our own proposal for integrated governance within the Spanish port system.

2.2. Digitalization and Smart Port Strategies

Digital transformation is reshaping the maritime and port sector, enabling the redesign of operational, logistical, and institutional processes [3]. The "Smart Port" paradigm has emerged as a response to increasing demands for efficiency, traceability, and sustainability, supported by the deployment of enabling technologies such as the Internet of Things (IoT), artificial intelligence (AI), blockchain, big data analytics, and 5G connectivity [23]. These technologies are not merely additive tools, but drivers of new organizational capabilities, decision-making logics, and service models. In this context, the digitalization of port operations is no longer optional; it is becoming a prerequisite for competitiveness in a global logistics environment increasingly governed by automation, data-driven optimization, and real-time responsiveness [24].

Numerous European ports have initiated digitalization projects aligned with this trend. In Spain, ports such as Valencia, Barcelona, and Bilbao have deployed initiatives ranging from smart access control systems and cargo tracking platforms to AI-based predictive maintenance and port community systems (PCSs) [2]. National programs such as the Port 4.0 Fund (launched by Puertos del Estado) have played a key role in supporting technological innovation, financing over 150 pilot projects across the port-logistics chain. At the European level, initiatives like the Digital Transport and Logistics Forum (DTLF) and the European Maritime Single Window environment (EMSWe) have further pushed for harmonization, interoperability, and the standardization of digital services across borders [3].

However, despite this momentum, the digitalization of Spanish ports remains fragmented. Each port authority tends to develop its own digital tools, procurement models, and technological priorities, often without alignment or interoperability with the rest of the system [25]. The academic literature has shown that this institutional dispersion results in duplicated efforts, uneven innovation capacities, and inefficiencies in data sharing. Moreover, while many studies have focused on the technical dimensions of smart ports (sensor networks, terminal automation, or cybersecurity protocols), fewer works have examined the strategic, organizational, or governance implications of digital transformation (Figure 1). Without systemic alignment, smart port projects may remain isolated upgrades, falling short of catalyzing ecosystem-wide transformation [26].

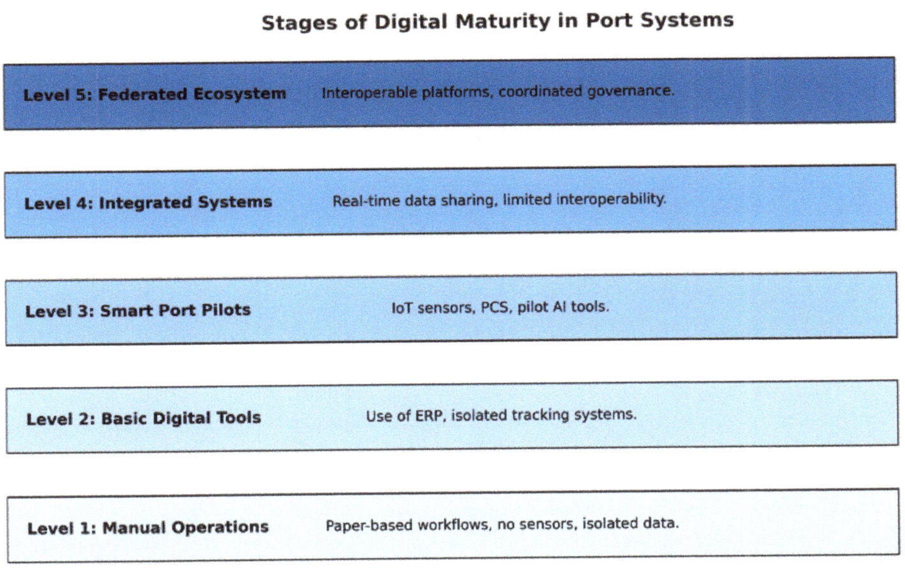

Figure 1. Stages of digital maturity in port systems.

Visual representation shows the progressive transition from manual, fragmented port operations to a fully integrated and interoperable digital ecosystem. Each stage reflects a qualitative leap in both technological deployment and strategic alignment.

To fully realize the potential of digitalization, ports must transition from isolated smart initiatives to a coordinated digital ecosystem [27]. This requires the development of shared standards, federated digital infrastructures, and integrated innovation policies at the national level. The concept of a smart port should therefore be reframed from a purely technical or operational perspective to one that incorporates institutional collaboration, cross-sectoral integration, and long-term strategic planning [28]. Such a shift would not only enhance efficiency, but also enable ports to serve as innovation platforms for urban mobility, energy transition, and smart logistics, aligning with broader public policy goals and creating a foundation for collective intelligence and resilience across the maritime-port system.

The conceptual model in Figure 2 shows a federated digital infrastructure linking autonomous port authorities through a shared interoperability layer. This layer enables coordinated interaction with urban systems, logistics platforms, public institutions, and academic stakeholders [10].

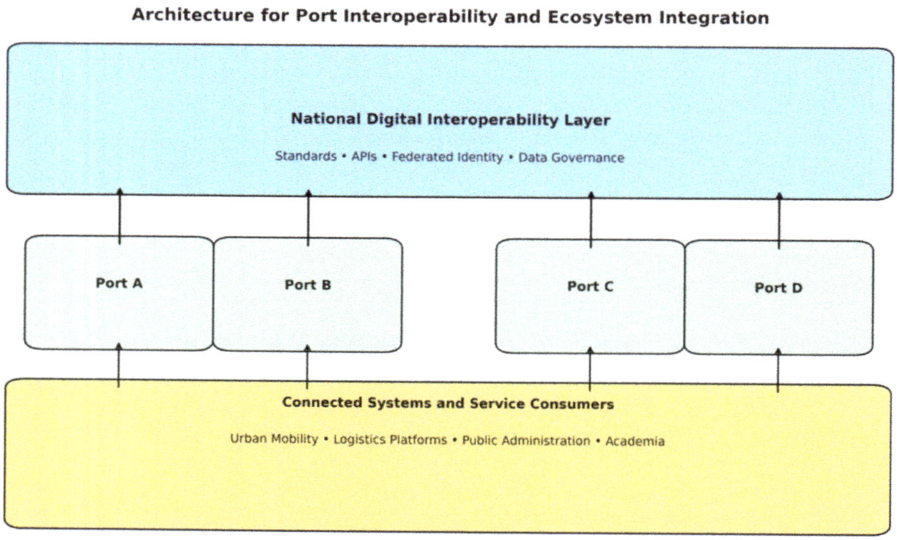

Figure 2. Architecture for port interoperability and ecosystem integration.

2.3. Port Competition, Cooperation, and Territorial Impact

Ports operate within a framework of economic rivalry, often competing for traffic, investment, and strategic positioning. In Spain, this competition has been formalized through financial autonomy and performance-based evaluation criteria [25]. However, the literature increasingly questions the long-term sustainability of this model, especially when it fosters a duplication of services, underutilized infrastructures, and fragmented innovation. Many authors have argued for a shift from inter-port competition to "coopetition", where ports cooperate within regional or national systems to generate mutual gains [14].

Ports are also increasingly analyzed in relation to their territorial and urban contexts. The evolution of port–city relations, including land use conflicts, environmental externalities, and the role of ports in metropolitan mobility, has become a critical area of study [29]. This body of work supports the idea that ports should be repositioned not merely as transport hubs, but as integrative nodes in regional innovation systems and sustainable development agendas [30]. However, while the territorial implications have been well-documented, fewer studies have explored how these factors can be operationalized in a strategic framework that aligns governance, innovation, and urban policy [31].

The dynamics of competition and cooperation among ports have long been discussed in the academic literature [15,32,33]. While competition is often seen as a driver of efficiency and innovation, excessive rivalry (especially within national systems) can lead to redundant infrastructure, disjointed investments, and diminished collective impact. The financial and managerial autonomy granted to each port authority often leads to internal competition, which undermines system-wide coordination [25]. This model, while aligned with neoliberal governance trends, has generated structural inefficiencies, particularly in the duplication of services and the absence of coordinated long-term planning. Scholars such as [4] have called for a shift toward "coopetition", wherein ports compete externally while collaborating internally through shared platforms, policy alignment, and harmonized investment strategies [34].

Simultaneously, ports are no longer understood solely as logistical infrastructures, but as actors embedded within complex territorial systems. Port–city relations, environmental conflicts, and intermodal connectivity have become central themes in port governance. The expansion of port hinterlands and the integration of ports into urban mobility and logistics networks require a reconceptualization of the port as a multiscalar, multifunctional node [35]. While several studies have addressed the challenges of port–city coexistence, few have offered strategic frameworks that operationalize this complexity into actionable models. The present paper contributes to this gap by linking governance reform and strategic innovation to the broader objective of integrating ports into sustainable territorial development processes [36].

2.4. Research Gap: Systemic Strategic Transformation of Port Ecosystems

Despite the growing body of literature on digitalization, governance, and port–city integration, there remains a significant gap regarding the systemic strategic transformation of entire port ecosystems, particularly at a national level. Most research has focused on individual ports or clusters, rarely addressing the design of coordinated strategies that span multiple autonomous authorities [37]. Moreover, while the Blue Ocean Strategy has been widely applied in business innovation, its adaptation to public infrastructure systems, especially ports, remains underexplored [38].

There is also limited academic guidance on how to reconstruct market boundaries, reach non-customers, or structure innovation portfolios within the port sector. Concepts such as value curves, ERRC matrices, or PEC maps are virtually absent from the port literature, despite their relevance for addressing fragmentation and unlocking new value propositions [39]. As a result, port systems continue to pursue incremental modernization without fully harnessing strategic design as a transformation tool [40].

This paper addresses this gap by applying the full BOS methodology to the Spanish port system. It offers not only a theoretical adaptation of strategic innovation tools to a public-sector setting, but also a prescriptive model grounded in diagnostic analysis, methodological rigor, and alignment with real-world policy frameworks (e.g., Port 4.0, EU Green Deal). In doing so, it contributes to reimagining the role of ports in national innovation strategies and territorial development.

3. Methodology

This study adopted a conceptual and strategic methodology based on the framework of the Blue Ocean Strategy (BOS), developed by Kim and Mauborgne [41], as a structured approach to redefining the Spanish port system under the principles of Port 4.0. The method was applied sequentially in four distinct stages: diagnosis, application of BOS tools, strategic formulation, and the construction of a new operational model (Figure 3).

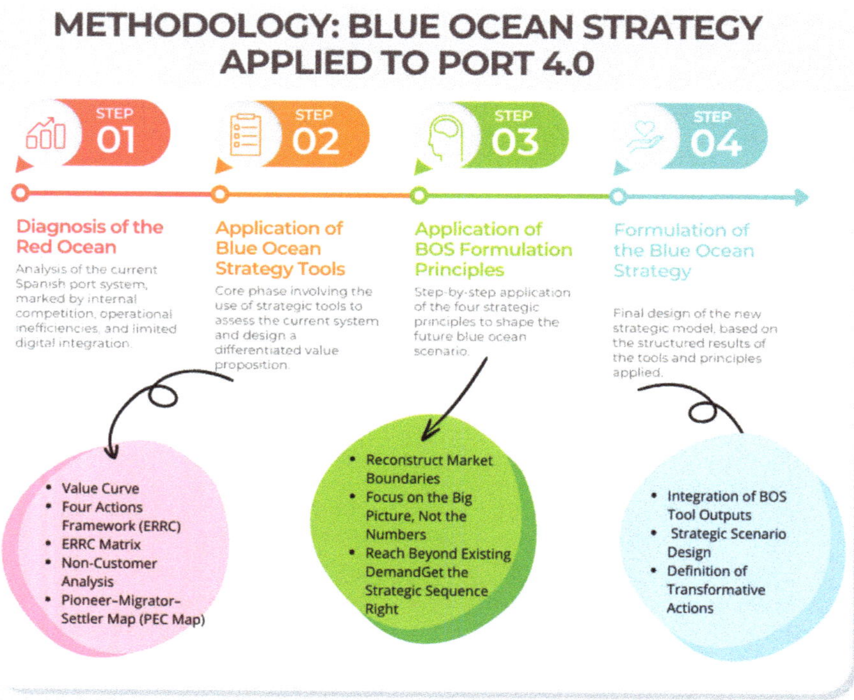

Figure 3. Methodology.

3.1. *Diagnosis of the Red Ocean*

The first stage consists of a critical analysis of the current Spanish port landscape, conceptualized as a "red ocean", which is characterized by intense internal competition, fragmented governance, limited technological integration, and uneven operational performance across port authorities. This diagnostic phase aims to identify inefficiencies, overlaps, and systemic limitations that hinder the sector's ability to transition toward sustainability and digital transformation.

3.2. *Application of Strategic BOS Tools*

The second stage focuses on the practical deployment of analytical tools derived from the BOS framework to structure the current competitive environment and support the formulation of a differentiated value proposition.

1. The Value Curve was used to map the key variables on which the Spanish port system currently competes, providing a visual summary of service features and investment intensity across authorities.
2. The Four Actions Framework (Eliminate–Reduce–Raise–Create) guides strategic thinking around which elements of the current system should be restructured to break away from the existing competition.
3. The ERRC Matrix, derived from the previous framework, summarizes the proposed changes by categorizing them into those that should be removed, scaled down, enhanced, or newly introduced.
4. Non-Customer Analysis identifies users and stakeholders not currently engaged by the port ecosystem—such as urban actors, smart mobility services, or digital

platform providers—and explores how they can be incorporated into the new value proposition.

5. The Pioneer–Migrator–Settler (PEC) Map classifies innovation opportunities according to their disruptive potential and market maturity, supporting the prioritization of strategic actions based on impact and feasibility.

These tools are not used in isolation but as a coherent system to extract actionable insights and define the foundations of a new, innovation-driven strategic scenario.

3.3. Application of the BOS Formulation Principles

Building upon the structured outputs from the previous stage, the methodology continues with the application of the four BOS formulation principles. These principles provide a logical and sequential pathway for transitioning from conceptual design to a coherent strategic blueprint.

1. Reconstruct market boundaries: The port sector is redefined as a cooperative and interoperable digital ecosystem rather than a collection of competing infrastructures. Market boundaries are expanded to include integrated logistics chains and data-sharing environments.
2. Focus on the big picture, not the numbers: The analysis shifts from isolated operational metrics to a system-wide perspective that considers long-term value creation, interoperability, and environmental impact.
3. Reach beyond existing demand: Strategic attention is directed toward actors and needs not currently addressed including digital service providers, intermodal urban mobility agents, and non-traditional port users.
4. Get the strategic sequence right: The proposed model is validated following a logical progression—ensuring utility for stakeholders, designing an accessible price strategy, optimizing cost structures, and addressing barriers to adoption such as technological resistance or regulatory inertia.

3.4. Formulation of the Blue Ocean Strategy

The final stage consists of synthesizing the information, insights, and strategic elements developed in the previous phases to formulate an integrated Blue Ocean Strategy tailored to the Spanish port system. This involves:

1. Consolidating the results of the BOS tools: The value curve, ERRC matrix, non-customer analysis, and PEC map are combined to generate a clear and actionable set of priorities for transformation.
2. Designing the strategic scenario: A conceptual model is articulated that emphasizes collaboration between port authorities, a shared digital platform for real-time operations and data exchange, and the implementation of sustainable technologies.
3. Defining transformative actions: Specific strategic initiatives are outlined such as unified governance frameworks, standardization of digital systems, investment in smart energy solutions, and the development of port–city synergies.

This methodology enables the construction of an uncontested market space aligned with the principles of Port 4.0 and sustainability while also proposing an operational transformation applicable to other port ecosystems facing similar challenges.

4. Results and Discussion

The methodological framework outlined in the previous section was applied to systematically analyze the current Spanish port system and explore a strategic alternative based on the Blue Ocean Strategy. This section presents the results obtained in each of the four methodological stages, discussing their implications in terms of operational transformation, strategic positioning, and the creation of a new value-driven, sustainable port ecosystem. The findings demonstrate how this approach enables the sector to move beyond competition and toward collaborative innovation.

4.1. Diagnosis of the Red Ocean

The initial assessment of the Spanish port system confirmed a number of structural and operational inefficiencies that justify the need for a strategic transformation. Among the key findings were the following:

- High internal competition among the port authorities, leading to inefficiencies in service delivery and price-based rivalry;
- Digital fragmentation, with unequal technological maturity levels across ports and no national interoperability;
- Lack of a unified vision, resulting in redundant investments and duplicated infrastructure;
- Environmental lag, as many ports still rely heavily on fossil fuels, lacking energy-efficient alternatives.

These conditions are typical of a "red ocean", where ports operate under conventional assumptions, compete for limited demand, and fail to create new market spaces.

4.2. Application of BOS Tools

To move beyond these limitations, several core tools of the Blue Ocean Strategy were applied. These tools allow for a structured diagnosis of the value dynamics in the current system and support the formulation of a differentiated strategic vision.

4.2.1. Value Curve

The Value Curve is one of the central tools in the Blue Ocean Strategy methodology. It enables a visual representation of the current competitive profile of an industry and serves as a basis for identifying how to differentiate strategically.

This study constructed a value curve to compare the traditional Spanish port model (Red Ocean) with the proposed strategy aligned with the principles of Port 4.0 and sustainable innovation (Blue Ocean). Seven key strategic variables were identified based on the literature, policy documents, and the analysis conducted in the diagnostic phase (Table 1):

1. Operational efficiency;
2. Inter-port collaboration;
3. Digitalization;
4. Energy sustainability;
5. Data interoperability;
6. Attraction of non-customers;
7. Shared governance.

Each variable was assessed on a 1–5 scale, with 1 indicating low or fragmented implementation and 5 representing a high and consistent level of value delivery across the system.

The traditional model was positioned with modest performance in efficiency, collaboration, and governance, and very low levels in areas such as digitalization and sustainability. In contrast, the proposed Blue Ocean model reflected high value creation across all dimensions, especially those related to innovation, interoperability, and environmental impact.

Table 1 presents the strategic variables considered in the value curve including their definitions and evaluation criteria. The ERRC Matrix, discussed in Section 4.2.2, summarizes the strategic actions derived from these variables.

Table 1. Strategic variables: Evaluation criteria.

Strategic Variable	Justification	Evaluation Criteria
Operational efficiency	Reflects the capacity of ports to manage flows efficiently, reduce turnaround times, and optimize internal processes.	Scored from 1 (fragmented, manual, or inefficient) to 5 (fully optimized, automated, and integrated).
Inter-port collaboration	Assesses the degree of coordination and cooperation among Spanish ports to avoid duplicated services and enable synergies.	Scored from 1 (isolated and competitive) to 5 (highly coordinated and collaborative).
Digitalization	Captures the extent to which ports have adopted digital systems for logistics, administration, and decision-making.	Scored from 1 (analog/manual operations) to 5 (full implementation of Port 4.0 technologies).
Energy sustainability	Measures the commitment to green practices, electrification of infrastructure, and reduction in carbon emissions.	Scored from 1 (heavy fossil fuel dependence) to 5 (clean energy adoption and electrification).
Data interoperability	Evaluates the availability and quality of data exchange across systems, platforms, and port stakeholders.	Scored from 1 (no interoperability or real-time data flow) to 5 (standardized, open digital infrastructure).
Attraction of non-customers	Indicates the effort to integrate new users and services (e.g., urban mobility actors, tech providers) into the port ecosystem.	Scored from 1 (ports serving only traditional clients) to 5 (integration of new actors and services).
Shared governance	Assesses whether port governance is unified and strategically aligned at the national level rather than fragmented per authority.	Scored from 1 (autonomous port-level governance) to 5 (shared national strategy and management alignment).

Figure 4 shows the resulting value curve. The Red Ocean strategy revealed a system heavily focused on competition, cost fragmentation, and limited collaboration. In contrast, the Blue Ocean strategy shifted the focus toward energy efficiency, transparency, synergies, data flow, and a unified logistics chain, suggesting a reconfiguration of the sector's value architecture.

Figure 4 illustrates the Value Curve, comparing the current competitive profile of Spanish ports with the proposed strategic repositioning, highlighting the shift from fragmented operations to integrated value creation.

The values assigned to each strategic variable in the curve were estimated on a 1–5 scale based on expert judgment, the analysis of institutional documents, and a comparative evaluation of the port authority strategies and capacities.

The values assigned to each strategic variable in the value curve were determined using a qualitative scale from 1 (low strategic value) to 5 (high strategic value). These estimations were based on the comparative analysis of port authority practices, institutional reports, and expert judgment. The Red Ocean line reflects the prevailing conditions in the Spanish port system, while the Blue Ocean line represents the expected value delivery under the proposed strategic transformation.

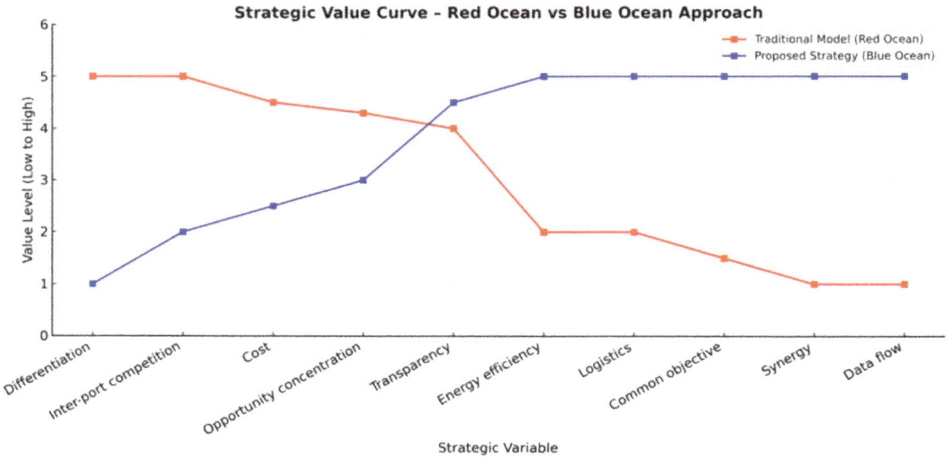

Figure 4. Strategic value curve—traditional model vs. Blue Ocean Strategy.

This reorientation implies:

- A reduction in redundant infrastructures and competitive behavior;
- The creation of new value through digital platforms and shared governance;
- The amplification of collective efficiency by aligning objectives across ports.

This curve laid the foundation for the subsequent application of the Four Actions Framework and the broader strategic transformation proposed.

4.2.2. Four Actions Framework (ERRC)

The Four Actions Framework (ERRC—Eliminate, Reduce, Raise, Create) provides a structured tool to analyze how the current value proposition of the Spanish port system can be reconfigured.

Following the construction of the value curve, the next step in the application of the Blue Ocean Strategy methodology was to implement the Four Actions Framework, also known as the ERRC grid. This tool enables a strategic rethinking of the industry by pushing organizations to systematically eliminate, reduce, raise, and create elements in their current value proposition. It is designed to break away from incremental innovation and instead foster bold, value-driven transformation.

In the context of the Spanish port system, the ERRC framework was used to derive actionable insights based on the limitations and opportunities identified in the diagnostic phase and the value curve analysis. The results were summarized in the ERRC Matrix (see Table 1), which proposed the following strategic directions:

Eliminate:

The analysis highlighted the need to eliminate two core aspects of the current system that no longer deliver strategic value:

- Price-based competition among port authorities, which leads to inefficiencies and a race to the bottom in service provision;
- Redundant port services and infrastructures, especially in areas where duplication adds operational cost but not value to users.

Reduce:

Several practices, while not entirely obsolete, were identified as misaligned with the long-term goals of sustainability and integration:

- Isolated infrastructure investments carried out independently by each port, without national coordination or shared priorities;
- Manual administrative procedures that persist in documentation, customs, and operational workflows, thus increasing the delays and error rates.

Raise:

To modernize the port system and enhance its strategic positioning, the following elements should be strengthened:

- Digital maturity standards across all port authorities, ensuring a minimum level of interoperability and integration;
- Adoption of green energy sources such as on-shore power supply (OPS) and the electrification of port operations;
- Deployment of IoT and sensor technologies to enhance monitoring, maintenance, and real-time decision-making.

Create:

The transformation proposed by the Blue Ocean approach also requires the creation of new capabilities and infrastructures:

- A national data-sharing platform that enables seamless interoperability between ports, logistics operators, and urban mobility systems;
- A shared governance framework, ensuring strategic alignment and coordinated action across the entire Spanish port system.

The ERRC Matrix (Table 1) synthesizes these strategic directions and serves as a bridge between diagnostic tools and the subsequent formulation of the Blue Ocean Strategy.

4.2.3. Non-Customer Analysis

A key principle of the Blue Ocean Strategy is the ability to expand market reach by engaging not only existing customers, but also with those who are currently outside the conventional demand spectrum—so-called non-customers. In the context of the Spanish port system, this strategic lens is essential to unlock latent value, foster cross-sector innovation, and enhance the port–city relationship.

The non-customer analysis in this study followed the classic BOS structure of three tiers of non-customers:

First Tier: "Soon-to-be" Non-Customers

These are stakeholders who interact occasionally or marginally with the port system but do not perceive it as a strategic ally or platform for development. In the Spanish case, this includes:

- Urban mobility providers (e.g., municipal transport agencies, shared mobility platforms);
- Local governments and metropolitan planning bodies, which typically operate independently from port decision-making.

These actors could become allies through joint planning, shared digital infrastructures, and logistics–mobility integration strategies.

Second Tier: "Refusing" Non-Customers

These stakeholders have consciously chosen not to engage with ports due to a lack of incentives, accessibility, or technological compatibility. Examples include:

- Digital solution providers (e.g., startups and SMEs in IoT, AI, Blockchain) excluded from port procurement due to rigid, traditional contracting models;
- Energy tech companies not integrated into port electrification strategies.

By adapting procurement models and aligning with innovation priorities, ports can attract these actors and enrich their digital and sustainability ecosystems.

Third Tier: "Unexplored" Non-Customers

These actors have never been considered potential users or partners of the port system. In this study, this group included:

- Inland logistics hubs and intermodal platforms, which currently operate in silos;
- Academic and research institutions specializing in AI, robotics, energy systems, or governance, rarely engaged beyond consulting.

By embracing these groups, ports can open up new pathways for R&D collaboration, training, and knowledge-driven governance.

Figure 5 could illustrate these three tiers in concentric circles or a stepped funnel, making it easier for readers to visualize the expansion of strategic scope.

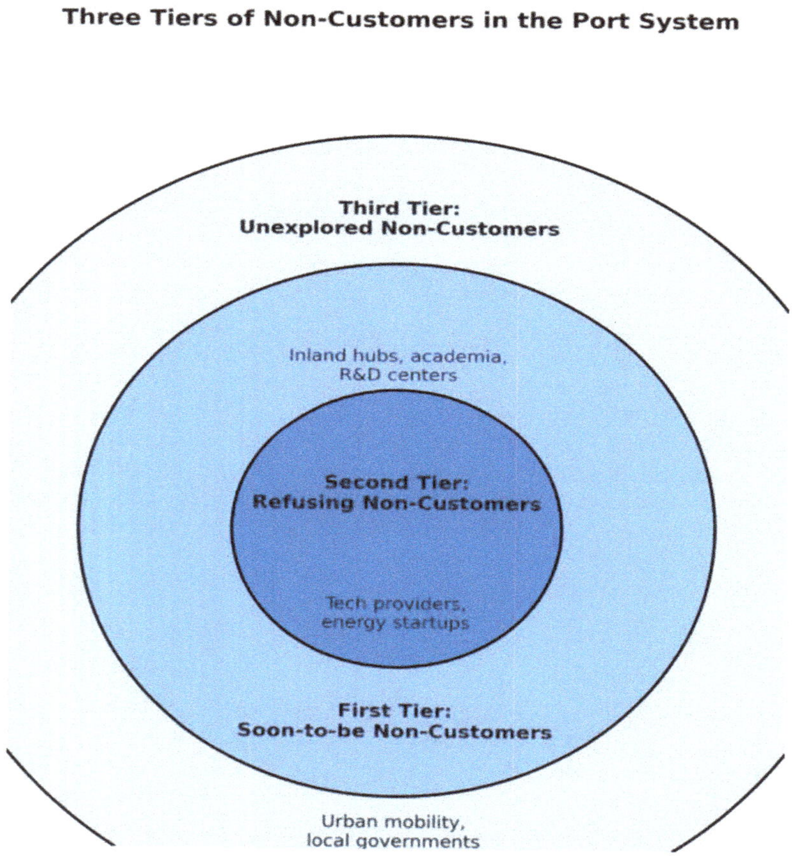

Figure 5. Three tiers of non-customers.

The incorporation of non-customers into the port strategy directly contributes to:

- The creation of new services and demand spaces;
- Increased innovation density around port infrastructure;
- Stronger ties between ports and their urban, social, and digital environments.

This analysis reinforces the idea that ports must transition from infrastructure managers to platform orchestrators in an increasingly interconnected and intelligent ecosystem.

4.2.4. PEC Map (Pioneers, Migrators, Settlers)

The Pioneer–Migrator–Settler (PEC) Map is a strategic tool within the Blue Ocean Strategy framework used to evaluate the innovation profile of an organization or sector. It classifies initiatives or capabilities based on their degree of originality, value creation potential, and market maturity, offering a portfolio view of current and future strategic assets.

In the context of the Spanish port system, the PEC Map was applied to classify the various initiatives and technologies identified throughout this study. The classification helps to prioritize investments, sequence transformation efforts, and balance risk and return in the implementation of a national Blue Ocean Strategy.

Pioneers

These are initiatives that represent true value innovation and open up new market space. They are forward-looking, high-impact, and mostly unexplored in the current port landscape. In this case:

- Interoperable digital infrastructure across all port authorities;
- Real-time data-sharing platforms connecting ports, logistics, and cities;
- Port-wide electrification strategies integrated with national energy transition plans;
- Collaborative governance frameworks that replace competition with alignment.

Pioneers should be prioritized for immediate strategic development as they anchor the creation of the Blue Ocean.

Migrators

These are initiatives that offer moderate innovation and currently exist in some ports but lack full deployment or integration across the system. Examples include:

- Smart port platforms developed in individual ports (e.g., Barcelona, Valencia);
- 5G experimentation projects, AI for port operations, and digital twins;
- Energy pilot projects such as shore power for docked vessels.

Migrators can be scaled and harmonized to support systemic transformation.

Settlers

Settlers are legacy systems and practices that, while still functioning, do not offer differentiation or strategic value in the current context. These include:

- Isolated administrative systems used by individual port authorities;
- Manual or paper-based customs and clearance procedures;
- Redundant investments in non-shared infrastructures.

Settlers should be restructured, integrated into broader systems, or gradually phased out.

A PEC Map diagram (Figure 6) visually plotted these three categories as a portfolio bubble chart, helping the readers see the distribution of innovation maturity across the port ecosystem.

The application of the PEC Map in this study ensured that the proposed Blue Ocean Strategy was not only visionary, but was also sequenced and grounded in realistic implementation pathways aligned with technological readiness and organizational capacity.

The PEC Map (Pioneers–Migrators–Settlers) offers a strategic portfolio view that classified the components of the Spanish port system according to their degree of innovation and future potential. By organizing these elements across two dimensions—level of innovation and time horizon (Today vs. Tomorrow)—the map provides insight into the balance between legacy systems, current development trajectories, and disruptive opportunities.

The Pioneers, clustered predominantly in the upper-right quadrant of the map, represent high-impact innovations with long-term potential. These include the information cluster, the proposed governance model, and digitalization initiatives envisioned as system-wide and fully interoperable. Their position in the "Tomorrow" column reflects that, although these strategies are being formulated or piloted, they have not yet been fully implemented. Nevertheless, they hold the greatest transformative capacity. For example, the governance model is intended to address institutional fragmentation and support shared strategic planning, while the information cluster would enable real-time interoperability across port systems, logistics providers, and urban mobility networks. These initiatives are pivotal to unlocking the full value of Port 4.0 and aligning with the European data and climate directives.

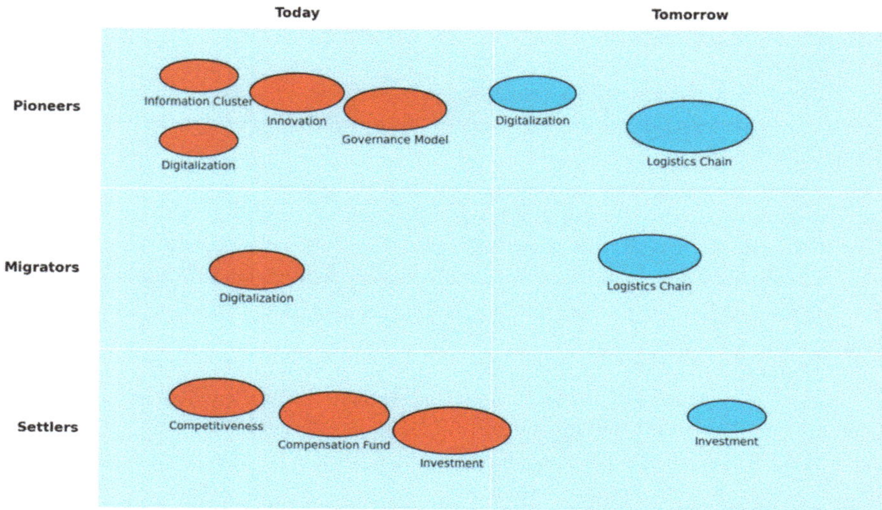

Figure 6. PEC Map diagram.

The Migrators occupy a transitional space and include digitalization and logistics chain improvements that are already in motion in some port authorities—particularly those with stronger innovation agendas—but have not yet reached the national scale or cross-system integration. Their classification reflects both their moderate originality and partial diffusion. These initiatives can serve as foundational steps that, if scaled properly, may evolve into Pioneers. However, if not harmonized or supported by national coordination, they risk stagnating as isolated best practices.

Finally, the Settlers—such as traditional investment practices, fragmented competitiveness, and the compensation fund—reflect legacy mechanisms that, although still functional, no longer offer strategic differentiation. They absorb resources and institutional attention but do not contribute meaningfully to the system's repositioning. Their placement in the "Today" column indicates that they continue to shape operational decisions but must be re-evaluated or restructured in light of the proposed Blue Ocean Strategy. Importantly, these elements are not necessarily obsolete, but their strategic value is low in a context that demands alignment, transparency, and innovation.

The visual distribution of initiatives in the PEC Map underscores a critical message: the Spanish port system is at a crossroads. While it maintains operational tools and governance logic rooted in the past, it simultaneously harbors a set of forward-looking, high-potential innovations that could redefine its role in the national economy. Moving toward the Blue Ocean requires deliberate investment in the Pioneer zone, strategic migration of scalable practices, and the gradual phasing out or transformation of legacy systems. This transition must be supported by coordinated governance, aligned incentives, and a clear roadmap linking short-term operational reforms with long-term structural transformation.

4.3. Application of BOS Formulation Principles

The formulation principles of the Blue Ocean Strategy provide a structured and sequential pathway for turning diagnostic insights into a coherent and future-ready strategic model. Based on the results from the diagnostic and tool-based phases, these principles were applied to guide the design of a Blue Ocean Strategy for the Spanish port system. Each principle contributes to redefining the logic of competition and unlocking new value creation spaces.

4.3.1. Reconstruct Market Boundaries

One of the foundational principles of the Blue Ocean Strategy is the need to redefine the boundaries of the market in which the organization operates. Rather than competing within the established limits and traditional segmentation, this principle encourages actors to look across sectoral, functional, and customer boundaries to discover new value spaces.

In the context of the Spanish port system, this principle is particularly relevant due to the rigid structure defined by the 28 independent port authorities, which have historically functioned under a logic of territorial fragmentation and institutional silos. The diagnostic phase revealed how this structure fosters internal competition, redundancies in infrastructure and services, and an underutilization of strategic synergies.

To reconstruct the market boundaries (Figure 7), the proposed Blue Ocean Strategy suggests a fundamental shift from a port-centric competitive model to a networked, interoperable port ecosystem, built on collaboration and digital integration. This reframing affects several layers:

- From territorial isolation to functional integration: The strategy proposes treating the port system as a national logistics and innovation platform rather than as a sum of disconnected local nodes.
- From sectoral specialization to cross-sectoral innovation: Ports are repositioned as connectors of maritime, urban, digital, and energy systems, enabling interactions with actors that were previously considered outside the port's operational scope.
- From infrastructure providers to platform orchestrators: Port authorities are encouraged to adopt a new role that goes beyond infrastructure management and embraces data coordination, intermodal integration, and ecosystem leadership.

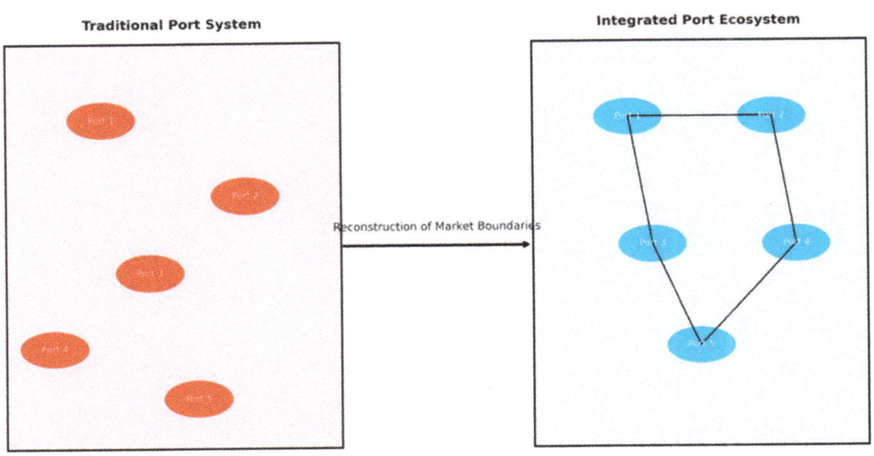

Figure 7. Reconstruct market boundaries.

This redefinition allows the port system to access new sources of demand, generate joint value with other sectors (e.g., smart cities, green energy, digital services), and eliminate internal inefficiencies. It also forms the strategic base for the principles that follow, all aimed at aligning the port ecosystem with the logic of the Blue Ocean.

4.3.2. Focus on the Big Picture, Not the Numbers

The second principle of Blue Ocean Strategy formulation urges strategists to move beyond fragmented metrics and conventional benchmarking and instead adopt a holistic, system-level vision. In traditional planning processes, excessive focus on isolated performance indicators—such as cargo throughput, cost per ton, or port turnaround times—tends to reinforce incremental thinking and sustain existing competitive structures.

In the Spanish port context, this pattern is evident: each port authority operates with its own KPI framework, investment priorities, and technological roadmap, often with limited alignment to national or collective objectives. This localism leads to a proliferation of uncoordinated digital projects, redundant infrastructure, and fragmented user experiences across the system.

The Blue Ocean approach recommends refocusing the strategic lens on the broader ecosystem, emphasizing structural transformation over short-term performance improvements. Concretely, this principle was applied in the following ways:

- System-level mapping: Instead of analyzing ports individually, the strategy evaluated the port system as a single digital and operational entity, considering the flows, synergies, redundancies, and collective externalities.
- Value-oriented logic: Strategic choices were guided not by isolated efficiency gains but by their contribution to collective value creation such as enhanced interoperability, shared data governance, and sustainable mobility integration.
- Visual tools to drive dialogue: Instruments such as the Value Curve, ERRC Matrix, and PEC Map not only serve analytical purposes, but also facilitate communication among decision-makers, shifting discussions from defensive benchmarking to strategic design.

The approach in Figure 8 enables the port system to move away from zero-sum logic and toward shared strategic horizons. It provides the foundation for joint innovation, efficient resource allocation, and the alignment of port transformation with broader national and European sustainability and digitalization goals.

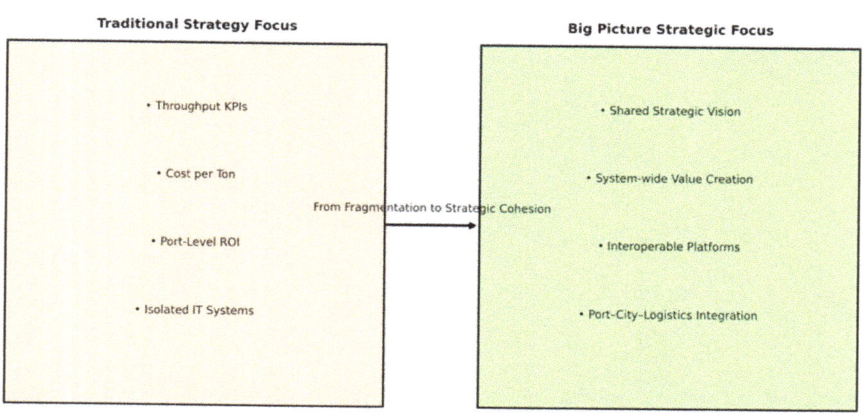

Figure 8. Blue Ocean approach: Focus on the big picture, not the numbers.

4.3.3. Reach Beyond Existing Demand

A key principle of the Blue Ocean Strategy is to move beyond the existing demand base and explore how to create value for non-customers—actors who, for various reasons, are not currently engaged with the port system but represent an untapped opportunity. Instead of segmenting and competing for the same users served by every port, this principle invites us to rethink who the system is truly for, and how its value proposition can be expanded.

In the Spanish port context, this principle is particularly relevant due to the historical focus on traditional maritime users (shipping lines, freight forwarders, and exporters/importers), while urban, technological, and mobility stakeholders have remained largely outside the strategic scope.

This study applied this principle through the structured three-tier non-customer analysis (as previously discussed in Section 4.2.3), identifying key expansion zones:

Tier 1—Soon-to-be non-customers:

These include actors such as local governments, urban mobility platforms, and metropolitan planning agencies who interact marginally with port operations but could benefit from joint digital and logistical solutions.

Tier 2—Refusing non-customers:

These are technology providers—especially in AI, IoT, blockchain, and green energy—that currently see the port system as too rigid, closed, or unattractive to collaborate with.

Tier 3—Unexplored non-customers:

This tier includes inland logistics hubs, academic institutions, and data-oriented research centers, which are not traditionally seen as clients or partners of port authorities, but could become key players in a data-driven, intelligent logistics ecosystem.

By targeting these three tiers (Figure 9), the proposed Blue Ocean Strategy enables the Spanish port system to:

- Open new service lines and partnerships beyond traditional maritime logistics;
- Increase the density and diversity of its innovation network;
- Reinforce its connection with urban and territorial development strategies;
- Become a platform for mobility-as-a-service, digital experimentation, and circular economy initiatives.

This expansion of demand not only increases the system's long-term value creation capacity, but also strengthens its legitimacy and resilience in a rapidly evolving economic and regulatory environment.

4.3.4. Get the Strategic Sequence Right

The final formulation principle of the Blue Ocean Strategy emphasizes the importance of following a logical and coherent sequence when moving from idea to implementation. Rather than designing a strategy based solely on feasibility or intuition, this principle recommends validating proposals through a four-step sequence:

1. Buyer utility;
2. Price;
3. Cost;
4. Adoption.

This sequence ensures that the strategy not only creates value, but does so in a way that is accessible, economically viable, and socially implementable.

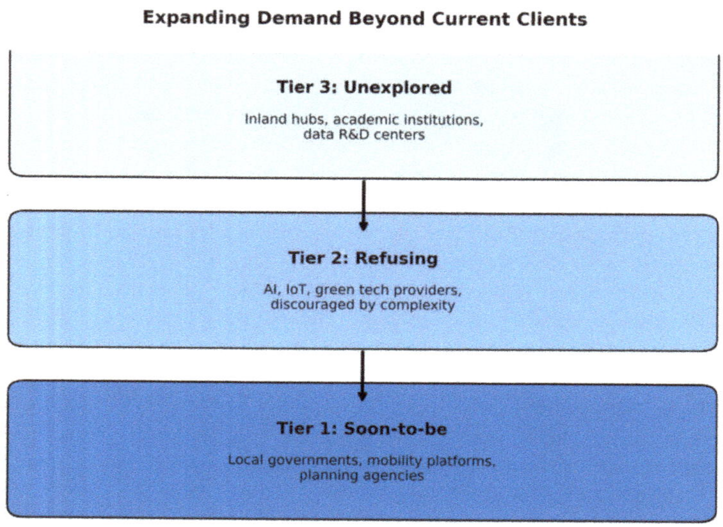

Figure 9. Reach beyond existing demand.

1. Buyer Utility

The strategy must deliver clear, compelling utility to the intended users. In the case of the Spanish port system, the proposed Blue Ocean Strategy addresses unmet needs from both traditional users (e.g., shipping lines, logistics operators) and new actors (e.g., tech companies, local governments). Utility is delivered through:

- Reduced operational complexity;
- Easier access to real-time data;
- Streamlined logistics and mobility coordination;
- Shared services that lower entry barriers for non-traditional users.

2. Price

The value must be offered at a price point that is acceptable to users and aligned with the public character of port operations. This implies:

- Promoting open-access digital infrastructure;
- Ensuring that participation in shared platforms is affordable and transparent;
- Avoiding hidden costs or proprietary lock-ins that discourage adoption.

3. Cost

Once the utility and pricing have been confirmed, the next step is to validate the cost-efficiency. The proposed strategy relies on:

- Reuse and standardization of digital tools across authorities;
- Phased implementation to reduce capex intensity;
- Elimination of redundant infrastructure and duplicated IT projects.

This cost rationalization strengthens the long-term sustainability of the strategic shift.

4. Adoption

The final test is whether the ecosystem can and will adopt the new strategy. In this case, key enablers for adoption include:

- Institutional alignment through governance frameworks;
- EU and national digitalization and decarbonization agendas;

- Demonstration pilots (e.g., Port 4.0 initiatives) that reduce the risk perception;
- Engagement of public and private stakeholders through co-design.

Applying this sequence ensures that the proposed transformation is not only ambitious, but also grounded and executable. It avoids the common pitfalls of tech-first or policy-first strategies by integrating user needs, pricing logic, operational feasibility, and stakeholder dynamics into a single roadmap.

4.4. Strategic Formulation of the Blue Ocean

The culmination of the Blue Ocean Strategy application lies in the formulation of a new strategic model for the Spanish port system—one that moves beyond fragmentation, competition, and reactive policy, and toward collaboration, interoperability, and shared innovation.

This final phase integrates the outcomes from previous analyses (Value Curve, ERRC Matrix, Non-Customer Mapping, PEC Classification, and strategic principles) into a coherent, future-oriented vision. The proposed Blue Ocean model is articulated around four fundamental strategic pillars.

4.4.1. A Unified Digital Infrastructure

The model proposes the deployment of a shared, interoperable digital platform across all 28 Spanish port authorities. This infrastructure allows for:

- Real-time coordination of port operations;
- Standardized data exchange between ports, logistics operators, and urban mobility systems;
- Reduction in duplicated technological efforts and enhanced system transparency.

This platform is not conceived as a top–down imposition, but rather as a federated architecture, allowing for modular implementation and localized adaptation.

4.4.2. Collaborative Governance

Instead of inter-port rivalry, the strategy encourages the adoption of a shared governance framework aligned with the national and EU strategic priorities. This includes:

- Harmonized investment planning;
- Joint digital and green transition strategies;
- Establishment of common standards for innovation and sustainability.

By aligning incentives and objectives, ports can operate as nodes of a coordinated network rather than competitors in a zero-sum game.

4.4.3. Expansion of the Value Network

The Blue Ocean model reframes the port from a static infrastructure to a dynamic innovation platform that engages new actors across sectors. This includes:

- Urban mobility providers;
- Tech and energy startups;
- Academic and research institutions;
- City governments and local communities.

The integration of these actors opens new demand spaces, supports circular economy models, and positions ports as active drivers of territorial innovation.

4.4.4. Commitment to Energy Transition

The strategy embeds energy efficiency and climate action as core elements of value creation. This involves:

- Electrification of port operations and shore power supply (OPS);

- Integration of renewable energy sources and green hydrogen pilots;
- Data-driven energy optimization strategies across logistics flows.

Environmental performance becomes a competitive and collaborative advantage, aligning the port system with the broader European Green Deal and global decarbonization efforts.

The proposed Blue Ocean Strategy for the Spanish port system does not simply suggest a set of innovations, it proposes a structural transformation from fragmented actors to an integrated ecosystem, from reactive investments to proactive value design, and from isolated performance to collective impact. The model is scalable, transferable, and designed to serve as a reference for other national port systems undergoing similar pressures.

In a future implementation scenario, the proposed strategy could enhance operational efficiency by reducing duplicated systems, standardizing data protocols, and improving coordination between ports. Although this study is conceptual, future empirical validation may involve performance indicators such as vessel turnaround time, inter-port cargo flows, energy consumption per operation, and stakeholder engagement in digital governance platforms.

The proposed model integrates specific green energy solutions such as shore power systems (OPSs), the electrification of cargo handling equipment, and the incorporation of renewable energy sources including solar and wind. These measures are aligned with national objectives under Spain's Integrated National Energy and Climate Plan (PNIEC) as well as European initiatives such as the Green Deal, Fit for 55, and the FuelEU Maritime package.

5. Conclusions

This paper applied the Blue Ocean Strategy (BOS) framework to the Spanish port system as a method for redefining its strategic trajectory in the context of Port 4.0. Through a structured four-phase methodology—diagnosis of the current competitive environment, application of BOS analytical tools, deployment of the formulation principles, and development of a new strategic model—this study proposed an alternative to the entrenched logic of inter-port competition, infrastructural duplication, and fragmented innovation.

The resulting strategy articulates a vision of the port system as a collaborative, interoperable, and innovation-oriented ecosystem. The Value Curve, ERRC Matrix, PEC Map, and Non-Customer Analysis collectively revealed a latent opportunity: to shift from siloed management toward shared governance, transition from isolated IT investments to a unified digital infrastructure, and expand the system's relevance by engaging actors beyond traditional maritime logistics.

Among the key contributions of this research is the formulation of a Blue Ocean Strategy tailored to a national port system—something that has been rarely explored in the existing literature, which has tended to focus on individual port case studies. Moreover, this proposal places digital and green transition at the core of value creation, aligning port reform with broader European Union strategies and climate commitments.

For the Spanish port system, the implications are significant. By adopting a Blue Ocean approach, port authorities could reduce inefficiencies, attract new partnerships, and reinforce their role as facilitators of sustainable, smart, and resilient logistics chains. The proposed model promotes cooperation over competition and positions ports as active agents in urban innovation and territorial development.

Naturally, this study had limitations. It was based on strategic and conceptual tools, supported by public documentation and the existing academic literature. Although consistent with recent institutional trends, the model has not yet been empirically validated through pilot implementation or stakeholder feedback mechanisms. The impact of such a transformation, in terms of cost savings, emissions reduction, or service performance, remains to be quantified.

Future research should focus on operationalizing the strategic pillars outlined here. This includes designing governance prototypes, testing data-sharing platforms in real environments, and developing energy transition indicators specific to port ecosystems. Additionally, comparative studies across European port systems could enhance the transferability of this framework and contribute to the construction of a broader Blue Ocean logic for maritime innovation.

In conclusion, the Blue Ocean Strategy offers not only a methodology, but a mindset—one capable of unlocking new value spaces in complex, mature sectors. Applied to the Spanish port system, it becomes a tool for structural transformation, guiding the sector toward a horizon of sustainable growth, digital interoperability, and long-term competitiveness.

Author Contributions: Conceptualization, N.G.-C., J.J.G.L., J.V.-C. and A.C.-O.; Methodology, N.G.-C. and J.J.G.L.; Validation, J.J.G.L.; Formal analysis, N.G.-C. and J.J.G.L.; Investigation, N.G.-C., J.J.G.L., J.V.-C. and A.C.-O.; Resources, J.J.G.L.; Data curation, N.G.-C. and J.J.G.L.; Writing—original draft, N.G.-C.; Writing—review & editing, J.V.-C.; Visualization, N.G.-C.; Supervision, A.C.-O. All authors have read and agreed to the published version of the manuscript.

Funding: This research received no external funding.

Data Availability Statement: The original contributions presented in this study are included in the article. Further inquiries can be directed to the corresponding author.

Conflicts of Interest: The authors declare no conflicts of interest.

References

1. Barbieri, E.; Capoani, L. Renewable Energy, Resilience, Digitalization, and Industrial Policies in Seaborne Transport. *Energies* **2025**, *18*, 1089. [CrossRef]
2. Vaca-Cabrero, J.; Domínguez Rastrojo, J.; González-Cancelas, N.; Camarero-Orive, A. Enhancing Sustainability in Port Infrastructure Through Innovation: A Case Study of the Spanish Port System. *Sustainability* **2025**, *17*, 2593. [CrossRef]
3. González-Cancelas, N.; Molina Serrano, B.; Soler-Flores, F.; Camarero-Orive, A. Using the SWOT Methodology to Know the Scope of the Digitalization of the Spanish Ports. *Logistics* **2020**, *4*, 20. [CrossRef]
4. Notteboom, T.E. Concentration and the Formation of Multi-Port Gateway Regions in the European Container Port System: An Update. *J. Transp. Geogr.* **2010**, *18*, 567–583. [CrossRef]
5. Sklyar, A.; Kowalkowski, C.; Tronvoll, B.; Sörhammar, D. Organizing for Digital Servitization: A Service Ecosystem Perspective. *J. Bus. Res.* **2019**, *104*, 450–460. [CrossRef]
6. Christodoulou, I.; Langley, P.A. A Gaming Simulation Approach to Understanding Blue Ocean Strategy Development as a Transition from Traditional Competitive Strategy. *J. Strateg. Mark.* **2020**, *28*, 727–752. [CrossRef]
7. Van Buuren, A.; Loorbach, D. Policy Innovation in Isolation?: Conditions for Policy Renewal by Transition Arenas and Pilot Projects. *Public Manag. Rev.* **2009**, *11*, 375–392. [CrossRef]
8. Lind, M.; Ward, R.; Jensen, H.H.; Chua, C.P.; Simha, A.; Karlsson, J.; Göthberg, L.; Penttinen, T.; Theodosiou, D.P. The Future of Shipping: Collaboration Through Digital Data Sharing. In *Maritime Informatics*; Lind, M., Michaelides, M., Ward, R., Watson, R.T., Eds.; Progress in IS; Springer International Publishing: Cham, Switzerland, 2021; pp. 137–149; ISBN 978-3-030-50891-3.
9. Henríquez, R.; de Osés, F.X.M.; Marín, J.E.M. Technological Drivers of Seaports' Business Model Innovation: An Exploratory Case Study on the Port of Barcelona. *Res. Transp. Bus. Manag.* **2022**, *43*, 100803. [CrossRef]
10. Su, Z.; Liu, Y.; Gao, Y.; Park, K.-S.; Su, M. Critical Success Factors for Green Port Transformation Using Digital Technology. *J. Mar. Sci. Eng.* **2024**, *12*, 2128. [CrossRef]
11. Heikkilä, M.; Saarni, J.; Saurama, A. Innovation in Smart Ports: Future Directions of Digitalization in Container Ports. *J. Mar. Sci. Eng.* **2022**, *10*, 1925. [CrossRef]
12. Parola, F.; Risitano, M.; Ferretti, M.; Panetti, E. The Drivers of Port Competitiveness: A Critical Review. *Transp. Rev.* **2017**, *37*, 116–138. [CrossRef]
13. Wilmsmeier, G.; Monios, J.; Pérez-Salas, G. Port System Evolution–the Case of Latin America and the Caribbean. *J. Transp. Geogr.* **2014**, *39*, 208–221. [CrossRef]
14. Notteboom, T.E.; Rodrigue, J.-P. Port Regionalization: Towards a New Phase in Port Development. *Marit. Policy Manag.* **2005**, *32*, 297–313. [CrossRef]

15. Bhattacharyya, O.; Khor, S.; McGahan, A.; Dunne, D.; Daar, A.S.; Singer, P.A. Innovative Health Service Delivery Models in Low and Middle Income Countries—What Can We Learn from the Private Sector? *Health Res. Policy Syst.* **2010**, *8*, 24. [CrossRef]
16. Cowie, J.; Heathcott, J. *Beyond the Ruins: The Meanings of Deindustrialization*; Cornell University Press: Ithaca, NY, USA, 2003.
17. Verhoeven, P. A Review of Port Authority Functions: Towards a Renaissance? *Marit. Policy Manag.* **2010**, *37*, 247–270. [CrossRef]
18. Debrie, J.; Raimbault, N. The Port–City Relationships in Two European Inland Ports: A Geographical Perspective on Urban Governance. *Cities* **2016**, *50*, 180–187. [CrossRef]
19. Szpak, A.; Gawłowski, R.; Modrzyńska, J.; Modrzyński, P.; Dahl, M. The Role of Cities in International Relations: The Third-Generation of Multi-Level Governance? In *The Role of Cities in International Relations*; Edward Elgar Publishing: Cheltenham, UK, 2022.
20. Tang, M.; Luo, X.; Ying, W. Multi-Level Governance in the Uneven Integration of the City Regions: Evidence of the Shanghai City Region, China. *Habitat Int.* **2022**, *121*, 102518. [CrossRef]
21. Monios, J. Identifying Governance Relationships Between Intermodal Terminals and Logistics Platforms. *Transp. Rev.* **2015**, *35*, 767–791. [CrossRef]
22. Wang, S.; Li, Q.; Khaskheli, M.B. Management Economic Systems and Governance to Reduce Potential Risks in Digital Silk Road Investments: Legal Cooperation between Hainan Free Trade Port and Ethiopia. *Systems* **2024**, *12*, 305. [CrossRef]
23. Yarali, A. *Intelligent Connectivity: AI, IoT, and 5G*; John Wiley & Sons: Hoboken, NJ, USA, 2021.
24. Dinh, G.H.; Pham, H.T.; Nguyen, L.C.; Dang, H.Q.; Pham, N.D.K. Leveraging Artificial Intelligence to Enhance Port Operation Efficiency. *Pol. Marit. Res.* **2024**, *31*, 140–155. [CrossRef]
25. Gómez Díaz, C.; González-Cancelas, N.; Camarero Orive, A.; Soler Flores, F. Digital Governance Approach to the Spanish Port System: Proposal for a Port. *J. Mar. Sci. Eng.* **2023**, *11*, 311. [CrossRef]
26. Cortes-Murcia, D.L.; Guerrero, W.J.; Montoya-Torres, J.R. Supply Chain Management, Game-Changing Technologies, and Physical Internet: A Systematic Meta-Review of Literature. *IEEE Access* **2022**, *10*, 61721–61743. [CrossRef]
27. Gerlitz, L.; Meyer, C. Small and Medium-Sized Ports in the Ten-t Network and Nexus of Europe's Twin Transition: The Way towards Sustainable and Digital Port Service Ecosystems. *Sustainability* **2021**, *13*, 4386. [CrossRef]
28. Almeida, F. Challenges in the Digital Transformation of Ports. *Businesses* **2023**, *3*, 548–568. [CrossRef]
29. Jansen, M.; Hein, C. Port City Symbiosis: Introduction to the Special Issue. *Marit. Econ. Logist.* **2023**, *25*, 211–229. [CrossRef]
30. Gravagnuolo, A.; Angrisano, M.; Fusco Girard, L. Circular Economy Strategies in Eight Historic Port Cities: Criteria and Indicators towards a Circular City Assessment Framework. *Sustainability* **2019**, *11*, 3512. [CrossRef]
31. de Langen, P.W.; Sornn-Friese, H.; Hallworth, J. The Role of Port Development Companies in Transitioning the Port Business Ecosystem; the Case of Port of Amsterdam's Circular Activities. *Sustainability* **2020**, *12*, 4397. [CrossRef]
32. Mdanat, M.F.; Al Hur, M.; Bwaliez, O.M.; Samawi, G.A.; Khasawneh, R. Drivers of Port Competitiveness among Low-, Upper-, and High-Income Countries. *Sustainability* **2024**, *16*, 11198. [CrossRef]
33. De Martino, M. Value Creation for Sustainability in Port: Perspectives of Analysis and Future Research Directions. *Sustainability* **2021**, *13*, 12268. [CrossRef]
34. Mirghaderi, S.A.; Aboumasoudi, A.S.; Amindoust, A. Developing an Open Innovation Model in the Startup Ecosystem Industries Based on the Attitude of Organizational Resilience and Blue Ocean Strategy. *Comput. Ind. Eng.* **2023**, *181*, 109301. [CrossRef]
35. Hesse, M. Approaching the Relational Nature of the Port-City Interface in Europe: Ties and Tensions Between Seaports and the Urban. *Tijdschr. Econ. Soc. Geogr.* **2018**, *109*, 210–223. [CrossRef]
36. Gurzhiy, A.; Kalyazina, S.; Maydanova, S.; Marchenko, R. Port and City Integration: Transportation Aspect. *Transp. Res. Procedia* **2021**, *54*, 890–899. [CrossRef]
37. Liao, H.-T.; Lo, T.-M.; Pan, C.-L. Knowledge Mapping Analysis of Intelligent Ports: Research Facing Global Value Chain Challenges. *Systems* **2023**, *11*, 88. [CrossRef]
38. Raimbault, N. From Regional Planning to Port Regionalization and Urban Logistics. The Inland Port and the Governance of Logistics Development in the Paris Region. *J. Transp. Geogr.* **2019**, *78*, 205–213. [CrossRef]
39. Anwar, M.; Henesey, L.; Casalicchio, E. Digitalization in Container Terminal Logistics: A Literature Review. In Proceedings of the 27th Annual Conference of International Association of Maritime Economists, Athens, Greece, 25–28 June 2019; pp. 1–25.
40. Martínez Torres, M.F.; Ruiz Aguilar, J.J.; Turias Domínguez, I.J.; Cerbán Jiménez, M.d.M. A Cost-Optimized Model for Dynamic Port Hinterland Delineation: An Application to the Spanish Port System. *Appl. Sci.* **2024**, *15*, 96. [CrossRef]
41. Kim, W.C.; Mauborgne, R. *Blue Ocean Strategy, Expanded Edition: How to Create Uncontested Market Space and Make the Competition Irrelevant*; Harvard Business Review Press: Cambridge, MA, USA, 2014.

Journal of
Marine Science and Engineering

MDPI

Article

Decarbonation Effects of Mainstream Dual-Fuel Power Schemes Focus on IMO Mandatory Regulation and LCA Method

Zhanwei Wang [1], Shidong Fan [1,*] and Zhiqiang Han [2,*]

1 School of Transportation and Logistics Engineering, Wuhan University of Technology, Wuhan 430063, China; 335161@whut.edu.cn
2 School of Naval Architecture, Ocean and Energy Power Engineering, Wuhan University of Technology, Wuhan 430063, China
* Correspondence: sdfan@whut.edu.cn (S.F.); hzq910413@zjou.edu.cn (Z.H.)

Abstract: Recently, the IMO has completed the guidelines on the life cycle greenhouse gas intensity of marine fuels to accelerate the application of alternative fuels. Low-carbon fuels may persist for decades and have become a key transitional phase in replacing marine fuels. A more comprehensive methodology for evaluating the carbon emission levels of marine fuels was explored, and the carbon emissions and environmental impacts of a 150,000-ton shuttle tanker under 19 dual-fuel power scenarios were evaluated using the Energy Efficiency Design Index (EEDI) and life cycle assessment (LCA) method. The results show that liquefied natural gas (LNG) has a higher carbon control potential level compared to liquefied petroleum gas (LPG) and methanol (MeOH), while LPG is superior to MeOH based on EEDI evaluation. LCA analysis results show that MeOH (biomass) has the best carbon control potential considering the carbon emissions of the well-to-tank phase of the fuel, followed by LNG, LPG, MeOH (natural gas, NG), and MeOH (coal). However, MeOH (NG) and MeOH (coal) had greater negative environmental impacts. This study provides method support and a direction toward improvement for revising related technical specifications and regulations for dual-fuel vessel performance evaluation, considering the limitations of various maritime regulations.

Keywords: life cycle assessment; carbon emissions; global warming potential; energy efficiency design index; dual fuel

1. Introduction

1.1. Background

The shipping industry is an important part of the international trade and logistics system, as well as a significant contributor to greenhouse gas (GHG) emissions and marine ecological damage [1,2]. To address the associated environmental issues, the International Maritime Organization (IMO) has successively introduced the Energy Efficiency Design Index (EEDI), Ship Energy Efficiency Management Plans (SEEMPs), and tier series standards to limit ship emissions during operation. According to Chapter 4 of the Annex VI amendment to the International Convention for the Prevention of Pollution from Ships (MARPOL), all newly built ships with a gross tonnage of ≥400 have been required to comply with the EEDI since 2013 [3–5]. Meanwhile, calculating and reporting the Energy Efficiency Existing Ship Index (EEXI) and Carbon Intensity Indicator (CII) for existing ships have been mandatory since 2023 [6]. Subsequently, the 2023 IMO Strategy on the Reduction of GHG Emissions from Ships was adopted, based on which the global shipping industry strives to achieve net-zero emissions by 2050 [7].

1.2. Literature Review

1.2.1. Evaluation of Decarbonization Effect Based on Different Marine Fuels and Traditional Technology Applications

The shipping industry's most important conventional propulsion systems are marine engines fueled with marine diesel oil (MDO) and heavy fuel oil (HFO). Due to their high carbon contents, the combustion of these fuels produces large amounts of GHGs and contributes to the acceleration of global warming [8]. Researchers have conducted a series of studies to effectively respond to IMO regulatory standards to reduce carbon emissions during ship operations and improve energy efficiency. More specifically, Adland et al. [9] used the performance parameters and weather data from 2012 to 2016 to study the effects of periodic hull cleaning on tanker energy efficiency. Cleaned hull and propeller surfaces were found to reduce the daily fuel consumption by decreasing the drag affecting the hull. Meanwhile, Norlund and Gribkovskaia [10] used a supply ship as the research object. By adding a suitable speed optimization strategy to the ship schedule and utilizing the waiting time to reduce the speed, they achieved energy savings and emission reduction. Jia [11] systematically analyzed the structural optimization design process of 4500 deadweight tonnage bulk carriers and the impact of structural lightweighting on ship energy efficiency. The results showed that by decreasing the structural weight, the ship's EEDI value could be moderately lowered and carbon emissions can be reduced; additionally, the load capacity can be increased or the main engine power can be reduced. Key optimization technologies include ship shape optimization, propeller modification, and the development and application of new energy-saving equipment can also further achieve the effect of carbon control and reduction in carbon emissions [12–15].

However, based on recent research trends, reducing carbon emissions based on existing optimization techniques and the single use of MDO/HFO remains challenging. Moreover, the mandatory requirements for new shipbuilding are difficult to meet, particularly in EEDI Phase III. Therefore, existing marine fuels are gradually shifting from single MDO/HFO to dual fuel, making studying new energy applications to replace traditional fuels an important research topic.

Alternative fuel selection prioritizes low-carbon options such as biofuels, methanol (CH_3OH, MeOH), liquefied natural gas (LNG), and liquefied petroleum gas (LPG), etc., and zero-carbon alternatives such as hydrogen and ammonia [16]. The literature shows that compared to conventional fuels, LPG, LNG, and MeOH can significantly reduce SOx and NOx emissions, reduce CO_2 and produce minimal particulates [17–22]. Specifically, LNG achieves 90–99% SOx reduction, 20–30% NOx abatement, and approximately 25% carbon reduction potential; LPG shows even more complete SOx elimination (90–100%) with 10–15% NOx reduction and 13–18% lower carbon emissions; while MeOH provides comparable SOx control (90–97%) but superior NOx reduction (30–50%), albeit with a more modest 5% decarbonization potential. Most notably, all three fuel alternatives achieve over 90% reduction in particulate matter emissions. Jafarzadeh et al. [23] conducted a study on the application of LNG to fishing vessels. They found that, as a clean and efficient energy source, LNG can be effective in improving the environmental footprint to some extent. Moroianu and Postolache [24] analyzed the performance of marine engines using MDO, HFO, and MeOH, identifying MeOH as an attractive alternative fuel given its reduced pollutant emissions and cost-effectiveness. Carbon emissions can also be reduced by using similar alternative fuels such as hydrogen, ammonia, and biofuel [25–32]. While considering factors such as energy availability, technological readiness, economic feasibility, environmental performance, fuel energy density, and regulatory frameworks, alternative fuels for ships are dominated by low-carbon fuels, with LNG, LPG, and MeOH being the primary type for ships, compared to zero-carbon fuels like hydrogen and ammonia. LPG, LNG, and MeOH are currently the main choice for newly built ships; even considering the future evolution of the shipping industry from low- to zero-carbon technologies, a long transition period remains for low-carbon fuels.

1.2.2. Evaluation of Decarbonization Effects Under Mainstream Methods

Figure 1 shows that consistent with the application of technologies to reduce carbon emissions, quantitative evaluation methods have been applied to assess the effects of carbon emissions on the shipping sector. Based on IMO regulations, the EEDI represents the carbon dioxide emissions per ton of cargo transported by a maritime vessel over one nautical mile. Initially, it was proposed as a predictor of the energy efficiency of newly built ships [33]. A lower EEDI indicates higher energy efficiency. However, the EEDI only utilizes a single value, while the effects of multiple conditional factors (including changes in sailing speed and improvements in the design of key ship equipment) are not considered. This index can only be used for the operational phase of a large ship, not for quantitatively evaluating the degree of carbon emissions during the design, construction, or dismantling phases of a ship's life cycle.

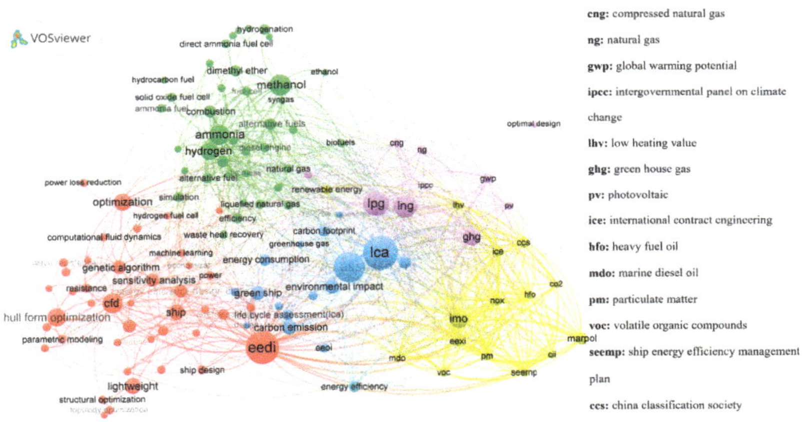

Figure 1. Keyword frequency statistics for research on methods to control carbon emissions over the past two decades.

The principles of the newly introduced EEXI and CII are consistent with the EEDI. Other indicators and methodologies must be considered to assess ships' carbon emissions more comprehensively and promote sustainable development throughout the ship cycle. Life cycle assessment (LCA) has recently been gradually applied to the marine sector [16,31,34,35]. Specifically, the IMO has developed 2024 guidelines on the life-cycle GHG intensity of marine fuels (2024 LCA Guidelines), focusing on the life-cycle GHG/carbon intensity of various types of low- and zero-carbon fuels [36]. IMO is increasingly emphasizing the importance of LCA in quantifying carbon emissions. The overall framework and content of LCA is based on ISO 14040 [37] and ISO 14044 [38]. Fang [39] used the LCA method to establish a spatiotemporal network model of the carbon footprint of a ship's life cycle and analyzed the movement paths and spatial trajectories of the bulk carrier "HUA YUN 5" during its life cycle. The analysis confirms that the carbon footprint of this type of ship is most significant during the shipping phase. Moreover, Han et al. [40] sought to further determine the impact of the structural design margins of fiber-reinforced polymer (FRP) fishing vessels on air pollution. To this end, LCA was used to conduct an environmental impact assessment of FRP materials. The results show that the increase in raw construction materials due to the hull structure design allowance negatively impacted atmospheric pollution; hence, the design rules must be further revised and improved. Chatzinkolaou and Ventikos [41] proposed a new framework for assessing gas emissions from ocean-going vessels in conjunction with the LCA and conducted a case study with Panamax tankers. During the operational phase of a ship, gas emissions from machinery and equipment accounted for a large proportion. The LCA can also be an important basis for comparing and selecting ship power and propulsion systems [42–44]. Currently, LCA applications in the marine sector primarily target the entire life cycle of small vessels such as tugboats, yachts, fishing boats, and ferries; the carbon emissions per gross ton of small ships are generally much higher than those of large commercial ships, likely exerting a greater environmental impact [45–47].

Compared with the EEDI, LCA can be used for the operational phase of a ship as well as for a full life cycle assessment. Therefore, LCA has been proposed as a statutory measure to decrease carbon emissions in the shipping industry [48–51]. Accurately assessing the environmental impacts of ships throughout their life cycles, including energy consumption, carbon emissions, and water pollution, is of great significance to determining the environmental impacts of ships. Moreover, LCA results are relatively conservative compared to experimental results, such as main engine emission tests. This is primarily due to current mainstream databases, such as Ecoinvent 3, defining marine fuel as a production, with the production stage critical in life cycle inventory data [34,35,46].

1.3. Research Gap and Objectives

As mentioned previously, although the EEDI can effectively evaluate the energy efficiency and the carbon emission level of newly built ships, it only applies to the operational phase. Furthermore, the issue of methane slips and their impact on carbon emissions cannot be further considered. While the LCA can quantitatively evaluate ships' carbon emissions and environmental impact performance from a complete life-cycle perspective, certain limitations remain. In particular, even the more comprehensive LCA databases are unable to accurately consider the carbon emissions of marine fuels during the combustion phase.

To address the research gap, a carbon emission level assessment method for marine fuels based on the EEDI and LCA is proposed. Taking a typical ship as the research object, different groups of dual-fuel power design schemes were developed from the perspective of marine fuels, particularly low-carbon fuels. Subsequently, EEDI and LCA evaluations were conducted, and a solution was proposed to address the lack of data in the LCA database. Meanwhile, the limitations of the EEDI in quantifying marine fuel carbon emissions during the application phase and the key role of LCA in the sustainable development of the shipping industry are discussed. The study provides a comprehensive quantitative framework for ship carbon emission assessment, and the findings can provide guidance for selecting dual-fuel solution for ships and promote the implementation and formulation of IMO regulations.

This paper is structured as follows. Section 2 examines the results of prior research and methodology, including target vessel type selection, dual-fuel power scheme development, and EEDI application. Section 3 analyzes the results of the LCA simulation, encompassing the entire process, from the definition and scope of the objectives to the interpretation of the results. Additionally, the results of prior EEDI research are discussed. Section 4 provides a multifaceted analysis and discusses the EEDI and LCA results. Finally, the main conclusions are summarized, while the limitations and future perspectives of the study are further discussed in Section 5.

2. Examination of Prior Research and Methodology

2.1. Target Ship Selection

Figure 2 shows the research object of this study, which was a 150,000-ton shuttle tanker built by COSCO Shipping Heavy Industry (Zhoushan) Co., Ltd., China. As a representative vessel from the company's core product line, this shuttle tanker series specializes in crude oil transportation from offshore fields to onshore terminals. The vessel class, certified by Det Norske Veritas (DNV) with unrestricted navigation certification, currently comprises six operational units in this series. The main engine model (MAN B&W 6G70ME-C10.5) uses MDO as the primary fuel. The current EEDI value of the N786 ship is 2.774, which is ~28.84% lower than the baseline value of 3.574. Despite meeting the second phase of the EEDI through technologies such as hull form optimization, main engine power reduction, speed optimization, energy-saving devices, and hull structure weight reduction, the N786 ship did not meet the third phase. Consequently, the N786 ship was selected as the research object for conducting comprehensive analysis of viable decarbonization solutions. Table 1 lists the main dimensions, engine, speed, EEDI values, and other characteristics.

Figure 2. 150,000-ton shuttle tanker considered in this study (N786).

Table 1. Main characteristics of the shuttle tanker.

Parameter	Value	Parameter	Value
Length (m)	278.90	75% MCR fuel consumption of main engine (g/kWh)	159.16
Distance between perpendiculars (m)	265.00	Number of generators and MCR (kW)	4000 for 5 units and 1600 for 1 unit
Molded breadth (m)	47.00	50% MCR fuel consumption of generators (g/kWh)	198.79
Molded depth (m)	24.90	Correction factor for reinforcement of ship voluntary structure	1.0105
Summer draft (m)	17.30	CSR correction factor	1.0150
Empty ship weight (t)	29,067	Design speed (kn)	14.5
Deadweight tonnage (t)	154,955	Baseline EEDI	3.574
Number of main engines and MCR (kW)	14,675 for 1 unit	Reached EEDI	2.774

MCR, maximum continuous rating; CSR, common structural rules.

2.2. Effect Analysis of Dual-Fuel Selection Scheme Based on EEDI

A dual-fuel power design solution is an excellent choice to decrease carbon emissions. Hence, an examination of the results of prior research must be prioritized. In recent years, COSCO Shipping Heavy Industry (Zhoushan) Co., Ltd. has successively provided design schemes for dual-fuel-powered vessels for this series of target vessels and LPG, LNG, and MeOH. All three design schemes have obtained Approval in Principle (AiP) certification from DNV, and research continues. EEDI evaluation was used to analyze the effect of improvements in various power schemes.

The study ship was designed for a cruising range of 18,000 nm and a speed of 14.5 kn. Assuming that the range of low-carbon fuels is adjustable and controllable, different ranges were determined. A total of 19 groups of low-carbon fuel range design schemes were developed, and the difference in the low-carbon fuel range of each group of schemes was set to 1000 nm. The traditional MDO power scheme was used for a low-carbon fuel range of 0 nm. Table 2 lists the specifications of the main dual-fuel engines selected for the different design options, based on AiP certified designs. Based on the data in Table 2, the fuel oil consumption, pilot oil requirement, and gas consumption for different dual-fuel power schemes were subsequently calculated. The SPOC is determined by SPOC/SGC item in Table 2. For example, in LPG applications, the consumption of 140.1 g of SGC requires 6.34 g of SPOC.

Table 2. Design solutions for main engine selection.

Fuel	Main Engine Model	SMCR/NCR	Specific FOC	SPOC/SGC
LPG	MAN 6G70ME-C10.5 LGIP-HPSCR	15,128 kW × 74 rpm/ 13,615 kW × 74 rpm	157.3 g/kWh	6.34/140.1 g/kWh
LNG	MAN 6G70ME-C10.5-GI	14,675 kW × 74 rpm/ 13,208 kW × 74 rpm	161.9 g/kWh	2.93/130.7 g/kWh
MeOH	MAN 6G60ME-C10.5-LGIM-SCRBP	14,675 kW × 94.5 rpm/13,208 kW × 94.5 rpm	168.8 g/kWh	8.90/343.0 g/kWh

SMCR/NCR, specified maximum continuous rating/normal continuous rating; FOC, fuel oil consumption; SPOC/SGC, specific pilot oil consumption/gas consumption.

According to the guidelines for the calculation of EEDI for new ships [52], calculations for 19 groups of schemes were conducted. As shown in Equation (1), the EEDI is used as a design index to assess the energy efficiency of a ship, where the numerator represents the CO_2 emissions of the ship, and the denominator represents the ship's total transportation capacity at the design speed.

$$\text{EEDI} = \frac{CO_2 \text{ emission from ship}}{\text{Completed work}} = \frac{CO_{2\ \text{Main engine}} + CO_{2\ \text{Auxiliary engine}} - \text{Reductions by different energy efficiency techniques}}{Capacity \times V_{ref}}, \quad (1)$$

where *Capacity* is the deadweight tonnage (in tons, t) and V_{ref} is the ship speed (knots, kn).

The EEDI calculation for the dual-fuel power schemes is outlined in Figure 3, and the specific calculations are shown in Table 3. Table 3 shows a significant decrease in the LNG-MDO fuel type as the low-carbon fuel range passes from 10,000 nm to 11,000 nm. Meanwhile, the same phenomenon of MeOH-MDO and LPG-MDO occurred when the low-carbon fuel range passed from 10,000 nm to 11,000 nm and 11,000 nm to 12,000 nm, respectively. The main reason is that when the low-carbon fuel range exceeds the above-mentioned ranges, the low-carbon fuels are considered primary fuels, due to the f_{DFgas} is greater than 0.5 in these scenarios according to Figure 3. Therefore, the calculation of the EEDI only considers the fuel consumption of both SPOC and low-carbon fuels, no longer considering the impact of conventional fuel oil on carbon emissions.

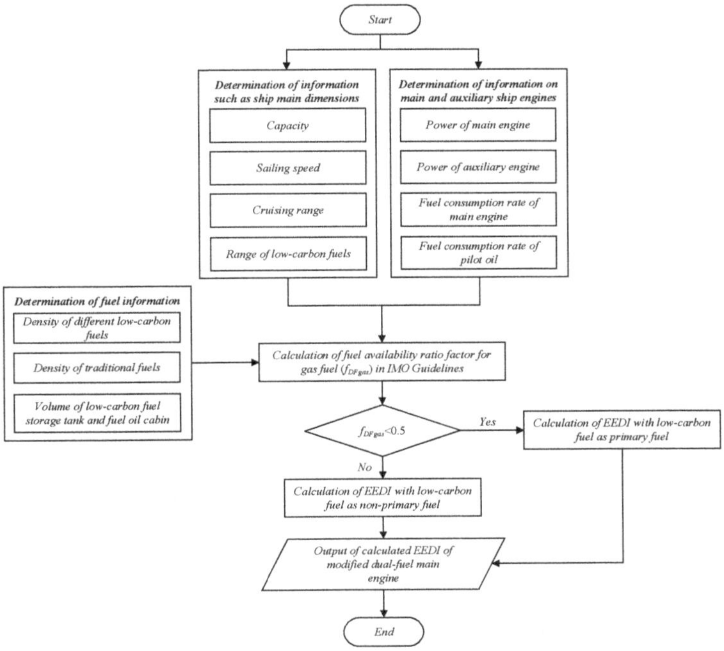

Figure 3. The EEDI calculation for the dual-fuel power schemes.

To further compare and analyze the optimization differences in different dual-fuel power schemes, the enhancement in the EEDI reduction factor is considered. The EEDI reduction factor is defined as the percentage change in the EEDI value achieved by the designed ship relative to the baseline EEDI value of the ship, as shown in Equation (2). Furthermore, the improvement of EEDI reduction factor is defined as the difference in the change in the reduction coefficient before and after optimization.

$$\text{EEDI reduction factor} = \frac{\text{Baseline EEDI} - \text{EEDI}_{attained}}{\text{Baseline EEDI}}, \tag{2}$$

Figure 4 compares the improvement effects of the EEDI reduction factor when the low-carbon fuel was used as non-primary and primary fuel. The low-carbon fuel range of 10,000 nm and 18,000 nm were selected for the comparison. This selection was based on the solution with the most substantial EEDI reduction effect. Utilizing low-carbon fuels, whether non-primary or primary, directly decreases the ship's carbon emissions. When using low-carbon fuels as non-primary fuels, the larger the range proportion of low-carbon fuels, the lower EEDI value. The LNG–MDO combination improved the EEDI reduction factor by 9.97% compared with the use of the MDO solution, outperforming the LPG–MDO and MeOH–MDO solutions by 3 and 7 times, respectively. When low-carbon fuels were used as primary fuels, the LNG–MDO solution improved the EEDI reduction factor by 20.80% compared with the use of the MDO solution, significantly outperforming the other two solutions. In addition, among low-carbon alternative fuels, whether used as non-primary or primary fuels, LNG has the highest EEDI reduction factor, followed by LPG and MeOH, respectively.

Table 3. EEDI values of 19 groups of dual-fuel power schemes.

Low-Carbon Fuel Range (nm)	Dual Fuel	Primary Fuel or Not (Y/N)	EEDI Value	Low-Carbon Fuel Range (nm)	Dual Fuel	Primary Fuel or Not (Y/N)	EEDI Value
0	LPG–MDO LNG–MDO MeOH–MDO	N	2.774	10,000	LPG–MDO LNG–MDO MeOH–MDO	N	2.666 2.418 2.724
1000	LPG–MDO LNG–MDO MeOH–MDO	N	2.838 2.738 2.769	11,000	LPG–MDO LNG–MDO MeOH–MDO	N Y Y	2.647 2.032 2.694
2000	LPG–MDO LNG–MDO MeOH–MDO	N	2.819 2.702 2.764	12,000	LPG–MDO LNG–MDO MeOH–MDO	Y	2.463 2.030 2.691
3000	LPG–MDO LNG–MDO MeOH–MDO	N	2.800 2.667 2.759	13,000	LPG–MDO LNG–MDO MeOH–MDO	Y	2.461 2.028 2.688
4000	LPG–MDO LNG–MDO MeOH–MDO	N	2.780 2.631 2.754	14,000	LPG–MDO LNG–MDO MeOH–MDO	Y	2.458 2.026 2.686
5000	LPG–MDO LNG–MDO MeOH–MDO	N	2.761 2.595 2.749	15,000	LPG–MDO LNG–MDO MeOH–MDO	Y	2.456 2.024 2.683
6000	LPG–MDO LNG–MDO MeOH–MDO	N	2.742 2.560 2.744	16,000	LPG–MDO LNG–MDO MeOH–MDO	Y	2.453 2.022 2.681
7000	LPG–MDO LNG–MDO MeOH–MDO	N	2.723 2.524 2.739	17,000	LPG–MDO LNG–MDO MeOH–MDO	Y	2.451 2.020 2.678
8000	LPG–MDO LNG–MDO MeOH–MDO	N	2.704 2.489 2.734	18,000	LPG–MDO LNG–MDO MeOH–MDO	Y	2.449 2.018 2.675
9000	LPG–MDO LNG–MDO MeOH–MDO	N	2.685 2.453 2.729				

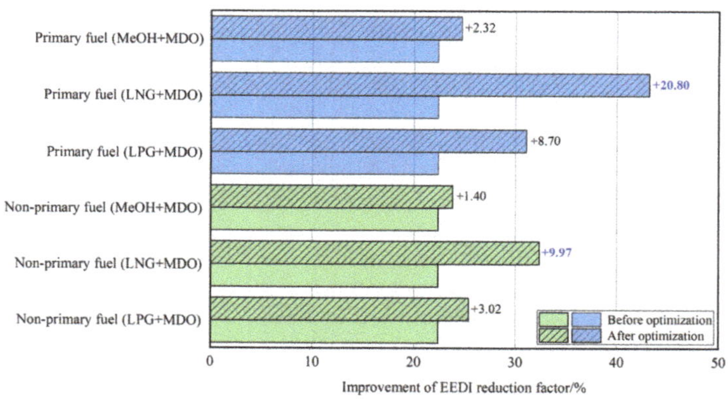

Figure 4. Improvement of the EEDI reduction factor provided by low-carbon fuels as primary and non-primary fuels.

2.3. Research Approach

In Section 2.2, different dual-fuel power options based on the EEDI were analyzed and discussed. The EEDI represents the CO_2 emissions per ton of transported cargo, which reflects the level of carbon emissions during ship operations. However, the results of prior studies showed that the EEDI evaluation technique does not consider the carbon emissions generated by the fuel during the production or transportation phases nor the issue of methane escape in the LNG power scenario. Meanwhile, when using low-carbon fuels as the primary fuel, there are some limitations and shortcomings in EEDI evaluation, especially the impact of the range of low-carbon fuels on carbon emissions reduction which cannot be quantitatively evaluated. Therefore, in this study, the N786 150,000-ton shuttle tanker was used as the research object and LCA was adopted to further study the three types of dual-fuel power scenarios in Section 2.2. The results were compared and analyzed using the EEDI assessment method.

In the marine sector, with a particular focus on marine fuel modules, LCA is used to assess the GHG emissions of the entire fuel chain involved in ship operations ("well-to-wake", as shown in Figure 5). From the initial production phase of the fuel, through extraction and cultivation, to processing and refining, transportation, storage, and ultimately use on board. Depending on the state of fuel bunkering, the whole process is mainly divided into the "well-to-tank" part and "tank-to-wake" part [36]. The environmental impacts of the well-to-tank fuel phase are influenced by several conditions such as mining, refining, and purification. The internationally dominant LCA-related Ecoinvent 3 database, which provides pre-combustion environmental impact emission factors, such as the mining of fuel products, was used to evaluate the environmental impacts in the well-to-tank phase. However, the database is not suitable for the combustion phase of the fuel [34,35,46], whereas the Intergovernmental Panel on Climate Change (IPCC) database provides detailed emission factors for the combustion phase ("tank-to-wake") of the fuel. The database was created and maintained by hundreds of experts and is supported by multiple corporate and non-profit organizations worldwide. The high credibility of the data provides an excellent solution for obtaining emission factors during the fuel combustion phase. Note that emission factors for the fuel combustion phase can also be obtained from actual main engine emissions tests or other databases. However, the Ecoinvent 3 database is also a comprehensive integration database of multiple data sources and methods, rather than being based entirely on actual testing of experimental methods. To ensure consistency in research methodology more effectively, the authoritative IPCC database was selected for the fuel combustion phase. Therefore, these two databases were used for LCA in this study and the IPCC method was used for environmental impact assessment.

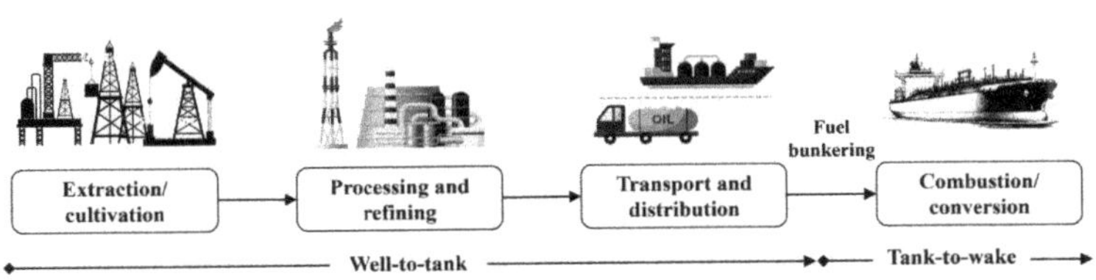

Figure 5. Methodology for life cycle assessment of marine fuel.

Figure 6 shows the evaluation method used in this study for carbon emission levels of marine fuel. This method mainly covers the following three parts:

1. Research preparation: Complete the data statistics of the target ship and develop corresponding dual-fuel power schemes based on the selection of the main engine.

2. Environmental impact assessment: Conduct environmental impact assessments for different scenarios based on the Ecoinvent 3 database and IPCC database.
3. Comparative analysis based on the EEDI and LCA: Considering the different focuses of the EEDI and LCA methods, the impact of different schemes on the results of both methods and the carbon emissions profile during the well-to-wake phase are considered.

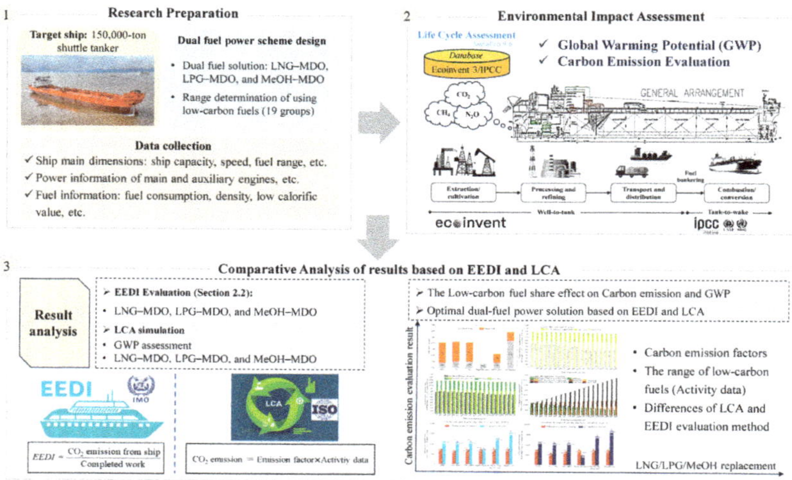

Figure 6. Evaluation method for carbon emission levels of marine fuel based on LCA [ISO 14001–ISO 14004] and EEDI.

From Figure 6, the method mainly combines the EEDI and LCA methods, aiming to overcome some limitations of the two individual methods. Meanwhile, a 150,000-ton shuttle tanker was selected as the study object and three groups of dual-fuel power design schemes (LNG–MDO, LPG–MDO, and MeOH–MDO) were developed to compare the effects of different dual-fuel power design options on the carbon control potential and environmental performance of ships.

3. LCA Simulation Result Analysis

3.1. Methodological Framework

The methodological framework of LCA was used in this study to evaluate and compare the environmental impact performance of three types of dual-fuel power schemes, that is, LNG–MDO, LPG–MDO, and MeOH–MDO, with that of conventional marine fuel (MDO), focusing on changes in the Global Warming Potential (GWP). The entire evaluation process strictly follows the standard procedures of LCA, including the definition of goal and scope, analysis of life cycle inventory, assessment of environmental impact, and interpretation of result. Furthermore, the obtained LCA results are compared with the prior EEDI results to explore the carbon control potentials of different dual-fuel power schemes. This will be more useful to fully reveal the differences in the actual effectiveness and potential of different dual-fuel power schemes in controlling carbon emissions.

3.2. Goal and Scope Definition

In the LCA simulations of this study, the main goal was to prioritize the comparative analysis of the environmental impact of three types of dual-fuel power scenarios, that is, LNG–MDO, LPG–MDO, and MeOH–MDO, for a total of 19 groups of low-carbon fuel range design scenarios used to replace the conventional marine fuel MDO over the full cycle range of marine fuels ("well-to-wake"). Both "well-to-tank" and "tank-to-wake" phases were included. The functional unit of the system is defined as the entire process from production to application of marine fuel, covering the processing and production of crude oil to the energy conversion of the combustion process in the operational phase. However, the fuel transportation process was given less consideration, as its environmental impact in the well-to-tank phase is relatively insignificant compared to raw material extraction and production processes. Although fuel transportation of great significance to LNG supply chains, relevant studies have shown that the transportation phase accounts for only 2% of total well-to-wake emissions for LNG [16,53]. The analysis primarily focused on the tank-to-wake phase to compare the results with those of a prior EEDI study. The overall system boundary framework applied to the LCA simulation is shown in Figure 7.

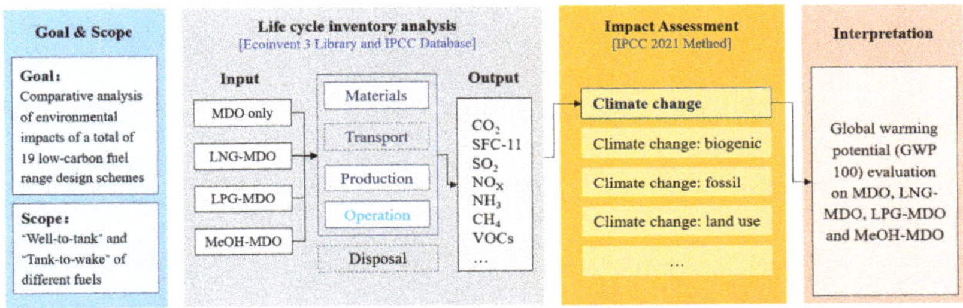

Figure 7. System boundary framework applied to the LCA simulation.

3.3. Life Cycle Inventory

The analysis of life cycle inventories is a key component that focuses on accurately quantifying all resource inputs and environmental outputs associated with the specific functional unit. In this study, a comprehensive and systematic comparison and analysis of the environmental impacts of marine fuels in all phases was achieved by examining in detail the data from the acquisition of raw materials up to the final application. Corresponding to the content of the 2024 LCA Guidelines developed by the IMO [36], both the "well-to-tank" and "tank-to-wake" phases of marine fuels were considered. The well-to-tank phase and fuel consumption derived from the 19 sets of dual-fuel power scenarios were used as inputs for the inventories. As previously mentioned, these inventories used the Ecoinvent 3 database provided by the LCA software SimaPro 9.6, which is one of the most comprehensive databases available. The tank-to-wake phase inventories utilized the IPCC results database [54–57]. This is because the emission factors for fuels in the application phase (i.e., fuel combustion phase) are not available in the current phase of Ecoinvent 3.10 database. The IPCC database, on the other hand, is regarded to be the authoritative repository of records on global climate change and includes a wealth of data on the rate of climate change, climate models, carbon emissions, and effects of climate change. Table 4 shows the key properties of the fuels used in this study. The preparation of MeOH was relatively diverse compared with that of the other two. Based on the database and research trends, the three main production and preparation processes of biomass, natural gas (NG), and coal were considered for methanol in this study. In addition, Table 5 lists the total consumption of low-carbon and conventional fuels for the 19 groups of dual-fuel power schemes. Based on the data provided in Tables 4 and 5, the GWP values were calculated using the rule of "activity data times by carbon emission factors".

Table 4. Fuel properties provided by Ecoinvent 3.10 implemented in SimaPro 9.6 and IPCC emission factor database.

Item		Unit	MDO	LNG	LPG	MeOH (Biomass)	MeOH (NG)	MeOH (Coal)
Low Calorific Value (LCV)		MJ/kg	42.7	48.0	46.3		19.9	
Density		kg/m^3	900	450	610		790	
Emission Factors	CO_2	g/kg	3212	2750	3000		1375	
	CH_4	g/kg	0.1	14.5	1		0.2	
	N_2O	g/kg	0.0798	0.0245	0.1770		0.01	
	well-to-tank	t CO_2-eq	0.84	1.07	1.06	0.68	0.88	4.46
	tank-to-wake	t CO_2-eq	3.24	3.16	3.07		1.38	1.38

Table 5. Consumption of alternative and traditional fuels for different design solutions.

	Fuel Consumption (t)					Fuel Consumption (t)			
No.	Low-Carbon Fuel		Diesel Fuel	Pilot Oil	No.	Low-Carbon Fuel		Diesel Fuel	Pilot Oil
	LPG					LPG	1452.52		66.15
1	LNG	0	2899.46	0	11	LNG	1431.24	1288.65	29.65
	MeOH					MeOH	3472.43		90.07
	LPG	145.25		6.61		LPG	1597.77		72.76
2	LNG	143.12	2738.38	2.97	12	LNG	1574.36	1127.57	32.62
	MeOH	347.24		9.01		MeOH	3819.67		99.08
	LPG	290.50		13.23		LPG	1743.02		79.38
3	LNG	286.25	2577.29	5.93	13	LNG	1717.49	966.49	35.58
	MeOH	694.49		18.01		MeOH	4166.91		108.09
	LPG	435.76		19.84		LPG	1888.28		85.99
4	LNG	429.37	2416.21	8.90	14	LNG	1860.61	805.40	38.55
	MeOH	1041.73		27.02		MeOH	4514.16		117.10
	LPG	581.01		26.46		LPG	2033.53		92.60
5	LNG	572.50	2255.13	11.86	15	LNG	2003.73	644.32	41.52
	MeOH	1388.97		36.03		MeOH	4861.40		126.10
	LPG	726.26		33.07		LPG	2178.78		99.22
6	LNG	715.62	2094.05	14.83	16	LNG	2146.86	483.24	44.48
	MeOH	1736.21		45.04		MeOH	5208.64		135.11
	LPG	871.51		39.69		LPG	2324.03		105.83
7	LNG	858.74	1932.97	17.79	17	LNG	2289.98	322.16	47.45
	MeOH	2083.46		54.04		MeOH	5555.89		144.12
	LPG	1016.76		46.30		LPG	2469.28		112.45
8	LNG	1001.87	1771.89	20.76	18	LNG	2433.10	161.08	50.41
	MeOH	2430.70		63.05		MeOH	5903.13		153.13
	LPG	1162.02		52.92		LPG	2614.54		119.06
9	LNG	1144.99	1610.81	23.72	19	LNG	2576.23	0	53.38
	MeOH	2777.94		72.06		MeOH	6250.37		162.13
	LPG	1307.27		59.53					
10	LNG	1288.11	1449.73	26.69					
	MeOH	3125.19		81.07					

3.4. Evaluation of Impact Assessment and Result Analysis

Converting the environmental loads of marine fuels in both the well-to-tank and tank-to-wake phases into specific quantitative indicators is essential for comprehensively and accurately assessing the actual impact of these phases on the environment. In other words, quantifying the environmental loads during the different phases is a key component in the scientific assessment of the environmental effects. The IPCC 2021 method on climate change, that is, GWP100 was adopted. The GWP100 is a measure of the intensity of the greenhouse effect of a GHG relative to carbon dioxide over a 100-year timeframe and is one of the more authoritative assessment techniques available. Based on the CO_2 equivalent values, the GHG emissions were converted to GWP values using the latest GWP coefficients listed in Table 6 [56]. Simultaneously, the numerical conversion was completed to obtain the carbon emissions of the ship during the operation phase based on the EEDI values of the 19 groups of dual-fuel power options listed in Table 3.

The carbon emission assessment results for the conventional MDO (zero low-carbon fuel range) and three dual-fuel power schemes (LNG–MDO, LPG–MDO, and MeOH–MDO) are shown in Figures 8–10. The LCA results show that the GWP values of LNG–MDO, LPG–MDO, and MeOH (biomass)–MDO gradually decrease with increasing range, whereas the GWP values of MeOH (NG) and MeOH (coal) increase, which raises the load on the environment. The observed variations primarily result from the combined effects of the specific carbon emission factors associated with each low-carbon fuel and differentiated fuel/gas consumption patterns among the three dual-fuel power schemes. These mechanisms will be thoroughly examined in the Discussion section. In contrast, using CO_2 emissions as a benchmark, the EEDI shows a "broken" trend based on the low-carbon fuel range. The main reason is that when the low-carbon fuel is considered as a primary fuel, the calculation of the EEDI only considers the fuel consumption of both SPOC and low-carbon fuels no longer considering the impact of conventional fuel oil on carbon emissions. Generally, when low-carbon fuels are used as non-primary fuels (i.e., LNG, MeOH $<$ 11,000 nm; LPG $<$ 12,000 nm), the range and CO_2 emissions exhibit a positive correlation, with emissions decreasing with increasing range. When low-carbon fuels are used as primary fuels (i.e., LNG, MeOH $\geq$ 11,000 nm; LPG $\geq$ 12,000 nm), the EEDI cannot quantify the correlation between CO_2 emissions and range. Further comparative analysis revealed that, compared with the EEDI evaluation method, LCA quantitatively analyzes the changes in the GWP values of the well-to-tank and tank-to-wake phases simultaneously, which reflects the environmental impacts of GHGs. The EEDI can only quantitatively analyze CO_2 emissions during the tank-to-wake phase and does not include low-carbon fuels as the main fuel. In addition, the analysis of the GWP composition ratios of marine fuels under different substitution scenarios shows that CO_2 contributes the most, followed by CH_4 and, finally, N_2O (Figure 11). Other GHG contributions, which are very small and have little influence on GHG effects, were not considered here.

Table 6. Global Warming Potential (GWP) values of major greenhouse gases.

Industrial Designation or Common Name	GWP Values for 100-Year Time Horizon: Fifth Assessment Report (AR5)
Carbon dioxide (CO_2)	1
Methane (CH_4)	28
Nitrous oxide (N_2O)	265
CO_2-eq	$gCO_2\text{-}eq_{(100years)} = 1 \times gCO_2 + 28 \times gCH_4 + 265 \times gN_2O$

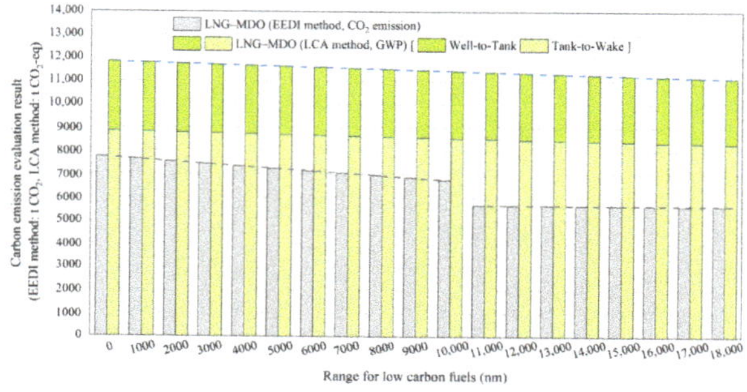

Figure 8. Carbon emission evaluation of the LNG–MDO dual-fuel power scheme.

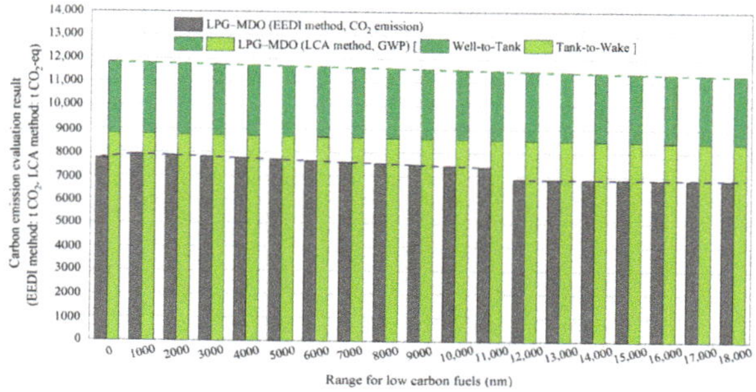

Figure 9. Carbon emission evaluation of the LPG–MDO dual-fuel power scheme.

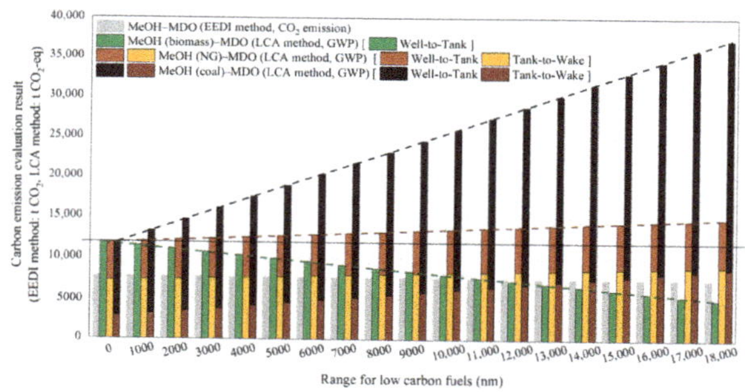

Figure 10. Carbon emission evaluation of the MeOH–MDO dual-fuel power scheme.

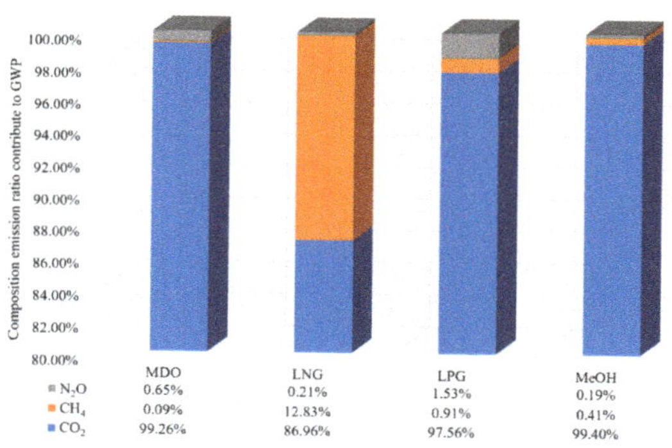

Figure 11. GWP composition ratio of marine fuels under different alternative scenarios.

4. Discussion

In this section, the previous research results of EEDI and LCA simulations were combined to further analyze and discuss the inventory inputs, carbon emissions, and life-cycle impacts of the research process. To determine the current applicable dual-fuel power scheme, three aspects were considered: the effects of emission factors on GWP, the carbon control potential of different alternative schemes, and the prospect analysis of low-carbon alternative fuels.

4.1. Effects of Emission Factors on GWP

Emission factors directly affect the estimation of GHG emissions, which represents a significant influence on the results of GWP calculations. Specifically, a higher factor means that producing the same amount of energy will release more GHG emissions, resulting in a higher GWP. The focus was placed on the full well-to-wake cycle of marine fuels including both the well-to-tank and tank-to-wake phases. The emission factors of the different marine fuels throughout the well-to-wake cycle are shown in Figure 12. Overall, the effects of the emission factors on the GWP can be ranked from smallest to largest as follows: MeOH (biomass) < MeOH (NG) < MDO < LPG < LNG < MeOH (coal). The integrated well-to-tank process for marine fuel oil covers the fuel process of crude oil procurement, subsequent processing, and production. In this phase, the effects of the emission factors on the GWP can be ranked from smallest to largest as follows: MeOH (biomass) < MDO < MeOH (NG) < LPG < LNG < MeOH (coal). For the tank-to-wake phase, which considers the use of marine fuels (i.e., process of fuel combustion), the emission factors can be ranked from smallest to largest as follows: MeOH (biomass) < MeOH (NG) = MeOH (coal) < LPG < LNG < MDO. MeOH (biomass) is a type of "renewable carbon dioxide". Although carbon dioxide is produced during subsequent methanol combustion, carbon emissions are captured in a loop; therefore, the loop is closed and carbon emissions are zero [58]. In addition, methane escape during this phase was considered. Data from the IPCC database were used. The methane escape ratios were generally similar to the results of actual dual-fuel mainframe emission tests [59]. The conventional methane slip ratio of 2% was prioritized. When the escape ratio was reduced, the LNG emission factor was further reduced. Based on analyzing the effect of the emission factors on the GWP, MeOH (biomass) is the most environmentally friendly alternative fuel for marine use. However, because of the effect of the low calorific value of the fuel (Table 4), the lower the calorific value of the fuel is, the less energy is supplied and more consumption is required for the same number of nautical miles of operation. Therefore, low-carbon fuels require further comparative analysis. The carbon emission factor in the EEDI calculation mainly considers the carbon emission during fuel combustion. However, through the analysis of this subsection, it can be seen that the impact of the extraction, refining, and production stages of fuel cannot be ignored, and it also shows the improvement direction of the carbon factor in carbon emission accounting of the existing EEDI and other relevant regulations so as to more comprehensively evaluate the carbon control potential of ships from the perspective of the whole life cycle of fuel.

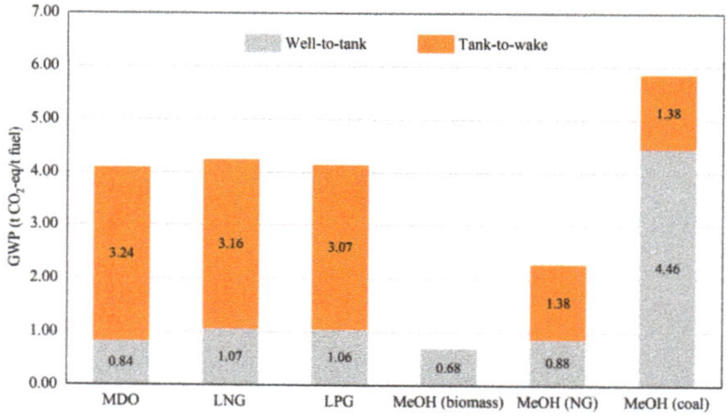

Figure 12. Emission factors of different marine fuels throughout the well-to-wake cycle.

4.2. Carbon Control Potential of Different Alternative Schemes

To further compare and analyze the carbon control potential of different low-carbon fuel alternatives, the results of the EEDI and LCA simulations of the conventional MDO power scheme were used as a baseline. Two typical working conditions of low-carbon fuels, that is, non-primary fuels (with a range of 10,000 nm) and primary fuels (with a range of 18,000 nm), were selected for further discussion, as shown in Figures 13 and 14.

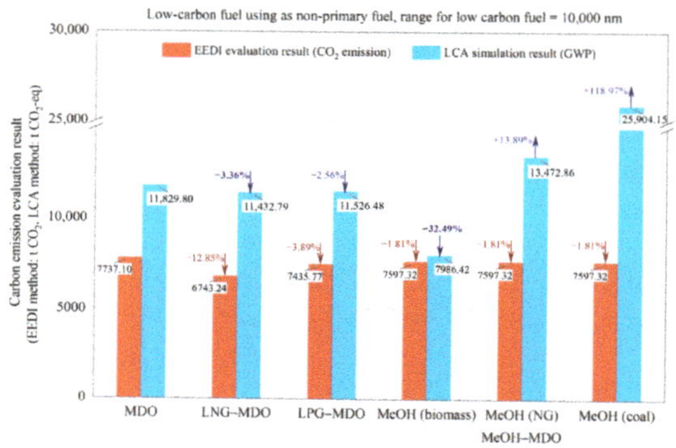

Figure 13. Carbon emission assessment results for low-carbon fuels as non-primary fuel under different alternative schemes.

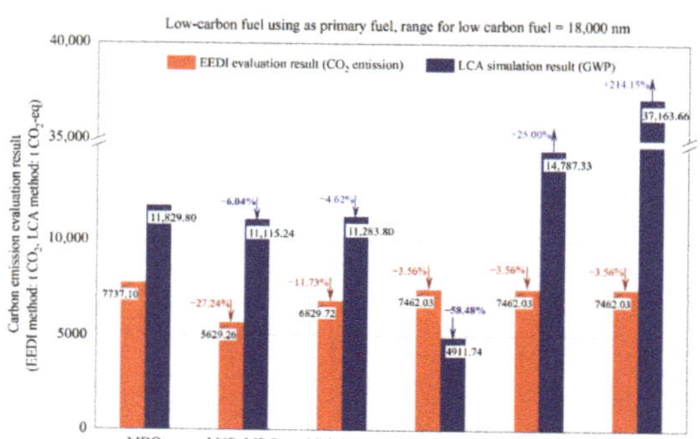

Figure 14. Carbon emission assessment results for low-carbon fuels as primary fuel under different alternative schemes.

Figure 13 shows the carbon emission assessment results for low-carbon fuels as non-primary fuels under different alternative schemes. The results show that the application of LNG, LPG, and MeOH reduced the carbon emissions by ~12.85% (993.86 t CO_2), 3.89% (301.33 t CO_2), and 1.81% (139.79 t CO_2), respectively, compared to the EEDI results of the traditional MDO case. Based on the LCA simulation results, compared to the traditional MDO case, the application of LNG, LPG, and MeOH (biomass) reduced the GWP by ~3.36% (397.01 t CO_2-eq), 2.56% (303.31 t CO_2-eq), and 32.49% (3843.38 t CO_2-eq), respectively. In contrast, the application of MeOH (NG) and MeOH (coal) increased the GWP by ~13.89% (1643.06 t CO_2-eq) and 118.97% (14,074.35 t CO_2-eq), respectively.

Similarly, Figure 14 shows the carbon emission assessment results for low-carbon fuels as primary fuel. The EEDI results shows that compared to the traditional MDO case, the application of LNG, LPG, and MeOH reduced the carbon emissions by ~27.24% (2107.84 t CO_2), 11.73% (907.38 t CO_2), and 3.56% (275.06 t CO_2), respectively. The LCA simulation results shows that compared to MDO case, the application of LNG, LPG, and MeOH (biomass) reduced the GWP by ~6.04% (714.56 t CO_2-eq), 4.62% (545.99 t CO_2-eq), and 58.48% (6918.05 t CO_2-eq), respectively. The application of MeOH (NG) and MeOH (coal) increased the GWP by ~25.00% (2957.53 t CO_2-eq) and 214.15% (25,333.87 t CO_2-eq), respectively. Regardless of the use of low-carbon fuel as non-primary or primary fuel, the application of LNG, LPG, and MeOH (biomass) has great potential for carbon control, reducing carbon emissions while minimizing the effects of GHGs on the environment. The results indicate that MeOH (biomass) is optimal, followed by LNG and LPG. In addition, MeOH (NG) and MeOH (coal) lead to a greater environmental load because of the production process.

4.3. Prospect Analysis of Low-Carbon Alternative Fuels

In Section 4.2, the carbon control potential of different fuel alternatives is discussed. Overall, based on the analysis of the evaluation results, MeOH (biomass) has a much better carbon control potential than LNG and LPG; however, many challenges remain. Fuel calorific value, fuel volume, and carbon emission factors are key in determining the total quantity of carbon emission. As mentioned before, MeOH has the lowest carbon emission factor during the operational phase among the evaluated alternatives, accounting for approximately about 43% and 44% of the LNG and LPG factors, respectively (excluding MeOH (biomass)). However, due to its smaller LCVs, MeOH may impact payload capacity. In contrast, LNG and LPG have higher LCVs, approximately 2.4 times that of MeOH, combined with their lower density and moderate emission factors, resulting in a smaller impact on payload. Therefore, under the same design conditions, MeOH consumption is higher, requiring more volume for fuel storage to meet the power demands of vessel operations. Additionally, the current state of raw materials poses certain challenges for MeOH production. Figure 15 shows the proportion of feedstock and resources used to generate global MeOH. Currently, the main sources of MeOH include natural gas, coal, and coke oven gas [58]. All of these raw materials need to go through complex processes and strict treatment steps in the production process to finally obtain qualified MeOH products. At the same time, insufficient supply of raw materials and complex production processes are also the main problems faced by current MeOH production. Therefore, measures are needed to improve the feasibility and economy of MeOH (biomass) production such as developing new biomass resources and optimizing pretreatment and conversion technologies [60–62].

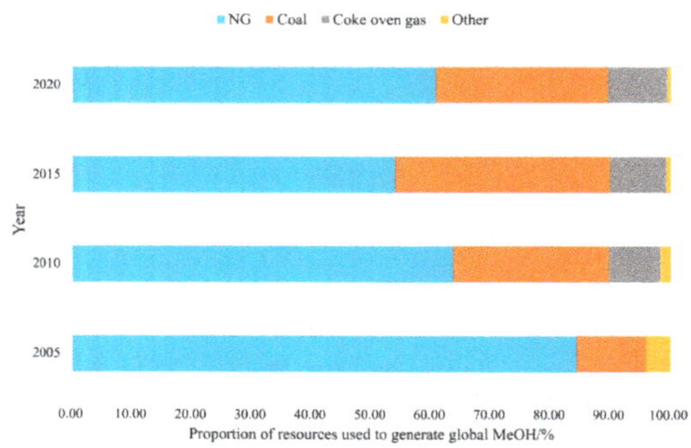

Figure 15. Global MeOH feedstock trend.

The results of this study show that the carbon control potential of LNG and LPG is second only to that of MeOH (biomass). LNG is increasingly favored as a low-carbon fuel that is easy to obtain and cost competitive. Its popularity is due to a number of significant advantages [63,64]. For example, LNG has a very high thermal efficiency, which makes the energy conversion process more efficient and helps reduce energy consumption. Secondly, LNG produces far fewer NOx during combustion than conventional fuels, which has a significant effect on improving air quality and reducing environmental pollution. Furthermore, its lower risk of bursting brings higher safety to the transportation and usage process, reducing the occurrence of accidents. In addition, the high compression ratio of LNG means that more energy can be stored in the same amount of storage space, improving the economy of storage and transportation. Related studies suggest that more NG reserves may be discovered in the future, which will further contribute to lowering the cost of LNG. Note that potential emission taxes could increase the interest in LNG [23,65].

Although LPG is widely recognized globally as a viable alternative fuel option for all types of vessels operating on local voyages and international routes, its popularity in practical applications in the maritime sector has not yet met widespread expectations [66]. The main reasons can be summarized as follows [67]: when LPG is used, ships must be refueled within a few weeks, which is inconvenient for long sea voyages and operations; despite the fact that the global infrastructure for LPG exhibits a high degree of completeness and is in good condition, this guarantees the safe and efficient storage and transportation of LPG. However, it should not be overlooked that current production capacity and supply facilities are relatively limited. This limitation is not only reflected in the fact that the expansion of production capacity may not be able to keep up with the rapid growth of market demand, but also in the lack of flexibility and scalability of the existing supply facilities in responding to sudden demand or large-scale applications. In addition, compared with LNG, LPG faces more serious challenges in terms of safety. Due to the higher density of LPG compared to air, once a leak occurs, the gas often spreads along the ground and is difficult to detect in a timely manner, greatly increasing the potential risk of accidents. Meanwhile, considering the low flash point temperature of LPG, it is extremely easy for it to burn rapidly once it encounters an ignition source or high temperature, thus triggering a fire or even a more serious explosion.

The world's famous shipping company Maersk previously led the charge in making methanol fuel an important development strategy, and more recently appears to be shifting toward LNG as a marine fuel. This may be due to the dwindling supply and high cost of green methanol compared with LNG based on energy density and extraction costs. Therefore, considering the carbon emission reduction effect based on the GWP of fuels from well-to-wake and cost issues, etc., LNG remains a good choice for the current low-carbon fuel transition phase. However, with the gradual maturity of biofuel preparation technology and the gradual expansion of biofuel production capacity, the optimal choice for the distant future is the promotion of green fuels, such as bio-methanol, which can help achieve net-zero emissions.

5. Conclusions

The shipping industry must adopt measures to achieve the ambitious goals of the IMO's strategy for reducing GHG emissions. In this study, a 154,000-ton shuttle tanker was used as the research object, and the design of three types of mainstream dual-fuel power schemes, that is, LNG–MDO, LPG–MDO, and MeOH–MDO, was prioritized. An in-depth analysis was focused on the carbon emissions profile and GWP of different dual-fuel power generation scenarios, providing a more environmentally friendly dual-fuel power solution for the main engine of ships. Furthermore, the results provide an effective analytical argument for the development and revision of the corresponding key IMO guidelines, which is of great significance for the decision-making process of the IMO in regulating and guiding the sustainable development of the global shipping industry.

Developed by the IMO, the EEDI standard mainly quantifies the carbon emissions of marine fuel during the operational phase, and the carbon emissions generated during fuel production cannot be considered. However, especially for certain fuels such as MeOH (coal) in this study, the emission factors during the production phase are much higher than those during the combustion phase, resulting in more carbon emissions. This reveals the limitation of the EEDI of only considering carbon emissions during the operational phase. Meanwhile, when low-carbon fuels are used as the primary fuel, the EEDI cannot accurately quantify their impact on carbon emission reduction. In contrast, LCA can be applied to evaluate the carbon control potential of low-carbon fuels quantitatively throughout the well-to-wake phase while also considering the environmental impact of additional GHGs. When benchmarked against conventional MDO's EEDI, LNG can achieve up to approximately 27.24% reduction in carbon emission potential. Comparatively, when evaluated relative to conventional MDO's GWP, MeOH biomass demonstrates a maximum decarbonization potential of about 58.48%. Therefore, with the increasing implementation of carbon control policies in the shipbuilding and dismantling phases in various countries, the carbon emission analysis model considering the full life cycle will have a significantly important role. In addition, because of the fundamental similarities between the EEXI and EEDI in terms of calculation formulae and assessment objectives, the methodological ideas and analyses of this study are equally applicable to existing ships (i.e., ships under the EEXI framework). For small vessels such as ferries and yachts that fall outside IMO regulatory jurisdiction, this study methodology can still be effectively applied to assess their environmental impacts through LCA, without being limited by ship type.

The main research results provide basic support for the revision of carbon emission factors in the relevant regulations of the shipping industry, such as the EEDI, that is to say, not only the carbon emissions in the combustion stage of fuel oil, but also the carbon emissions in the extraction, refining, and production stages of fuel oil should be considered, especially in the important period of transformation from traditional fuels to low-carbon and zero-carbon fuels. However, this study has certain limitations. First, although LCA was conducted for the three types of low-carbon fuels, zero-carbon fuels and their manufacturing processes were not considered. Second, although emission factor data provided by the internationally authoritative IPCC database were used, actual main engine emission tests were not conducted; thus, the conclusions of the comparative analysis may contain discrepancies. Therefore, zero-carbon fuels should be considered, and the reliability of the method provided in this study should be verified experimentally.

Author Contributions: Conceptualization, Z.W. and Z.H.; methodology, Z.H.; resources, S.F.; data curation, Z.W.; writing—original draft preparation, Z.W.; writing—review and editing, Z.W., Z.H. and S.F.; visualization, Z.W.; supervision, S.F.; funding acquisition, S.F. All authors have read and agreed to the published version of the manuscript.

Funding: This research was funded by the National Natural Science Foundation of China, grant number 52071240.

Data Availability Statement: All data used in this study are presented in the paper.

Acknowledgments: We appreciate the partial data sources provided by Zhoushan COSCO Shipping Heavy Industries Co., Ltd.

Conflicts of Interest: The authors declare no conflicts of interest.

References

1. Borén, C.; Grifoll, M.; Castells-Sanabra, M. Emissions assessment of container ships sailing under off-design conditions. *J. Mar. Sci. Eng.* **2023**, *11*, 1983. [CrossRef]
2. Elkafas, A.G.; Elgohary, M.M.; Shouman, M.R. Numerical analysis of economic and environmental benefits of marine fuel conversion from diesel oil to natural gas for container ships. *Environ. Sci. Pollut. Res. Int.* **2021**, *28*, 15210–15222. [CrossRef] [PubMed]
3. Hu, Q.; Zhou, W.; Diao, F. Interpretation of the initial IMO Strategy on Reduction of GHG Emissions from Ships. *Shipbuild. China* **2019**, *60*, 195–201.
4. International Maritime Organization. *Report of the Marine Environment Protection Committee on Its Sixty-Second Session*; MEPC 62; MEPC: London, UK, 2011.
5. International Maritime Organization. *International Maritime Organization's Initial Greenhouse Gas Strategy*; IMO Marine Environment Protection Committee (MEPC): London, UK, 2018.
6. International Maritime Organization. *IMO Regulations to Introduce Carbon Intensity Measures Enter Into Force*; IMO Marine Environment Protection Committee (MEPC): London, UK, 2022.
7. International Maritime Organization. *IMO Strategy on Reduction of GHG Emissions from Ships*; IMO Marine Environment Protection Committee (MEPC): London, UK, 2023.
8. Ait Allal, A.; Mansouri, K.; Youssfi, M.; Qbadou, M. Toward an evaluation of marine fuels for a clean and efficient autonomous ship propulsion energy. *Mater. Today Proc.* **2019**, *13*, 486–495. [CrossRef]
9. Adland, R.; Cariou, P.; Jia, H.; Wolff, F.C. The Energy efficiency effects of periodic ship hull cleaning. *J. Clean. Prod.* **2018**, *178*, 1–13. [CrossRef]
10. Norlund, E.K.; Gribkovskaia, I. Reducing emissions through speed optimization in supply vessel operations. *Transp. Res. Part D Transp. Environ.* **2013**, *23*, 105–113. [CrossRef]
11. Jia, T. *Research on the Influence of Lightweight Structure to the Energy Efficiency Design Index*; Dalian University of Technology: Dalian, China, 2016.
12. Feng, Y.; el Moctar, O.; Schellin, T.E. Parametric Hull Form Optimization of Containerships for Minimum Resistance in Calm Water and in Waves. *J. Mar. Sci. Appl.* **2021**, *20*, 670–693. [CrossRef]
13. Kang, J.G.; Kim, M.C.; Kim, H.U.; Shin, I.R. Study on Propulsion performance by varying rake distribution at the propeller tip. *J. Mar. Sci. Eng.* **2019**, *7*, 386. [CrossRef]
14. Kumar, Y.H.; Kumar, R.V. Development of an energy-efficient stern flap for improved EEDI of a typical high speed displacement vessel. *Def. Sci. J.* **2020**, *70*, 95–102. [CrossRef]
15. Ren, H.; Ding, Y.; Sui, C. Influence of EEDI (Energy Efficiency Design Index) on ship–engine–propeller matching. *J. Mar. Sci. Eng.* **2019**, *7*, 425. [CrossRef]
16. Perčić, M.; Vladimir, N.; Fan, A. Life-cycle cost assessment of alternative marine fuels to reduce the carbon footprint in short-sea shipping: A case study of Croatia. *Appl. Energy* **2020**, *279*, 115848. [CrossRef]
17. Zhao, Z.; Cui, W.; Qi, Y. Prospects of various ship power fuels under the transformation of marine energy. *Marit. China* **2022**, *6*, 42–48.
18. Tu, H. Comprehensive analysis on the adaptability of clean energy for marine use. *China Ship Surv.* **2022**, *1*, 58–62.
19. International Maritime Organization. Implementation of Sulphur 2020 Limit—Carriage Ban Adopted. Available online: https://www.imo.org/en/MediaCentre/PressBriefings/Pages/19-Implementation-of-sulphur-2020-limit-.aspx (accessed on 16 April 2025).
20. Burel, F.; Taccani, R.; Zuliani, N. Improving sustainability of maritime transport through utilization of Liquefied Natural Gas (LNG) for propulsion. *Energy* **2013**, *57*, 412–420. [CrossRef]

21. Li, C.; Hu, Y. Development of clean alternative fuel LPG in marine engines. *Mar. Equip. Mater. Mark.* **2022**, *30*, 68–73.
22. Balcombe, P.; Brierley, J.; Lewis, C.; Skatvedt, L.; Speirs, J.; Hawkes, A.; Staffell, I. How to decarbonise international shipping: Options for fuels, technologies and policies. *Energy Convers. Manag.* **2019**, *182*, 72–88. [CrossRef]
23. Jafarzadeh, S.; Paltrinieri, N.; Utne, I.B.; Ellingsen, H. LNG-fueled fishing vessels: A systems engineering approach. *Transp. Res. D Trans. Environ.* **2017**, *50*, 202–222. [CrossRef]
24. Moroianu, C.; Postolache, F. Marine propulsion engine behaviour using fossil fuel and methanol. *J. Phys. Conf. Ser.* **2018**, *1122*, 012018. [CrossRef]
25. Boretti, A.; Osman, A.; Aris, I. Direct injection of hydrogen, oxygen and water into a novel two stroke engine. *Int. J. Hydrogen Engery* **2011**, *36*, 10100–10106. [CrossRef]
26. Fernández-Dacosta, C.; Shen, L.; Schakel, W.; Ramirez, A.; Kramer, G.J. Potential and challenges of low-carbon energy options: Comparative assessment of alternative fuels for the transport sector. *Appl. Energy* **2019**, *236*, 590–606. [CrossRef]
27. Gilbert, P.; Walsh, C.; Traut, M.; Kesieme, U.; Pazouki, K.; Murphy, A. Assessment of full life-cycle air emissions of alternative shipping fuels. *J. Clean. Prod.* **2018**, *172*, 855–866. [CrossRef]
28. Labeckas, G.; Slavinskas, S. Performance and emission characteristics of a direct injection diesel engine operating on KDV synthetic diesel fuel. *Energy Convers. Manag.* **2013**, *66*, 173–188. [CrossRef]
29. Matzen, M.; Alhajji, M.; Demirel, Y. Chemical storage of wind energy by renewable methanol production: Feasibility analysis using a multi-criteria decision matrix. *Energy* **2015**, *93*, 343–353. [CrossRef]
30. Tang, Y.T.; Ma, X.Q.; Lai, Z.Y.; Chen, Y. Energy analysis and environmental impacts of a MSW oxy-fuel incineration power plant in China. *Energy Policy* **2013**, *60*, 132–141. [CrossRef]
31. Yan, X.; He, Y.; Fan, A. Carbon footprint prediction considering the evolution of alternative fuels and cargo: A case study of Yangtze River ships. *Renew. Sustain. Energy Rev.* **2023**, *173*, 113068. [CrossRef]
32. Zubi, G.; Dufo-López, R.; Carvalho, M.; Pasaoglu, G. The lithium-ion battery: State of the art and future perspectives. *Renew. Sustain. Energy Rev.* **2018**, *89*, 292–308. [CrossRef]
33. International Maritime Organization. *MEPC 65/22/Annex 14. Guidelines for Calculation of Reference Lines for Use with the Energy Efficiency Design Index (EEDI)*; IMO Marine Environment Protection Committee (MEPC): London, UK, 2013.
34. Han, Z.; Jang, J.; Souppez, J.-B.R.G.; Maydison Oh, D. Environmental implications of sandwich structure of glass fiber-reinforced polymer ship. *Ocean Eng.* **2024**, *298*, 117122. [CrossRef]
35. Maydison, K.H.; Lim, H.K.; Heo, J.; Choe, S.-B.; Kim, J.-S.; Jang, J.; Oh, D. Sustainable retrofitting for shipping: Assessing LNG dual fuel impact on global warming potential through life cycle assessment. *Results Eng.* **2024**, *23*, 102484. [CrossRef]
36. International Maritime Organization. *Guidelines on Life Cycle GHG Intensity of Marine Fuels (2024 LCA GUIDELINES)*; IMO Marine Environment Protection Committee (MEPC): London, UK, 2024.
37. *ISO 14040*; Environmental Management—Life Cycle Assessment—Principles and Framework. British Standards Institution: London, UK, 2006.
38. *ISO 14044*; Environmental Management—Life Cycle Assessment—Requirements and Guidelines. British Standards Institution: London, UK, 2006.
39. Fang, L. *Research on Ship's Carbon Footprint Based on Theory of Life Cycle*; Ningbo University: Ningbo, China, 2015.
40. Han, Z.; Jang, J.; Oh, D. Design margin of composite structures of Korean fishing vessel and assessment of its environmental impact. *Shipbuild. China* **2020**, *61*, 90–100.
41. Chatzinikolaou, S.D.; Ventikos, N.P. Holistic framework for studying ship air emissions in a life cycle perspective. *Ocean Eng.* **2015**, *110*, 113–122. [CrossRef]
42. Dong, D.T.; Cai, W. A comparative study of life cycle assessment of a Panamax bulk carrier consideration of lightship weight. *Ocean Eng.* **2019**, *172*, 583–598. [CrossRef]
43. Jeong, B.; Wang, H.; Oguz, E.; Zhou, P. An effective framework for life cycle and cost assessment of marine vessels with the aiming to select optimal propulsion systems. *J. Clean. Prod.* **2018**, *187*, 111–130. [CrossRef]
44. Ling-Chin, J.; Roskilly, A.P. Comparative life cycle assessment of marine power systems. *Energy Convers. Manag.* **2016**, *127*, 477–493. [CrossRef]
45. Greer, K.; Zeller, D.; Woroniak, J.; Coulter, A.; Winchester, M.; Deng Palomares, M.L.; Pauly, D. Global trends in carbon dioxide (CO2) emissions from fuel combustion in marine fisheries from 1950 to 2016. *Mar. Policy* **2019**, *107*, 103382. [CrossRef]
46. Jeong, S.; Woo, J.H.; Oh, D. Simulation of greenhouse gas emissions of small ships considering operating conditions for environmental performance evaluation. *Int. J. Nav. Arch. Ocean Eng.* **2020**, *12*, 636–643. [CrossRef]
47. Olmer, N.; Comer, B.; Roy, B.; Mao, X.; Rutherford, D. *Greenhouse Gas Emissions from Global Shipping, 2013–2015 Detailed Methodology*; International Council on Clean Transportation: Washington, DC, USA, 2017; pp. 1–38.
48. Blanco-Davis, E.; Zhou, P. Life Cycle Assessment as a complementary utility to regulatory measures of shipping energy efficiency. *Ocean Eng.* **2016**, *128*, 94–104. [CrossRef]

49. Burman, M.; Kuttenkeuler, J.; Stenius, I.; Garme, K.; Rosén, A. Comparative Life Cycle Assessment of the hull of a high-speed craft. *Proc. Inst. Mech. Eng. M* **2016**, *230*, 378–387. [CrossRef]
50. Wang, H.; Boulougouris, E.; Theotokatos, G.; Zhou, P.; Priftis, A.; Shi, G. Life cycle analysis and cost assessment of a battery powered ferry. *Ocean Eng.* **2021**, *241*, 110029. [CrossRef]
51. Oh, D.; Lee, D.; Jeong, S. Environmental impact evaluation on lightweight structure design of a composite ship by LCA (Life Cycle Assessment). *J. Korean Soc. Precis. Eng.* **2019**, *36*, 875–881. [CrossRef]
52. International Maritime Organization. *Guidelines on the Method of Calculation of the Attained Energy Efficiency Design Index (EEDI) for New Ships*; MEPC: London, UK, 2022; p. 364.
53. Fan, A.; Xiong, Y.; He, Y.; Yang, L.; Yin, Q. Lifecycle Carbon Footprint Study for Alternative Fuel Power of Yangtze River Ships. *Ship Eng.* **2022**, *44*, 70–75.
54. Ewemoje, T.A.; Oluwaniyi, O.O. Mechanised shea butter production in south-western Nigeria using Life Cycle Assessments (LCA) approach from gate-to-gate. *Agric. Eng. Int.* **2016**, *18*, 230–242.
55. Huang, J.; Fan, H.; Xu, X.; Liu, Z. Life cycle greenhouse gas emission assessment for using alternative marine fuels: A very large crude carrier (VLCC) case study. *J. Mar. Sci. Eng.* **2022**, *10*, 1969. [CrossRef]
56. Myhre, G.; Shindell, D.; Bréon, F.-M.; Collins, W.; Fuglestvedt, J.; Huang, J.; Koch, D.; Lamarque, J.-F.; Lee, D.B.; Nakajima, T.; et al. Anthropogenic and Natural Radiative Forcing. In *Climate Change 2013: The Physical Science Basis. Contribution of Working Group I to the Fifth Assessment Report of the Intergovernmental Panel on Climate Change*; Stocker, T.F., Qin, D., Plattner, G.-K., Tignor, M., Allen, S.K., Boschung, J., Nauels, A., Xia, Y., Bex, V., Midgley, P.M., Eds.; Cambridge University Press: Cambridge UK; New York, NY, USA, 2014; pp. 659–740.
57. Seithe, G.J.; Bonou, A.; Giannopoulos, D.; Georgopoulou, C.A.; Founti, M. Maritime transport in a life cycle perspective: How fuels, vessel types, and operational profiles influence energy demand and greenhouse gas emissions. *Energies* **2020**, *13*, 2739. [CrossRef]
58. Tabibian, S.S.; Sharifzadeh, M. Statistical and analytical investigation of methanol applications, production technologies, value-chain and economy with a special focus on renewable methanol. *Renew. Sustain. Energy Rev.* **2023**, *179*, 113281. [CrossRef]
59. Comer, B.; Beecken, J.; Vermeulen, R.; Sturrup, E.; Paschinger, P.; Osipova, L.; Gore, K.; Delahaye, A.; Verhagen, V.; Knudsen, B.; et al. *Fugitive and Unburned Methane Emissions from Ships (FUMES): Characterizing Methane Emissions from LNG-Fueled Ships Using Drones, Helicopters, and On-Board Measurements*; International Council on Clean Transportation (ICCT): Washington, DC, USA, 2024.
60. Ghiasirad, H.; Skorek-Osikowska, A. Biomethanol production via electrolysis, oxy-fuel combustion, water-gas shift reaction, and LNG cold energy recovery. *Int. J. Hydrogen Energy* **2024**, *52*, 610–626. [CrossRef]
61. Khan, S.; Jain, G.; Srivastava, A.; Verma, P.C.; Pande, V.; Dubey, R.S.; Khan, M.; Haque, S.; Ahmad, S. Enzymatic biomethanol production: Future perspectivee. *Sustain. Mater. Technol.* **2023**, *38*, e00729. [CrossRef]
62. Thakare, M.; Pandit, S.; Pandit, C.; Ray, S.; Alkahtani, A.M.; Alabbosh, K.F.; Ranjan, N.; Geetha, S.; Joshi, S.J. A concise review on biorefinery remedies for biomethanol production and processing. *Biocatal. Agric. Biotechnol.* **2024**, *57*, 103061. [CrossRef]
63. Mehra, R.K.; Duan, H.; Juknelevičius, R.; Ma, F.; Li, J. Progress in hydrogen enriched compressed natural gas (HCNG) internal combustion engines—A comprehensive review. *Renew. Sustain. Energy Rev.* **2017**, *80*, 1458–1498. [CrossRef]
64. Sapra, H.; Godjevac, M.; De Vos, P.; Van Sluijs, W.; Linden, Y.; Visser, K. Hydrogen-natural gas combustion in a marine lean-burn SI engine: A comparitive analysis of Seiliger and double Wiebe function-based zero–dimensional modelling. *Energy Conserv. Manag.* **2020**, *207*, 112494. [CrossRef]
65. Acciaro, M. Real option analysis for environmental compliance: LNG and emission control areas. *Transp. Res. Part D Transp. Environ.* **2014**, *28*, 41–50. [CrossRef]
66. Xing, H.; Stuart, C.; Spence, S.; Chen, H. Alternative fuel options for low carbon maritime transportation: Pathways to 2050. *J. Clean. Prod.* **2021**, *297*, 126651. [CrossRef]
67. Bilgili, L. A systematic review on the acceptance of alternative marine fuels. *Renew. Sustain. Energy Rev.* **2023**, *182*, 113367. [CrossRef]

Journal of
*Marine Science
and Engineering*

MDPI

Article

Data-Driven Analysis of the Causal Chain of Waterborne Traffic Accidents: A Hybrid Framework Based on an Improved Human Factors Analysis and Classification System with a Bayesian Network

Xiangyu Yin [1], Yan Yan [2], Jiahao Wang [2], Hongzhuan Zhao [3], Qingyan Wu [4] and Qi Xu [3,*]

[1] China Waterborne Transport Research Institute, Beijing 100088, China; yxy730622@163.com
[2] Guangxi Key Laboratory of ITS, Guilin University of Electronic Technology, Guilin 541004, China; 19126550340@163.com (Y.Y.); 19106444843@163.com (J.W.)
[3] School of Architecture and Transportation Engineering, Guilin University of Electronic Technology, Guilin 541004, China; zhaohongzhuan@guet.edu.cn
[4] Department of Infrastructure, Guilin University of Electronic Technology, Guilin 541004, China; peterwuqy@163.com
* Correspondence: xuqi@guet.edu.cn

Abstract: In the context of economic globalization, waterborne transportation plays an important role in international trade and logistics. However, waterborne traffic accidents pose a severe threat to life, property safety, and the environment. To gain a deeper understanding of the causal mechanisms behind waterborne traffic accidents, we conducted a data-driven analysis of the causal chain of waterborne traffic accidents. By constructing a hybrid framework integrating an improved HFACS (Human Factors Analysis and Classification System) with a Bayesian network model, we conducted a multi-dimensional analysis of accident causes. The constructed model was quantitatively analyzed by applying genie software to the accident samples collected from the China MSA. The results indicate that there are 12, 3, 6, 2, 4, and 7 causal chains leading to collisions, contact, fires/explosions, windstorm accidents, sinking, and other types of accidents, respectively. These research results can serve as a reference for the enhancement of the safety of waterborne transportation.

Keywords: waterborne traffic accidents; causal chain analysis; HFACS; Bayesian network; data-driven analysis

1. Introduction

With the acceleration of economic globalization and the expansion of global trade, waterborne transportation has assumed an irreplaceably important role. This form of transportation offers significant advantages, such as allowing for high carrying capacities, low costs, and high long-haul transport efficiency. Nevertheless, accidents often affect the waterborne transportation system, particularly since the 1990s, when changes in marine shipping and the international environment sharply highlighted the importance of shipping safety [1]. These accidents pose significant threats to lives, property, and the environment [2].

Waterborne traffic accidents are not caused by a single factor but occur as the result of the interaction of multiple factors. Therefore, it is especially important to analyze the causal chain of waterborne traffic accidents so as to reduce the likelihood of their occurrence. The HFACS (Human Factors Analysis and Classification System) model, an effective system for analyzing and classifying human factors contributing to accidents, can help identify and categorize human factors that lead to accidents. Furthermore, the HFACS can integrate various methods to explore the causal relationships between factors leading to accidents [3]. As a probabilistic graphical model, a Bayesian network can learn the probabilistic dependencies between variables by means of statistical data. By constructing a Bayesian network model, one can identify and analyze the influences of various factors in accidents, providing a new method for accident risk analysis and prediction [4]. Therefore, constructing a hybrid framework combining the HFACS and Bayesian networks, especially in a data-based setting, is a feasible and useful method of comprehensively understanding the mechanisms of accident occurrence, providing a scientific basis for accident prevention and quick responses.

2. Literature Review

2.1. Identification of Causality of Waterborne Accidents

Research on the causality of waterborne traffic accidents has made remarkable progress. Such studies, through the comprehensive application of various methods and theories, have deeply analyzed the root causes of accidents. Wang et al. [5] conducted an in-depth analysis of multiple factors affecting the severity of marine accidents, including human, vessel, environmental, and management factors. They used global accident data and statistical models to reveal the independent and interactive effects of these factors on accident severity. Liu et al. [6] employed machine learning techniques to comprehensively analyze the causes of marine accidents in China's coastal waters. They utilized algorithms, including decision trees and random forests, and neural networks to uncover the complex relationships between human, vessel, environmental, and management factors and accident severity. Sun et al. [7] employed complex network theory to analyze the causes of waterborne traffic accidents. They constructed a causal network model, analyzed its characteristics, and identified critical causal factors, such as human negligence and management loopholes. Wu et al. [8] analyzed ship collision accidents in the Yangtze River that occurred from 2013 to 2017 through text-mining techniques, identifying 33 causal features, covering human, vessel, environmental, and management factors. They used a Bayesian network model and text-mining methods to predict marine risks and discovered that human factors were the primary causes. Using GIS technology, Wang et al. [9] explored the spatial distribution patterns of global maritime accidents, including accident frequency and severity, through density analysis and clustering analysis. Ho Namgung et al. [10] developed a collision risk inference system for Maritime Autonomous Surface Ships (MASSs) adhering to COLREGs. Addressing gaps in existing methods, their innovative approach defines avoidance actions based on danger levels, sets risk indices using expanded ship domains, extracts key factors from overlapping domains, and employs an adaptive neuro-fuzzy system for learning. This system enhances collision warning timing and positioning, offering more decision-making time for prevention.

The HFACS model, an effective tool in the field of aviation safety, has been applied to the waterborne transportation field by many scholars. Kaptan et al. [11] applied the HFACS to the maritime sector to uncover the root causes and causal chains of waterborne traffic accidents. Yildiz et al. [12] applied the HFACS-PV method to explore HOFs in maritime accidents. Their analysis revealed this method's applicability to collisions, groundings, sinkings, and other types of accidents. Wang et al. [13] used the HFACS-FCMs model to investigate the human factors in ship grounding accidents. They found that inadequate ship safety management organizations constitute the most significant factor in grounding accidents, followed by organizational influences and unsafe behavior antecedents. Yıldırım et al. [14] utilized the HFACS and statistical methods to analyze human factors in collisions and groundings. GIS and FTA of Black Sea accidents revealed that COLREG violations and communication lapses were the primary causes of collision, while grounding accidents stemmed from watchkeeper errors and inadequate bridge resource management communication. Huang [15] utilized the HFACS framework combined with expert scoring methods and gray theory to analyze the human factors in marine traffic accidents. The HFACS-MTAI system achieved precise identification, systematic classification, and quantitative ranking of human factors. Chen et al. [16] analyzed the causes of marine traffic accidents using the HFACS-MA framework to comprehensively assess human and organizational factors. Through case studies and public accident reports, they validated this framework's effectiveness and identified key causal factors. Chauvin et al. [17] improved the HFACS model using the classification tree method to analyze the causal factors at each level of the collision accidents.

The Bayesian network model has also been widely applied in analyzing the causes of waterborne traffic accidents. Fan et al. [18] employed a data-driven Bayesian network model to focus on human factors, exploring their impact on the probability and outcomes of various waterborne traffic accidents when interacting with non-human factors. Antão et al. [19] utilized Bayesian Belief Network (BBN) models to investigate the impact of human error on coastal shipping accidents under various sea conditions. Their findings revealed that human error and variations in risk acceptance and perception among the crews of different vessel types had significant influences. Wang et al. [20] used a Bayesian network model to investigate the severity of global waterborne transportation accidents in order to reveal the complex relationships between weather, ship characteristics, human factors, and accident severity. Meng et al. [21] adopted a data-driven Bayesian network model integrating physical knowledge to explore risk factors influencing ship collision accidents. The research results demonstrated the model's effectiveness in identifying and analyzing key risk factors of ship collisions. Fan et al. [22] employed a Bayesian network approach to explore risk factors influencing maritime transportation accidents. They constructed a network model reflecting the interdependencies among the influencing factors. Tian et al. [23] utilized a Bayesian network model to investigate ship collision accidents recorded by the Zhejiang Maritime Safety Administration of China. Their research revealed that human factors such as improper lookout, inadequate collision assessments, and improper collision avoidance measures are the primary causes of accidents and can be used to predict accident probabilities. Hänninen et al. [24] conducted a statistical analysis of ship collisions in the Gulf of Finland and utilized a Bayesian network to investigate the causal relationships between human factors and the final collision outcomes. Meng et al. [25] employed a combined N-K model and Bayesian networks to assess coupled risks in Chinese ship collisions. They found that multifactor coupling was more effective than bifactor coupling, with human and management factors being vital and ship/environment factors' impacts becoming more significant with varying probabilities.

Wang et al. [26] integrated the HFACS with a BN to investigate the root causes of marine accidents, emphasizing the importance of human and organizational factors. Their results showed that combining the HFACS with a BN can effectively uncover complex human factor chains and organizational deficiencies in marine accidents. Rostamabadi et al. [27] developed an FBN-HFACS model to analyze human and organizational factors (HOFs) in process accidents. The model serves as a robust tool for accident prevention and safety management. Jiang et al. [28] applied the HFACS, Bayesian network, and path analysis models to analyze the causal pathways of waterborne traffic accidents. Their research revealed the key causal pathways for accidents such as collisions, sinkings, and contacts. Li et al. [29] studied the impact of HOFs on ship collision accidents using the HFACS and a Bayesian network model. The HFACS model identified multi-level causes of accidents, while the Bayesian network revealed the probabilistic dependencies between causes and identified critical causal pathways. Their study revealed that crew operational errors and equipment failures were the primary causes of ship collision accidents. Wang et al. [30] employed the HFACS-BN model to explore HOFs in collisions between merchant ships and fishing vessels. Their study revealed the interactions between various human and organizational factors in collision accidents and their degrees of influence on accident occurrence. Özkan et al. [31] analyzed nearly two decades of ship accidents in the Black Sea region using the HFACS and a Bayesian network model. They found that the frequent accidents in the coastal areas of the Black Sea were primarily due to crew operational errors, equipment failures, and inadequate management.

2.2. Research Gap Analysis

Scholars in the field of causation analysis have conducted extensive research on waterborne traffic accidents. Nevertheless, despite significant progress in this field, there is still room for improvement. Firstly, the research conducted so far is mainly based on data from a specific region or a single type of waterborne traffic accident, and the applicability of the aforementioned method under different circumstances requires further validation. Furthermore, waterborne traffic accidents often involve the intricate interplay of multiple complex factors, and current research efforts in deciphering these interactions are still inadequate. Lastly, the tendency to use single-framework methodologies in existing research has somewhat limited the breadth and depth of inquiry.

Therefore, we conducted a data-driven analysis of the causal chains of waterborne traffic accidents, which can be applied more generally to different situations. Firstly, a hybrid framework integrating an improved HFACS and a Bayesian network model was constructed to conduct a multi-dimensional analysis of accident causes. Secondly, based on the constructed framework, we analyzed the underlying causes from multiple dimensions and conducted a chain analysis of the causal factors, quantitatively assessing the correlations between factors and their degrees of influence on accident occurrence. Thirdly, we identified the global causal chain of waterborne traffic accidents for different types of accidents, providing information that can serve as a general reference for practical waterborne traffic accident prevention.

3. Data-Driven Hybrid Framework Based on the HFACS and a Bayesian Network

3.1. Basic Data-Driven Analysis Framework

In this paper, we propose a data-driven hybrid framework that can be used to analyze the causal chain of waterborne traffic accidents in order to attain a better understanding of the interaction mechanisms between potential causal factors in the waterborne transportation system. The basic framework is shown in Figure 1. As Figure 1 shows, the main procedure includes the following steps.

***Step 1*: Initialize the accident dataset.** Add all the accident samples collected at the beginning of the procedure into the accident dataset one by one.

***Step 2*: Calculate the number of occurrences for causal factors.** The number of occurrences for causal factors is calculated according to the novel MTAACS (Maritime Traffic Accident Analysis and Classification System, an improved Classification method based on HFACS) proposed in this paper.

Step 3*: For every causal factor pair, conduct a correlation analysis.** This analysis is conducted using the chi-square test. If the causal factor pair passes the test and the pair is the last factor pair, then proceed to ***Step 4; otherwise, test the next causal factor pair.

***Step 4*: Construction and application of Bayesian networks.** This step includes five sub-steps, including constructing a conditional probability table, identifying key factors, performing a sensitivity analysis of the key causes, identifying cause paths, and conducting a global cause chain analysis.

***Step 5*: Update the accident dataset and return to Step 2 if there are any new accident data; if not, output the analysis result.**

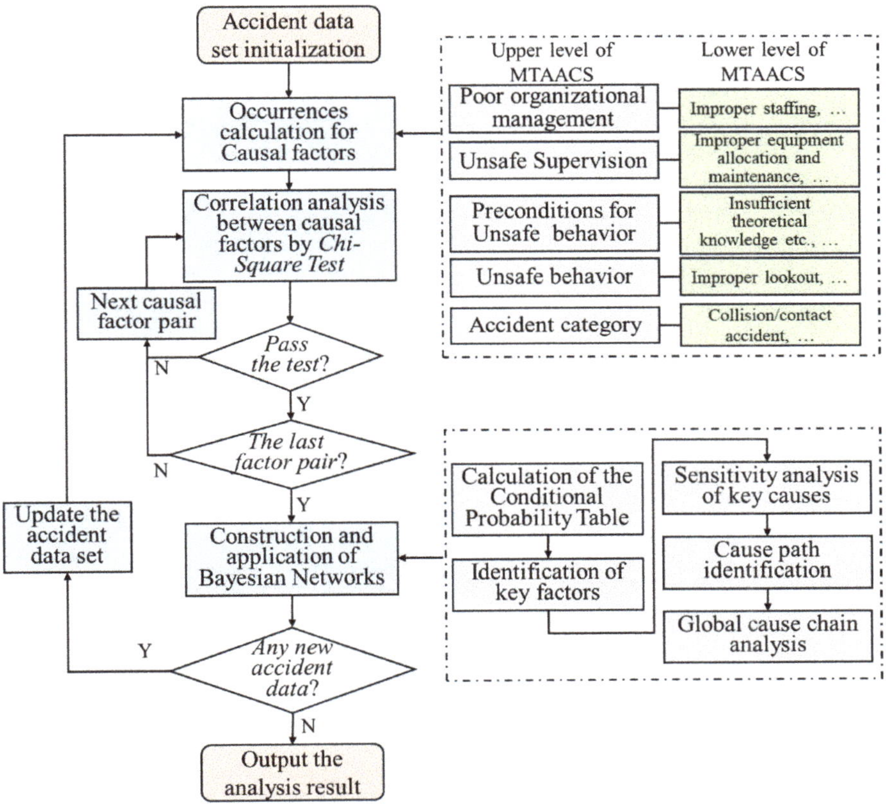

Figure 1. Illustration of the basic data-driven analysis framework.

With regard to the field of exploring analytical frameworks for the causes of waterborne traffic accidents, our study significantly distinguishes itself from the existing research through our innovative combination of the Human Factors Analysis and Classification System with a Bayesian network model integrated in a data-driven framework, which should renew the analysis results according to the updating of the datasets. Different from traditional research methods, which tend to be confined to single-dimensional analyses focusing solely on human factors or technical faults, the framework constructed in this study systematically identifies and categorizes human factors in accidents using the HFACS. These factors comprehensively encompass multiple dimensions, such as individuals, organizations, and the environment, thereby providing more comprehensive and in-depth insights. Furthermore, we introduce an advanced probabilistic graphical model of the Bayesian network. This model not only effectively handles complex dependencies between variables but also quantifies the uncertainty of accident causes through probabilistic reasoning, more accurately capturing the ambiguity and dynamic characteristics of accident causation. This provides solid data support for formulating accident prevention strategies.

3.2. Data Collection and Basic Analysis

To ensure the accuracy and representativeness of the data, we primarily sourced our information from the water traffic accident investigation reports publicly released on the official website of the China Maritime Safety Administration from June 2015 to February 2023. These reports comprehensively document various types of accidents, including collisions, contact, fires/explosions, windstorms, sinkings, etc. However, upon conducting an in-depth analysis of the causal chains of various accidents, it was observed that there were relatively few sample data for strandings and contact. Given the impact of sample size on research accuracy, strandings and contact were excluded in this study. The sample data for the other types of accidents are relatively abundant in the publicly available reports, sufficient to provide the necessary information for an in-depth analysis and understanding of their causal chains.

After systematic data collection and screening, we gathered a total of 756 investigative reports on water traffic accidents along China's coast, serving as the core data foundation for this research. For collision accidents, each vessel was considered an independent research sample. We classified several common types of accidents and recorded their respective occurrence frequencies, as shown in Table 1.

Table 1. Description and frequency of accident types.

Accident Type	Accident Description	Accident Frequency
Collision accident	An accident in which two or more ships are directly impacted and cause damage at the same time to the same area.	315
Contact Accident	An accident in which a ship collides with structures above or below the water, such as quay walls, docks, navigation aids, bridge piers, floating facilities, or obstacles to navigation (such as sunken ships, sunken objects, and wooden piles), causing damage.	49
Fire/explosion accident	A fire or explosion on a ship caused by an uncontrolled ignition source due to several potential reasons during navigation, berthing, or operation.	34
Windstorm accident	An accident in which a ship suffers losses due to a strong storm.	32
Sinking Accident	An accident in which a ship sinks or capsizes of its own accord.	135
Other Accidents	An accident that does not fall under the specific categories of collision, contact, fire/explosion, windstorm, or sinking but still results in the sinking of or damage to a ship, crew member and passenger casualties, and environmental pollution.	191

The statistical data in Figure 2 clearly show that there are significant differences in the proportions of different types of accidents in the overall dataset. Collision accidents, other accidents, and sinking accidents account for a relatively high percentage, indicating that these types of accidents are more likely to occur. In contrast, the proportions of contact accidents, fire/explosion accidents, and windstorm accidents are relatively low, suggesting that these types of accidents are less likely to occur. This data pattern provides us with an intuitive overview of accident occurrences.

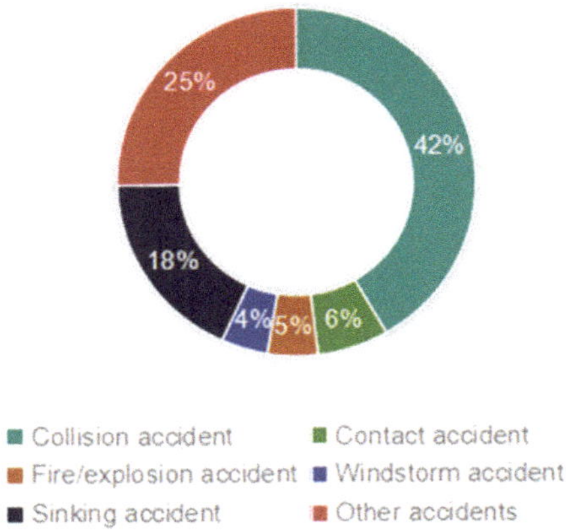

Figure 2. Accident-type distribution chart.

3.3. Improved HFACS (MTAACS) for Causes of Waterborne Traffic Accidents

In the process of accident analysis, the HFACS model logically provides a basic framework for the causal chain as well as a fundamental pathway with which to identify the root causes of accidents, namely, management factors: direct cause–indirect cause (precursor)–root cause. Although the HFACS model provides a systematic logical framework for causal linkages, there are several limitations to its direct application to classifying factors. Firstly, the factor hierarchy set in the HFACS model may overlook the unique nature of causal structures in waterborne transportation accidents. Secondly, given that the HFACS model lacks clarity in terms of accident classification—a feature that is particularly crucial for the analysis of waterborne traffic accidents in this paper, wherein causal analysis of various accidents is the key research content—it is necessary to incorporate a specific "accident level" as an indispensable part of the framework structure to explicitly classify various accident types. Furthermore, at the "preconditions for unsafe behavior" level, the influence of the external environment must be analyzed in addition to internal factors. It is worth noting that the impact of adverse external environments does not necessarily have a direct causal relationship with upper-level factors such as "adverse organizational influences" and "inadequate supervision". Lastly, the HFACS model assumes there is a linear causal relationship between adjacent levels, which may be different from the nonlinear causal relationships in waterborne traffic accidents. This complexity necessitates a more nuanced examination of the interactions and associations between factors in the analysis of waterborne traffic accidents.

Thus, in this paper, we propose a novel Maritime Traffic Accident Analysis and Classification System (MTAACS) based on an improved HFACS. The proposed system fully accounts for the unique characteristics and complexities of waterborne traffic accidents. It categorizes the causative factors of waterborne traffic accidents into five hierarchical levels: organizational influences, inadequate supervision, preconditions for unsafe behaviors, unsafe behaviors, and the accident layer. Factors within the same level are considered to be in parallel relationships, implying there are no direct causal linkages. This arrangement is designed to avoid oversimplifying the complexity of waterborne traffic accidents and allows analysts to comprehensively consider the interactions between various factors at the same level. Concurrently, the relationships between factors at adjacent levels are described as nonlinear. This indicates that a single factor may simultaneously impact multiple lower-level factors or that multiple factors may collectively act on the same lower-level factor. This nonlinear relationship can more accurately reflect the complexity and diversity of the various factors involved in waterborne traffic accidents, providing a deeper and more comprehensive analytical perspective for accident prevention.

Based on waterborne traffic accident investigation reports, causative factors relevant to each dimension were extracted. The accident risk causative factors extracted in this study stem from descriptions given in accident investigation reports and can be classified into 38 factors, as shown in Table 2 and Figure 3.

Table 2. Integrated factors causing water transportation accidents.

Factor Level	Num *	Causal Factor	OC *
Poor organizational management	A1	Improper staffing	94
	A2	Inadequate education and training	87
	A3	Lack of communication within the team	41
	A4	Incomplete safety management system	46
	A5	Inadequate implementation of safety management	224
	A6	Improper provision of chart information	10
	A7	Incomplete or invalid ship certificates	52
Unsafe supervision	B1	Improper equipment allocation and maintenance	81
	B2	Lack of supervision and guidance	76
	B3	Improper voyage planning	74
	B4	Improper selection of navigation areas/improper operation and management	87
	B5	Failure to correct mistakes	11
Preconditions for unsafe behavior	C1	Insufficient theoretical knowledge, work experience, skill level, etc.	275
	C2	Insufficient safety awareness	285
	C3	Poor health status	43
	C4	Incomplete ship operation, navigation performance, information, and equipment	117
	C5	Improper cargo securing and stowage	70
	C6	Insufficient visibility	67
	C7	Excessive wind, waves, and currents	135
	C8	Complex navigation environment	79
Unsafe behavior	D1	Improper lookout	393
	D2	Improper watchkeeping	141
	D3	Misjudgment of danger	286
	D4	Improper collision avoidance actions	183
	D5	Failure to detect collision objects early	2
	D6	Improper use of signals	117
	D7	Improper use of equipment	90
	D8	Failure to travel at a safe speed	150
	D9	Failure to take early action	220
	D10	Improper location of anchorage	5
	D11	Improper emergency measures	73
	D12	Improper typhoon prevention measures	35
Accident category	E1	Collision accident	521
	E2	Contact accident	50
	E3	Fire/explosion accident	34
	E4	Windstorm accident	32
	E5	Sinking accident	135
	E6	Other accidents	192

* "Num" is an abbreviation of "Numbering", and "OC" is an abbreviation of "Occurrences".

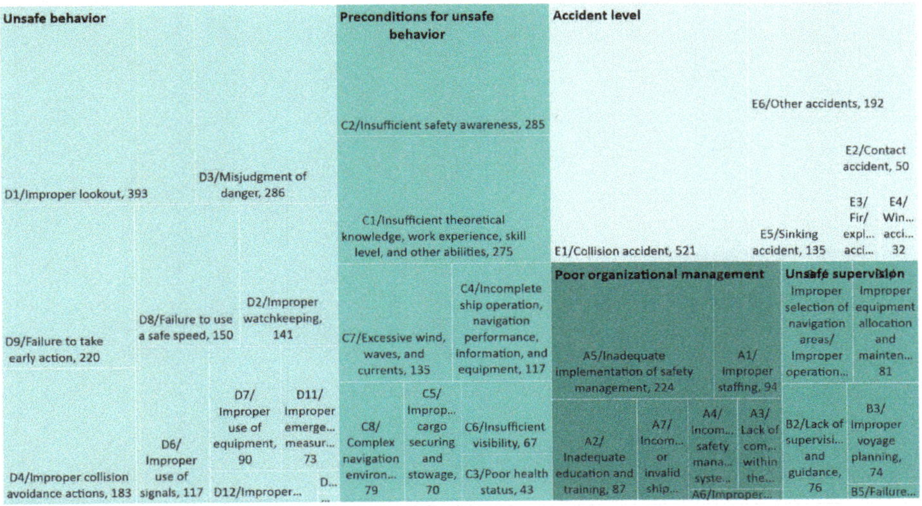

Figure 3. Statistical chart regarding the layered causes of water traffic accidents.

3.4. Correlation Analysis for Causal Factor Pairs

In analyzing the causal factor chain of waterborne traffic accidents, structural learning becomes a crucial initial stage, with the core objective of clarifying the topological structure of the Bayesian network. As the foundation for constructing the Bayesian network model, structural learning determines how various nodes in the model are interconnected and ascertains their dependency relationships.

In this process, the constraint-based search method plays a decisive role. In this method, various techniques from statistics and information theory are employed, such as the chi-square test, mutual information, and the G2 test. In this study, the chi-square test was selected as the core analytical tool for structural learning. By analyzing the statistical correlations in accident data through the chi-square test, potential links between different causal factors were identified, providing a basis for constructing the topological structure of the Bayesian network model.

Taking C2 (insufficient safety awareness) and D1 (improper lookout) as examples, the chi-square test was applied to examine whether there was a significant statistical correlation between these two factors. The results of applying the chi-square test to analyze the potential link between C2 (insufficient safety awareness) and D1 (improper lookout) are shown in Figure 4. This figure indicates that the expected count (116.19) was significantly greater than 5, and the total sample size reached 964, satisfying the conditions for conducting a Pearson's chi-squared test. The test results show that the p-value is less than 0.05. According to conventional statistical norms, this result indicates a rejection of the null hypothesis (i.e., C2 and D1 are independent of each other) and the acceptance of the alternative hypothesis, indicating that there is a statistically significant relationship between C2 (insufficient safety awareness) and D1 (improper lookout).

C2 * D1 Crosstabulation

			D1 0	D1 1	Total
C2	0	Count	323	356	679
		% within D1	56.6%	90.6%	70.4%
	1	Count	248	37	285
		% within D1	43.4%	9.4%	29.6%
Total		Count	571	393	964
		% within D1	100.0%	100.0%	100.0%

Chi-Square Tests

	Value	df	Asymptotic Significance (2-sided)	Exact Sig. (2-sided)	Exact Sig. (1-sided)
Pearson Chi-Square	129.361[a]	1	<.001		
Continuity Correction[b]	127.733	1	<.001		
Likelihood Ratio	143.592	1	<.001		
Fisher's Exact Test				<.001	<.001
Linear-by-Linear Association	129.227	1	<.001		
N of Valid Cases	964				

a. 0 cells (.0%) have expected count less than 5. The minimum expected count is 116.19.
b. Computed only for a 2x2 table

Figure 4. Chi-square test result chart for C2 × D1.

3.5. Construction of Bayesian Network for Causes of Waterborne Traffic Accidents

3.5.1. Basic Concepts and Formulas

A Bayesian network is a probabilistic graphical model used to represent dependencies between variables. It is presented in the form of a Directed Acyclic Graph (DAG), a graphical structure that enables an intuitive understanding and analysis of causal relationships and conditional dependencies among variables. In the model, nodes represent random variables, while directed edges represent the interrelationships between variables, and their relationships are quantified through conditional probabilities. The construction and analysis of Bayesian network (BN) models primarily involve the following three steps. The first step is to determine the BN's structure. In the context of complex networks, we integrated data-driven approaches with expert knowledge, employing structural learning methods and optimizing based on statistical test results and prior information to accommodate multi-node scenarios. Secondly, the BN parameters must be determined. Given the scale of the network, in this study, we trained the parameters based on sample data to ensure that parameter estimation was both efficient and robust. Finally, an analysis of the BN, particularly a sensitivity analysis, was conducted. Through this analysis, significant precursor variables influencing the target nodes can be identified, thereby allowing the extraction of key causal factors and pathways for water transportation accidents and providing scientific support for accident prevention strategies.

The theoretical basis of Bayesian networks lies primarily in probability theory and graph theory. The basic concepts and formulas for the construction of a Bayesian network are as follows.

(1) Prior Probability

Prior probability is the initial assessment of the probability of a random event or a variable taking on a specific value before the observation of any relevant data or evidence. In Bayesian statistics and Bayesian networks, prior probability reflects an initial belief about the state of an event or a variable based on historical data, expert knowledge, or background information.

(2) Posterior Probability

Posterior probability is the reassessment of the probability of a random event or a variable taking on a specific value after the observation of new data or evidence. Posterior probability is calculated based on Bayes' theorem, which combines the prior probability (the probability before the observation of new data) and the likelihood function (the probability of observing the data given that the event has occurred). Posterior probability represents an updated belief about the state of an event or a variable given the new evidence or observations.

(3) Conditional Probability

The conditional probability formula is a fundamental concept in probability theory that describes the probability of one event occurring given that another event has already occurred. The conditional probability formula is defined as follows: Let A and B be two events, and $P(B) > 0$ (i.e., the probability of event B occurring is not zero). Then, the probability of event A occurring given that event B has already occurred is denoted as $P(A \mid B)$.

(4) Joint Probability

The joint probability formula describes the probability of two or more events occurring simultaneously. For two events A and B, the joint probability formula is as follows:

$$P(A \cap B) = P(A,B) \tag{1}$$

However, if events A and B are independent, their joint probability can be obtained by multiplying their individual probabilities:

$$P(A \cap B) = P(\mathrm{A}) \cdot P(B) \tag{2}$$

In Bayesian networks, joint probabilities are not usually given directly but instead calculated indirectly through conditional probability tables. These conditional probability tables describe the probability of a child node taking on each possible value given the values of its parent nodes.

(5) Total Probability Formula

The total probability formula is an important theorem in probability theory that describes the probability of a complex event occurring in terms of the probabilities of a series of simpler events and their conditional probabilities.

Let the full sample space be S. If the events $E_1, E_2, E_3, \ldots E_n$ form a complete event group (i.e., they are mutually exclusive and exhaustive, and $E_1 \cup E_2 \cup E_3 \cup \ldots \cup E_n = S$) and $P(E_i) > 0 (i = 1, 2, 3, \ldots, n)$, then for any event A, we have

$$P(A) = P(A|E_1)P(E_1) + P(A|E_2)P(E_2) + \ldots + P(A|E_n)P(E_n) = \sum_i^n P(A|E_i)P(E_i) \tag{3}$$

(6) Bayes' Formula

Bayes' Theorem provides a formula that describes the relationship between two conditional probabilities.

Let the full sample space be S and A be an event within S. If $E_1, E_2, E_3, \ldots E_n$ is a complete event group (i.e., they are mutually exclusive and exhaustive, and $E_1 \cup E_2 \cup E_3 \cup \ldots \cup E_n = S$) and $P(A) > 0$, $P(E_i) > 0 (i = 1, 2, 3, \ldots, n)$, then Bayes' Theorem states that

$$P(E_i|A) = \frac{P(A|E_i)P(E_i)}{\sum_i^n P(A|E_i)P(E_i)} \tag{4}$$

After the topological structure of the Bayesian network is determined through structural learning methods, the next crucial step is to solve for the network parameters. In this process, the Maximum Likelihood Estimation (MLE) method, a commonly used statistical approach, is widely applied in parameter learning for Bayesian networks.

Based on the available accident data, MLE was used to compute the likelihood function of the data and find the parameters that maximize this function to estimate the conditional probability table (CPT) for each node. The results of parameter estimation provide the CPTs for each node in the Bayesian network model, representing the probability distribution of each node given its parent nodes' states. Further analysis of these parameters can reveal the dependency relationships between different factors and their degrees of influence.

3.5.2. Bayesian Network for Causes of Water Traffic Accidents

Based on the established conditional probability table (CPT), we could obtain the Bayesian network for the causes of water traffic accidents using GeNle software Academic 4.1. The network is shown in Figure 5. Therein, five different colors means different categories of factor levels. The blue and yellow bars are calculated based on the provided data, with blue indicating the probability of the accident's cause occurring and yellow indicating the probability of not occurring.

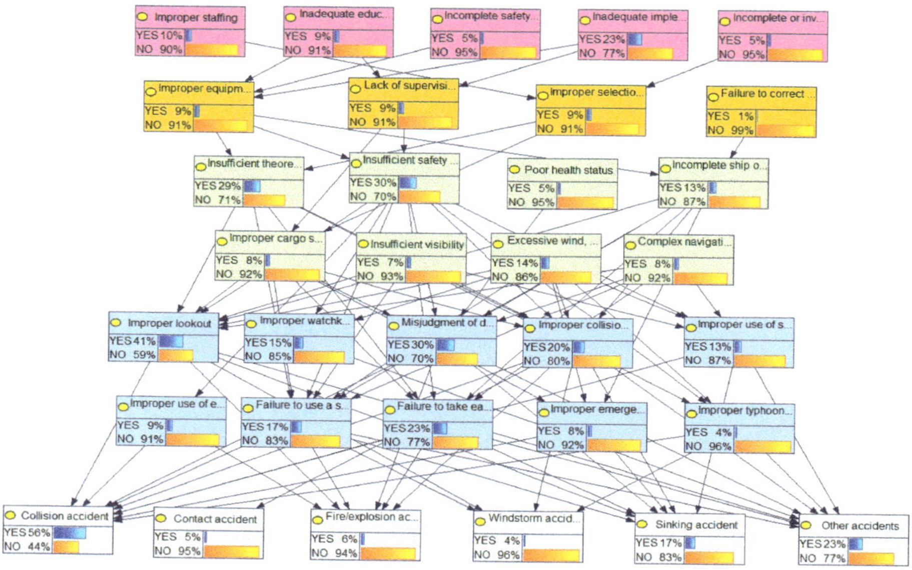

Figure 5. Chart depicting the Bayesian network calculation results for causes of water traffic accidents.

4. Analysis of the Causal Chain of Waterborne Traffic Accidents

4.1. Identification of Key Factors in Waterborne Traffic Accidents

Identifying and analyzing the key factors are crucial for understanding the causes of waterborne traffic accidents, predicting accident risks, and formulating effective preventive measures and emergency response strategies. Based on the node probabilities presented in Figure 5, the top three factors with higher occurrence probabilities in each level were extracted and compiled into a table of key factors, as shown in Figure 6.

Based on Figure 6, at the organizational management level, inadequate safety management execution (A5), improper staffing (A1), and inadequate education and training (A2) were identified as the three core factors affecting waterborne traffic safety. In particular, the high incidence of inadequate safety management execution emphasizes a significant lack of safety management. Together, these three factors constitute potential root causes of waterborne traffic safety accidents, creating hidden dangers and increasing risks during ship operations.

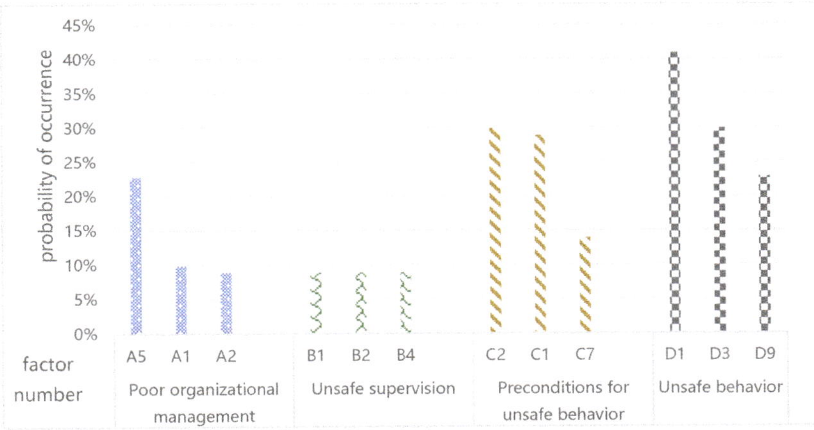

Figure 6. Top three causes with the highest occurrence probabilities at each level.

At the unsafe-supervision level, inadequate equipment provisioning and maintenance (B1), a lack of supervision and guidance (B2), and improper navigation area selection and operational management (B4) are the three most problematic areas. These deficiencies not only have a direct impact on decision making and execution at the next level but also indirectly amplify safety risks regarding waterborne traffic.

At the preconditions-for-unsafe-behaviors level, insufficient safety awareness (C2); inadequate theoretical knowledge, work experience, skill levels, and other competencies (C1); and external factors such as excessive wind, waves, and currents (C7) were recognized as the primary factors inducing waterborne traffic accidents. These preconditions foster the occurrence of unsafe behaviors, increasing the likelihood of accidents.

At the unsafe-behavior level, improper lookout (D1), misjudgment of hazards (D3), and failure to take early action (D9) are the behaviors that most directly lead to accidents. These behaviors involve misjudgments and faulty decisions made by crew members during navigation, highlighting the importance of enhancing crew safety awareness and operational skills.

4.2. Sensitivity Analysis of Causes of Waterborne Traffic Accidents

Sensitive factors can be identified through sensitivity analysis of the Bayesian network. The aim of this analysis is to determine the degrees of influence of network parameters on the posterior probability of specific target nodes, thereby identifying the key factors contributing to various types of accidents. The core algorithm of Bayesian network sensitivity analysis involves calculating the differential of the target node's posterior probability with respect to network parameters, thus quantifying the impact of parameter changes on the posterior probability. In GeNIe, by setting the state of the target node and conducting sensitivity analysis, the importance of upper-level factors with respect to lower-level factors can be visually displayed. Combined with the calculation results for the sensitivity parameters, further quantitative analysis can be conducted.

Taking a windstorm accident as an example, we set E4 (windstorm accidents) in the accident layer as the target node and conducted a sensitivity analysis as shown in Figure 7. During this process, the colors of the Bayesian network nodes were changed to visually indicate the locations of sensitive factors. Red nodes represent the locations of relevant parameters, with darker shades indicating the higher sensitivity of the associated parameters. In Figure 7, it is evident that insufficient theoretical knowledge, work experience, skill levels, and other competencies (C1); excessive wind, waves, and currents (C7); and inadequate typhoon prevention measures (D12) are the factors most relevant to windstorm accidents. The causal strength of each factor needs to be combined with sensitivity parameters for quantitative assessment. GeNle provided the maximum, minimum, and average values of the sensitivity of each node's relevant parameters. Among them, the maximum value is consistent with the shade of the associated node's color, indicating that the maximum value of the sensitivity analysis is the primary criterion for assessing the sensitivity of associated nodes.

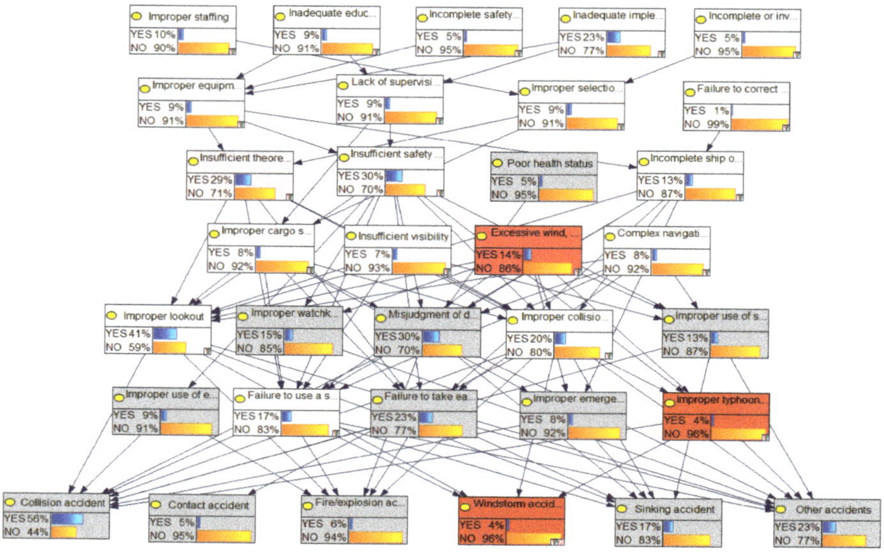

Figure 7. Sensitivity analysis network diagram for wind disaster accidents.

Through the sensitivity analysis of various factors contributing to waterborne traffic accidents, sensitivity parameter data were obtained, as shown in Figures 8–13, and Figure 14 lists the top three sensitivity factors for different types of accidents. These parameters reveal the relative importance of causal factors in different types of accidents.

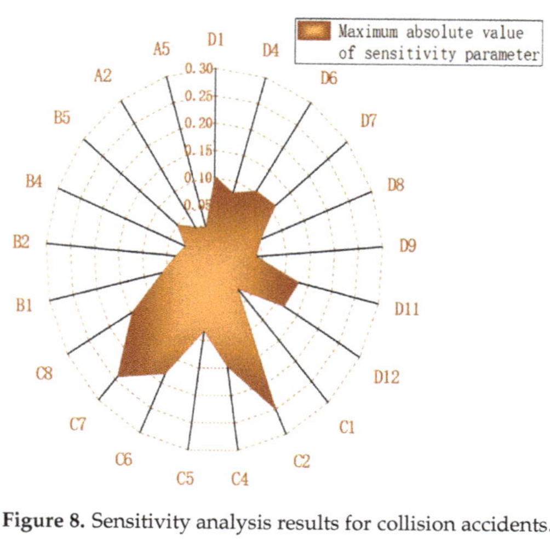

Figure 8. Sensitivity analysis results for collision accidents.

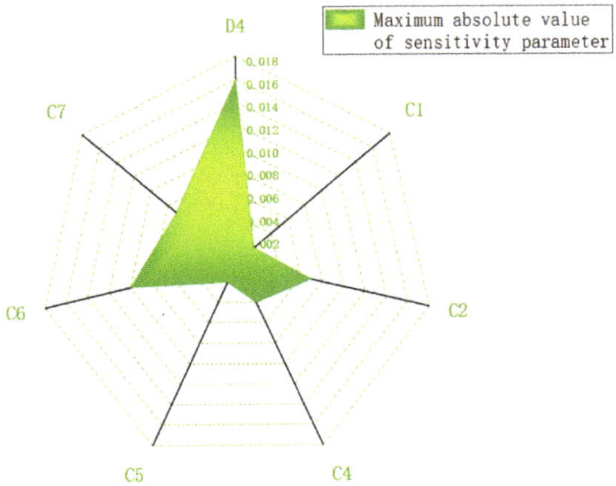

Figure 9. Sensitivity analysis results for contact accidents.

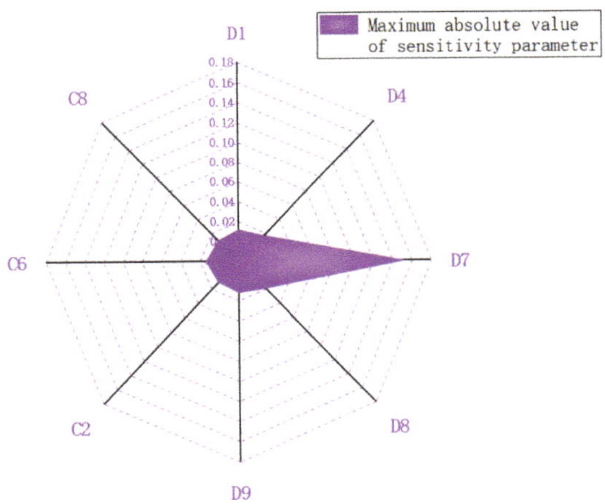

Figure 10. Sensitivity analysis results for fire/explosion accidents.

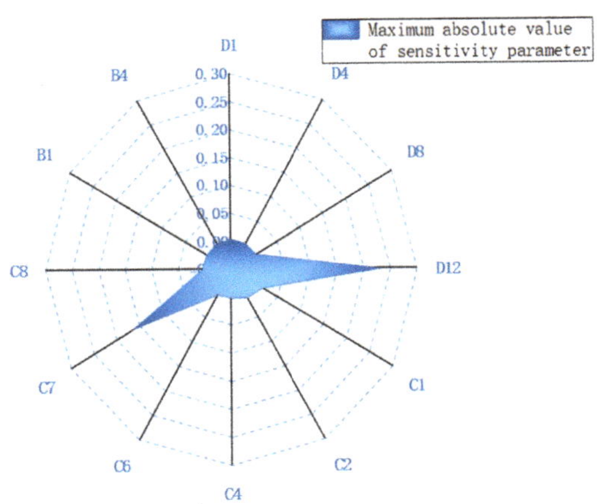

Figure 11. Sensitivity analysis results for windstorm accidents.

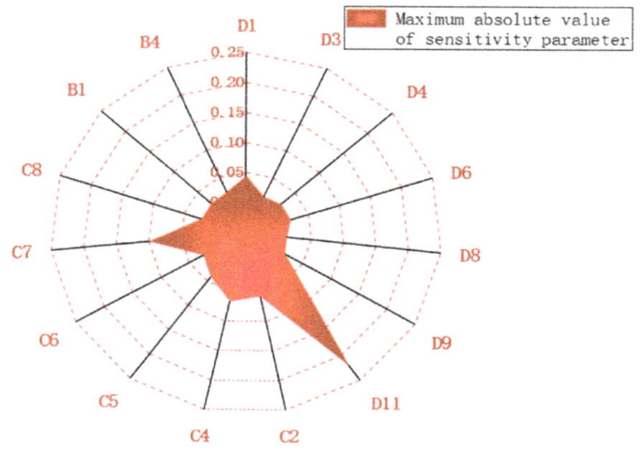

Figure 12. Sensitivity analysis results for sinking accidents.

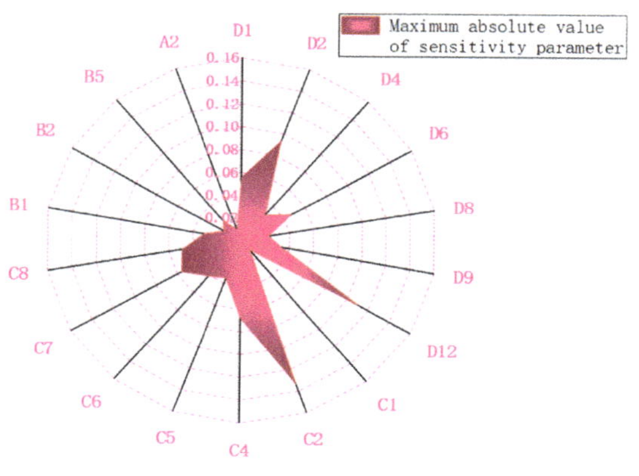

Figure 13. Sensitivity analysis results for other accidents.

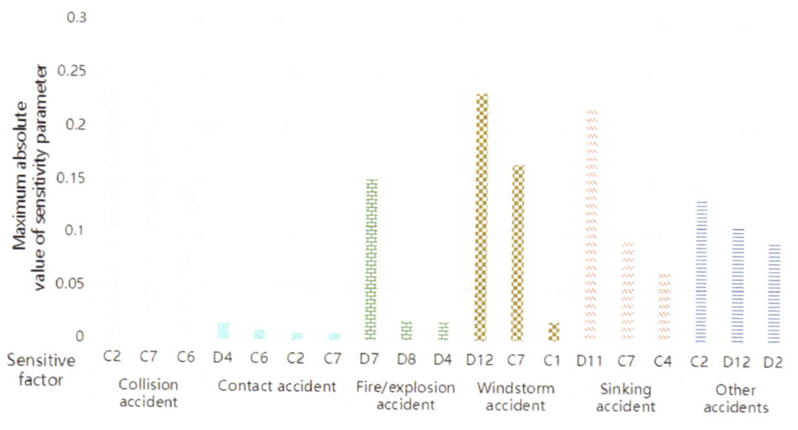

Figure 14. Top three sensitivity factors for different types of accidents.

Based on Figures 8 and 14, for collision accidents, the analysis results show that insufficient safety awareness (C2) has the largest absolute value for the sensitivity parameter, indicating that a crew's level of safety awareness has the most significant impact on the occurrence of collision accidents. Therefore, enhancing crew safety awareness is of great significance for reducing collision accidents.

According to Figures 9 and 14, for contact accidents, improper collision avoidance behavior (D4) has the largest absolute value for the sensitivity parameter among the causal factors. This result indicates that during navigation, proper collision avoidance behavior is crucial in preventing contact accidents.

Based on Figures 10 and 14, for fire/explosion accidents, the absolute value of the sensitivity parameter for improper equipment use (D7) is the highest, indicating that improper use of equipment is one of the primary contributors to fire/explosion accidents. Therefore, enhancing the standardization and safety of equipment use is crucial for reducing fire/explosion accidents.

According to Figures 11 and 14, for windstorm accidents, the absolute value of the sensitivity parameter for inadequate typhoon prevention measures (D12) is the highest, indicating that the adequacy of typhoon prevention measures taken by ship staff directly relates to the incidence of windstorm accidents under severe weather conditions such as typhoons. Therefore, strengthening the formulation and implementation of typhoon prevention measures for ships is crucial for reducing the risk of windstorm accidents.

As shown in Figures 12 and 14, for sinking accidents, the absolute value of the sensitivity parameter for improper emergency measures (D11) is the highest, indicating that the adequacy of emergency measures taken by ship staff in emergency situations plays a decisive role in preventing sinking accidents. Therefore, improving crews' emergency response capabilities and formulating effective emergency measures are important ways of reducing sinking accidents.

As shown in Figures 13 and 14, for other types of accidents, insufficient safety awareness (C2) has the largest absolute value for the sensitivity parameter, which again emphasizes the universal importance of enhancing crews' safety awareness for reducing waterborne traffic accidents.

4.3. *Cause Paths of Waterborne Traffic Accidents*

After deeply exploring the key and sensitive factors of waterborne traffic accidents, the next step was to construct the corresponding cause paths. Based on the results of the sensitivity analysis, a rigorous screening process was conducted for the causal factors in the chain analysis model of waterborne traffic accident causes. Specifically, the causal factors that had insignificant impacts on various accidents (i.e., those with node sensitivity parameters less than 0.01) were eliminated.

However, in the analysis of causal factors for contact accidents, a special situation was identified: the sensitivity of most causal factors was less than 0.01. To comprehensively and thoroughly analyze the cause chains of contact accidents, a special strategy was adopted: when constructing the cause paths, the three nodes with the highest sensitivity at each level were selected for focused analysis. The results are shown in Table 3 and Figure 15. In Figure 15, five different colors with numbers mean different categories of factor levels.

As evident in Table 3 and Figure 15, the number of cause paths for various types of waterborne traffic accidents varies: there are 12 cause paths for collision accidents, while there are 3 for contact accidents, 6 for fire/explosion accidents, 2 for windstorm accidents, 4 for sinking accidents, and 7 for other types of accidents.

Table 3. Cause paths for different types of accidents.

Accident Type	Cause Paths
Collision accidents (E1)	A2(A5)-B1-C2-D1-E1; A2(A5)-B1-C2-D6-E1; A2(A5)-B1-C2-D7-E1; A2(A5)-B1-C2-D12-E1; B5-C4-D1-E1; B5-C4-D11; C6-D1-E1; C6-D6-E1; C7-D1-E1; C7-D6-E1; C7-D11-E1; C7-D712-E1
Contact accidents (E2)	C2-D4-E2; C6-D4-E2; C7-D4-E2
Fire/explosion accidents (E3)	C2-D4-E3; C2-D7-E3; C2-D8-E3; C6-D4-E3; C8-D7-E3; C8-D8-E3
Windstorm accidents (E4)	C1-D2-E4; C2-D2-E4
Sinking accidents (E5)	B1-C2-D1-E5; B1-C4-D1-E5; B5-C4-D1-E5; C7-D11-E5
Other accidents (E6)	A2(A4)(A5)-B1-C2-D1-E6; A2(A4)(A5)-B1-C2-D2-E6; A2(A4)(A5)-B1-C2-D12-E6; A2(A4)(A5)-B1-C7-D12-E6; A2(A5)-B2-C2-D1-E6; A2(A5)-B2-C2-D2-E6; A2(A5)-B2-C2-D12-E6

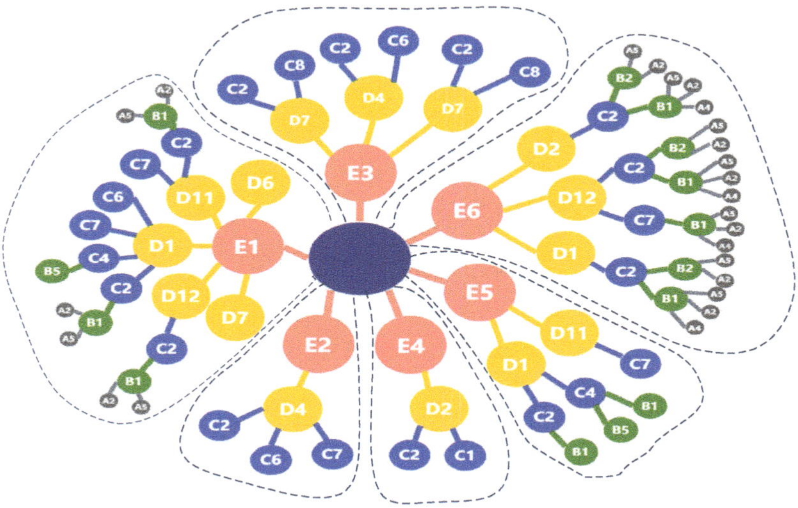

Figure 15. Cause paths of different types of accidents.

The cause paths of collision accidents show complex diversity, encompassing the intertwined influences of human errors (such as improper lookout and operation) and environmental factors (such as insufficient visibility and complex current conditions). In contrast, the cause paths of contact accidents are relatively concise, centered on improper collision avoidance behavior while also being influenced by certain environmental factors. The cause paths of fire/explosion accidents reveal potential risks across multiple links, ranging from equipment failures to improper human operations, demonstrating these accidents' unique complexity. The cause paths of windstorm accidents are closely related to extreme weather conditions, especially insufficient wind resistance capabilities of ships under severe weather conditions. The cause paths of sinking accidents involve improper cargo stowage and ship operational performance issues. The cause paths of other accident categories are more extensive and complex, reflecting the diversity and complexity of the causes of waterborne traffic accidents.

4.4. Global Causal Chain Analysis of Waterborne Traffic Accidents

Using the causal reasoning method of Bayesian networks, we analyzed the global causal chains of the causal factors of waterborne traffic accidents. Via the causal reasoning method, the target node (i.e., the occurrence status of different accident types) was set as a known condition, and the posterior probabilities of evidence nodes were calculated to infer the possible causes of an accident.

Taking windstorm accidents (E4) as an example, the occurrence status of the target node was set to 1, and the posterior probabilities of its four parent nodes—failure to travel at a safe speed, improper lookout, improper collision avoidance behavior, and inadequate typhoon prevention measures—were derived. The results show that the occurrence probability for inadequate typhoon prevention measures (D12) was the highest, reaching 69%. Further analysis of the three other parent nodes of inadequate typhoon prevention measures—a lack of theoretical knowledge/work experience/skills/other abilities, insufficient safety awareness, and excessive wind/waves/currents—revealed that the posterior probability of excessive wind, waves, and currents (C7) was relatively high. Using the same method, a global causal chain analysis was conducted for other types of accidents, yielding the global causal chains for all the accident types, as shown in Table 4.

Table 4. Global causal chains for different types of accidents.

Accident Type	Global Causal Chain
Collision accidents (E1)	A1-B4-C1-D1-E1
Contact accidents (E2)	A5-B2-C2-D1-E2
Fire/explosion accidents (E3)	A5-B1-C2-D7-E3
Windstorm accidents (E4)	C7-D12-E4
Sinking accidents (E5)	A1-B4-C1-D3-E5
Other accidents (E6)	A1/A5-B4/B2-C1/C2-D3

The results given above have the following implications.

(1) Collision accidents occur frequently due to a complex array of causes. These include improper staffing; improper selection of navigation areas/improper operation and management; insufficient theoretical knowledge, work experience, skill levels, and other abilities; and improper lookout. Measures including enhancing crew safety training, optimizing navigation environments, improving crew management systems, and enhancing vessel lookout capabilities are essential to prevent collision accidents.

(2) Contact accidents are mainly caused by inadequate implementation of safety management, a lack of supervision and guidance, insufficient safety awareness, and improper lookout. Based on these results, strengthening crew training (especially regarding the standardized process of safety management), enhancing supervision normalization, and enhancing vessel lookout capabilities are crucial to reduce the probability of contact accidents.

(3) Fire/explosion accidents often result from inadequate implementation of safety management, improper equipment allocation and maintenance, insufficient safety awareness, improper use of equipment, etc. To prevent these accidents, regular inspections and maintenance of fire-fighting and explosion-proof equipment are essential. Additionally, strict fire source management systems must be implemented.

(4) Windstorm accidents frequently occur due to crew members' lack of theoretical knowledge, work experience, and skills, leading to ineffective supervision, guidance, and inadequate plans under severe weather conditions. Hence, improving crew competence, enhancing decision making under extreme conditions, and implementing scientific navigation plans are crucial for safe navigation during wind disasters.

(5) Sinking accidents are mainly caused by improper staffing; improper selection of navigation areas/improper operation and management; insufficient theoretical knowledge, work experience, skill levels, etc.; and misjudgment of danger. To prevent

these accidents, crew safety training should be enhanced, navigation environments should be optimized, and the supervision of ship maintenance should be strengthened. Other accidents often result from various combinations of factors, such as inadequate education/training, leading to ineffective safety management and improper equipment handling.

Furthermore, among all kinds of the accidents investigated in this paper, human errors such as improper lookout and misjudgment of hazards are influential factors which may lead to the occurrence of the waterborne accidents.

5. Conclusions

Water traffic accidents not only result in casualties and losses but also pollute waterways and damage ecosystems. In this study, we focused on investigating the causal chains of waterborne traffic accidents in a data-driven context, delving into their underlying factors and chain reactions. Based on China's nationwide investigative reports on waterborne traffic accidents, we systematically extracted the causal factors leading to accidents and constructed a hybrid framework based on an improved HFACS and a Bayesian network. Through quantitative analyses, including chi-square tests and sensitivity analyses, the causal chains of different types of accidents were thoroughly explored, revealing the fundamental mechanisms behind accident occurrences. The results indicate that there are 12, 3, 6, 2, 4, and 7 causal chains leading to collisions, contact, fires/explosions, windstorm accidents, sinking, and other types of accidents, respectively. These research results can serve as a reference for enhancing the safety of waterborne transportation.

Future research will focus on the following: (1) comparative analysis with different maritime regions, so as to explore the similarities and differences in causal chains across different maritime environments; (2) the application of new technologies, such as artificial intelligence, to deepen our understanding of the causal chains of waterborne traffic accidents. More specifically, with the advent of MASs (Maritime Autonomous Ships), advanced sensors and algorithms have significant potential to reduce human errors, in turn enhancing the safety and reliability of water transportation. Additionally, MASs, equipped with AI and machine learning, exhibit improved emergency response capabilities, enabling autonomous judgment and timely action in critical situations. However, as MASs evolve, it is imperative to strengthen regulation and legal frameworks and establish safety standards, operational norms, and robust monitoring mechanisms to ensure their safe and reliable operation. We aim to explore these technological advancements in accident prevention, monitoring, and emergency responses, thereby driving the development of water transportation safety management towards greater intelligence and automation.

Author Contributions: Conceptualization, X.Y.; Methodology, Q.X.; Software, Y.Y.; Formal analysis, J.W.; Data curation, Q.W.; Writing—original draft, X.Y.; Writing—review & editing, H.Z. and Q.X. All authors have read and agreed to the published version of the manuscript.

Funding: This research was funded by the Natural Science Foundation of China (52362055), Guangxi Science, and the Technology Major Program of China (Grant No. AA23062053).

Institutional Review Board Statement: Not applicable.

Informed Consent Statement: Not applicable.

Data Availability Statement: The data presented in this study are available on request from the corresponding author.

Conflicts of Interest: The authors declare they have no conflicts of interest.

References

1. Formela, K.; Weintrit, A.; Neumann, T. Overview of definitions of maritime safety, safety at sea, navigational safety and safety in general. *Int. J. Mar. Navig. Saf. Sea Transp.* **2019**, *13*, 285–290. [CrossRef]
2. Peng, Z.; Jiang, Z.; Chu, X.; Ying, J. Spatiotemporal Distribution and Evolution Characteristics of Water Traffic Accidents in Asia since the 21st Century. *J. Mar. Sci. Eng.* **2023**, *11*, 2112. [CrossRef]
3. Wu, B.; Kou, L.; Ma, Q. Research on HFACS based on accident causality diagram. *Open J. Saf. Sci. Technol.* **2017**, *7*, 77–85. [CrossRef]
4. Zhang, G.; Thai, V.V. Expert elicitation and Bayesian Network modeling for shipping accidents: A literature review. *Saf. Sci.* **2016**, *87*, 53–62. [CrossRef]
5. Wang, H.; Liu, Z.; Wang, X.; Graham, T.; Wang, J. An analysis of factors affecting the severity of marine accidents. *Reliab. Eng. Syst. Saf.* **2021**, *210*, 107513. [CrossRef]
6. Liu, K.; Yu, Q.; Yuan, Z.; Yang, Z.; Shu, Y. A systematic analysis for maritime accidents causation in Chinese coastal waters using machine learning approaches. *Ocean Coast. Manag.* **2021**, *213*, 105859. [CrossRef]
7. Sun, J.; Li, M.; Xiu, X. Analysis of the Causes of Water Transportation Accidents Based on Complex Network Theory. *J. Dalian Marit. Univ.* **2023**, *49*, 80–90+160.
8. Wu, Y.; Jiang, F.; Yao, H.; Huang, M.; Ma, Q. Analysis of Causal Factors and Risk Prediction for Inland River Ship Collision Accidents Based on Text Mining. *J. Transp. Inf. Saf.* **2018**, *36*, 8–18.
9. Wang, H.; Liu, Z.; Liu, Z.; Wang, X.; Wang, J. GIS-based analysis on the spatial patterns of global maritime accidents. *Ocean Eng.* **2022**, *245*, 110569. [CrossRef]
10. Namgung, H.; Kim, J.S. Collision risk inference system for maritime autonomous surface ships using COLREGs rules compliant collision avoidance. *IEEE Access* **2021**, *9*, 7823–7835. [CrossRef]
11. Kaptan, M.; Sarıalioğlu, S.; Uğurlu, Ö.; Wang, J. The evolution of the HFACS method used in analysis of marine accidents: A review. *Int. J. Ind. Ergon.* **2021**, *86*, 103225. [CrossRef]
12. Yildiz, S.; Uğurlu, Ö.; Wang, J.; Loughney, S. Application of the HFACS-PV approach for identification of human and organizational factors (HOFs) influencing marine accidents. *Reliab. Eng. Syst. Saf.* **2021**, *208*, 107395. [CrossRef]
13. Wang, Q.; Sha, Z.; Zhang, J.; Ma, J. Human Factor Analysis of Ship Grounding Accidents Based on the HFACS-FCMs Model. *J. Shandong Jiaotong Univ.* **2024**, *32*, 103–109+123.
14. Yıldırım, U.; Başar, E.; Uğurlu, Ö. Assessment of collisions and grounding accidents with human factors analysis and classification system (HFACS) and statistical methods. *Saf. Sci.* **2019**, *119*, 412–425. [CrossRef]
15. Huang, H. Analysis and Application of Human Factors in Marine Traffic Based on the HFACS Method. *Shandong Ind. Technol.* **2018**, *05*, 216–218+238.
16. Chen, S.; Wall, A.; Davies, P.; Yang, Z.; Wang, J.; Chou, Y.H. A Human and Organizational Factors (HOFs) analysis method for marine casualties using HFACS-Maritime Accidents (HFACS-MA). *Saf. Sci.* **2013**, *60*, 105–114. [CrossRef]
17. Chauvin, C.; Lardjane, S.; Morel, G.; Clostermann, J.P.; Langard, B. Human and organizational factors in maritime accidents: Analysis of collisions at sea using the HFACS. *Accid. Anal. Prev.* **2013**, *59*, 26–37. [CrossRef] [PubMed]
18. Fan, S.; Blanco-Davis, E.; Yang, Z.; Zhang, J.; Yan, X. Incorporation of human factors into maritime accident analysis using a data-driven Bayesian network. *Reliab. Eng. Syst. Saf.* **2020**, *203*, 107070. [CrossRef]
19. Antão, P.; Soares, C.G. Analysis of the influence of human errors on the occurrence of coastal ship accidents in different wave conditions using Bayesian Belief Networks. *Accid. Anal. Prev.* **2019**, *133*, 105262. [CrossRef]
20. Wang, L.; Yang, Z. Bayesian network modelling and analysis of accident severity in waterborne transportation: A case study in China. *Reliab. Eng. Syst. Saf.* **2018**, *180*, 277–289. [CrossRef]
21. Meng, X.; Li, H.; Zhang, W.; Zhou, X.Y.; Yang, X. Analyzing risk influencing factors of ship collision accidents: A data-driven Bayesian network model integrating physical knowledge. *Ocean Coast. Manag.* **2024**, *256*, 107311. [CrossRef]
22. Fan, S.; Yang, Z.; Blanco-Davis, E.; Zhang, J.; Yan, X. Analysis of maritime transport accidents using Bayesian networks. *Proc. Inst. Mech. Eng. Part O J. Risk Reliab.* **2020**, *234*, 439–454. [CrossRef]
23. Tian, Y.; Qiao, H.; Hua, L.; Yan, S.; Zhang, Q. Bayesian Network Model for Maritime Ship Collisions and Its Application. *J. Nav. Univ. Eng.* **2023**, *35*, 28–33.
24. Hänninen, M.; Kujala, P. The effects of causation probability on the ship collision statistics in the Gulf of Finland. *Mar. Navig. Saf. Sea Transp.* **2010**, *4*, 79–84.
25. Meng, X.; Li, H.; Zhang, W.; Zhou, X.Y.; Yang, X. Analyzing ship collision accidents in China: A framework based on the NK model and Bayesian networks. *Ocean Eng.* **2024**, *309*, 118619. [CrossRef]
26. Wang, H.; Jin, G. Cause analysis of marine accidents based on HFACS and Bayesian networks. In Proceedings of the International Conference on Smart Transportation and City Engineering, Chongqing, China, 16–18 December 2023; Volume 13018, pp. 1022–1030.

27. Rostamabadi, A.; Jahangiri, M.; Zarei, E.; Kamalinia, M.; Banaee, S.; Samaei, M.R. A novel fuzzy bayesian network-HFACS (FBN-HFACS) model for analyzing human and organization factors (HOFs) in process accidents. *Process Saf. Environ. Prot.* **2019**, *132*, 59–72. [CrossRef]
28. Jiang, Y.; Wan, Z.; Chen, J. Path Analysis of the Causes of Coastal Water Transportation Accidents in China. *J. Dalian Marit. Univ.* **2024**, *50*, 76–84.
29. Li, Y.; Cheng, Z.; Yi, T.; Fan, X.; Wu, B. Use of HFACS and Bayesian network for human and organizational factors analysis of ship collision accidents in the Yangtze River. *Marit. Policy Manag.* **2022**, *49*, 1169–1183. [CrossRef]
30. Wang, H.; Chen, N.; Wu, B.; Soares, C.G. Human and organizational factors analysis of collision accidents between merchant ships and fishing vessels based on HFACS-BN model. *Reliab. Eng. Syst. Saf.* **2024**, *249*, 110201. [CrossRef]
31. Özkan, U.; Serdar, Y.; Sean, L.; Wang, J.; Kuntchulia, S.; Sharabidze, I. Analyzing Collision, Grounding, and Sinking Accidents Occurring in the Black Sea Utilizing HFACS and Bayesian Networks. *Risk Anal. Off. Publ. Soc. Risk Anal.* **2020**, *40*, 2610–2638.

MDPI AG
Grosspeteranlage 5
4052 Basel
Switzerland
Tel.: +41 61 683 77 34

Journal of Marine Science and Engineering Editorial Office
E-mail: jmse@mdpi.com
www.mdpi.com/journal/jmse